CW01626874

ANALYSES CONCERNING PASSIVE AND ACTIVE SYNTHESIS

Lectures on Transcendental Logic

EDMUND HUSSERL

COLLECTED WORKS

EDITOR:
RUDOLF BERNET

VOLUME IX

ANALYSES CONCERNING PASSIVE AND ACTIVE SYNTHESIS
Lectures on Transcendental Logic

TRANSLATIONS
PREPARED UNDER THE AUSPICES OF
THE HUSSERL-ARCHIVES (LEUVEN)

A list of titles in this series can be found at the end of this volume.

EDMUND HUSSERL

ANALYSES CONCERNING PASSIVE AND ACTIVE SYNTHESIS

Lectures on Transcendental Logic

TRANSLATED BY

ANTHONY J. STEINBOCK
Southern Illinois University at Carbondale
Carbondale, Illinois, USA

KLUWER ACADEMIC PUBLISHERS
DORDRECHT / BOSTON / LONDON

Library of Congress Cataloging-in-Publication Data

Husserl, Edmund, 1859-1938
Analyses concerning passive and active synthesis : lectures on transcendental logic / Edmund Husserl ; translated by Anthony J. Steinbock.
p. cm. -- (Edmund Husserl collected works ; vol. 9)
Includes bibliographical references and index.
ISBN 0-7923-7065-1 (hb : alk. paper)
1. Transcendental logic. I. Title.

B3279.H93 A53 2001
193--dc21

2001035837

ISBN 0-7923-7065-1 (hb)
ISBN 0-7923-7066-X (pb)

Published by Kluwer Academic Publishers
PO Box 17, 3300 AA Dordrecht, The Netherlands

Sold and distributed in North, Central and South America
by Kluwer Academic Publishers,
101 Philip Drive, Norwell, MA 02061, U.S.A.

In all other countries, sold and distributed
by Kluwer Academic Publishers, Distribution Center,
PO Box 322, 3300 AH Dordrecht, The Netherlands

Printed on acid-free paper

Printed in the Netherlands.

For Joseph and Samara

TABLE OF CONTENTS

MAIN TEXTS

PART 1: PRELIMINARY CONSIDERATIONS FOR THE LECTURE ON TRANSCENDENTAL LOGIC

PART 2: ANALYSES CONCERNING PASSIVE SYNTHESIS: TOWARD A TRANSCENDENTAL AESTHETIC

Division 1: Modalization

Division 2: Evidence

Division 3: Association

Division 4: The In-Itself of the Stream of Consciousness

PART 3: ANALYSES CONCERNING ACTIVE SYNTHESIS: TOWARD A TRANSCENDENTAL, GENETIC LOGIC

SUPPLEMENTARY TEXTS

SECTION 1. FIRST VERSION OF MAIN TEXT PART 2 (1920/21)

SECTION 2: APPENDICES

SECTION 3. RELATED ESSAYS

SECTION 4. ON STATIC AND GENETIC PHENOMENOLOGICAL METHOD

TRANSLATOR'S INTRODUCTION

The *Analyses Concerning Passive and Active Synthesis* was Edmund Husserl's phenomenological investigation into the *origin of truth*. We find here an early indication of an historical reflection and the identification of a "crisis," the description of primordial dimensions of experience, the genealogy of judgment, and the employment of a new, genetic phenomenological method. While a large portion of the material comprised under this heading is a translation of Husserliana XI, *Analysen zur passiven Synthesis*, it also includes essential additions to the main text of Husserl's lecture, some supplements, and a partial reorganization of the material.

The "Translator's Introduction" is offered as an orientation to this work. This Introduction is divided into four sections. Section 1 situates the work historically and conceptually, discusses its composition and revised title, and provides a basic overview of material making up this lecture. Section 2 situates the *Analyses* in the context of a genetic phenomenology, since it is this methodological approach that enables the description of phenomena treated in the *Analyses*. Section 3 elaborates upon the novel and significant themes in these lectures, such as passivity, affective allure, association, motivation, the unconscious, etc. Section 4 includes final editorial notes on the translation and my acknowledgements. Rather than reserving a special section to explain the translation of various key terms, I integrate this clarification into the course of the explications of sections 2 and 3, and on occasion, discuss them in footnotes appended to the translated text.

1. The Historical and Conceptual Context

Presented here as *Analyses Concerning Passive and Active Synthesis: Lectures on Transcendental Logic* is one of Edmund Husserl's most renowned series of lectures presented in the 1920s.

Offered three times, Winter Semester 1920/21, Summer Semester 1923, and Winter Semester 1925/26, Husserl's lectures are virtually contemporaneous with writings devoted to the problem of "intersubjectivity" and "individuation" (1921–1927) his reflections on the reduction from *Erste Philosophie* (1923/24), and his considerations of cultural crises and its potential for renewal in the *Kaizo* articles (1922-24). As such, the *Analyses* occupy both an historical and a conceptual "middle point" of his work.

Historically speaking, the *Analyses* are situated between major, well-known published works. On the one hand, they arise twenty years after Husserl's ground-breaking *Logical Investigations* (1900/01), a decade and a half after his first lectures on time-consciousness (1905), and nearly ten years following his *Ideas* (1913); on the other, they precede by several years his *Formal and Transcendental Logic* and his *Cartesian Meditations* (both from 1929), and they anticipate his *Crisis* (1934–37) by more than a decade.

While the major insights, novel notions, as well as the import and contribution of these lectures will be explained below, it is possible to say provisionally that these lectures also occupy a center point conceptually. As expressive, even exemplary of his genetic method, they succeed Husserl's earlier phenomenology of consciousness by surpassing both the Cartesian static analysis peculiar to the *Ideas* and the formalism of his early time-consciousness lectures, and they anticipate his generative investigations into intersubjectivity, history, and the lifeworld by initiating a regressive style of inquiry into origins that becomes the hallmark of Husserl's later undertakings in the *Crisis*.

Husserl's fame was well established by the time of these lectures. According to the *Quästurakten* or the "registrar's list" at Albert-Ludwigs-Universität Freiburg where Husserl held these lectures, Husserl had 176 persons in attendance the first time he gave them under the title of "Logik" in 1920/21, 133 enrolled in 1923 (now entitled "Ausgewählte phänomenologische Probleme" ["Selected Phenomenological Problems"]), and the numbers tallied 65 in 1925/26 in lectures newly entitled "Grundprobleme der Logik" ["Fundamental Problems of Logic"].[1] A survey of these registrar's lists reveal a number of names familiar to those acquainted with the phenomenological tradition: Alfred Adler, Oskar Becker, Franz-Josef Brecht, Käthe Hamburger, Max Horkheimer, Fritz Kaufmann, Paul Landsberg, Walther Marseille, Arnold Metzger, Fritz Neumann, Hans Reiner, Wilhelm Szilassi (1920/21); Marvin Farber, Karl Hanser, Ludwig Landgrebe, Hasime Tanabe (1923), and Eugen Fink, (again, Ludwig Landgrebe), Walter Sachs (1925/26).

1. Passive Synthesis and Transcendental Logic

In recent years, these lectures have achieved a near legendary status under the shorthand rubric of "passive synthesis." How does a lecture series preoccupied with the general problem of logic win its world-wide renown as the "passive synthesis" lectures? There are at least two reasons for this, one editorial (a), one philosophical (b). After discussing these reasons, I explain the composition of this English edition and the reasons for its revised title.

A. One reason these lectures have come to be known as the "passive synthesis" lectures—a reason almost too obvious to mention—is due to the title assigned to them by the editor of Husserliana XI, Margot Fleischer, namely, *Analysen zur passiven Synthesis* (1966) [*Analyses Concerning Passive Synthesis*]. Why this title? The original titles Husserl gave to the lectures—"Logic," "Selected Phenomenological Problems," and "Fundamental Problems in Logic"—she notes, were simply too broad for the collection of texts that she assembled in the Husserliana volume. While the title "Transcendental Logic," which Husserl assigned to the lectures on the folders containing the manuscripts, did give them more specification, this was to her mind still too imprecise. Instead, she wished to capture the sense attributed to these investigations by Husserl himself, to wit, *Urkonstitutionen* or the analyses of primordial modes of constitution. And while she

[1] I am grateful to the Albert-Ludwigs-Universität Freiburg's Universitätsarchiv for providing me with the *Quästurakten* of these three semesters in question. I would also like to thank Sebastian Luft and Matthias Haenel for transcribing the lists from the Sütterlin handwriting.

could have also chosen the title "Transcendental Aesthetic" to evoke this sense of the investigations—a title suggested by the occurrence of this expression both in the *Analyses* and in *Formal and Transcendental Logic*—she thought that in the wake of Kant it would have given the reader a false impression of what was to be expected from this work. For these reasons, Fleischer settled on the expression "passive synthesis" for the title of this collection, uniting the main portion of the lectures she collated and the supplementary material. This expression is not unwarranted, for it occurs at least a half a dozen times throughout the work. It has *de facto* proved itself to be a title suited to the material selected for publication in Husserliana XI.

B. The title, however, is not the sole reason for these lectures to have acquired their acclaim as the "passive synthesis" work. While the issue of passive synthesis is a fundamental one and does occupy a large portion of Husserl's investigations in Husserliana XI, the context in which the lectures unfold is a broader one. This context, as intimated above, is *transcendental logic*.

Husserl's *Formal and Transcendental Logic* (published in 1929) was conceived as an "Introduction" to phenomenology, and as such joins the *Logical Investigations*, *Ideas I*, and is later joined by *Cartesian Meditations* and the *Crisis*. In distinction to, e.g., *Ideas I*, the way into phenomenology takes place *via* the natural attitude, in particular, as it is functional in the mathematician and logician. While formal logic—understood both as the apophantic science of propositions and deductive relations as well as the formal ontology of individual objects—serves as the starting point of analysis, it cannot be seen as self-sufficient; it requires an investigation into subjective accomplishments that constitute mathematical and logical truths; it requires a "transcendental logic." But even this, writes Husserl, demands a deeper founding. For as a "critique" of the limits and capacities of logical reasoning, a transcendental logic must understand how a streaming egoic life of consciousness can be constituted as a true being, and it must do this by appealing to a theory of experience and actuality that founds active cognition and its ideal objects (pp. 112, 259–60, 386).[2]

Thus, when considering the function of the *Analyses* in this broader context, we are witness to a peculiar, but almost typical phenomenological movement, a "zig-zag," if you will. Even though Husserl understood his *Formal and Transcendental Logic* as another "introduction" to phenomenology, and even though this work followed his lectures making up the *Analyses*, Husserl's *Formal and Transcendental Logic* itself can be read as an introduction to the project of the *Analyses*. Let me explain.

Husserl's actual "Introduction" to these lectures given in 1920/21 (included here in the English edition as "Main Text, Part 1," but published only as an appendix to Husserliana XVII, *Formale und Transcendentale Logik*) begins with a preliminary consideration of the term "logic." Tracing the term "logic" back to its Platonic founding and to its Greek roots in "*logos*," and then to the more original "λέγω" as "gathering together," and "expounding upon," Husserl detects in logic a vocation of the critical justification of reason, and as such, a vocation to be the science of all sciences (pp. 1, 8, 387). As a radical and universal *a priori* theory of science, logic is not to be understood merely as an axiomatic and formalistic deductive system, formed by abstracting general traits from existing or past sciences; for intrinsic to all factual sciences at our disposal is an animating teleological orientation. Even if we never encounter this teleological idea as such, it nonetheless functions guidingly and efficaciously—even if implicitly—when we practice science or operate from theoretical interest. If we find today that the sciences treat their objects of study in a detached, particularized, and fragmented manner, this would only be an expression of the way in which the particular sciences themselves become detached from "the aim, sense, and possibility of genuine science." They have lost the sense of *their own* orientation that ultimately gives them meaning and to which they refer back as indexes.

[2] All references to the *Analyses* in the "Translator's Introduction" will given to this English edition.

Yet despite the fact that the particular sciences have abandoned their own normative sense, a phenomenological investigation will not simply do away with the sciences in their current cultural forms; they cannot simply be passed over in a fundamental analysis. For as scions of the instituting idea of logic, the special sciences still harbor their internal sense even in their self-forgetfulness. By examining them, and more specifically, by examining the science of logic as it has been handed down to us, we can gain a clue, a leading clue, to logic's vocation of critical self-justification and as the universal theory of principles and of norms of all sciences.

This self-forgetfulness and possibility of recovery, however, is not as innocent and facile as it seems. The tragedy we currently face, laments Husserl, is that the sciences have *inverted* the original relation between logic and science such that (1) the sciences have made themselves autonomous; in this ostensible, mystifying self-sufficiency and groundlessness, (2) they have become splintered in relation to each other; and in this process, (3) logic has been transformed into a sub-discipline of the sciences, a pragmatic technology borrowing its methods from mathematics, becoming a limited theoretical instrument brushed aside with scorn.

It is precisely through this inversion and its ramifications that the sciences have lost their internal sense and landed in a kind of self-forgetfulness of scientific objectivism. "In other words, logic, which was originally the torchbearer of method and which claimed to be the pure doctrine of principles of possible knowledge and science, lost this historical vocation and, understandably, remained far behind in its development" (p. 4). The paradox here is that the autonomy of the sciences from logic (logic as a justifying system of principles of all objective justification) has only a putative autonomy, one which exacerbates the sciences inability to emerge as self-sufficient, since they are completely ignorant of their own sense and without foundation. So, writes Husserl, while at first we novices are filled with enthusiasm in engaging in the positive sciences, we end up being deeply dissatisfied because we do not become wiser and better through them, as is clearly their pretension (p. 6).

For Husserl, the fact that we were no longer moving in the same direction sketched out by this optimal idea of logic suggests that a rupture, a constitutive abnormality, has ensued, one which we might call a crisis in the "spiritual common good of humanity" (p. 28). But provided that we want to be more than mere professionals, specialists, and academics, provided that we want to take ourselves as human beings "in the full and highest sense," we are called upon to "raise ourselves above the self-forgetfulness of the theoretician ... who knows nothing of his accomplishment and of the motivations compelling them, who lives in them, but does not have a thematic view of this accomplishing life itself" (pp. 5).[3]

Phenomenology as transcendental philosophy wants to recover the philosophical spirit of logic. The way proposed to do so in these lectures is a genetic one. Though I will say more in section 2 of the "Translator's Introduction" regarding genetic method, let me remark here in a general manner that by clarifying its *origins*, not as something static, but as origins that are originating, we can recover the lost sense of logic, a sense that remains obfuscated in the present sciences. In this way we are in a position to discover the presuppositions of logic by investigating the genetic formations of sense.

Only a transcendental logic can be an ultimate theory of science, for it treats the objects of thought precisely as accomplishments of the activity of thinking. Transcendental phenomenology makes such a theory of science possible because it inquires back from ready-made propositions, from theories "already there," to thinking life in which these formations are accomplished; it goes back still more deeply from the givenness of all types of objects that underlie possible theories to the experiencing life in which those objects are pregiven, and most radically, it understands how the life of consciousness itself can be

---

[3] It is not a mere coincidence that the issues of self-forgetfulness and the call for us to become more profoundly human are echoed in other articles Husserl penned during the early 1920s, namely, his "Kaizo" articles (1922-24). The Kaizo articles (Hua XXVII), which also presuppose a genetic method, are concerned specifically with the crisis of human culture and its "renewal" as an ethical humanity. See my "The Project of Ethical Renewal and Critique: Edmund Husserl's Early Phenomenology of Culture," *The Southern Journal of Philosophy*, Vol. 32, No. 4 (1994), 449-64.

constituted as a "true being," as an ideal correlate of possible verifications (p. 259–60).

In order to undertake a transcendental logic, "tremendous transcendental-phenomenological preliminary work must be accomplished" (p. 7). This preliminary work entails, in part, tracing the accomplishments of thinking to their genetic origins in *passive*, pre-cognitive syntheses. In moving from the dimension of the constituted to the constituting, Husserl incorporates a regressive, archeological movement from the *active* cognitive dimensions to the passive kinaesthetic ones. It is in this sense that the project as it actually took shape in *Formal and Transcendental Logic* becomes an introduction to and preparatory for the *Analyses*. But equally, this beginning regressive movement also has to be understood as preparatory for the inverse direction that the *Analyses* will take for their explicative method.

Once we have regressed back to the origins of the great world of constituting life, we describe this life "by beginning from below and ascending upward, to show how genuine thinking in all its levels emerges here, how it is motivated and is built-up in its founded accomplishment" from the most basic structures of consciousness (pp. 32 and 607 fn. 93), tracing the "storied structure of constitution" (p. 270). The *Analyses* undertake the task of a "transcendental-phenomenological aesthetic" as founding for a transcendental-phenomenological logic, thus investigating the systematic connections of passive sense formation. Only from the sphere of passivity, contends Husserl, can we grasp the most fundamental of all shortcomings in the foundation of traditional logic, one that concerns the validity of logical norms, and the ultimate principle of logical norms, namely, the principle of contradiction and the law of the excluded middle (pp. 143, 149, 386). A genetic method allows us to elucidate the dynamic formations of sense in the passive sphere as foundational for logic. For this reason, "paradoxically," a critique of the ideal structures of logical reason which takes as its point of departure the investigations into a formal and transcendental logic cannot be limited to the sphere of logic; it demands a transcendental aesthetic.[4]

Described as a "transcendental aesthetic," the tremendous preliminary work mentioned above entails not merely recovering the foundation for active syntheses and cognitive operations, but of describing the passive sphere of experience in its own integrity, its own essential laws and contributions in the constitution of evidence, and the modalizations of evidence peculiar to it. A transcendental aesthetic within a genetic methodological register will bracket all judicative knowing, determinative and predicative thought, and focus on the occurrences of apperception in general, the objects of possible perception that have the sense-form of time and the sense-shape of spatiality, and investigate how sense unities are constituted through associative syntheses. It will require investigations into the structure of sensibility as the continual constitution of space and time through self-temporalization in time-consciousness and lived-bodily kinaesthesis (pp. 444–45).

Extending to all features of space-time constitution, a transcendental aesthetic will broach even a generative analysis of the constitution of space and time in terms of earth-ground and world-horizon, investigating lifeworlds in terms of their normative significance as "home" and "alien."[5] The *Analyses* from the 1920s dealing with "passive synthesis" did not go this far, and stays for all practical purposes on the level of genetic phenomenology, that is, within the span of individual facticity or the intragenerational constitution of community.

On the one hand, by understanding the tenor of this genetic methodological movement that underlies the *Analyses*, we have

[4] In a manuscript belonging to the *Analysen* (F I 37, 68b), Husserl referred to his phenomenology of experience—in distinction to Kant—as a "new transcendental aesthetic." See the "Introduction" to Edmund Husserl, *Lezioni sulla sintesi passiva*, trans., Vincenzo Costa (Milano: Edizioni Angelo Guerine e Associati, 1993) p. 29, fn. 14.

[5] In a manuscript stemming from 1930 or 1931 that deals with the generative phenomena of homeworld and alienworld, Husserl notes: "Important for the method of the constitutive, correlative structure of the transcendental aesthetic—that is, of the correlative system of validity of the world as the world of experience" (Hua XV, p. 214 fn. 1). And in a similar context of elucidating the constitution of a homeworld, Husserl writes: "This becomes the task of a *transcendental aesthetic* ..." (Hua XV, pp. 234 ff). See also p. 632 fn. 102.

some further philosophical justification for these lectures having acquired their fame as the "passive synthesis" lectures. But the explication of passive synthesis does not complete the *Analyses*, since a transcendental aesthetic must ascend upwards to a transcendental logic, and thus is situated concretely within the problematic of a transcendental logic. To this extent, the rubric of passive synthesis though accurate also misses the broader context of his lectures. This is especially poignant with the inclusion of new manuscript materials that belong to this lecture series. As I will note below, the new material concerns the role of active synthesis and motivation for the constitution of formal ontologies.

In addition to Husserl's actual "Introduction," the material that is published here as Part 3, bearing on active synthesis, demands situating these passive synthesis texts in their original, proper, and broader context, and modifying the title of the edition to reflect this framework. To keep a continuity with the Husserliana edition, its title, and the recognition it has attained internationally under the rubric of "passive synthesis," but also to reflect its context and the content of the new material that completes the lecture series, I have, in consultation with the Husserl-Archives in Leuven, modified the title of the English edition: *Analyses Concerning Passive and Active Synthesis: Lectures on Transcendental Logic*. Let me now turn to the composition of this edition and what appears under this title.

2. The Composition of the Analyses

What is published as Husserliana XI is not the complete lecture series Husserl delivered in the 1920s, but a portion of it, supplemented by relevant appendices and essays. The English edition includes all the material published in Husserliana XI, along with four new manuscript texts, two of which complete the lecture series (Parts 1 and 3 of the Main Text), and another two that complement the supplementary materials. I will mention these texts in the course of explaining the composition of the *Analyses*.

(i) *Main Texts*. The first new text included in this edition is Husserl's actual "Introduction" to these lectures already mentioned above. Entitled here, "Preliminary Considerations of a Transcendental Logic," Part 1 sets up the framework for Husserl's phenomenology, his genetic investigations into the context of transcendental logic, and shows the necessity of beginning with an inquiry into the associative laws of passive synthesis or a transcendental aesthetic. This text from 1920/21, stemming from the manuscript F I 37 was published only in 1974 as an appendix to the Husserliana edition of *Formale und transzendentale Logik.*

Part 2, as noted, represents the main text of Husserliana XI and stems partly from manuscript F I 37, but mostly from F I 38. To fit into the comprehensive framework of Husserl's reflections, I have given this Part the subtitle, "Passive Synthesis: Toward a Transcendental Aesthetic." Here, Husserl discusses the problem of evidence, of modalization, articulates his phenomenology of association, and describes the roles of affection and attention for the constitution of sense, and eventually for the constitution and genesis of a phenomenological in-itself.[6] Since I will discuss some of its main themes below in section 3 of this "Introduction," let me highlight the second new addition to this volume.

The second new addition, included here as Part 3: "Active Syntheses: Toward a Transcendental, Genetic Logic," has been recently published as *Aktive Synthesen: Aus der Vorlesung "Transzendentale Logik." 1920/21. Ergänzungsband zu den "Analysen zur passiven Synthesis"* (Kluwer, 2000). Edited by Roland Breeur, this manuscript stems primarily from the signature F I 39 and originally belonged to the same series of lectures that make up the *Analyses* from 1920/21.[7] As Breeur notes, the fact that Husserl both continued to rework this Part and to integrate it into the new pagination of his lecture material (even though he presumably had to leave it out of his subsequent lecture due to

[6] A very small portion, approximately 28 out of 220 Husserliana printed pages of this Part 2 (from the Division on "Modalization") was taken by Landgrebe and used for *Erfahrung und Urteil,* though the order of presentation in the later does not often correspond to the presentation in Husserliana XI. See D. Lohmar, "Zur Entstehung und den Ausgangsmaterialien von Edmund Husserls Werk *Erfahrung und Urteil,*" *Husserl Studies,* Vol. 13 (1996), 31-71.

[7] When Landgrebe edited *Erfahrung und Urteil,* he also drew from Ms. F I 39 that makes up this Part 3. The portion taken equals approximately 31 pages of *Erfahrung und Urteil.* See Lohmar, "Zur Entstehung." Again, Landgrebe selected and published the material in a different order.

lack of time) supports the view that the entire lecture series demands being treated as an integral whole, and accordingly demands being presented under the same cover.[8]

Part 2 traces the constitution of sense through passive associative syntheses leading up to the constitution of the object as such through acts of remembering and expectation. Part 3 picks up on this transition from the passive to the active spheres and describes various levels of "objectivation." Thus, we have the movement from the perceptual to the judicative, from the pre-predicative to the predicative realms, or the genesis of the in-itself or true being. Active objectivation, as I will note below, can take shape with a mix of passive and active syntheses, although the overwhelming contribution here is by means of active, categorial syntheses, culminating in conceptualizing judgment, in particular, the realm of universal judging for all possible objectlike formations, the classification of their relations, and thus, the transcendental grounding of formal logic.

It is within this framework of Husserl's lectures on "Transcendental Logic" that the manuscripts in question rejoin their appropriate context. These three Parts (with two new additions) give the most complete presentation of Husserl's lecture series to date. As a whole, it makes up the Main Text of this volume.

(ii) *Supplementary Texts*. The second half of this edition entitled "Supplementary Texts" is divided into four sections. While Parts 1 and 2 of the Main Text date back to 1920/21, Husserl—in his own words—"unfortunately reworked" it in 1923 and "partly improved, partly spoiled" §§12 - 40 of Part 2 in 1925/26.[9] Section 1 of the Supplementary Texts reproduces the earliest original version of this segment of lectures from 1920/21; the passages printed here in italics are the passages left in tact by Husserl and reappropriated in the subsequent lecture of 1925/26.

Section 2 of the supplementary materials are the "Appendices." They include all the appendices originally edited by Fleischer and all those edited by Breeur in the *Ergänzungsband* that accompany

[8] See Breeur, "Vorwort des Herausgebers," p. i.

[9] From the "Textkritische Anmerkungen" Hua XI, 445.

Part 3 of the Main Text.[10] By placing all the appendices directly after both versions of the lecture series, I depart from the order given them in Husserliana XI; I place them here because the first two sections of the Supplementary Texts make up a subsection of manuscripts that either have a direct connection to the Main Text (i.e., its earlier draft) or make direct editorial references to the Main Text.

Also included under the heading of Supplementary Texts, but without the same kind of direct reference to the Main Text are two further sections. Section 3 presents two supplementary essays provided by Fleischer that stem primarily from F I 37 and F I 38, with a date of 1920/21, with some pages possibly originating from 1923. These texts, "Perception and its Self-Giving," and "Consciousness and Sense—Sense and Noema," concern the nature of perception, time-consciousness, and the constitution of an objective sense.

The last section of the Supplementary Texts supplies texts that bear generally on phenomenological method, and in particular on the relation between "static and genetic" phenomenology and the phenomenon of genesis. The presentation of this section departs from the original German edition in two ways. First, whereas Fleischer included only one text on static and genetic method from 1921, the English edition includes two additional ones that are companions to the first, stemming from the same manuscript B III 10, but published separately in 1973 in Husserliana XIV, ed., Iso Kern.[11]

---

[10] The appendices to Part 2 of the Main text are taken primarily from manuscripts D 19, F I 37, F I 38, and F I 29, but also from A VI 32, A VI 33, D 6, D 13, B III 12; those to Part 3 stem principally from F I 39, but with excerpts from F I 37 and A III 11.

[11] The first two longer essays Husserl wrote at St. Margen in 1921; the third shorter one carries a possible date of 1923 and makes an allusion to the lectures mentioned above entitled, "Einleitung in die Philosophie" from F I 29. The two not originally published in Hua XI (the second and third essays) are both taken from Hua XIV.

There is one other manuscript on static and genetic method belonging to this cluster of writings on static and genetic method, B III 10, entitled, "Statische und genetische phänomenologische Methode. Eingeborenheit. Genesis von Apperzeptionen. Allgemeinster Begriff von Apperzeption" ["Static and Genetic Phenomenological Method. Innateness. Genesis of Apperceptions. The Most General Concept of Apperception."]; it is, however, unavailable for this edition.

Finally, there are other texts dealing with static and genetic method dating from the 1930s, but they are beyond the scope of the material covering this translation.

Second, rather than being placed in the middle of the edition, preceding the appendices, the English edition positions them at the very end. The reasons for this are two-fold. First, as indicated, these writings on method stem from a different set of manuscripts, the B III 10 series and are not part of the lecture series on "Transcendental Logic." Second, these manuscripts focus explicitly on the question of genesis as a theme for phenomenology and on the difference between static and genetic phenomenological methods. In fact, they represent the *first explicit formulation* of this difference. To locate them in the middle of the volume risks losing them among *thematic* matters, when the question of *method*, genetic method, is essential to the undertaking of the *Analyses Concerning Passive and Active Synthesis* and actually makes them possible. Hence, though they are placed at the end of the edition, they deserve their own prominence in the constellation of issues raised in the *Analyses*.

Because genetic method is at the heart of these lectures, and in order to compensate for the sketchiness of Husserl's presentation of the ideas surrounding static and genetic phenomenology, I devote the next section of this "Introduction" to clarifying the question of method, with a particular emphasis on the difference between static and genetic methods.

2. Genesis and Genetic Phenomenological Method

It would be misleading to characterize phenomenological method only as a way of circumscribing modes of givenness, since the phenomenal field on its own part can overstep the bounds of a pronounced or presupposed methodological orientation, demanding the formulation of a new methodology. This is the position in which we find Husserl and his phenomenological philosophy by 1921. For it was at this time that Husserl was led to formulate explicitly the difference between static and genetic phenomenological methods.

Husserl's writings on static and genetic method not only mark Husserl's explicit effort to formulate systematically a difference internal to phenomenological method in terms of the static and the genetic, they also show the distinctive traits of each method, how the methods are to be organized in terms of the motivational descriptor of guiding threads or "leading clues." To question back is to question after founding relations of validity, and this for Husserl means an inquiry into genesis.

To be sure, Husserl was not the first to distinguish between static and genetic elements of experience. Husserl himself suggests this by referring to the difference between static and genetic method in the same terms Dilthey used for psychology, namely, as "descriptive" [*beschreibende*] and "explanatory" [*erklärende*].[12] Lurking in the background is not only Dilthey, but also Brentano and his distinction between descriptive psychology and genetic and physiological psychology.[13]

To cite these historical precedents is to acknowledge that phenomenology did not develop in a vacuum; but it does not mitigate the originality of Husserl's own phenomenological distinctions no matter how tardy they may seem to the contemporary reader. Because Husserl had described genetic matters that exceeded the scope of static constitution, including phenomena like apperception, normality and abnormality, kinaesthesis, association, etc.—phenomena that came under the general title of "primordial constitution"—Husserl was provoked by the very matters themselves to catch up reflectively with his own descriptions. This means that Husserl had undertaken genetic analyses implicitly without phenomenology having been explicitly cognizant of itself as having this genetic methodological dimension, and that the distinction between static and genetic methods is internal to the movement of phenomenology.[14]

---

12 See Wilhlem Dilthey, "Ideen über eine beschreibende und zergliedernde Psychologie" (1894) in *Gesammelte Schriften: Band 5. Die Geistige Welt*, ed., Georg Misch (Göttigen: Vandenhoeck & Ruprecht, 1957), 139-240. Whereas Dilthey takes description as interpretive description and explanation as something the natural sciences do, Husserl takes descriptive phenomenology in a narrower, "static" sense in order to contrast it with a genetic phenomenological research perspective that takes up an interpretative position with respect to the teleological genesis of sense. See Robert Scharff, "Non-Analytic, Unspeculative Philosophy of History: The Legacy of Wilhelm Dilthey," *Cultural Hermeneutics* (1976): 295-331.

13 Franz Brentano, *Psychologie vom empirischen Standpunkt*, second edition (Leipzig: Meiner, 1924).

14 T. Sakakibara gives us one example of this, locating Husserl's turn to a genetic phenomenology in Husserl's concept of the pure ego with its habitualities both from the

Looking back from our privileged perspective, with the distinction between static and genetic method and matters already in hand, we can say that Husserl's initial preoccupation was with matters and an approach that are "static." By static we understand two things: first, a constitutive approach that is concerned with *how* something is given or *modes* of givenness, and second, a concern with *essential structures*. In Husserl's terminology, a static method can address both strictly "phenomenological" (i.e., constitutive) as well as "ontological" (i.e., essential) dimensions of experience. Thus, a static approach can interrogate the interplay of intention and fulfillment, the meant features of an object, the noetic qualities of an act, etc., as well as the structural or essential possibilities of the particular object or act within the intentional correlation. Here one would examine the structures and the being of these structures (for example, formal and material essences, typicalities, regions, etc.).

The fact that Husserl actually began from a static research perspective betrays the following two-fold methodological prejudice: First, it was assumed that it is better to begin with constitutive questions rather than taking the being of things for granted, that is, it is more helpful to see how sense as constitut*ed* is given to the constitut*ing* pole of experience, and *then* to proceed to structural or ontological questions. Second, it was assumed that it is better, constitutively, to proceed with something at rest rather than something in process; it is advantageous to begin with the "simple," and then advance to the "complex." Accordingly, Husserl granted a methodological priority to an investigation into constitutive problems that did not broach the question of temporal genesis.

---

pencil manuscripts of *Ideas II* (1912) and in the main manuscript for Division 3 of *Ideas II* (1913) concerning the constitution of the spiritual world. See Tetsuya Sakakibara, "Das Problem des Ich und der Ursprung der genetischen Phänomenologie bei Husserl," in *Husserl Studies* Vol. 14, No. 1 (1997), 21-39.

Moreover, at least on Husserl's own account, his distinction between static and genetic matters pre-dates even this. For example, in June, 1918, Husserl writes to Paul Natorp that "... already, for more than a decade, I have overcome the level of static Platonism and have situated the idea of transcendental genesis in phenomenology as its main theme." Edmund Husserl, *Briefwechsel. Band V: Die Neukantianer*, ed., Karl Schuhmann (Boston: Kluwer, 1994), 137.

By genesis Husserl understands three variations of experience: (1) genesis within the purely active sphere of experience where the ego functions in rational acts, (2) genesis between the active and passive spheres of experience, where one traces the origins of activity in passivity (or between the judicative in the perceptual spheres of experience), and finally, (3) "primordial constitution" as a phenomenology of passive experience, including apperception, motivation, affection and association, kinaesthesis, etc. Ultimately these three dimensions of genesis are bound to the analysis of the genesis of the monad (a point that I address below).

The fact that Husserl began with static structures and a static model of constitution and not genesis does not mean that genesis was absent from the horizon of his thought, for *to prefer* stasis, even if it be in the form of ignorance, is already to acknowledge the problem of genesis for later work, implicitly rooting the problem of stasis in that of genesis—something that Husserl himself came to see. It is for this reason that Husserl provocatively asks whether one could even undertake a static phenomenology with the genetic dimension being fully suspended (see pp. 630f. and 633).

As noted, Husserl thought that the best way to handle more complex matters in phenomenology (like the problem of self-temporalization, or later the problem of cultural communities and historicity, in short, "generativity") was to prepare the groundwork with static investigations. Following such "preparatory" work, it would be suitable to proceed to higher constitutive levels of analysis.

Yet it was only after explicitly tackling the problems of genesis and more "complex" features of experience that Husserl *retroactively* understood the problem of genesis not to be more complex than that of stasis, but rather, more *concrete* and more fundamental. Likewise, static matters were no longer seen to be "simple," but now more *abstract*. This inversion was only discerned after having arrived explicitly at genesis through the leading clue of stasis, even though one could in no way derive genesis from stasis. Genesis has to be seen as more fundamental than stasis, though pedagogically, stasis guided us to the problem of genesis.

Ontological questions, questions concerning the being of things as they can be presented in the natural attitude, function as "leading clues" to constitutive questions, both static and genetic ones. While Husserl was initially wary of genesis (at least in the *Logical Investigations*, since empirical psychology imputed to ideal objects a subjective genesis in consciousness instead of taking logical entities as self-given to consciousness), he did confront the problem of genesis in a forceful manner after 1915. This is not to say that there are no themes peculiar to the problem of genesis earlier, say, as early as the *Logical Investigations* (1900-1) with his notion of motivation or association, or his *Thing and Space* (1907) with descriptions of the kinetic syntheses of perception and kinaesthesis. Husserl's own point is that these and similar analyses are still too implicit and abstract. Even Husserl's work on time-consciousness from his *On the Phenomenology of the Consciousness of Internal Time* is not really a full-fledged genetic analysis because it is too formal: "Mere form is admittedly an abstraction, and thus from the very beginning the analysis of the intentionality of time-consciousness and its accomplishment is an analysis that works on [the level of] abstractions" (pp. 173).

Remaining solely on the level of the temporal modes of givenness like impression, retention, protention, is still too formal, too abstract, and it is not until we get to the problems of association and affection, and the individuation of the monad that the problem of genesis really comes into play. This is due to the fact that an inquiry into the question of constitution is not necessarily an inquiry into the problem of genesis: "... attending to constitution is not attending to genesis, which is precisely the genesis of constitution and operates as genesis in a monad" (see pp. 644 and 639). The matter of genetic phenomenology, then, concerns monadic individuation and its genesis. What is monadic genesis?

The monad is a process of becoming in one unique time with one unique ego. As temporally enduring, the monad is not confined to a Now-point or a collocation of Nows, but exists as having been, a having been that transcends the past toward a futural becoming. As a uniform temporal form by virtue of horizons everything is related and interconnected to everything else in the dynamic unity of the monad. The monad is a "living unity" capable of having dispositions that are "unconscious." It is true that habitualities as the precipitations of acts are no longer actively conscious since they have receded from the living present and can become sedimented to the null-point of active vivacity; nevertheless, there is a dynamic interplay between act and affection because as expressing an "abiding style" or "abiding *habitus*," the habitual character of the monad can affectively provoke sense, prefiguring a perceptual or even judicative world from the density of that personal character that is not egoic. This density, by which the monad retains its identity passively and that is in part formative of the active ego, points to the concrete individuation of the monad: the fact that the monad is a unique "unity of its living becoming," a unity of its "sedimented history" that it bears as a heritage of the past (see pp. 635 and 637f.). Phenomenology of genesis then is the phenomenology of the primordial becoming in time, of the genesis of one shape of consciousness emerging from another, acquiring a temporal opacity. In short, it is a phenomenology of what Husserl calls at this time, "facticity."

In an important and revealing appendix to the *Analyses* (Appendix 8)—the longest one in this collection—Husserl describes the peculiar temporality of the individual within a genetic register. Transcendental subjectivity (i.e., the "natural" individual as clarified in terms of its constitutive powers and limits—neither the mundane nor the phenomenologizing ego), Husserl writes, neither is born nor dies. This immortality of the individual is due to two things. First, as transcendental subjectivity, the individual is constitutive of time, is the source of temporalization, and to this extent cannot be contained within time.[15]

Second, it presupposes that the phenomenological investigations are limited precisely by genetic parameters, parameters articulated by the former constitutive issues. Static phenomenology, let us recall, cannot account for any constitutive disruptions beyond the living present; here something like sleep would be a

[15] Making explicit an earlier insight in the time-consciousness lectures that consciousness as self-temporalizing cannot be understood as "temporal." See PCIT, 88, 345f.

constant barrier to analysis. Genetic phenomenology, however, can give an account of how remembering can potentially span constitutive gaps like sleep, fainting spells, etc. (pp. 151f.), but it cannot cover more radical breaks like birth and death. Only when phenomenology is broadened and deepened to a generative dimension do birth and death become integrated into constitutive world features because now one investigates the constitutive sense of the individual being constituted within generative homeworlds and alienworlds, hence, admitting of transcendental successors and progenitors, the phenomena of sharing births and deaths, being constituted by another (say, *as* "father" by a child), etc.[16] Within a genetic phenomenology, however, it is quite correct to insist on the immortality of transcendental subjectivity.

Transcendental subjectivity, as individuated in its personal orientation, is the concrete monad. But this concreteness is not tantamount to independence. Husserl writes, for example, that in contrast to his contention in the Third Logical Investigation (and in *Ideas I*, § 15, as well), what is concrete should be regarded as non-independent; only an analysis that makes abstractions can view "phases" as if they were concrete and independent. This holds not only for the temporal phases of impression, retention, and protention in relation to the concrete unity of the living present, or living presents as phases in relation to the concrete monad, but *mutatis mutandis* for the phases of individuated monads in relation to an intermonadic community (see p. 639). Through a genetic account of monadic genesis, a static, one-sided account of intersubjectivity is implicitly called into question.

Although Husserl distinguished between two different methodological orientations, simply naming two different methodological dimensions is not sufficient for describing the (structural) differences between stasis and genesis, for this would still remain static. Rather the very formulation of static and genetic methods and matters itself demands articulating the relation between static and genetic methods, that is, it itself requires a *genetic* description. The questions to be handled now concern "how the investigations

---

[16] See my "Liminality and Liminal Experience" in *Alter: revue de phénoménologie*, 6 (1998), 275-296.

are to be ordered," and working out the order of these "necessary phenomenological investigations" entails addressing "the leading clues of the system" (see pp. 633).

Expressing the differences between static and genetic methods in a relation of *leading clue* produces a ripple effect within transcendental method. First, one does not only move progressively from constitutive phenomenology to eidetic considerations, but now regressively from the natural attitude and essential structures (and the sciences of those structures) to constitutive matters. "Beginning with the natural attitude, one can also take the 'natural concept of the world' [i.e., the lifeworld] as a leading clue" (p. 633).[17] One begins with static method as eidetic analysis, which can take place "naively" within the natural attitude and all its rich implications, and then submit these results to a constitutive, properly speaking phenomenological analysis. And he writes: "*Is not static phenomenology precisely the phenomenology of leading clues*, the phenomenology of the constitution of leading types of objects in their being ..." (p. 644)? Within a static register now, one moves regressively to constitutive phenomenology.

Second, although we can find such a methodological reconfiguration implicitly at the conclusion of *Ideas I* as Husserl prepares to launch into a regional ontology of *Ideas II*, this reconfiguration of transcendental method comes into sharper focus for Husserl when the entire static method of investigation is placed in a relation of leading clue to genetic method. Here static ontology is not merely a leading clue to constitutive problems in general, but static constitution also becomes a leading clue to genesis. "Another constitutive phenomenology" named "phenomenology of genesis" is one that works from results of static constitutive phenomenology; a genetic phenomenology follows the histories of the constitution of objects that are there for the concrete monad as well as traces the genetic "history" of the monad itself (see p. 634).

---

17 This is evocative of what Husserl calls in his *Crisis* writings an "ontology of the lifeworld"; see esp. § 51.

however, does not mean that no ego is present, but only that the ego is not engaged in active participation.[18] This leads Husserl to a paradoxical formulation of the process as an "active passivity" in order to characterize a constitution and acquisition of sense that is, on the one hand, not nothing and is also somehow "subjective" (since there is in the intentionality of drive, or tendencies of desire toward an objectlike formation), but which, on the other hand, does not stem from "an activity proceeding from the ego."[19]

Seeking some historical anchorage for his concept of passivity, Husserl appeals to the notion of "productive synthesis" in Kant's First Edition of the *Critique of Pure Reason*. A productive synthesis, he writes, "is nothing other than what we call passive constitution, nothing other than the team-work ... of the constantly higher developing intentionalities of passive consciousness ..." (p. 410). Husserl cites the A edition of the *Critique* because Kant speaks of a faculty, the power of imagination, that is independent of the Understanding rather than being subordinate to and a function of the Understanding. Whereas the understanding has the spontaneous character of *active syntheses* that hold together and connect the sensuous manifold according to rules, sensibility has the character of *passivity*, since the inner and outer senses merely receive sense data. In this case, of course, the expression, "passive synthesis" is oxymoronic from a Kantian perspective.

But Husserl's analyses of "*passive* synthesis" challenge this schism between the sensibility and the understanding by describing intentionality as the interplay of intention and fulfillment as they both pertain to the perceptual and the cognitive spheres of experience. If truth is not alien to the sphere of sensibility (any more than intuition is to judgment), then passive syntheses are not without epistemic import, and a transcendental aesthetic cannot be foreign to the problems of truth, evidence, and their modalizations. Sensibility does make a contribution to the acquisition of knowledge, and an enterprise that wants to determine the limits, powers, and conditions of human cognition

---

[18] It is for a lack of available terms, writes Husserl, that we have recourse to the appositive "passive" even though this dimension of sense constitution is functionally operative and effective (p. 119).

[19] See Ideas II, 349.

(i.e., critical philosophy) must not only address active syntheses discernible in a transcendental logic, but it must be attentive to the unique and irreducible sphere of passive syntheses peculiar to a transcendental aesthetic.

3. The meanings of passivity are also determined *ex negativo*. Active processes include but do not exhaust making judgments, predicating something of an object, reflection, and linguistic acts. On this score, Husserl tends to regard passivity as basically equivalent to *perceptual, pre-predicative, pre-reflective, and pre-linguistic* experience, and thus with an implicit teleological orientation to reason. Because the constitution of space and time genetically and noetically understood also implies the presence of various bodily predispositions, passivity suggests a realm of bodily habits that were once actively acquired but subsequently have become *sedimented* into a style of comportment, and yet are accessible pre-reflectively. While one must not be too hasty in identifying the perceptual with the pre-linguistic (for language, too, becomes sedimented, and can become a passive acquisition without being distinctively aesthetic), Husserl does understand by passive, "aesthetic" experience, appealing both to the sense of the Greek *aesthesis*, and the constitution of spatio-temporal nexus through the kinaesthetic character of experience. This does not mean, as it would for Kant, that pre-meaningful accomplishments of sensibility are apprehended by the understanding in order then to constitute objects of knowledge, but rather that the origin of meaning must be sought in the sphere of passive intentionality, that aesthesis (passivity) is foundational for logos (activity), and echoing this relation of foundation, that a transcendental aesthetic should prepare the work for a transcendental logic.

4. Passivity is the sphere of *pregivenness and objectlike formations*. It is no coincidence that Husserl's use of the expression pregivenness [*Vorgegebenheit*] comes into full bloom in the years of the *Analyses*, and continues to flourish through out his subsequent investigations. For it is with the genetic explication of a level of experience (namely, passivity) in which objects are not (yet) constituted by an active ego that Husserl must speak in a way that evokes a mode of self-presentation that is not the givenness of an object in the genuine sense, namely, constituted

through egoic participation in temporal *acts*. And although this anticipates the discussion below of "affection," it is unavoidable to mention it here: According to the *Analyses*, something is *pregiven* insofar as it exercises an affective allure on me without being grasped by me as such, responsively or egoically.[20] Here, this "something" generally receives the appellation, objectlike formation [*Gegenständlichkeit*], that is, something that exhibits the basic structure of an object (including an object-phase), but is more "elementary" than an object in the full-fledged sense or has *not* (yet) exhibited objectivity.[21] This is not to say that the objectlike formation cannot have its own internal continuity maintained in passive time-consciousness, but only that it is not the result of active processes that give it an identity such that it becomes a theme of cognitive interest.

On the other hand, something is said to be *given* insofar as the *ego* yields to the allure and has turned toward it attentively, laying hold of it in egoic acts of interest, cognition, explication, or examination, etc. In any case, however, in order for an object to become an "object," it must first be constituted with the intervention of remembering.[22]

5. Passivity as the realm of perceptual and aesthetic experience is the *basis for* activity as the realm of cognitive and logical experience. The emphasis on the phenomenon of passivity in Husserl's writings arises about the same time as that of genesis, not only because there is a lawful regularity in the becoming of sense, but also because the inquiry into genesis implies the emergence of activity from passivity. In this respect, Husserl is

---

[20] This is certainly one sense of pregivenness. Another sense also occurring around the time of the *Analyses* is what is always already there for the individual who can in turn appropriate it either passively or actively.

[21] This is why I translate the term *Gegenständlichkeit* as objectlike formation, rather than objectivity or even objecthood. The expression *gegenständlich* I render "objectlike," and depending upon the context, "with objects," since Husserl sometimes uses the expression to indicate it as qualifying the noun: e.g., *gegenständliche Feld*. Here he does not mean a field that is like an object, but rather, a field filled with objects that can potentially become thematic.

[22] Cf. pp. 230, 298 and p. 615; and see Appendix 26: Repetition and Essential Identity of Rememberings.

In this work, Husserl will use both *Gegenstand* and *Objekt* interchangeably, and for this reason I translate them both as "object."

attempting to give an account of the genesis of reason through ascending modes of attentiveness. By passivity, then, Husserl means that which makes the active levels of experience possible; passivity is a "founding" level of experience, the "subsoil" of creative life, provides the "basic, essential conditions of the possibility of a subjectivity itself," and readies what can then be taken up in specifically egoic cognition (cf. pp. 163, 169). Accordingly, the perceptual, passive sphere has the character of being "originary" in relation to the judicative, active sphere.

In sum, one can say that passivity indicates a primordial regularity of sense-genesis in which the ego does not participate; it characterizes a pre-reflective dimension of experience of pregivenness of objectlike formations, a dimension that is founding for activity. Discerning such a sphere of passivity (and its implicit relation to activity) Husserl is able to describe the motivation from the passive to the active spheres under the rubric of fundamental laws of association.

B. Affection and Association

In our ordinary, everyday attitude, something is present as if the whole object were already there, ready-made. A phenomenological shift in perspective will show that such a presentation is only a pretension, and a two-fold one at that. First, the thing does give itself in perception as the whole object, but by its very nature it does so "paradoxically," incompletely in and through perspectives or modes of givenness (p. 39). Second, the object in-itself is taken as such by us, which is to say, the in-itself is only constituted as in-itself-for-us. Underlying Husserl's analyses into affection is his attempt to describe fundamental regularities in the constitution of sense through passive syntheses of association that show *how* something can present it-self-to-us as in-itself, and to investigate the implications this has for a science worthy of its name.

From at least 1905 to the end of his life, Husserl devoted literally thousands of pages to his phenomenology of time-consciousness. A phenomenology of time-consciousness describes the most general and basic syntheses that connect particular reductionistic, or behavoristic connotations, I translate the term *Reiz* as "allure," rather then, say, as "excitation" or "stimulus."[28]

By affection, then, Husserl does not mean a contextless power or a third person force; rather, he understands an "enticement to be" on the part of the "object," a *motivational* solicitation or pull to attentiveness, eventually to respond egoically and epistemically, though the response does not have to be egoic; in fact, even if there is an egoic response in the form of an active turning toward, this not imply the cessation of the affection (e.g., pp. 91, 196, 523). Intrinsic to the notion of affective allure is the notion of motivation. Let me clarify this here.

2. *Motivation.* The concept of motivation was explained in *Ideas II* in contrast to a law of cause and effect peculiar to the "naturalistic" attitude; it is a "because-thus" relation of efficacy peculiar to the "natural" (natural in the broad sense) or "personalistic" attitude, a relation that concerns not things of a natural-scientific reality, but givennesses of experiential actuality.[29] With respect to *Ideas II* one finds the concept of motivation greatly deepened in the *Analyses*. The concept of motivation is deepened in three ways. First, motivation pertains to both passive and active spheres, and the relationship between the two. Second, motivation is described in the context of modalizations of being and belief. Third, it is brought into the framework of affection.

A. Since I have already discussed the significance of passivity above, and address the issue of active synthesis below, I will not dwell on this point here. Suffice it to say that motivations can be functional within the passive and active spheres in different ways, and in a particular instance, motivations in the passive sphere can

---

[28] R. Rojcewicz and A. Schuwer, for example, translate *Reiz* as "stimulus" in their translation, *Ideas II*. With *Reiz* in the sense of allure, however, one could also speak of attraction, appeal, or lure.

[29] Perhaps one of the best formulations of the concept of motivation in Husserl comes from Merleau-Ponty: one phenomenon *releases* another, not by virtue of some objective efficacy connecting natural events, but by the sense that it offers; there is a "*raison d'être*" that orients the flow of the phenomena without being explicitly posited in the "awakening" or the "awakened" and which functions as an operative reason rather than a judicative rationale. Maurice Merleau-Ponty, *Phénomènologie de la perception* (Paris: Gallimard, 1945), 61.

provoke an active response. Relations of motivation permeate all dimensions of intentional life: broadly speaking, with the active sphere in terms of egoic motivations of interest; within the passive sphere, e.g., in terms of a fusion or connection of matters within a sense-field, or again, in terms of kinaesthetic motivations; and in the transition from the passive to the active sphere, typified in the advent of egoic awakening through an attentive turning toward, a turning toward instigated by some pregivenness turned givenness.[30]

B. Among the various possible modalizations of evidence that can occur: doubt, annulment, repression and superimposition, disappointment, negation, affirmation, it is possibility that stands out as one of the most significant for the issue of motivation. Here Husserl distinguishes between open and enticing possibilities. An open possibility is putatively a possibility that can be fulfilled in an indeterminate scope of determinations, where nothing in particular is called for, just, say, the scope of "color," but not any color in particular. An enticing possibility, on the other hand, will exercise a demand for a particular determination. Some possibilities will have more "weight" than others, some will speak in favor of or against a possible sense. Some will be overpowering. Thus, motivation is clarified in terms of a demand on the part of the object that entices its very constitution in varying degrees of weightiness. From the side of the "subject" motivations are explicated as soliciting appearances in the world; the subject, too, has a kind of power to usher a world of appearances, but because this is a co-relative structure, the subject's prompting is not an absolute imposition (p. 152).

From Husserl's explication of enticing possibilities, we can glean the noetic and noematic components of motivation. Noematically speaking, motivations stemming from the objective vector of the intentional relation can be understood as a "propensity to be"; noetically speaking they can be understood as a propensity or enticement to believe and "to turn toward" attentively (see §§11–13). (It is at this point that we can see a

[30] See §13 and Division 2, Chapter 4.

transition from the passive sphere of givenness to the attentive turning toward in the lowest active level, namely, "receptivity.")

C. But all of this still takes on one further step of exposition, for now the notion of motivation, interpreted in terms of enticing possibilities, is articulated in the context of affection—the topic of Division 3 of the analyses. Although Husserl will not give detailed phenomenological descriptions of affection in Division 2, he nonetheless introduces affection in the framework of motivations that function as enticing possibilities. This is more implicit than explicit. Here Husserl speaks of the demand of the enticing possibility as an "affective force" that is "in relief" or "prominent" (pp. 85, 90–1), of the weight as being "strongly affective," and noetically, of the fact that I am motivated not only when the thing exerts an affection or enticement on the ego, but when I yield to the affection either passively without egoic attentiveness, or actively in an attentive turning toward (pp. 86, 134).

Given the connections that Husserl makes in Division 2 between motivation, enticing possibilities, and affection, and given Husserl's later conclusions on the status of affection in the constitution of sense, we would have to read the latter back into the former and determine that concretely and phenomenologically, there really are no purely open possibilities, so that sense is always exerting a demand to be with some degree of intensity. And although Husserl himself hardly makes such a claim, one would also have to assert that from the phenomena of protention to epistemic striving, all consciousness is to some degree already affectively colored (cf. p. 277).

3. Affective Prominence. Gestalt psychology has shown that something can only be perceived as a figure against a background, and that this figure/ground organization is an irreducible perceptual structure. What marks Husserl's contribution is not his appeal to a perceptual Gestalt, or figure/ground structure[31]; this is already

[31] Holenstein notes that the "figure-background" structure was systematically described for the first time by E. Rubin, *Visuell wahrgenommene Figuren* published in 1921 (Elmar Holenstein, *Phänomenologie der Assoziation: Zu Struktur und Funktion eines Grundprinzips der passiven Genesis bei E. Husserl* (The Hague: Martinus Nijhoff, 1972) 287, fn. 37). On the relation between Gestalt Psychology and Husserl's thought, see Holenstein, 275-302.

operative in his thought. Not presupposing ready-made objects or taking for granted their genetic constitution, as in a data-sensualism peculiar to certain psychologies and epistemologies of his time,[32] Husserl's genetic inquiry demands asking whether or not the formation of sense-unities is given prior to or only along with affective force: Are there unities of sense first constituted as prominent that only subsequently become affective, or does the actual emergence of these unities coming into relief depend upon the play of affective forces (cf. pp. 200ff.)? Because Husserl focuses not on an already determined world, but on the *emergence of prominence* in the genesis of passivity, on the very becoming of sense as is in the *process of becoming determinate*, Husserl's analyses lead him to the conclusion that something coming into relief is *an affectively charged relief; prominence is always already an affective prominence* (pp. 211, 221).

But in the living present there is not just one single ray of affective force on me, for *many things exercise affections and rival for my attention* to some degree or other. In a field of affective tendencies that rival one another for attention, some will be more or less significant than others, and these rivalries can occur within the same sense field or across sense fields. Even though there is an affective force, it is not necessary that it draw my attention to it; it may only be "perceived" in a passive attention as it knocks at the antechamber of the ego; it may not yet or ever achieve an active attention, cognitive or otherwise (pp. 214–15).[33]

Through affective interconnections, one thing is connected to another and charged with affective resonance, prior to active egoic involvement. Just because the ego does not turn attentively toward a particular allure does not mean that it is not exerting its pull on us; it may simply be that the ego is following out some other affective force, or something else is drowning out this affective allure. This is why Husserl rejects an all or nothing account of

---

It is interesting to note as a reminder that Husserl's first lecture of the *Analyses* was Winter Semester 1920/21.

32 Cf. FTL, 286-88

33 I encounter these rivaling forces with a *pre-reflective preferential directedness that is always selective/exclusive*, more precisely, a pre-reflective tendency toward the optimum as the principle of selection of practical possibilities (cf. Ms. D 13 3, 151a).

affection, and resorts to the analysis of the *gradation* of affective forces (cf. pp. 215ff.).

4. The Unconscious and Repression. A completely undifferentiated field of affective forces is only possible after something has exerted an affective force in some manner in the living present such that the zero of affective force takes place as *limes* in the retentional past, which in principle can be reawakened in a variety of ways. This subsequent dormancy of affective force corresponds to a zero-point of affective force, an affective zero-region that Husserl terms the "unconscious."[34]

When we hear the expression, "the unconscious," most of us immediately think of Freud. Husserl's phenomenology of the unconscious, however, must not be immediately assimilated to Freud's meta-psychological treatment of the unconscious. Briefly, for Freud the unconscious is formed by repressed contents that are affect laden drive or instinct representatives, governed by primary processes, e.g., condensation and displacement; while they are not given straightforwardly to consciousness, they can make it circuitously to consciousness by passing the "censor" in disguised form; they can emerge in jokes, dreams, slips of the tongue; in this way, what is repressed can return to consciousness and become accessible in a new way.[35]

But for Husserl the unconscious is the nil of the vivacity of consciousness, a nil, which he clarifies, is not a nothing; it is only a nothing of affective force that has arisen from original constitution, a nothing with respect to those accomplishments that presuppose an affectivity above the zero-point. The unconscious, then, is the realm in which objectlike formations are no longer affective. Still, they can be reawakened because the constituted sense is implied in the unconsciousness. Awakening occurs through various passive and active associative syntheses, an awakening produced by virtue of an affective communication.

Now, when Husserl uses the term "repression" he means generally the affective eclipse of one thing by another within a

[34] See, for example, pp. 201, 216ff.

[35] See, for example, Sigmund Freud, "Das Unheimliche," in *Gesammelte Werke*, Vol. 12, ed., Anna Freud (Frankfurt am Main: Fischer, 1947), 229-68

field. Something emerges with more force than something else, thus repressing it, even though what is repressed can continue to exert its allure. In essence, the repression that takes place in this instance is not due to one element being in the present as opposed to something else being in the past; rather, the key to Husserl's epistemology of repression lies in affective force: Something represses another because it exercises a stronger affective force in this particular context. It is only because he tends to identify the strongest affective force with the present that the present is understood to repress the past, which it in fact can do. And in this case, a dormant retentional affection can surface in the present once more by provoking an active turning toward, namely, through remembering.

Still, the identification of the present with affective force is called into question when, for example, we have two or more things *in the present* beginning to exercise an affective force, issuing in a conflict in the perceptual field. The eclipse or repression of affective forces, where the repressed element still exerts an affective force, triumphs not because it is present *per se*, but because of the power that the affection exercises in this situation. As a consequence (and genetically speaking) something becomes present in and through its affective force, winning out. Insofar as both the repressed and the repressing objectlike formations are still affectively charged, and not the nil of affective vivacity, they cannot be ascribed to the unconscious. Accordingly, this epistemic "region" would warrant the title of the unconscious, not because it is not actively awakened—for this would be to equate the passive sphere (the "pre-consciousness," "pre-givenness," even the habitual lived-body) with the unconscious—rather, there are still affective forces in passivity that are not nil, as well as modes of attention, even if they have not surpassed the threshold of active awakening.[36]

[36] For a more detailed discussion of themes relating to passivity, affection, and the phenomenology of the unconscious, and on the general problem of the genesis of logic, see Bruce Bégout's insightful work, *La généalogie de la logique*: *Husserl, l'anté-prédicatif et le categorial* (Paris: Vrin, 2000). See also, Anne Montavont, *De la passivité dans la phénoménologie de Husserl* (Paris: PUF, 1999).

5. Self-Givenness and the Object It-Self. The insight that something coming into relief is always already an affective relief suggests that an affective force is always exerted on the ego, or at least, on a "subject" that can become egoic and exercise an egoic response. It confirms that intentionality is not a structure that stems one-sidedly from consciousness; it is a dynamic co-operative structure, *a "constitutive duet"* (cf. p. 52).[37] Since both sides of the intentional correlation are operative, it is no coincidence that Husserl not only attributes ipseity to the subject, but also to the object. Legion throughout the *Analyses* are references made to the "self" [*Selbst*] of the object. This points not only to an identifiable core that makes up the object, but to the fact that the object is not reducible to consciousness, that it has its own kind of density and otherness that both solicits and evades us, and that one cannot arbitrarily prompt it into being. The object holds itself back, at a distance, which is precisely what allows it to give it-self in an intentional relation.

The more common expressions like *Selbstgebegenheit* (with which Husserl means both the giving of the self of the object on the part of the subject as a noetic process, and the self-giving of the self of the object from the object), I render as "self-givenness"; in this case, one should hear in the expression "self," then, not the subject, but the ipseity of the object, the self-givenness of object it-self in the intentional relation.[38] As such, the object it-self is never experienced in a neutral manner; it exercises an affectively significant allure on the perceiver to be constituted as such, that is, for the "ego" to turn toward it attentively and to constitute it as a theme of interest in an active manner. The fact that some*thing* is actually heard or seen or smelled, etc., is due to "affective rays" radiating from the object, drawing in its wake the horizonal referential implications.

But, what more precisely is the object it-self? The object it-self, the object in-itself-for-us, is the normatively significant

[37] It is because of this affective pull that the object can also call out to us from it horizonal referential implications, providing the subject with a kind of pre-understanding that is elicited by a particular side of the object beckoning us to examine it more closely.

[38] Where the context is not always clear when Husserl simply uses the expression "Selbst," I will occasionally translate it with the interpolation: self [of the object],

"optimal."[39] The genetic concept of the optimal (or optimum) replaces or rather deepens the static notion of the "noematic core" of *Ideas I* and clarifies the normative significance of the "objective sense." Not only is normality defined by concordance [*Einstimmigkeit*] as a series of harmoniously interlocking appearances, but as the advent of norms in terms of the maximum of richness and differentiation in a unity. Thus, the thing itself, its true being, is the optimal, or the system of optima, as it is experienced as "saturated" for a perceiver in the context of the lifeworld (p. 61–2). All other possibilities fluctuate around it; they become constituted in experience as "abnormal," i.e., indexes to the optimal as *their* norm of givenness (Hua XIV, 121-23, 134; Hua XIII, 379.)[40]

As the thing it-self, the optimal is not only what is or what was, but what will be. As such, it functions as a norm that orients and guides a concordant course of present perceptions from the future, which is to say, it prescribes a course of possible confirmation and of possible verification that the active ego can discover (pp. 266–69).

6. Association, Remembering, and the Object. Husserl's investigation into affection is grounded in his attempt to describe how something can present it-self-to-us as an in-itself, and to investigate the implications this has for phenomenology. The self of the object is constituted as such when the objectlike formation emerges from indeterminacy into prominence; it does this through the allure of the objectlike formation that rivals for determinacy by exercising an affective force on the perceiver as an invitation to be. One of the motivations for the whole Division on association is prefigured in Division 2, namely, the investigation into the origin of truth and into the role that remembering—as constituted through associative motivations and as a source of

[39] D 13 IV, 26: "Das Optimum ist also ein relatives Maximum des geübten Reizes, dem nun auch ein Interesse, eine Zuwendung des Ich, folgen mag. Das Ich ist nun bei dem es affizierenden ...." And D 13 IV, 28: "Jedes außer-zentrale Bild, das das Interesse erweckt, leitet dieses Interesse fort zu seinem Optimum und weckt ein Streben, den geraden Weg dahin zu durchlaufen, um es zu realisieren."

[40] D 13 I, 242b: "Das normale Ding fungiert im Leben zugleich in dem anderen Sinne als normal, als es zur Norm für die anomalen Erscheinungen dient."

definitiveness—plays in constituting the identical object beyond the momentary present. It is his phenomenology of association that then allows Husserl to return in Division 4 immediately to the issue of the in-itself as the true self of the object.

Although there are many different ways of becoming attentive, even within the sphere of passivity itself, the privileged case of becoming attentive for Husserl is the transition from passivity to activity through an egoic "turning toward" the pregiven affectively charged objectlike formation, in turn, constituting it as "object" and now as "given" to consciousness. These affective awakenings or affective interconnections are understood as "associations" by Husserl, and while associations include passive syntheses that accomplish unity and diversity (through pairing, similarity, uniformity, heterogeneity, fusion and contrast, etc.), Husserl favors the affective awakening or association as a specifically *reproductive* temporal act that confers an identity or an ipseity on the object over time, namely, remembering. Before discussing the concept of remembering here, and its yield, let me first clarify Husserl's phenomenological notion of association.

Needless to say, the theory of "association" has had a long history, and is perhaps most well-known from the English empiricists and the tradition of empirical psychology, especially Locke and Hume, but also Hartley, and a little later, Brown and Mill, and still later Spencer.[41] In connection to Husserl, one should not overlook the movement in the German psychology of the 19th and early 20th century, represented especially by Lipps, Herbart, Wundt, and Stumpf.[42] The notions of resemblance, contrast, contiguity, vivacity, coexistence and succession, similarity and

---

[41] John Locke, *An Essay Concerning Human Understanding* (1690), esp. Book II, Chapter 33. David Hume, *A Treatise of Human Nature* (1739), I, 1, 4f : "Of the Connection or Association of Ideas." Hartley, *Observations*, esp., Part. I, 65-83. Thomas Brown, *Lectures on the Philosophy of the Human Mind* in 3 Volumes (Andover: Mark Newman, 1822), esp., lectures 34-37; John Stuart Mill, *Analysis of the Phenomena of the Human Mind* (London: Baldwin and Cradock, 1829), esp., Vol. 1, Chapter 3; Herbert Spencer, *The Principles of Psychology* (London, 1899), §111 ff, 129 ff.

[42] Lipps, *Grundtatsachen*, esp., Division 3; Johann Friedrich Herbart, *Lehrbuch*, 30 ff. and 132 ff.; Wilhelm Wundt, *Grundriß der Psychologie*, esp., 264, 268 ff. Carl Stumpf, *Erscheinungen und psychische Funktionen*, (Berlin: Königli. Akademie der Wissenschaften, 1907), esp. 22 ff. Supplementary Texts, Appendix 17.

contrast, fusion, habit—all notions peculiar to Husserl's *Analyses*—are present in some form or other in these authors.

It would be fruitless to try to duplicate the historical documentation and philosophical acumen that Elmar Holenstein has brought to the problematic of association as it bears on Husserlian phenomenology and in particular, on the *Analyses*.[43] Here, I want to note two things. First, Husserl's own contribution to the theory of association is to have taken "association" out of both an objective and subjective explanatory framework by submitting it to a phenomenological clarification, which is to say, for Husserl association designates an *a priori* essential lawful regularity of *genetic*, "passive" constitution of sense, irreducible to habit, custom, or mental processes; he brings associative syntheses into an intentional and *motivational* (and not causal) nexus concerning affectively significant primordial phenomena and their role in the awakening of affective tendencies toward attentiveness (association even dovetails at times with "apperception" as the transference and implication of sense); thus, association is not a juxtaposition or succession of objective facts; rather, it designates an internal referring; and finally, association serves as a way of explicating how objects get constituted as such in the transition from distinctively perceptual functions to cognitive acts.

Second, Husserl's theory of association challenges a form/content dualism. Since an actual formation of sense necessarily presupposes affective force and affective differentiation, affection is an essential condition for every kind of synthesis, and without it there would not be a living present articulated with objects (pp. 212–13). It is primarily due to Husserl's phenomenology of affective force as co-eval with the emergence of prominence that Husserl's phenomenology of association is able to liberate his theory of constitution from a sensualism and a dualism. It is well-known, for example, that Husserl's earlier theory of constitution was articulated in terms of a form/content (hyle/morhpe) dualism whereby an interpretative intentional apprehending act is said to animate the intimately inherent (non-intentional) material sense-data (something that enabled Husserl to account for the fact that,

---

[43] Holenstein, *Phänomenologie der Assoziation.*

can only be an object when I can return to it again and again as "the identical," which presupposes remembering (pp. 141, 614f.). But as we noted, the object is also the futural object, and so the future too plays its role by projecting a norm around which perceptual fulfillment be achieved, and in this way expectation has a constitutive function in the constitution of the object it-self.

In the *Analyses*, Husserl does not generally differentiate between different active memorial levels[46] (only between retention and remembering) neither does he differentiate between various levels of active futural orientation (but only between protention and expectation, which he virtually equates with anticipation). Nevertheless, no matter how inchoate, remembering and expectation serve as transitions from passivity to activity; they can both be modes of active attentiveness, and they are formative of more active levels of objectivation.

C. Activity and Active Synthesis

If we were to live only in passivity, contends Husserl, and if it were not possible for us to carry out free activity, we would be blind to the sphere of true being. Part 3 of the *Analyses* picks up precisely here with the descriptions of higher and higher levels of attention and processes of active objectivation. In fact, having once moved through careful phenomenological analyses from the level of passivity "up" to that of activity, Husserl—in typical fashion—reinterprets this "higher" level now as "concrete" and understands that passivity and activity are really only phenomenologically designated internal differentiations of cognitive experience, and that we can only gain the level of experience of passivity for reflection by presupposing and *abstracting* from the accomplishments of activity. It is this movement within the concrete whole of experience that enables Husserl to consider explicitly in Part 3 the genetic transition from passivity to activity

[46] Though on one occasion he does speak of a necessary "self-forgetfulness" in memory or a non-egoic remembering where in memory, the ego is not "awake." See pp. 595–96.

and to describe the roles of affection and attention in levels of experience peculiar to active cognitive life.

Remembering and expectation are necessary but not sufficient conditions for an object to be constituted as such. In addition, what is required for something to be constituted as an "object" in the "complete and genuine sense" is that it be constituted as identical with itself and that it be originally constituted as the thematic object for the ego in identifying activity (p. 297–98).

A consciousness of the "object" is genuinely carried out only first in egoic acts, and is only there as object for the active ego. "Identical sense," "being," "modalities of being," "true being," "verification," get their genuine character only first within the framework of activity (p. 275). One of Husserl's goals in the *Analyses* is to understand active objectivation as a genetic gradation of thematic accomplishments. Since the intellect is a name for the constitutive accomplishments of objects that the ego has given to itself through activities of identification, and since there are as many modes of identification that we have for a theme as there are modes of objectivation, we can expect a spectrum of so-called "intellectual" or "spiritual" *acts* that differ in manners of generating the object. Each level entails a new "attitude," new active syntheses, and a corresponding alteration in thematization: From cognitive interest, explication, determinative identification, conceptualization, to as-such judgments. Further, each stage or "attitude" can admit of further differentiation. For example, just within the level of conceptualization, one would have to distinguish between concepts (operating apophantically in meanings and judgments) and essences (operating ontologically in simple and categorial objects), and further, between various levels of essences from empirical universals, to morphological essences and types, to pure *eide*. To make all these differentiations explicit is far beyond the scope of this Introduction. Let me merely sketch the broad parameters of movement mapped out by Husserl to higher levels of active synthesis.

If receptivity is the bridge, as it were, from passivity to activity, and if the awakening peculiar to this transition initially entails a submission that motivates an active turning toward, then what Husserl calls cognitive interest goes beyond receptivity by not merely focusing on a theme, but by taking it as an object that is in principle explicable as object: Cognitive interest naturally functions a spur to processes of explication, yielding an examination into the theme that a) delves into its content, and b) extends beyond the theme to other objects.

Examination is a special cognitive process in which several things are called to our attention in a unity, whereby one explicates a substrate or a subject in its properties: Now one says "S" is determined by the partial self-identification as α, β, γ, whereby the "S" remains one and the same "S" in the unity of this activity, as it progresses to newer and newer concentrations. What was merely a content of the theme (e.g., red), now becomes the thematic object; but here, what has gradually become the special theme enters into the partial identity with the S that is still retained. In this way, the interest in the object that goes from α to β is fulfilled in the concentration of each moment or special theme, and what we acquire is an enrichment of sense in this synthesis. Even though the special themes of the object are actively made explicit as a way of clarifying "S," the synthesis of coinciding that forms identity and determines their relation and the coinciding of S is *not active*, rather, it is a result of a special *passive* synthesis (pp. 339ff.). Moreover, this passing from object to object that is held together by passive syntheses are excitations for possible determinations, but it itself is not yet the activity of determinative synthesis, it is not a judging.

Judgment is a determining process that *actively* relates one theme to another such that S is determined as the form of the subject, and the α as the form of determination, etc.

If we focus on the "S" that is enriched in sense, "S" becomes an object of a new apprehension; now an active movement in the transition from, say, "S" to α is intent on generating the element accruing to "S." Now we have the active consciousness that the "S" undergoes determination by being expressed as "S is α" or "S contains α."

According to Husserl, the proposition is a correlate, the "What" of the judicative act, and in this regard is the "judicative proposition," the *propositio*. When we substantivize the proposition in the form of "that" statements ("that S is α"), we have a new level of objectivation that fulfills the proposition, the judgment maintained as valid, we have an state-of-affairs or *Sachverhalt* (cf. §62). This is the foundation for other possible determinative judgments: the relation between whole and parts, the combination of parts and their forms of connection, the relationships between the elements as connected in these forms, etc.[47]

In each case, it is an enrichment of sense that gives occasion to turn to activities of determination that form relations. The universal investigation into the categorial realm, according to its forms, and pure laws determined by these forms, laws of true existence of states-of-affairs, make up the discipline called "formal logic." This is the systematic disclosure of radically different genera of relations that are possible for the object, in formal generality, that is, forms and pure laws of states-of-affairs; it is a matter of a formal classification and axioms of relations. When we are motivated by the association of similarity and our direction of categorial interest is trained on the universal, when the universal and its relation to particulars becomes an explicit issue for thought, and when we judicatively assess the objects as such, we have that mode of synthetic cognition Husserl calls conceptualization.

Grasping the One or the universal has a unique field of interest. In this case, we do not merely attribute similar features to different objects: "The bookcase is brown," "The desk is brown," "The cup is brown"; rather, now we mean by "brown" the same color in all instances, subsuming the individuals under the *eidos*, "brown": Thus, no longer "S' is β'," "S'' is β''," etc., but rather, "S' is β," "S'' is β," "S''' is β," etc. "For now" writes Husserl, "the interest, the ray of attentiveness must pass clear through the individual objects that are already constituted, and by traveling along the path of uniformity and carrying out the coinciding, the One that is constituted becomes thematic" (p. 350). This can yield a still new active determination. By focusing on the particular aspects, they will be shown to all have a peculiar relationship to the universal,

---

47 A whole is not an state-of-affairs, but an state-of-affairs is a whole insofar as it can be divided up (§59).

namely, as a *particularization of the eidos*. Here, the new judgment arises, "This is an α," i.e., it is a particular, an instance of α.

In turn, the field of openness that is constituted with the universal, (the universal of repetition with unending scope) unleashes the "whatsoever" or *Beliebigkeit*. We can choose any "this" or "that" whatsoever as the determinative theme; not only do new species emerge, but new thematic formations. Accordingly, it is one thing to have this rope before us thematically, and another to make a judgment about it *as such*, about any rope whatsoever whereby this rope gets constituted as an example.

By assuming the as-such into a theme, we have a conceptualizing function of universal and particular conceptual grasping. We can also operate in the scope of pure concepts, in a manifold of possibility; here we gain pure as-such judgments, like judgments of laws, like "A triangle has three angles"; or possibility, "A triangle can have a right angle." We can modify this in free variation and find a lawful regularity. "With this, the realm of universal judging is opened up, the realm of law-giving, the knowledge of law-giving for itself and for all judgments as such, and therefore for all possible objectlike formations" (pp. 354–55).

As these syntheses and categorial objects occur in *natural experience*, they are not yet "critical," and must be submitted to a constitutive or transcendental clarification. Still, as we noted at the outset, the scientific and ontological attitudes, formal logic, mathematics and regional or material ontologies that are constrained by the objects in question, can become clues to a transcendental analysis as an inquiry into the origin of truth. Beginning with a transcendental aesthetic, we will have the foundation for a broad scope of transcendentally formed disciplines, like a transcendental physics, as the transcendental science of space and time, a transcendental psychophysics and psychology, a transcendental sociology.[48] But we will also have the foundation for the most

48 In his later writings, Husserl will regard not only the lived-body as constitutive of spatiality, but the Earth *qua* earth-ground as constitutive of the lived-body and the spatiality of nature. (See, for example: Edmund Husserl, "Grundlegende Untersuchungen zum phänomenologischen Ursprung der Räumlichkeit der Natur"

universal theory of norms of all sciences in principle, transforming them and giving to them their justification, namely, a transcendental logic.

4. Note on the Translation and Acknowledgments

The writings that make up the English edition of the *Analyses* draw from four different sources published in the German Husserliana Critical Edition. In the order of the presentation of the English edition and translation they are as follows: Hua XVII (*Formale und transzendentale Logik*), Hua XI (*Analysen zur passiven Synthesis*), Hua XXXI (*Aktive Synthesen*), and Hua XIV (*Zur Phänomenologie der Intersubjektivität*). The pagination given in the margins of the English text correspond to the pagination of these editions. All of the German editors' insertions are indicated by angled brackets < >, and the translator's insertions by square brackets [ ].

The *Analyses* are lectures. As such, they were never written with an eye to publication. The advantage is that the lecture format of these presentations provides the reader with a candor not seen in some of his other well-known writings intended for publication; we see Husserl addressing the participants in his lecture directly, and we observe Husserl working out insights and we witness a thinking in process. But this also means that much of what he wrote could be elaborated upon or modified in the course of his lecture. Its disadvantage is that this writing style does not make for the most cogent of presentations, and this is particularly dramatic in the case of Husserl. He is already known for his laconic style; on this score, the situation is only exacerbated.

(1934) in *Philosophical Essays in Memory of Edmund Husserl*, ed., M. Garber (Cambridge: Harvard University Press, 1940), pp. 307-25.) These analyses belong as well to a transcendental aesthetic begun here. Such transcendental problems will ultimately concern the co-generation of a generative nexus in terms of normatively, geo-historically, and socially significant "lifeworlds," i.e., homeworlds and alienworlds peculiar to a "generative phenomenology." See Anthony J. Steinbock, *Home and Beyond: Generative Phenomenology after Husserl* (Evanston: Northwestern University Press, 1995).

But this has also meant that many of the connections he makes to earlier sentences or references are ambiguous, connections that he could have easily supplemented in his oral presentations. Rather than try to duplicate the ambiguity or even incomprehensibility of Husserl's text, I have often had to become an overt interpreter of this work, either by substituting nouns for pronouns, or by intervening with explications and/or interpolations supplied in square brackets []. Where the difficulty in terminology in concerned, the reader should be reminded that Husserl was still in the process of inventing phenomenological language by borrowing from the resources of "natural language," which is at root not bothered with "phenomenological distinctions" (p. 536). Where a common, technical term is used, or where I want to suggest that multiple interpretations of a term are called for, I cite the German expression in a footnote. Finally, some of Husserl's sentences are laboriously long; at times a paragraph can go on for pages. In these instances, I have simplified the syntax by breaking down the sentences, and where the longer passages contain several distinctive thoughts, I have broken down the text into shorter paragraphs.

While undertaking the translation and edition of such magnitude I have benefited from the support, advice, and assistance of several sources and persons. I owe a particular note of thanks to the Southern Illinois University at Carbondale (SIUC) which provided me both with a grant covering three years of funding for this project in the form of travel monies, commodities, contractual services, and a research assistant line, as well as an additional grant in the form of a Summer Research Fellowship. I am also grateful for a stipend from the National Endowment for the Humanities (NEH) covering two months of Summer work on this translation project (1999). The assistants at the Husserl-Archives in Leuven, and the director of the Archives, Rudolf Bernet, have been supportive in providing me with material and editorial advice. I would also like to thank Roland Breeur for keeping me supplied with advance renditions of "Aktive Synthesen," which has now become "Part 3" of the *Analyses*. Thanks are also due to Rudolf Bernet for allowing me to cite from unpublished manuscripts.

It has also been my fortune to have had capable and amiable research assistants. If this translation has accomplished its goal of providing the reader with an intelligible English rendition that is both accurate and faithful to the sense of this celebrated work of Husserl's, it is due in no small part to the diligent readings by and enlightening conversations with my research assistants, Ms. Stephanie Windolph (1997), Ms. Tanja Stähler (1998), and Ms. Christiane Hochkeppel (1999), all distinguished visiting doctoral students from the Bergische Universität Wuppertal, Germany. Thanks are also due to several people who have read portions of this manuscript and made helpful suggestions on the translation: Jim Hart, Sebastian Luft, Ullrich Melle, and Donn Welton, as well as to those who have offered comments on this "Introduction": Rudolf Bernet, Roland Breeur, Natalie Depraz, Christian Lotz, and Art Luther. Finally, I would like to acknowledge John Hartmann for the arduous task of compiling the index to the English translation.

On occasion, I have consulted Costa's Italian translation of Husserliana XI, *Lezioni sulla sintesi passiva*[49], and Bégout's, Depraz's, and Kessler's French translation, *De la synthèse passive*[50] either for advice or in order to control various technical terms peculiar to the *Analyses*. Finally, I owe special acknowledgements to Shin Nagai, with whom I first read the *Analysen*, to Donn Welton for having suggested this translation project, and to Leslie Brown, my spouse and best friend, for having encouraged it. Any shortcomings in the translation are solely my responsibility.

This translation is dedicated both to my son, Joseph, who spent many hours as infant and toddler nestled on my lap "helping" me translate, and to my daughter, Samara, who immediately followed suit.

---

[49] See citation above, fn. 4.

[50] Edmund Husserl, *De la synthèse passive: Logique transcendentale et constitutions originaires*, trans., Bruce Bégout and Jean Kessler, with Natalie Depraz and Marc Richir (Grenoble: Jérôme Millon, 1998).

BIBLIOGRAPHY OF HUSSERL'S WORKS FOR TRANSLATOR'S INTRODUCTION AND TRANSLATOR'S NOTES

Hua I	*Cartesianische Meditationen*, ed., S. Strasser. The Hague: Nijhoff, 1950.
Hua III	*Ideen zu einer Phänomenologie und phänomenologische Philosophie: Erstes Buch*, ed., Walter Biemel. The Hague: Nijhoff, 1950.
Hua IV	*Ideen zu einer Phänomenologie und phänomenologische Philosophie: Zweites Buch*, ed., Marly Biemel. The Hague: Nijhoff, 1952.
Hua V	*Ideen zu einer Phänomenologie und phänomenologische Philosophie: Drittes Buch*, ed. Marly Biemel. The Hague: Nijhoff, 1984.
Hua VI	*Die Krisis der europäschen Wissenschaften und die transzendentale Phänomenologie: Einleitung in die phänomenologische Philosophie*, ed., Walter Biemel. The Hague: Nijhoff, 1954.
Hua VIII	*Erste Philosophie (1923/24): Zweiter Teil*, ed., Rudolf Boehm. The Hague: Nijhoff, 1959.
Hua X	*Zur Phänomenologie des inneren Zeitbewußtseins (1893–1917)*, ed., Rudolf Boehm. The Hague: Nijhoff, 1966.
Hua XIV	*Zur Phänomenologie der Intersubjektivität: Texte aus dem Nachlaß: Zweiter Teil: 1291-1928*, ed., Iso Kern. The Hague: Nijhoff, 1973.
Hua XV	*Zur Phänomenologie der Intersubjektivität. Texte aus dem Nachlaß. Dritter Teil: 1929-1935*, ed., Iso Kern. The Hague: Nijhoff, 1973.
Hua XVI	*Ding und Raum: Vorlesungen 1907*, ed., Ulrich Claesges. The Hague: Nijhoff, 1973.
Hua XVII	*Formale und transzendentale Logik: Versuch einer Kritik der logischen Vernunft*, ed., Paul Janssen. The Hague: Nijhoff, 1974.

Hua XXVII	*Aufsätze und Vorträge (1922-1937)*, ed., Thomas Nenon and Hans Reiner Sepp Dordrecht: Kluwer. 1989.
Hua XVIII	*Logische Untersuchungen: Erster Band*, ed. E. Holenstein. The Hague: Nijhoff, 1975.
Hua XIX	*Logische Untersuchungen (2 Vols.)*, ed., Panzer. The Hague: Nijhoff, 1984. English Translations
LI	*Logical Investigations*, trans., J. N. Findlay. London: Routledge & Kegan Paul, 1970.
Crisis	*The Crisis of European Sciences and Transcendental Phenomenology*, trans., David Carr. Evanston: Northwestern University Press, 1970.
CM	*Cartesian Meditations*, trans., Dorion Cairns. The Hague: Nijhoff, 1977.
Ideas I	*Ideas Pertaining to a Pure Phenomenology and to a Phenomenological Philosophy: First Book*, trans., Fred Kersten. The Hague: Nijhoff, 1982.
Ideas II	*Ideas Pertaining to a Pure Phenomenology and to a Phenomenological Philosophy: Second Book*, trans., Richard Rojcewicz and André Schwer. Dordrecht: Kluwer, 1993.
Ideas III	*Ideas Pertaining to a Pure Phenomenology and to a Phenomenological Philosophy: Third Book*, trans., Ted Klein and William Pohl. The Hague: Nijhoff, 1980.
FTL	*Formal and Transcendental Logic*, trans., Dorion Cairns. The Hague: Nijhoff, 1978.
PCIT	*On the Phenomenology of the Consciousness of Internal Time (1893-1917)*, trans., John Barnett Brough. Dordrecht: Kluwer, 1991.
TS	*Thing and Space: Lectures of 1907*, trans., Richard Rojcewicz. Dordrecht: Kluwer, 1997.

<Main Texts>

<PART 1: PRELIMINARY CONSIDERATIONS FOR THE LECTURE ON TRANSCENDENTAL LOGIC>[1] [351]

<1. Introduction>

It is my intention in these lectures to present a few fundamental considerations toward a phenomenological logic. By the word "logic" I do not understand a subordinated, theoretical, and normative special science in the sense that it is usually taken today, even, say, in the sense in which the modern mathematician has shaped logic as a special mathematical discipline. Logic in the full and universal sense, the sense that we will have in view, is the science that consciously reappropriates the task that was enjoined to logic in general from its historical origin in the Platonic dialectic: namely, the task to be a universal theory of science, and at the same time, a theory of science in principle. A theory of science in principle signifies a science that is in principle a science of all sciences as such.

Logic as a theory of science is then the science of the *a priori* of all sciences as such, the theory of what gives them sense as formations of practical reason, what they must necessarily fulfill if they are actually able to be what they want to be, namely, formations of practical reason. As a pure, *a priori* theory of science, logic wants to bring to light "pure" generalities according to the Socratic-Platonic method. Thus, it does not wish to follow

[1] Editor: Wintersemester 1920/21
Translator: The following pagination to the German text corresponds to Husserliana XVII.

empirically the same path as the pregiven so-called "sciences," the cultural forms that have emerged in fact and that bear the name "science," only then to abstract from them empirical types. Rather, free from all ties to factuality, it wants to bring to complete clarity the teleological idea one always has obscurely in mind when operating from purely theoretical interest. Steadily pursuing the pure possibilities of a cognitive life in general, it wants to bring to the light of day the essential forms of genuine knowledge and science in all their fundamental shapes, as well as the essential presuppositions to which they are bound, the necessary methods [352] that lead to them. In all of this, then, lie the necessary norms against which is to be measured how far a factual science (initially only a presumptuous science) conforms to the idea of science, the extent to which its particular modes of knowledge are genuine modes of knowledge, its methods, genuine methods—methods, that is, which according to their principle form do justice to a pure and formally general norm. The sense of "formal" here consists in precisely nothing other than this: The guiding question is not the one concerning a particular science with particular regions of science, but rather, the question concerning the aim, sense, and possibility of genuine science as such.

Historically, what we call science in the narrow sense today developed from logic, namely, it developed at first from the normative guidelines elaborated in the Platonic dialectic. The classical expression which says that all sciences have arisen from the maternal ground of philosophy fits especially well for logic and, on the other hand, for sciences in the particular sense that we all have in mind today.

In a broader sense, we likewise give the name "science" to the cosmological theories of the pre-Platonic era, to similar cultural formations of other peoples and times, even to astrologies and alchemies, and the like. But at best they are inchoate forms, preliminary stages of science—and this holds especially for pre-Platonic philosophy or the science of the Greeks no less then it does for ancient Egyptian mathematics, for ancient Babylonian astronomy.

Science in a new sense first arises from the Platonic founding of logic, from the radical and critical reflection on essence and eidetic exigencies of genuine knowledge and of genuine science, and from the disclosure of norms according to which a science arises that is henceforth consciously directed toward normative justification, a science consciously justifying its own method. In accordance with its intention, this is a justification from pure principles, that is, a logical justification. Science in the new sense, then, no longer wants to operate naively on the basis of purely theoretical interest. It strives to justify from principles every step it takes in its authenticity, in its necessary validity. Accordingly, in this case, the original sense is such that the logical insight pertaining to principles, taken from the pure idea of possible knowledge and of the method of knowledge in general, precedes the method undertaken in a factual manner as well as the factual formation of science, and guides it in an *a priori* manner; but the sense is not such that the fact of some arbitrary method and science arising naively, and the type read-off from the fact, would have to pose as a norm in order to provide a model for scientific accomplishments in general.

Plato's logic arose as a reaction to the universal denial of science—a denial characteristic of sophistic skepticism. If skepticism denied what is in principle the possibility of something like science in general then Plato had to consider precisely what is [353]
in principle a possibility of science, and he had to found it critically. If science as such was called into question, then of course one could not presuppose the fact of science. In this way Plato was led down the path of the pure idea. His purely ideal logic or theory of science that shapes pure norms (and not read-off from factual sciences), had the mission of only now making possible factual science and guiding it practically. And precisely by fulfilling this vocation it actually did help to fashion sciences in the precise sense: new mathematics and natural science, etc., whose further developments in higher levels are our modern sciences.

However, the original relation between logic and science has become curiously inverted in modern times. The sciences made themselves autonomous. They cultivated highly differentiated methods in the spirit of critical self-justification, a spirit that had now become second nature to them; the fruitfulness of these methods became evident and certain <through> experience or <through> the reciprocal ratification by all the specialists being in agreement. While they did not cultivate these methods in the naiveté of the everyday person, they did it in a naiveté of a higher level, in a naiveté that dispensed with justifying method from pure principles by having recourse to the pure idea in accordance with ultimate *a priori* possibilities and necessities. In other words, logic, which was originally the torchbearer of method and which claimed to be the pure doctrine of principles of possible knowledge and science, lost this historical vocation and, understandably, remained far behind in its development. Even the grand reformation of mathematics and of the natural sciences in the 17th Century by figures like Galileo, Descartes, and Leibniz was still determined by logical reflection on the nature and exigency of genuine natural knowledge, on their *a priori* necessary goals and methods. Thus, if perfecting logic in these beginnings still precedes perfecting science, and if they still go hand in hand, then this essential relationship is modified in the following epoch, in the epoch in which the sciences, rendered autonomous, turn into special branches of science that no longer bother with a logic and that even brush it aside with scorn. But even logic completely departs from its own proper sense and its inalienable task in most recent times. Instead of pursuing the pure essential norms of science according to all their essential formations in order thereby to be able to provide an orientation in principle, it is instead happy to copy norms and rules from the factual sciences, especially from the highly esteemed natural sciences.

Perhaps this signals a deeper and more consequential tragedy of modern scientific culture than what one is in the habit of lamenting in scientific circles. It is said that the number of special branches of science have grown so vast, and each particular one has become so copiously diffuse in their special field of [354]
knowledge and methods, that no one is any longer able to make full use of all this wealth, to enjoy having a command of all the epistemological treasures.

The shortcoming of our scientific situation appears to be a much more essential one, a more radical one in the literal sense of the term; it does not concern collective unification and appropriation, but the rootedness of the sciences, which is a rootedness in principle, and the unification of them from these roots. It would remain a shortcoming even if an incredible mnemonic technology and a pedagogy guided by it would make it possible for us to have an encyclopedic knowledge of the theoretically and objectively established facts in the sum-total of the respective sciences.

Lacking are the centrating ideas that would easily illuminate all thinking in the special branches of science and that would spiritualize all its particular results by relating them to eternal poles; lacking is what removes from all the special branches of science the blinders that are necessary only for their special work; lacking is the capacity to integrate them into the single universal nexus of actual and possible knowledge and thereby to understand this nexus as a nexus that is necessary in principle. But there is still much more that is lacking, namely, the reference back to the phenomenological primordial sources of all knowledge, the deepest founding of all objective sciences arising from the universality of knowing consciousness. Thus, lacking is a systematic fundamental science that would provide an ultimate understanding of all theory arising from the originally sense-giving sources of the subjectivity that accomplishes knowledge.

If the highest task of knowledge is not only calculating the course of the world, but understanding it—as Lotze characterized this task in a well-known dictum—then we have to take this dictum in the sense that we rest content neither with the way in which the positive sciences methodologically shape objective theories, nor with the way in which a theoretical logic directs the forms of a possible genuine theory to principles and norms. We must raise ourselves above the self-forgetfulness of the theoretician who in his theoretical accomplishments devotes himself to the matters, to the theories and methods, and who knows nothing of the interiority of his accomplishment and of the motivations compelling them—who lives in them, but does not have a thematic view of this accomplishing life itself.

We will understand what is accomplished as genuine theory and genuine science only through a clarification of principles that descends into the depths of the interiority that accomplishes knowledge and theory, i.e., into the depths of transcendental, phenomenological interiority; this is a clarification that investigates the theoretical sense-giving and accomplishment of reason in its essential necessity, a sense-giving and accomplishment that is carried out in the interplay of transcendental nexuses of motivation. But only through such a clarification will we also understand the true sense of that being, the sense that science wanted to bring to light in its theories as true being, as true nature, [355] as the true world of spirit.

Thus, only a transcendental science, that is, a science directed into the hidden depths of accomplishing cognitive life, and thereby a science that is clarified and justified—only this science can be the ultimate science; only a transcendental-phenomenologically clarified world can be a world that is ultimately intelligible, only a transcendental logic can be an ultimate theory of science, only it can be an ultimate, deepest, and most universal theory of principles and of norms of all sciences, and at the same time transform them into clarificatory and intelligible sciences. While the contemporary positive sciences, and even the exact sciences, initially fill the novice with enthusiasm and in fact spiritually enriches him, in the end they leave him deeply dissatisfied; *notabene*: provided that he wants to be more than a professional and a specialist, provided that he wants to understand himself as a human being in the full and highest sense and wants to understand the world, and wants to pose to himself and to the world questions of ultimate knowledge and conscience.

We sense this, and we are particularly sensitive in our unfortunate time to the fact that the sciences are lacking in philosophical spirit, lacking the spirit of ultimate and principle purity and clarity, and above all, the spirit of that clarity that we call phenomenological, transcendental clarity. And this is precisely the reason for the lamentation that we do not become wiser and better through them, as is certainly their pretension.

But if we seize upon the idea of logic once again as broadly and as broad-mindedly as it should be grasped in accordance with its original intention, and if we animate it with the transcendental spirit that has been awakened in the modern era but has not arrived at pure self-consciousness, then we will have to say that what is lacking in the modern sciences is true logic as the mother of their true method: logic that lights the way for them with a most profound self-knowledge of knowledge and renders them intelligible in all of their activities.

Accordingly, this logic does not want to be a mere technology for sorts of extremely pragmatic accomplishments of spirit that one calls scientific, a technology after all that one orients in an empirical manner toward practical results. It wants once more to precede all possible sciences as a justifying system of principles of all objective justification, a system of principles that understands itself through an absolute method, namely, for what should count as science and should be able to develop as a genuine science.

Just how much the sciences are in need of such a logic, or rather, just how little they are capable of emerging as sciences that are self-sufficient, and of persevering in such a self-sufficiency is seen in the conflict concerning the true sense of their foundations, a conflict that all the sciences share, no matter how exact they may be. And we see that in truth they are completely in the dark regarding their own sense. To be sure, only transcendental logic allows one to understand completely that the positive sciences can only bring about a relative, one-sided rationality, a rationality that leaves in its wake a complete irrationality as its necessary counterpart. But only a comprehensive rational science is science [356]
<in> the highest sense, like ancient philosophy originally wanted to be.

At the very least I would like to give you a look into some of the profound levels of this universal logic; and if I am not in the position of making this logic thematic in its entire universality, it is not only because of its magnitude and difficulty (as well as the multitude of its subordinated disciplines), but above all because it has become clear that in order to bring to light an actually intelligible transcendental logic, tremendous transcendental-phenomenological preliminary work must be accomplished from the very beginning. Even if, historically and subjectively speaking, the outlines of the positive sciences and of a positive or theoretical logic have been developed first, phenomenological investigations nevertheless form what is first in itself out of which all fundamental forms of logical structures must proceed in a general manner and according to an intelligible motivation. We will be exclusively occupied with such transcendental-logical foundations in these lectures.

2. "Thinking" as the Theme of Logic. Speaking[2], Thinking, What is Thought

The term, *logos*, from which the name "logic" is derived, has a great number of meanings that have arisen from quite intelligible modifications of the more original meanings of λέγω, that is, from the meanings of gathering together[3], expounding upon[4], then expounding upon by words, by means of discourse. In developed language, λόγος sometimes means "word" and "discourse" itself, other times what the discourse is about, the matter of the discourse. But it also means, further, thought that is put in the form of sentences and that is generated by a speaking subject for the purpose of communicating or even for the subject himself, thus, so to speak, the spiritual sense of the linguistic proposition, what one also calls simply the theorem without thinking of anything grammatical here, precisely as something meant by the grammatical propositional expression, likewise, the sense of names; in particular, *logos* means, in the case of universal words, the universal concept belonging to them as sense.

Further, in many expressions *logos* also refers to the intellectual[5] act itself, to the activity of stating, to asserting, or to other modes of thinking in which such a sense-content relative to the respective objects or states-of-affairs are generated.

But all these meanings of the word *logos* assume a specific sense—especially everywhere scientific interests are in play—by the idea of a norm of reason entering into this sense. Then *logos* means reason itself as a faculty, but also rational, i.e., evidential thinking or thought directed toward evident truth. More specifically, *logos* also means the faculty of forming legitimate

---

[2] *Reden*. Translator: Depending upon the context, I use the English expressions "speech," "speaking," and "discourse" to translate the German term "*Rede*" and "*Reden*."

[3] *zusammenlegen*

[4] *darlegen*

[5] *geistigen*

concepts, and even this signifies the rational formation of concepts as well as this legitimate concept itself.

Finally, we mention an even more specific coinage of these [357] meanings whereby the specifically "scientific" element of its sense also comes to the fore: We then mean a scientific concept, a scientific concept-formation, a scientific thinking or the corresponding intellectual faculty.

If we now take this evidently harmonizing manifold of meanings of the word *logos* as a clue to the formation of the first idea of a science of *logos*, rich and closely related themes are opened up for theoretical research and normative application. It is easy to find a natural course of research here. If we concern ourselves with the second and third groups of meanings, the theme of reason as the faculty of correct thinking to be justified in an evident manner as a conceptual, scientific faculty, will lead us beyond the more general question concerning how temporary acts of an ego ground[6] corresponding habitual faculties, and will lead immediately to the question concerning the nature of the "rational" acts of thought that are under consideration.

But now, before the specific quality of this rationality can be considered, the specific quality of thinking itself must naturally become a theme, prior to all distinctions of what is rational and irrational.

We are led by the sense of our talk of *logos* chiefly to conceptual thinking and conceptual thought. Conceptual thinking in general prior to the norm however does not encompass all thinking in general, at least when thinking is taken in the broadest sense of the word. So, let us come back to thinking in the broadest sense, considering it for the time being.

Now, since human thinking is normally carried out linguistically, and all rational operations are practically completely bound to discourse, since all critique, from which rational truth is said to result makes use of language as intersubjective critique, and as a result always leads to statements, then not only acts of thinking and thoughts are at issue initially, but discourse,

---

[6] Translator: Reading *begründen* for *bedürfen*. This reading is based on a correction by the Husserl Archives in Leuven of a transcription error.

statements, stated thoughts. Thus, we arrive at the first group of meanings of the term, *logos*.

The first group of meanings of "the logical" can be brought under three headings: speaking, thinking, what is thought. Naturally, we can likewise speak of their corresponding faculties: the faculty of discourse, to be conceived only along with speaking, and by means of thinking relating to what is thought. Thus, we are examining higher psychic beings, human beings, and are not referring to animals. Only human beings have language and reason, only human beings can carry out psychic acts such that they are subject to the normative regulation of reason, at least this is the general conviction.

Only the human being generates cognitive formations in the form of thinking, like those of a scientific culture, and is able to bring them to expression linguistically, documenting them; only [358]
human beings have something like literature.

3. The Ideality of Linguistic Phenomena

The three headings that we specified above, however, are still quite ambiguous; due to the abundant obscurity of the terms employed they require further distinction and clarification. First, we recognize that we must not overlook a certain distinction where the term speech or language is concerned. We distinguish the articulated word, speech that is spoken at present taken as a sensuous phenomenon, especially as an acoustical phenomenon, from the word and proposition itself or from a string of propositions itself making up a larger discourse. It is for good reason that we speak precisely of a repetition of the same words and sentences in the event we are not understood, repeating [what we have said]. In a treatise, in a novel, every word, every sentence is singularly unique, and it cannot be duplicated by a repeated reading, be it aloud or to oneself. Indeed, in this case it is not a matter of who reads it: each one having his own voice, intonation, etc. We distinguish not only the treatise itself (taken here in merely a grammatical sense of a composition of words and language) from the manifolds of uttered reproductions, but likewise from the manifold documentations that endure on paper and in print, or on pergament and in ink, on clay tablets in cuneiform, etc. The one and only linguistic composition is reproduced a thousand-fold, for instance, in book form, and we say without hesitation, "the same book" with "the same name," "the same treatise"; and to be sure, this sameness already holds in a purely linguistic regard, while it holds in yet another way when completely distilling the content of signification of which we will soon speak.

Language as a system of signs—signs with which thoughts are expressed in contrast to other types of signs—provides us in general and in many respects with subtle and wondrous problems. One of these problems is the ideality of language that we just encountered and which is usually completely overlooked. We can also characterize it in this way: Language possesses the objectivity of objectlike formations[7], of the so-called spiritual or cultural world and not that of mere physical nature. As an objective, spiritual formation, language has the same features as other spiritual formations: In this way, we distinguish from the thousand reproductions of an engraving, the engraving itself, and this engraving, that is, the engraved image itself, is intuitively read-off of every reproduction, and is given in every one of them in the same way as identically ideal. Likewise, when we speak of the Kreutzer sonata in distinction to any of its arbitrary reproductions. Even if the sonata itself consists of sounds, it is an ideal unity, and its sounds are no less an ideal unity; they are not for instance physicalistic sounds or even the sounds of external, acoustic perception: the sensuous, thing-like sounds, which are only really [359]
available precisely in an actual reproduction and intuition of them. Just as a sonata is reproduced over and over again in real reproductions, so too are the sounds reproduced over and over again with every single <sonata> sound[8] of the sonata in the corresponding sounds of the reproduction. The same holds likewise for all linguistic formations; indeed, what is at issue here does not concern what they express (no matter how great a role

[7] Translator: Reading *Gegenständlichkeiten* for *gegenständlichen*. This reading is based on a correction by the Husserl Archives in Leuven of a transcription error.

[8] Translator: <Sonate> added by the Husserl Archives in Leuven.

"expressing" and the dearth of clarity surrounding the relationships here. We can delimit this way of speaking about "expressing" in a preliminary fashion by noting that something is meant with every word and with every relation of words that is composed to form the unity of a discourse: namely, whenever speech is actually expressive speech, functioning normally. Of course, a parrot, a magpie does not really speak. We also exclude now deceptive speech or lying that means something other than what it says. Corresponding to the unity of discourse is a unity of meaning[11], and corresponding to the linguistic divisions and forms of discourse are the divisions and formations of meaning. But this is not external or juxtaposed to the words; rather, in speaking we continuously carry out an internal act of meaning[12] that melds with the words, as it were, animating them. The result of this animation is that the words and the entire discourse incarnate within themselves a meaning, and incarnated in them, bear it within themselves as sense.

We do not need to go any further now, and we can delimit the first and broadest sense of thinking provisionally, namely, that it should encompass those psychic lived-experiences in which this act of meaning consists, this act of meaning in which precisely the meaning is constituted for the speaking subject or, in a parallel manner, constituted for the listening, comprehending subject— thus, the signification, the sense that is expressed in discourse. For example, if we pronounce a judgment, if we assert that Germany will rise again in glory, we have carried out a unity of an internally "thoughtful" assertion with the very words of the statement being asserted. Whatever other psychic accomplishments that may be carried out so that the words themselves come about, and whatever role they may play for the fusion generating the "expression," we only pay attention to what is fused, to the judicative acts that function as sense-giving, as bearing sense within themselves, which therefore constitute within themselves the judicative meaning that finds its expression in the assertoric proposition. Many kinds of psychic lived-experiences accordingly remain out [361]

[11] *Meinung*

[12] *Meinen*

of consideration. Not considered are the indicative tendencies belonging to the words, like those belonging to all signs, the phenomena of pointing away-from-itself and into-the-meaning, the phenomena of pointing-toward what is meant. Also not considered are other psychic lived-experiences that emerge along with them, like those for instance in which we turn to an interlocutor to whom we wish to announce our judgment, etc., but naturally only insofar as the character of the address is not itself expressed in the discourse; for example, "I tell you ...." What we have learned with the example of the assertoric statement holds generally. If we express a wish like "God be with me!," we will have together with the articulated generation of the words a certain wishing, a wishing that is expressed precisely in the articulated organization of words, a wishing that for its part has an articulated content running parallel to it; likewise when we promulgate an order, pose a question, etc. Taken in such a broad manner, thinking means each lived-experience which, during the act of speaking, belongs to the primary function of the expression, precisely to the function of expressing something; thus, it is that lived-experience in which the sense to be expressed is constituted in consciousness; this is thinking, be it a judging or a wishing, willing, questioning, supposing.

Let us retain this broadest concept, which I would like to note at once is not the concept determinative of traditional logic; whether or not we hold fast to this generality, it is important initially to take a stand on it and to examine it scientifically. We hold fast at once to the universality of the coinciding of language and thinking. That designates for us now two parallel domains; they correspond to one another as the domain of possible expressions and as the domain of possible senses, of possible expressive intended meanings. They yield in their intertwined unity the two-sided domain of current and concrete discourse, of sense-fulfilled discourse. Thus, every assertion is at once speech and currently intended meaning, more precisely, judicative intended meaning; every pronounced wish at once optative speech and the current wish itself, the current wish-meaning, etc. It will become clear that there is more than a duality here, so that we must sharply distinguish between the act of intending and the intended meaning, the act of judging and the judgment, and in this manner for all cases, so that a three-fold relation results. We investigate, rather, [362] an important general character peculiar to all lived-experiences that exercise a sense-giving accomplishment, wherever expressions actually exercise their expressive function: thus, in normal discourse and listening that comprehends.

<5. Sense-Constituting Lived-Experiences as Egoic Acts>

All such lived-experiences are not only modes of consciousness in general, but egoic acts; and this is what we want to make clear now. In the course of our psychic life, waking life is only one type; there is another one besides this one, deep dreamless sleep, unconsciousness. We arrive at both these types in their contrast by presentifying actual lived-experiences of awakening, by a retrospective intuitive grasping of preceding phases of consciousness in comparison with wakefulness itself. Even if we cannot say anything in more detail at all about the content of what is past and about what is experienced in a torpid manner, we can describe with evidence the typical essence of the contrast. There is an experiencing taking place when in a stupor, as well. But there is not perception in the genuine sense or an experience of another sort; there is nothing of a cognitive theme; there is nothing of a judgment; there is nothing of an object of an emotional interest; there is nothing properly speaking of an object being loved or hated at present, there is nothing of a desiring or a willing. What is the distinguishing trait of such lived-experiences that, in the very broadest sense (a broadness that is certainly not customary), could be characterized as lived-experiences of interest, and that mark the wakeful psychic life with the character of wakefulness? We can respond: Psychic life is wakeful, that is, the ego is wakeful, and this is the case insofar as it carries out specific egoic functions in the present, that is, carries out in the present an "I perceive"; that is to say, I turn toward an objectlike formation, viewing it, seeing it, I am absorbed in it; likewise, I remember, I am absorbed in viewing the memorial object; likewise, I make comparisons and distinctions; in comparing I recognize that two different perceptual objects possess the same type; or I am inclined to love, to cherish,

to respect someone I have in mind, or I turn against him in hate and contempt; I carry out an act of striving, consider the means, make a decision and do it. Note well the way in which emphasis is placed on the "I," the ego. In such wakeful lived-experiences of experiencing, taking cognizance, drawing conclusions, valuing, willing, we find the ego as the peculiar center of the lived-experiencing, as the one being absorbed in it, or the one suffering it in a conscious manner; it is the identical pole, the center of actions and passions—the latter corresponds to conditions like: I am sad, I am delighted, I am happy. The term "I" here is not empty, and on the other hand, we mean "I" neither as a corporeal human being, nor as the entire psychic life, actually, nothing at all of life and living. Rather, here the ego is identified in reflection as the center of life and lived-experiencing, the center to which are [363]
related perceiving, judging, feeling, willing.

But we understand this in such a way that the expressions, "I perceive," "I judge," "I feel," "I want," designate at the same time an essential shape of these lived-experiences themselves that is essentially given through their ego-centration. Here, the ego is everywhere living in these acts as carrying them out, as being related to the perceptual object, the judged object, the willed object through these acts. The ego is not a box containing egoless lived-experiences, or a slate of consciousness upon which they light up and disappear again, or a bundle of lived-experiences, a flow of consciousness or something assembled in it; rather, the ego that is at issue here can be manifest in each lived-experience of wakefulness or lived-experiential act as pole, as ego-center, and thus as involved in the peculiar structure of these lived-experiences; it can be manifest in them as their outward radiating or inward radiating point, and yet not in them as a part or a piece. This can be seen by the fact that in order to grasp this outward radiating point thematically, we must exercise a peculiar reflection, one going in a opposite direction. We do not find it like a part, as something in the lived-experience or literally on it; rather, the structure of the lived-experience, its directional structure that goes toward what is presented, toward what is wished for, etc., points back to an outward radiating point and to the directedness of this ego toward its intentional theme. It is also evident here that all such lived-experiences that emerge in this distinctive shape, *ego cogito*, in the unity of a stream of lived-experience, manifest the identically same ego: I, who I perceive, am identically the same as the I who then judges or feels, desires, wants, and only by virtue of this identity can I say that all of these are my acts. A curious polarization of the stream of consciousness! All lived-experiential acts [are] centered in a single, fully identical pole. Only through a reflective grasping of this central ego—which however is only graspable as the subject of its acts, as a subject carrying them out—does every other concept of the ego, even that of the personal and psychophysical human ego get its sense, no matter the number of new determinative moments that may be taken up by these new concepts of the ego.

What we have said here will be further clarified when we note that the wakeful life of an ego does not only contain such egoic lived-experiences in which the central ego emerges as a present functional center, and thus gives to its lived-experiences the shape, *ego cogito* (to speak with Descartes). Wakeful life has, so to speak, a background of non-wakefulness, constantly and with eternal necessity. When I actually perceive an object, that is, look at it, take note of it, grasp it, regard it, it will never be without an unnoticed, ungrasped background of objects. In this case we distinguish what is secondarily noticed from what actually goes unnoticed. In general, in addition to the object that is primarily noticed, with which I am occupied in a privileged way while viewing it, there are still other single objects that are co-noticed, be they given in a second or third order co-grasping. This will take [364]
place in such a way that in passing over from the observation of one object to the observation of another, I am indeed no longer looking at the first one, I am no longer primarily occupied with it, properly speaking; but I still have a hold on it, I do not let it slip from my attentive and conceptual hold, and along with that, everything I had previously grasped: It continues to belong to me in a modified way, and in this way I still have a hold on it. I am still present there as the central, present ego; as a wakeful ego, I still have a relation to it in an *ego cogito*. But in contrast to it we have a broad lived-experiential field, or as we can also say, a field of consciousness that has not entered into such a relation with the

ego or with which the ego has not entered into such a relation: It may knock on the door of the ego, but it does not "affect" the ego, the ego is deaf to it, as it were.

The wakeful ego with its lived-experiences in the specific sense of wakefulness, lived-experiences of the *ego cogito*, thus has a constant, broad horizon of background lived-experiences to which the ego is not present and "in" which it does not reside: They may be sensations, like sensations of sound, but the ego is not alert to them; physical objects or beings with a lived-body may appear in the surrounding space as in transition or as in rest, but the ego does not carry out an "I perceive" or an "I take note" with regard to <them>; affects may be intertwined with these background lived-experiences or with their objects, spilling over into a general atmosphere of well-being or malcontentment; even tendencies, lived-experiences of drive, may be rooted in them, tendencies which for instance incline away from malcontentment, but the ego is not present there. Belonging here are also flashes of insight, imaginings that arise, memories, theoretical insights that emerge or even stirrings of the will, decisions that are not however taken up by the ego. Only when the ego carries them out do they get the shape of "*ego cogito*," of the "I am occupied in imagination with what is shaped by the imagination," "I think through the theoretical insight," "I carry out the stirring of the will," etc. Thus, the wakeful egoic life is distinguished from the egoic life that is not awake, from the ego that is "in a stupor" in the broadest sense, and the two are distinguished by the fact that in the latter, no lived-experience in the specific sense of wakefulness is there at all and no present ego is there at all as its subject, while in the other case, precisely such a wakeful ego is there as the subject of specific acts.

6. Foreground Lived-Experiences and Background Lived-Experiences

Every act in the specific sense has the fundamental character of being a consciousness of something, an "intentional lived-experience." The perceptual lived-experience is in itself a perception of something, for example, a house; the cognitive lived-experience is in itself a lived-experience of something known, like when the house is recognized as a residence; in every judgment there is something, a judged state-of-affairs, in every desiring is something desired, in every willing, something willed. This is the broad concept of the concept of intentional lived-experience. For even background lived-experiences are [365] intentional. A universal life that is a life of consciousness through and through encompasses the specific acts, those that are specifically ego-consciousness, as well as the background consciousness. Thus within wakeful life, for example, we constantly have a visual space that is filled up and that is present to consciousness. If we take note of a single tree in a landscape that is open before us, then this landscape as a spatial field with multiple objects is given to consciousness and is there for us. That is to say, as a whole and according to all single features, the background objects are objects for us through the fact that they appear, through the fact <that> in their corresponding lived-experiences, they have the character of intentional lived-experiences. Every appearance is the appearance of what is appearing in it, the lived-experience of an appearance of a house <in> the landscape is precisely the appearance of this house, whether we pay particular attention just to it or not. The egoic act in the specific sense is thus a special form of carrying out intentional lived-experiences. Let me mention that I have originally introduced the term "act" in my *Logical Investigations* for this broadest concept of intentional lived-experience; this is the sense in which the term is now customarily employed in the literature. For this reason I now say emphatically "egoic act" or "act in the restrictive sense," where it is a matter of acts having a distinctive form of execution.

Taking place in the life of consciousness is a constant transformation of the modalities of execution; foreground lived-experiences, egoic acts, lose this form of execution and then take on the altered form and *vice versa*. That holds for all types of lived-experiences of consciousness. Cognitive acts, acts of pleasure, volitional acts do not simply disappear when we no longer carry them out from the standpoint of the ego; they become background lived-experiences.

It is evident here that background lived-experiences, in contrast to the acts corresponding to them, are modified through and through, whereas they do indeed share something essentially in common so that we find it necessary to speak further of the same judgments, wishes, etc. Certainly, they are not actually the same. It is not like shoving things in a room away from the window into dark corners, where the things themselves remain unchanged. The moment a background lived-experience becomes present, that is, the moment the ego becomes an ego carrying out acts through it, it has, as lived-experience, become completely and essentially transformed. So too, *vice versa*. And yet even the judging that has been shoved into the background is still a judging of this and that; background perception is still perception of exactly the same thing.

It belongs to the very essence of a lived-experience that it is a consciousness of the same thing in the transition from one mode of execution to another. Purely through their own essence and in passing from one mode to the other, they found the consciousness of the unity and of the sameness of what is given to consciousness in them; a kind of coinciding sets in, the coinciding that ensues precisely according to their intentional content as the content given in them. This situation makes intelligible our way of speaking of acts, acts that become latent and then patent again. [366]

A counter argument that emerges in the background of consciousness during a conflict is initially a latent act; its intentionality (which brings it to the idea of such and such an argument) is a hidden intentionality, up to the point where we intervene[13], as it were, and now actualize it, that is, carry out an explicit argumentation, an argumentation of the corresponding content that issues from the ego-center.

---

[13] Translator: Reading *eingreifen* for *angreifen*. This reading is based on a correction by the Husserl Archives in Leuven of a transcription error.

<7. The Interconnection between Expressing and Signifying as the Unity of an Egoic Act>

Our theme was the clarification of thinking, of thinking together with speaking, of thinking that exercises the sense-giving function on speaking. Whenever we actually speak or actually take part in a discourse by listening to it and by understanding it, this actuality will consist in the execution of egoic acts in sense we have defined. This initially concerns thinking that gives sense to words. The speaking person means something in the spoken words, and this act of meaning, this "thinking" that belongs to speaking is an act (or a unitary nexus of acts) carried out by the ego. That toward which the ego is intentionally directed in these acts is what the ego means in speaking these words, what the words as discourse "express."

Further, we note that even those lived-experiences in which the words themselves are produced for us as the ones speaking, lived-experiences in which the words are given to consciousness and are there for us, have the character of egoic acts, and [we note] that our analysis has accordingly also taught us something with respect to the specific mode of linguistic consciousness. Words as actually spoken words do not emerge in a background distant from the ego; as speakers, we generate them, and through this generation we are directed toward them in acts, and not in a latent intentionality. Further, if one says that we mean or express this or that with words, then even this synthetic unity of the act of meaning with words belongs to the circle of the specific participation of the ego. In verbal consciousness, words have the character of signs; the character of indicating is inherent in them; radiating from them are indicative tendencies that aim at what is meant and terminate in the content of the meanings. This intertwining belongs to the intentional stock of the unity of verbal and linguistic consciousness, and this peculiarity obviously produces the following: that the expression and what is expressed, that verbal and semantic consciousness are not juxtaposed to one another, disjointed, but rather, make up a unity of consciousness in which the doubled unity of word and sense is constituted. The moment we conceptually do away with these indicative tendencies and rid ourselves of them, we no longer have words at all, we have meaningless sounds like a parrot makes in its so-called "talk"; we have signs that do not signify anything, in fact they are then no longer really signs, to say nothing of expressions.

Even this intentionality unifying the words themselves and the sense, the lived-experience of the word and thinking, has the character of patent intentionality; the pure ego is present there. The ego seizes the word in regarding it; it grasps its indicative tendency; it willingly allows itself to be guided by it, to be initiated into the execution of thinking; it allows itself to be [367] oriented by what is thought as what is meant by the words. But we do not intend the words themselves here! We can also intend the words in other acts; we can become interested in them, as we are wont to say, making them our "theme" in this interest, possibly making them our theoretical theme like we would do if we were grammarians. In this case, we carry out theoretical judgments and theoretical discourses corresponding to them, discourses that proceed with new words; obviously, then, the difference comes to light between the words that are our grammatical theme, and the words that we use in order to express ourselves on this theme, in order to express our thoughts with respect to them. Both words are essentially given differently to consciousness; in one case, the acts that are directed toward them are acts of interest, in the other, they are not. In the broadest sense, but not in a customary literal sense, we could speak of interest precisely in order to say that an act is altogether carried out, that is, that in it the ego is present to the respective intentional object, that the ego is directed toward something in the act. But the normal concept of interest says more, it means that peculiar mode of carrying out acts whereby what is given to consciousness in it is a theme for the ego. If the words themselves are not given to consciousness thematically in each current discourse, they will still necessarily have a theme, namely, one that resides in what is meant with the words. Thus, the sense-giving act is an intending act in the specific sense of a thematic act that, in the mode of interest, is directed toward a theme residing in the content of the act.

The actuality of indication that adheres to the word in the consciousness of an actual discourse gains a clarifying determination through our elucidation of the feature of thematic acts. The word points away from itself, as it were, [pointing] toward what is expressed as the thematic sense. This analysis obviously concerns every type of sign, or rather, actual signifying, be it a linguistic or non-linguistic one. A certain imperative, the firm indication of its thematic intending, firmly adheres to every sign according to its essence as sign.

If our interest is nevertheless turned toward the sign itself, then the thematic preference of the sign is one that runs counter to its function; it takes place in it despite this; and one senses that this "despite" has a phenomenological character.

We have arrived at the insight that neither any kind of intentional lived-experience, nor even any kind of act can be found in the sense-constituting function, neither with respect to signs in general, nor accordingly with respect to speaking. Only acts in the mode of thematic acts, acts of interest in a specific sense can function in such a way; only acts through which what is given to consciousness in them has for the ego the preferential character of thematic intending.

Naturally, this character even enters into background modifications, like everything we demonstrate in actual discourse [368] with respect to structures, but precisely as modified; modifications can be bestowed upon all acts, and hence those of discourse, too. In this respect it does not require any further special exposition. Let us remain in the sphere of wakeful activity which is alone fruitful for us. What I state, what I express in speaking is my theme, my "what I intend" in the moment of my current speaking.

If I assert something, then my thematic act is a judging, and I have my judicative theme, a judicative intending. Likewise, if I express a wish, then my wishing is the thematic act, the wish is made as my optative-intending, in interrogative discourse, the interrogative act has thematic form, etc. All in all, there is accordingly a multiplicity of acts carried out by the one presently speaking, acts synthetically connected to one another forming a unity of one act. Not only do we have a continual sequence of acts when moving laterally along discourse in its single words and sentences—a continual sequence of acts that are connected with one another and thereby constituting for the one speaking the unity of a discourse built up out of the sense-endowed words and sentences, and which is now a unitarily meaningful discourse. Not only, I say, do we have this multiplicity moving laterally along discourse, but it also belongs to every cross-section, so to speak, to a diversity; that is, it belongs to every part of the discourse and possibly to every part of the word insofar as it is still discourse, still endowed with sense. At each place we have the organization of the act according to word and sense, thus the synthesis itself resulting from an overarching act, namely, from an indicative act that assigns to the connected acts simultaneously a different place and function.

<8. Theme, Interest, Indication>

At the end of our last lecture, we characterized the entire nexus of expressing and signifying as a unity of an egoic act. Now we can immediately pick up this line of thought here in order to provide the character of thinking with a necessary depth, a depth with which a novel and at the same time especially distinctive mode of execution by egoic acts in general will come to light. If we compare the mode of execution in which the central speaking ego carries out the word-constituting act and, on the other hand, the sense-constituting act, we will encounter a sharp contrast. We also called the latter mode of thinking the act of meaning or intending. For example, what is expressed in propositional speech like "geometry is the science of space," is what the person speaking judicatively "means." But while he means the judgment "with" the words, in this case he does not mean the words themselves. They are given to the ego in a manner that is quite different from the way in which what is judged is given to the ego. Residing in the latter, so to speak, is the *terminus ad quem*, and through this it has a special priority in relation to the word, although the ego's regard was also directed toward the word.

[The fact that we] have in mind words (like objects in general) in specific acts is not yet to say that the acts are directed toward the words by intending them. To be sure, we can also attribute to them the distinction of an intending, but then [this would only [369]
occur] in correspondingly altered acts. We can become especially

interested in the words, as one also says, we can make them our "theme" through this interest, like we do as grammarians, for instance. Then they become our theoretical themes; in relation to them we carry out theoretical judgments and corresponding to them, theoretical discourses that are deployed in new words. Especially clear here is the difference between words that are our grammatical themes, and words that we use in order to pronounce our theoretical thoughts about these themes. Both are given to consciousness in an essentially different manner; in one case, the acts that are directed toward them are acts of interest in the restrictive sense of the term, in the other case, they are not. One could indeed speak in a broadest though unconventional sense of the interest of the ego with respect to each act. For the ego, as pure ego, is absorbed in each act and *interest*[14]; it is directed toward something and absorbed in it. But the normal concept of interest means more; it means a peculiar mode of carrying out an act, whereby in this concept of interest something is given to consciousness, given in the special manner of a theme, of a theme that one has one's eye on.

The expression, "intending" or "meaning" is used generally now and then for all acts in order to depict the directedness of the ego toward the intentional content, and for this reason one must distinguish a thematic intending or the thematic act from intending or act in general. In this way there is a theme with potentially many special themes in every discourse, only that precisely the words themselves of the discourse are not themes. The theme resides in what is meant in the words. The sense-giving act is not only a second interlaced act, but interlaced as a thematic act, an act of interest. The different manner of carrying it out, suggested by the words "interest" and "theme"—where the former points to the ego and its action, and where the latter does not—obviously belongs to the lived-experiential acts even outside of assertoric discourse. It is also apparent here that there are different degrees of interest and, on the other hand, modes of interest that are not merely differences of degree. Thus, an intuitive regard of objects and events of the environing-world can be of more or less interest;

[14] Translator: Husserl uses the Latin expression, taking the term in its literal sense.

the ego has its theme in these matters, but is absorbed in them more or less intensely. On the other hand, while the ego has its primary theme in these matters that the ego regards, it can still not only notice additional events, but can take an interest in them. But then they are secondary themes, interests of a second order.

If we return now for a moment to the special sphere of expression, we will be struck by the curious interconnection between the function of sense as a thematic one and the function of indication, an interconnection whose fundamental character becomes intelligible only now. Expressed in a more complete manner, the word points away from itself and to the sense in normal discourse, that is, the word directs interest. The word-sign [370] that is in itself not a matter of interest serves to draw attention to the sense as something that matters to the ego.

This analysis is obviously fitting for every kind of sign or for acts in which they exercise their present signicative function, be they linguistic signs or other types of signs like signals from a boatman. The moment our interest is directed toward the signs themselves and is arrested there (rupturing this normal function), like when it is directed toward the written signs or toward the flag that serves as a signal, abnormality shows up in the lived-experience itself. One feels that it goes against the grain, so to speak, and that one is not only violating a habit, but a habitual determinative end, a practical imperative.

In this way we have thus also gained a deeper insight into the essential structure of living speech, above all the knowledge that sense-giving thinking cannot be just any act, but only one that has the general character of a thematically intending act, be it in other respects a judicative intending, a presumptive intending, an intending of doubt, an intending of a wish or an intending of volition.

<9. The Regression from Theoretical *Logos* to the Pre-theoretical Sense-Giving Life of Consciousness>

Now we want our investigation to go beyond the narrow realm to which it has been restricted, [namely,] the realm of thinking as the sense-giving function peculiar to statements. Actually, every step of our preceding analyses of interiority functioning in linguistic thinking has already elicited the demonstration of the general character of consciousness, which points beyond the narrow field. Now we want to enter the broadest generality in which it is no longer a matter of words and statements, although it does concern, in an expanded sense, sense-givenness and furthermore the distinctions between what is rational and irrational—distinctions that belong to the special theme of every logic.

As we said right at the beginning, the genuine theme of logic alludes to the group of significations most rich in content and so to speak magnified: the group of significations of the word *logos* related to reason, specifically as scientific reason, and to the achievements that are accomplished in it, accordingly, to the entirety of the linguistically marked structure, the structure which the rubric "scientific theory," "scientific system" expresses. Thus <logic> should encompass the principles and theorems, the deductions and proofs in their entire systematic interconnections, just like they would be objectively presented in an ideal textbook, namely, as the spiritual common good of humanity. As I already said earlier, the term "rational" is a normative term. What is rational is the true, the genuine; it is what even irrationality, the human being insofar as he thinks irrationally, strives toward, but what he lacks in his unclarity and confusion because of an ingenuine, irrational method. We can accordingly say that logic bears on science in the genuine sense, or as we can also put it, it wants and wanted from the very beginning to be the universal [371] theory of science, the science of the essence of genuine science in general. Under the rubric of science, humanity wanted to know the world in a systematic manner, or in the specialization of investigative interests, wanted to know some type of special unending-open region of the world. This guiding idea of science which was initially unclear was supposed to have been made sharp and clear. The essential features of genuine science, those to which the truth of its methods and its theories are bound in regulative necessity, were supposed to be brought to light, and because of their clarity were supposed to be recognized in this pressing necessity. Thus, the goal was simultaneously to gain an evident norm for all procedures of practical reason in the foundation of genuine science, and building upon this, to ever higher accomplishments of truth. Since it constantly concerns here the modes of the accomplishing and the results of the accomplishment, the subjective activity of the scientists and the structure of objective spiritual formations following from it, namely theories, the efforts of elucidation and of scientific knowledge pertaining to the theory of science or logic would have to be two-fold: subjectively directed toward the activity of knowing and, on the other hand, objectively directed toward theory.

However, only in the modern era has one seen, or rather first merely felt obscurely and then seen with evidence, how this two-fold structure demands deep and tremendously encompassing investigations if one really wants to reach an understanding of the essence of scientific accomplishment as the essence of an accomplishment of reason. As soon as systematic portions of sciences were won in a certain naive evidence (as already in antiquity, Euclidean mathematics, the beginnings of an astronomy and mechanics, and from there, certain firm and precisely formed theories were given in evidence, theories whose epistemic value seemed incontestable because of this naive evidence), one understandably held onto these models conceptually, and the focus was predominantly fettered by what was objectively available, by the manifold formations of theory. One initially held that theories consist of propositions, they progress from true propositions to true propositions; insight grasps the truth, and thus also justifies the pretension toward truth.

Propositions, whose truth is immediately evident, lead through deductions to conclusions that become evident in their dependent truth. The entire nexus composed of elementary deductions, produced in its unity, is itself the unity of truth as theory. These whole formations that are built up from single propositions are indeed linguistically expressive formations, but the linguistic element in it (for instance, varying according to the national language), is irrelevant here. Prominent in this variation of the mere linguistic element is the pure thought, the pure significance, the identical proposition, or as one also says, the judgment. It is only to the latter that evidence and the predication of truth or possibly falsity adheres. In this sense not only is a single proposition only a very complex judgment, but so too is the whole unity of a theory.

In this way logic was directed toward a theory of theory; it [372] looked at these pure unities of significance; one examined it in a kind of descriptive and classificatory manner. One systematically distinguished the general forms of these significances, the forms of judgment and of their elements, and the form of the connections through which complex judgments arise; elementary forms of judgment like: *S is P, all S are p, some S are p, if S is p then Q is r*, etc. The systematic production of those forms of judgment-complexes that are called deductions also belonged here. Following this, one could then examine these forms to see to what extent they yield general conditions of possible truth and falsity of judgments formed in such a way. If one examines the forms of deduction in this way, it seems evident that one cannot arbitrarily tie propositions to deductions, or, forms of proposition to forms of deduction, namely, insofar as it is evident that deductions of certain forms are in principle false and that from the standpoint of truth, only certain forms of deduction are admissible. Every deduction with the form "*given all A are B and all B are C, all A are C*" is correct with respect to the consequence, but if it would read "*not all A are C*" the deduction would be false. From there one could see that belonging to the form of judgments as to the form of pure propositional thoughts are laws of form which, depending upon the circumstance, say that judgments and formations of judgments of such and such pure forms are contradictory once and for all, they are in principle false; others are not contradictory, they can be true according to their form.

The Aristotelian syllogistic form, and the later, more or less purely shaped formal logic arose in this way. According to the core which is alone useful, it offers in effect the beginnings of a doctrine of forms and a doctrine of validity of judgments related to the pure form, and thus the beginnings of a theory of possible forms of theories. Traditional logic did not bring about anything more with respect to a theory of theory; and on the other hand, with respect to the investigations subjectively directed to the essence of scientific thinking, scant little was accomplished, that is, in relation to a critique of knowledge. One has sought in vain since Locke to make headway by means of a psychology of knowledge and a theory of rationally valid cognition grounded in it. But the naturalism of this psychology was unable to grasp consciousness and the accomplishment of consciousness from within, and although it had pretensions of being grounded in inner experience, it was even unable to see into this peculiarity of consciousness at all; the naturalism of this psychology got paid back in its absurd theories of knowledge emerging here—absurd in the strongest sense; one indeed felt the absurdity of these theories, but strived in vain to clarify it. What was completely unintelligible in the modern era was this marriage between pure ideal theories of the formal logic of signification, on the one hand, and the theories of epistemological investigations, on the other. The propositions, the theories emerge from the interiority of accomplishing thought, somehow; but just what this interior [373] thought looks like, and what it is, and what it accomplishes as so-called "evidence"—that remains obscure.

It was only with phenomenology that we first had avenues of access, methods, and insights that make possible an actual theory of science, namely, through its radicality in going back to sense-giving consciousness and the whole of conscious life. It is phenomenology that seriously inquires back from the ready-made propositions, theories, to thinking consciousness and to the broader nexus of the life of consciousness in which these formations are constituted; and it inquires back, going still more deeply from all types of objects as the substrata of possible theories, to experiencing consciousness and its essential characteristics which make the experiencing accomplishment intelligible. It has allowed us to see in a presuppositionless manner the feature of intentionality as the very feature that makes up the fundamental essence of consciousness. It has generated methods of developing the hidden implication of one consciousness in another, an implication that is given everywhere with this feature, and therefore of making intelligible how objectivity as a true being of every kind is shaped as an accomplishment in the subjectivity of the life of consciousness, and is then shaped as a higher level of accomplishment which is there as theory. If one goes back from theory that is dead, so to speak, and has become objective, to the living, streaming life in which it arises in an evident manner, and if one reflectively investigates the intentionality of this evident judging, deducing, etc., one will immediately be lead to the fact that what stands before us as the accomplishment of thought and was able to show itself linguistically rests upon deeper accomplishments of consciousness. Thus, for example, in order to be able to emanate from actual evidence every theory that refers to nature presupposes natural experience—what we call outer experience. In this way, all theoretical knowledge in general ultimately leads back to an experience.

We see upon closer inspection that already under this rubric, "experience," a sense-giving accomplishment is carried out, indeed, a highly ramified, intricate one, and one that is even covered by a broadly apprehended rubric of reason and unreason, whereby the rational operation alone, one that has taken shape in a certain free spontaneity, can function as a verifying foundation of a genuine theory.

It is impossible to understand what thinking (which is a highly built-up accomplishment) is in the specific sense in order to be able to be expressed by language and universal words and in order to provide a science, a theory, if we do not go back prior to this thinking, back to those acts and accomplishments that make up the most expansive part of our life. For not only does a pre-theoretical life reside in this expansiveness, but a pre-linguistic life as well, one that immediately ceases to be in its original, primitive peculiarity with every expression.

And thus I set the task of our further lectures to open up this expansive, great world of the interiority of consciousness and [374] under the guiding viewpoint of a theory of science, and by beginning from below and ascending upward, to show how genuine thinking in all its levels emerges here, how it is motivated and is built-up in its founded accomplishment.

We want to deal with the great, universal theme of sense-giving. We called thinking sense-giving. And we had already distinguished this sense-giving thinking from what is thought in it, or as one can also say in this correlation, from the thought. Thus, for example, judicative thinking, judicative intending and on the other hand the judgment itself, the optative intending and the wish itself, the volitional intending and what is willed itself in the intending. The word "intending" or "meaning"[15] is used for both; likewise, the special words judgment, wish, decision, question, etc., are ambiguous. In the psychological, logical, ethical language of the modern era, both of them are indistinguishably muddled, although the clarity and distinctness of the differences that are necessary here are not only useful but fundamental for all these disciplines; the differences are also of decisive importance especially for the pure distinctions concerning the investigative areas of the logical disciplines.

That there are constantly temptations for confusion shows from the very start how important the clarification of the distinction is. By engaging in such a clarification important insights are immediately opened up. Thus, we distinguish the intending and the intended meaning, the sense-giving act and the sense itself (which is given to consciousness thematically in the sense-giving act). This holds generally. When a thematic act is attached to words, what is meant in the act is called the sense of the word, or even, its significance[16] because the word signifies[17]. But independently of the fact whether an act has such a function of lending words significance, and perhaps being able to lend words significance, it has in itself a sense-content. Accordingly, we must liberate the concept of sense from its relation to expressions. Put in a quite general manner, every intentional lived-experience possesses as such its intentional sense; the latter becomes precisely a specifically meant sense when the ego becomes a subject who carries out acts thematically and becomes the subject of thematic interest. Let us now enter this realm of greater generality, the general realm of sense-giving and sense; without an encompassing study of this realm, all attempts to clarify logic in the specific sense are hopeless.

---

[15] *Meinung*

[16] *Bedeutung*

[17] *deutet auf*

<10. Perception and Perceptual Sense>

Let us begin with any external perception. If we observe an unchanging object at rest, for example, a tree standing before us, we pass over it with our eyes, now we step closer to it, now back away from it, now here, now there, we see it now from this, now from that side. During this process the object is constantly given to us as unchanged, as the same; we see it as such; and yet a slight turn of our attention teaches us that the so-called perceptual images, the modes of appearance, the aspects of the object [375] constantly change. In a constant variation of modes of appearance, perspectives, that is, during a constant variation in the actual lived-experience of perception, we have a consciousness that runs through them and connects them up, a consciousness of the one and the same object. This variation is given to consciousness, and yet it is hidden in a certain way; in the normal attitude, the natural attitude that is turned outward toward things, we do not consciously notice the variation.

I spoke of a turning of attention. More specifically, I spoke of a turning of the thematic regard and even more precisely of a reflection. In fact, we speak of a reflection in all cases, where in any kind of conscious lived-experience a direction of the thematic regard is prefigured from the very start as normal, that is, as a necessary, thematic attitude that serves as a starting point from which we must turn away in order to get hold of something new in our conscious lived-experience. This is how it works for external perception.

Belonging to external perception is a thematic basic attitude, namely, the attitude directed toward the external object, which without further ado we call the object of perception. Customarily and from the very outset, we consider attentive perceiving,[18] that is, this normal thematic directedness toward the external object, as belonging to the concept of external perception. But a reflective conversion of the thematic regard is possible at any time and in an evident manner, and then our perceptual images themselves become graspable and grasped. In and through their variation, we

---

[18] *das Gewahren*

also see then, evidently, a thoroughgoing unity of the perceiving lived-experience. However we may conceive of it as temporally articulated and temporally partitioned, we find it composed of perceptions (and it is not conceivable in any other way). Each one has its content of appearance as well as a content of appearance that is constantly different, and each one has its object that appears "there in the flesh." But this object is the same object in all these stretches and phases of the continual unitary perception; it is the same thanks to the thoroughgoing "coinciding" of the appearances being carried out in perception itself. And it is the same for consciousness! It is not the appearances themselves according to their contents of appearance that coincide; to be sure, they are always different and temporally spaced out; and yet there is a certain "coinciding" that is expressed in this evidence, to wit, that in every one of these modified appearances, the same tree appears, and the perceptual intending, the constantly thoroughgoing thematic intending, intends this object that is overall the same. We now name this same object given to consciousness as identical in the continuity of appearance using a preliminary concept: the sense or objective sense of perception. Let me say in advance: In precisely such a manner, every conscious lived-experience possesses its sense within itself. This is to say that instead of carrying out a conscious lived-experience naively, we can make any kind of conscious lived-experience thematic by reflecting on it; and then—be it with respect to the temporal stretches of its variable continuity, be it in comparison with other such separated lived-experiences—we can always find they make possible an evident consciousness of the identity of the content, that what two [376]
consciousnesses intend is the same. In each case we call this meant same object the objective sense of these lived-experiences. To be sure, we are remaining for the time being with perception. The objective sense in our example is thus the perceptually appearing tree as such; it is meant in and through all perceptions in an evident manner.

But now it is extremely important to avoid a misunderstanding. The perceived tree is naturally and simply there for us as an existing reality as we live naively in perception: at least in the normal case of perception which is presupposed here, namely, in which an experiential motivation for a doubt or a negation is not operative. Of course, this does not rule out the fact that we are still subject to deception. Were that not the case, if perception did not have its incontestable legitimacy that can also be demonstrated through further experience, the tree would exist as an actual part of nature. And *vice versa*: If the tree actually exists, perception has its demonstrable legitimacy in the form of possible acts of legitimating attestation. Both are evidently equivalent. Note now that the objective sense of perception is nothing less than or means the same thing as the actual perceptual object, the sense of perception of a tree nothing less than or means the same thing as the actual natural object, tree. When we speak of sense, we are not at all concerned whether or not the perceiver has carried out a legitimate perception that the perceiver or anyone else can ratify through new experiences. We only inquire into what perceptual lived-experiences bear in their very essence and what they as perceptions irrevocably bear within themselves, no matter how a judgment of legitimacy of it may run, attributing legitimacy to it or contesting its legitimacy. This is to say in other words that we do not inquire into whether this tree, the tree that the perceiver naively sees (and not merely given to the perceiver in a general manner, but posited by the perceiver in the certainty of its existence), has a place in nature, in the totality of realities that are to be posited with legitimacy.

It is a matter of indifference to us whether, in the realm of possible positings of the object to be grounded as legitimate, one of the positings comes about that accords or does not accord with our perception in its very sense-content. Be that as it may, it is beyond doubt that perception possesses in itself what appears to it as such, possesses its perceptually meant object, and that several perceptions with different perceptual contents accord within it in an evident manner and according to an evident identity. We can also put it in this way: Perception is an intentional lived-experience and has immanently, within itself, an intentional object as an inseparable sense. If we make a judgment about this sense, we thus judge something that is demonstrable in an evident manner and therefore has being, but immanent being, even if it also turns out later to be that the perception was a deceptive one.

By conceptually shifting our way of speaking, one speaks of the perceptual object only where one makes the claim to judge reality, [377] like in all normal perceptual judgments about the surrounding things, and <not> merely about purely immanent objects, for example, about the perceived tree as such. No one would disagree <in this case> that nothing in reality corresponds to this tree that I see before me, for instance, in a dream as actually there and in the flesh.

What is designated there as the "tree" is obviously the immanent sense-content of perception itself, and a sense-content is not a tree, not a thing as such; that is, it is not a thing in factual nature. Therefore, a shift of significance has occurred here and—since all similar shifts are customarily alluded to in written form by quotation marks—I also make a habit of expressing the shift as the difference between tree in quotation marks and the tree simpliciter. It is quite similar to the way in which we say, for example, "Socrates is a philosopher," and at another time, "Socrates is a proper name." In the latter case we use quotation marks to show more clearly that we do not speak of Socrates himself, but of the word, Socrates.

Our considerations have thus yielded a fundamental distinction, initially in relation to the very special fundamental shape of consciousness that we call perception:

(1) The full, concrete lived-experience of perceiving. We did not have the occasion to speak of many things that belong here; for example, when it was a thematic perceiving, the moment of the directedness toward the object that results from the pure ego. In particular then,

(2) the variable manifold appearances, aspects necessarily belonging to every phase of perception, but combined in the continuity of perceiving through a peculiar synthesis, a kind of "coinciding," a kind of synthesis, insofar as through it the phenomenally distinguished and possibly completely distinguished aspects form a unity in the evident consciousness of the same object. I—the one experiencing—know about the being of this lived-experience and about the different modifications of it only by a reflective change in perspective through which I grasp it thematically and then judge it thematically.

(3) This same object, the object in quotation marks itself; that which is the same appearing object in each of these appearances, what each appearance means, the intentional object as such.

The introduction of the concept of sense is unclear. Sense is initially introduced as the intentional object, what is meant or intended as such. That is ambiguous, as is the object in quotation marks. When I carry out the phenomenological reduction, I have for every "act" its meant object, the intentional object that contains in it all modalities of being, which is "being."

But then it amounts to this schism between the intentional content and the intentional modal-character, which initially looks [378]
like a distinction between two components. The intentional content in this sense, the "material," the "quality," is also the "intentional object," what is merely presented[19], what is qualified there modally.

That is an entirely different concept of sense and intentional object. All of that will come to light in the following exposition, but it must be pursued in the correct manner from the very start, and the distinctions <must> be made, even if they are only provisional ones.

[19] *das bloß Vorgestellte*

<PART 2: [3]
ANALYSES CONCERNING PASSIVE SYNTHESIS: TOWARD A TRANSCENDENTAL AESTHETIC>[20]

<SELF-GIVING IN PERCEPTION>

<§1. Original Consciousness and the Perspectival Adumbration of Spatial Objects>

External perception is a constant pretension to accomplish something that, by its very nature, it is not in a position to accomplish. Thus, it harbors an essential contradiction, as it were. My meaning will soon become clear to you once you intuitively grasp how the objective sense exhibits itself as unity <in> the unending manifolds of possible appearances; and seen upon closer inspection, how the continual synthesis, as a unity of coinciding, allows the same sense to appear, and how a consciousness of ever new possibilities of appearance constantly persists over against the factual, limited courses of appearance, transcending them.

Let us begin by noting that the aspect, the perspectival adumbration through which every spatial object invariably appears, only manifests the spatial object from one side. No matter how completely we may perceive a thing, it is never given in perception with the characteristics that qualify it and make it up as a sensible thing from all sides at once. We cannot avoid speaking of such and such sides of the object that are actually perceived. Every aspect, every continuity of single adumbrations, regardless how far this continuity may extend, offers us only sides. And to our mind this is not just a mere statement of fact: It is

[20] Translator: The following pagination to the German text corresponds to Husserliana XI.

inconceivable that external perception would exhaust the sensible-material content of its perceived object; it is inconceivable that a perceptual object could be given in the entirety of its sensibly intuitive features, literally, from all sides at once in a self-contained perception.

Thus, this fundamental division between what is genuinely [4] perceived and what is not genuinely perceived belongs to the primordial structure of the correlation: External perception and bodily "object." When we view the table, we view it from some particular side, and this side is thereby what is genuinely seen. Yet the table has still other sides. It has a non-visible back side, it has a non-visible interior; and these are actually indexes for a variety of sides, a variety of complexes of possible visibility. That is a very curious situation peculiar to the very essence of the matter at hand. For proper to the very sense of every perception is perception's perceived object as its objective sense[21], that is, this thing, the table that is seen. But this thing is not [merely] the side genuinely seen in this moment; rather (according to the very sense of perception) the thing is precisely the full-thing that has still other sides, sides that are not brought to genuine perception in this perception, but that would be brought to genuine perception in other perceptions.

Generally speaking, perception is original consciousness. We have, however, a curious schism in external perception: Original consciousness is only possible in the form of an actually and genuinely original conscious-having of sides and a co-conscious-having of other sides that are precisely not originally there. I say co-conscious, since the non-visible sides are certainly also there somehow for consciousness, "co-meant" as co-present. But they do not appear as such, genuinely. They are not there like reproductive aspects are, as intuitions that exhibit them; we can nevertheless produce such intuitive presentifications[22] any time we like. Viewing the front side of the table we can, whenever we like, orchestrate an intuitive presentational course[23], a reproductive

[21] *gegenständlicher Sinn*

[22] *Vergegenwärtigungen*. See translator's note, p. 110, fn. 64.

[23] *Vorstellungsverlauf*. Translator: The term, "*Vorstellung*" is translated throughout as "presentation," and not, for example, as representation. Whereas the latter term suggests an

course of aspects through which the non-visible side of the thing would be presented to us. But here we are doing nothing more than presentifying a course of perceptions to ourselves in which we would see the object—passing from a perception to new ones—from ever new sides in original aspects. Still, that only happens in exceptional circumstances. It is clear that a non-intuitive pointing beyond or indicating is what characterizes the side actually seen as a mere side, and what provides for the fact that the side is not taken for the thing, but rather, that something transcending the side is intended in consciousness as perceived, by [5]
which precisely *that* is actually seen. Noetically speaking, perception is a mixture of an actual exhibiting that presents in an intuitive manner what is originally exhibited, and of an empty indicating that refers to possible new perceptions. In a noematic regard, what is perceived is given in adumbrations in such a way that the particular givenness refers to something else that is not-given, as what is not given belonging to the same object. We will have to understand the meaning of this.

Let us first note that every perception, or noematically speaking, every single aspect of the object in itself points to a continuity, to multifarious continua of possible new perceptions, and precisely to those in which the same object would show itself from ever new sides. In every moment of perceiving, the perceived is what it is in its mode of appearance [as] a system of referential implications[24] with an appearance-core upon which appearances have their hold. And it calls out to us, as it were, in these referential implications: "There is still more to see here, turn me so you can see all my sides, let your gaze run through me, draw closer to me, open me up, divide me up; keep on looking me over again and again, turning me to see all sides. You will get to know me like this, all that I am, all my surface qualities, all my inner sensible qualities," etc.

You understand what I mean to convey with this suggestive manner of speaking. In the particular present perception I have just

---

active cognitive operation, "*Vorstellung*" is functional on a passive level of experience as well.

[24] *Verweisen*

these aspects and their modifications, and no others, just these aspects that are always limited ones. In each moment the objective sense is the same with respect to the object as such, the object that is meant; and it coincides with the continual course of momentary appearances, as for instance this table here. But what is identical is a constant *x*, a constant substrate of actually appearing table-moments, but also of indications[25] of moments not yet appearing. These indications are at the same time tendencies, indicative tendencies that push us toward the appearances not given. They are, however, not single indications, but entire indicative systems, indications functioning as systems of rays that point toward corresponding manifold systems of appearance. They are pointers into an emptiness since the non-actualized appearances are neither [6]
consciously intended as actual nor presentified. In other words, everything that genuinely appears is an appearing thing only by virtue of being intertwined and permeated with an intentional empty horizon, that is, by virtue of being surrounded by a halo of emptiness with respect to appearance. It is an emptiness that is not a nothingness, but an emptiness to be filled-out; it is a determinable indeterminacy. For the intentional horizon cannot be filled out in just any manner; it is a horizon of consciousness that itself has the fundamental trait of consciousness as the consciousness of something.

In spite of its emptiness, the sense of this halo of consciousness is a prefiguring that prescribes a rule for the transition to new actualizing appearances. Seeing the front side of the table, I am also conscious of the back side, along with everything else that is non-visible, through an empty pointing ahead, even though it be rather indeterminate. But no matter how indeterminate it may be, it is still a pointing ahead to a bodily shape, to a bodily coloring, etc. And only appearances that adumbrate things of that kind and that determine more closely what is indeterminate in the framework of this prefiguring can be integrated concordantly; only they can stay the course of an identical *x* of determination as the same, being determined here newly and more closely. This holds time and again for every perceptual phase of the streaming process

[25] *Hinweisen*

of perceiving, for every new appearance, only that the intentional horizon has altered and shifted. Proper to every appearing thing of each perceptual phase is a new empty horizon, a new system of determinable indeterminacy, a new system of progressing tendencies with corresponding possibilities of entering into determinately ordered systems of possible appearances, of possible ways that the aspects can run their course, together with horizons that are inseparably affiliated with these aspects. In the concordant coinciding of sense, they would bring the same object as being ever newly determined to actual, fulfilling givenness. To our mind, the aspects are nothing for themselves; they are appearances-of only through the intentional horizons that are inseparable from them.

We thereby distinguish further between an inner horizon and an outer horizon of the respective aspect-appearance. It should be recognized that the division applying to what is genuinely perceived and what is only co-present entails a distinction between determinations with respect to the content of the object [a] that are [7] actually there, appearing in the flesh[26], and [b] those that are still ambiguously prefigured in full emptiness. Let us also note that what actually appears is, in itself, also laden with a similar distinction. Indeed, the call resounds as well with respect to the side that is already actually seen: "Draw closer, closer still; now fix your eyes on me, changing your place, changing the position of your eyes, etc. You will get to see even more of me that is new, ever new partial colorings, etc. You will get to see structures of the wood that were not visible just a moment ago, structures that were formerly only viewed indeterminately and generally," etc. Thus, even what is already seen is laden with an anticipatory intention. It—what is already seen—is constantly there as a framework prefiguring something new; it is an x to be determined more closely. There is a constant process of anticipation, of preunderstanding. In addition to this inner horizon there are then also outer horizons, prefigurings for what is still devoid of any intuitively given framework that would require only more differentiated ways of sketching it in.

[26] *leibhaft*

<§2. The Relationship of Fullness and Emptiness in the Perceptual Process and the Acquisition of Knowledge>

In order to gain a deeper understanding we must pay attention to how fullness and emptiness stand in relation to one another at each moment, how emptiness adopts fullness in the flow of perception, and how fullness becomes emptiness again. We must understand the structure of interconnections for every appearance as well as the structure that unites all series of appearances. In the continual progression of perception, as in the case of every perception, we have protentions that are continuously fulfilled by what occurs anew, occurring in the form of the primordial-impressional Now. And here as well. As each external perceiving progresses, the protention has the shape of continuous anticipations becoming fulfilled. That is to say, out of the indicative systems of the horizons, certain indicative lines are continually being actualized as expectations; the latter are continuously fulfilled in aspects that are being determined more closely.

In the previous lecture we approached the unity of each external perception from different directions. External perception is a temporal run-off of lived-experience where appearances [8] concordantly pass into one another and form the unity of coincidence corresponding to the unity of sense. We came to understand this flux as a systematic network of progressive fulfillment of intentions that obviously, when viewed from the other side, goes hand in hand with an emptying of intentions that are already full. Every momentary phase of perception is in itself a network of partially full and partially empty intentions. For, in every phase we have genuine appearances, that is, a fulfilled intention, albeit only gradually fulfilled, since there remains an inner horizon that is unfulfilled and an inner horizon of indeterminacy that is still determinable. Moreover, proper to every phase is a completely empty outer horizon that tends toward fulfillment and, in the transition toward a definite direction of progress, strives toward it in the manner of empty anticipation.

Viewed more precisely, we now have to describe the process of perception as a process of acquiring knowledge; and we have to distinguish further (in the following manner) between fulfillment and the process of determining more closely. While the empty horizon—both inner and outer—fashions its next fulfillment in the march of perception, this fulfillment does not merely consist in tracing over in intuition the prefigured sense of which one is emptily conscious. Indeterminacy, as we said, belongs essentially to the empty premonition which is, as it were, the presentiment of what is to come. We also spoke of determinable indeterminacy. Indeterminacy is a primordial form of generality whose nature it is to be fulfilled in the coincidence of sense only by "specification." As long as this specification itself has the character of indeterminacy (the specific indeterminacy as opposed to the general indeterminacy just mentioned), it can attain further specification, etc., in new steps. But now we should consider that this process of fulfillment, which is a specifying fulfillment, is also a process of knowing something more closely; it is not only a momentary acquisition of knowledge, but at the same time a process of acquisition [that takes place] within knowledge that is abiding and that becomes habitual. This will become clearer shortly.

Let us note in advance that the primordial place of this accomplishment is the continuously co-functioning retention. First, let us recall that the continually progressing fulfillment is at [9]
the same time a continually progressing emptying. For, as soon as a new side becomes visible, a side that has just been visible gradually disappears from sight, becoming finally completely non-visible. But what has become non-visible is not cognitively lost for us. Thematic perception does not merely drive at continually possessing some new aspect of the object that would be intuitively grasped from moment to moment, as if what was formerly given would slip away from the grasp of [perceptual] interest. Rather, in passing over the object, perception drives at fashioning a unity of originary acquisitions of knowledge through which the object, according to its specific content, would reach an original acquisition, and through it would become an abiding epistemic

possession.[27] And in fact we understand the original acquisition of knowledge by observing the following situation: The process of determining more closely, which comes about with fulfillment, imparts afresh a specific moment of sense. While it vanishes from the field of genuine perception in the progression to new perceptions, it remains held retentionally. (That already takes place prethematically, already in background perceiving. In thematic perception, retention has the thematic character of keeping-a-hold-of.) Accordingly, the empty horizon (into which what is new enters by virtue of retention) has a character other than the empty horizon peculiar to the expanse of perception, that is, before the latter originarily appeared. Having already once seen the back side of an unfamiliar object and, turning back to perceive the front side, the empty premonition of the back side now has a determinate prefiguring that it did not have previously. The unfamiliar object is thereby transformed in the perceptual process into a familiar object; in the end, I have exactly what I had started with, namely, a perspectival appearance. And if the object has moved entirely out of our field of perception, then we have an altogether completely empty retention of it. Nevertheless, we still have the entire epistemic acquisition of it, and we still have a hold on it in thematic perceiving. Our empty-consciousness now has an articulated, systematic sense that is sketched in—something that did not exist previously and especially at the beginning of the [10]
perception. What was previously a mere framework of sense, a wide ranging generality, is now specified meaningfully in an articulated manner; to be sure, it awaits further experience in order to take on still richer epistemic contents as contents of determination. If I turn back again to the perceptions of the earlier determination, they will issue in the consciousness of recognition, in the consciousness: "I already know all that." Now a mere bringing to intuition takes place, and with it, a fulfilling confirmation of the empty intentions, but no longer the process of determining more closely.

[27] Every content of an unaltered thing can be reached time and again through perception; I can go around the surface; ideally the thing can be divided, and can be viewed time and again from all surface sides, etc.

<§3. The Possibility of Our Acquired Knowledge Being Freely at Our Disposal>

By acquiring knowledge originally, perception also acquires permanent, lasting possession of what it has acquired; it is a possession that is at our disposal any time. How is something freely at our disposal? Although this thing that is already familiar to me has become empty, it is freely at my disposal insofar as the empty retention remaining behind can be freely filled up at any time; it can be made present at any time by a re-perception in the sense of a re-cognition. By walking around it, drawing nearer to it, touching it with my hands, etc., I can once more see all the sides that are already familiar to me; I can experience them again, they are ready for perception. And this holds true likewise for the next time. The fact that a re-perception, a renewed perception of the same thing, is possible for transcendence characterizes the fundamental trait of transcendent perception, alone through which an abiding world is there for us, a reality that can be pregiven for us and can be freely at our disposal.

To this we must add yet another essential observation. If we have become familiar with a thing and a second thing appears in our field of vision, and if, with respect to the side genuinely seen, it accords with the earlier and familiar thing, then according to an essential law of consciousness (by virtue of an inner coinciding with the earlier thing awakened through the "association of similarity"), the new thing receives the entire epistemic prefiguring from the earlier one. It is apperceived, as we say, with the same non-visible qualities as the previous one. And even this prefiguring, this acquisition of an inner tradition, is also freely at [11]
our disposal in the form of actualizing perception.

But how does this having something freely at our disposal look now upon closer inspection? What makes possible the free foray into our world that is thoroughly interwoven with anticipations; what makes all existing knowledge and new knowledge possible? Let us privilege here the normal and basic instance of the constitution of external existence, namely, the constitution of unaltered spatial things. Whether alterations of things can occur without being perceived and yet can be known in all their unperceived elements in a variety of perceptions and experiences that follow—this is a theme for a clarification existing on a higher level, a theme that already presupposes clarifying the possibility of knowing existence in rest.

Thus, in order to understand at least this basic feature of the constitutive problematic, we ask what having acquisitions of knowledge freely at our disposal looks like—acquisitions I already have, however incompletely; what does it look like specifically in the case of unaltered thingliness? What makes it possible?

From what we have said above, we see that every perception *implicite* invokes an entire perceptual system; every appearance that arises in it implies an entire system of appearance, specifically in the form of intentional inner and outer horizons. We cannot even imagine a mode of appearance in which the appearing object would be given completely. No final presentation in the flesh[28] is ever reached in the mode of appearance as if it would present the complete, exhausted self of the object. Every appearance implies a *plus ultra* in the empty horizon. And since perception does indeed pretend to give the object [completely] in the flesh in every appearance, it in fact and by its very nature constantly pretends to accomplish more than it can accomplish. In a peculiar way, every perceptual givenness is a constant mixture of familiarity and unfamiliarity, a givenness that points to new possible perceptions that would issue in familiarity. And that will continue to hold in a new sense, differently from what has come to light up to now.

Let us now take a look at the formation of unity through coinciding as it pertains to sense by examining the transition of appearances, for instance, when approaching or walking around an object, or in eye movement. The fundamental relationship in this dynamic transition is that of intention and fulfillment. The empty [12] pointing ahead acquires its corresponding fullness. It corresponds to the more or less rich prefigured possibilities; but since its nature is determinable indeterminacy, it also brings, together with the fulfillment, a closer determination. Thus, here we have a new "primordial-institution," or as we can say here again, a primordial-impression, since a moment of primordial originality emerges.

---

[28] *Leibhaftigkeit*

What is already given to conscious in a primordial-impressional manner points to new modes of appearance through its halo which, when occurring, emerge as partly confirming, partly determining more closely. By virtue of inner intentions—unfulfilled and those now in the process of being fulfilled—what has already appeared itself becomes enriched. In this progression, moreover, the empty outer horizon that was intertwined with the appearances achieves its next fulfillment, at least a partial one. The part of the horizon that remains unfulfilled passes over into the horizon of the new appearance, and it goes on like this continually. That aspect of the object which has already appeared is partially lost again as it moves away from givenness, i.e., the appearance; the visible becomes non-visible again. But it is not lost. I remain conscious of it retentionally and in such a way that the empty horizon of the appearance present at this time receives a new prefiguring that points in a determinate manner to what has already been given earlier as co-present. Having seen the back side and having turned back to the front side, the perceptual object has kept a determination of sense for me; likewise in emptiness, it points to what was previously seen. They all belong now abidingly to the object. The process of perception is a constant process of acquiring knowledge that holds on to what was acquired epistemically in sense; it thereby fashions an ever newly altered and ever more enriched sense. During the ongoing perceptual process, this sense is added to the grasped object itself in its presumed [complete] presentation in the flesh.

Now, it depends upon the direction of the perceptual processes as to which lines are brought to fulfillment from the system of unfulfilled intentions, that is, which continuous series of possible appearances will be realized out of the entire system of possible appearances of the object. Advancing along this line, the empty [13]
intentions are transformed respectively into expectations. Once this line is pursued, the series of appearances run their course in the sense of continuously arousing and steadily fulfilling expectations that stem from the current kinaestheses, while the remaining empty horizons are left in dead potentiality.

Lastly, we still have to mention that the integral harmony [taking place] in the coinciding of adumbration-appearances, which pass over into one another by way of intention and fulfillment, not only concerns the appearances taken as wholes, but also all their moments and parts that can be differentiated. Thus, there is something corresponding to every filled spatial point of the object in the entire series of appearances; they continuously pass into one another such that this point in the appearance exhibits itself as a moment of the appearing spatial form.

If we ask, finally, what gives unity within every temporal point of the momentary appearance—unity considered as the entire aspect in which the particular side is exhibited—we will also come across reciprocal intentions that are fulfilled simultaneously and reciprocally. The transition of appearances following one after the other are all in dynamic displacement, enrichment and impoverishment.

The object appearing constantly new, constantly different, is constituted as the same in these exceedingly intricate and wondrous systems of intention and fulfillment that make up the appearances. But the object is never finished, never fixed completely.

We must point here to a side of the noematic constitution that is essential for the objectivation of the perceptual object, [namely,] to the side of kinaesthetic motivation. We mentioned in passing time and again that the courses of appearance go hand in hand with the orchestrating movements of the lived-body[29]. But that must not remain something that we only mention haphazardly in passing. The lived-body is constantly there, functioning as an organ of perception; and here it is also, in itself, an entire system of compatibly harmonizing organs of perception. The lived-body is in itself characterized as the perceiving-lived-body. We recognize it then purely as a lived-body, subjectively movable and in perceiving activity, as subjectively self-moving. In this regard it does not come into consideration as a perceived spatial thing, but rather with respect to the system of so-called "movement-sensations" that run their course during perception, in eye [14] movements, head movements, etc. And they do not simply run

---

[29] *Leib*

parallel to the flow of appearances there; rather the kinaesthetic series under consideration and the perceptual appearances are related to one another through consciousness. By viewing an object I am conscious of the position of my eyes and at the same time—in the form of a novel systematic empty horizon—I am conscious of the entire system of possible eye positions that rest at my disposal. And now, what is seen in the given eye position is so enmeshed with the entire system that I can say with certainty that if I were to move my eyes in this direction or in that, specific visual appearances would accordingly run their course in a determinate order. If I were to let the eye movements run this way or that in another direction, different series of appearances would accordingly run their course as expected. This holds likewise for head movements in the system of these possibilities of movement, and again for the movement of walking, etc., that I might bring into play.

Every series of kinaestheses proceeds in its own way, in a manner totally different from the series of sensible data. It runs its course in such a way as to be freely at my disposal, free to inhibit, free to orchestrate once again, as an originally subjective realization. Thus, the system of lived-body movements is in fact characterized with respect to consciousness in a special way as a subjectively free system. I run through this system in the consciousness of the free "I can." It may happen that I involuntarily dwell upon something, that for instance my eyes involuntarily turn this way or that. But at any time, I can capriciously pursue such a path of movement or whatever path of movement I like. As soon as I have an appearance of the thing in such a situation, a system of internally coherent manifold appearances of the same thing is thereby prefigured in the original consciousness of the sequence of appearances.

A propos the appearances I am not free: When I undertake a series of movements in the free system, "I move myself," the appearances that are arriving are already prefigured. The appearances form dependent systems. Only as dependent upon kinaestheses can they continually pass into one another and constitute a unity of one sense. Only by running their course in these ways do they unfold their intentional indicators. Only through this interplay of independent and dependent variables is [15] what appears constituted as a transcendent perceptual object, precisely as an object that is more than what we directly perceive, as an object that can completely vanish from my perception and yet still persist. We can also say it is constituted as such only by the fact that its appearances are kinaesthetically motivated, and consequently that it is in my freedom, in accordance with the knowledge that I have acquired, to let the appearances run their course randomly as original appearances in their system of concordance. Through the appropriate eye movements and other lived-bodily movements I can, in the case of a familiar object, turn back at any time to the old appearances that give me back the object from the same sides. Or, by freely returning to the appropriate place, I can once again perceive and identify the object no longer perceived.

Thus, in every perceptual process we see a constitutive duet being played: (1) The system of my free possibilities of movement is intentionally constituted as a practical, kinaesthetic horizon. This system is actualized each time I run through single paths of movements with the character of familiarity, that is, of fulfillment. We are not only thereby conscious of every eye position that we have at the moment, every position of the thing-body as the momentary sensation of movement, but we are also conscious of them as a place in a system of places; thus we are conscious of them with an empty horizon which is a horizon of freedom. (2) Every visual sensation or visual appearance that arises in the visual field, every tactile appearance that arises in the field of touch is ordered with respect to consciousness, to the current situation of the consciousness of the parts of the lived-body, creating a horizon of further possibilities that are ordered together, creating a horizon of possible series of appearances belonging to the freely possible series of movement.

In relation to the constitution of transcendent temporality we should note here that every path of actualization that we would *de facto* enter down in realizing this freedom would yield continuous series of appearances of the object. All of these series would exhibit the object for one and the same expanse of time; they would all exhibit the same object in the same duration, only from different sides. In accordance with the sense of the constituted object, all determinations that would be known through this process would be co-existent.

<§4. The Relation of *esse* and *percipi* in Immanent and Transcendent Perception> [16]

All this holds only for transcendent objects. An immanent object, like a lived-experience of black, offers itself as a lasting object, and in a certain way through "appearances" as well. But it only does so like any temporal object in general. The temporally extending duration requires the constant modification of the modes of givenness in accordance with the modes of appearance of the temporal orientation. Now, the spatial object is also a temporal object, so the same holds for it, too. But it still has a second, special way to appear. By directing our attention to the temporal fullness and especially to the primordial-impressional phases, we come up against the radical difference between the appearance of transcendent and immanent objects. The immanent object has only one possible way to be given in the original in every Now, and therefore every mode of the past also has only one single series of temporal modifications: to wit, that of presentification, with the changing past objects being constituted in it. But the spatial object has infinitely many ways [to be given in the original] since it can appear in the Now, that is, in an original way from its different sides. Though it appears *de facto* from this side, it could have been able to appear from other sides, and accordingly every one of its past phases have infinitely many ways in which it could exhibit its past fulfilled points of time. We can also say: The concept of appearance has a new and unique sense for the transcendent object.

If we consider exclusively the Now phase, then in the case of the immanent object, appearance and that which appears cannot be separated in the Now phase. What arises anew in the original is the particular, new black-phase itself, and without being exhibited. And appearing means here nothing other than a to-be devoid of any exhibiting that points beyond, and a to-be-conscious-of in the original.

But on the other hand, with respect to the transcendent object, it is clear that the thing of which we are conscious in the flesh as a thing in the new Now is given to consciousness only in and through an appearance; that is, exhibiting and that which is exhibited, adumbration and that which is adumbrated are to be [17] distinguished. If we exchange the noematic attitude that we have privileged up to now with the noetic attitude in which we turn our reflective regard toward the lived-experience and its "intimately inherent"[30] components, we can also say that a transcendent object such as a thing can only be constituted when an immanent content is constituted as substratum. Now, this immanent content for its part is substituted, as it were, for the peculiar function of the "adumbration," of an exhibiting appearance, of a being exhibited in and through it. When we do not regard the appearing thing-object, but the optical lived-experience itself, the thing-appearance that arises anew in each Now—as we say, the optical appearance—is a complex of surface color moments that are extended in this way or that; these surface color moments are immanent data, and we are thus conscious of them in themselves just as originally as, say, red or black. The manifold changing red-data in which, e.g., any surface side of a red cube and its unaltered red is exhibited, are immanent data.

Yet, on the other hand, the matter does not rest with this mere immanent existence. In the immanent data, something is exhibited in the unique manner of adumbration, which the immanent data themselves are not; in the visual field, a sameness, an identical spatially extended body-color is exhibited in the alteration of the immanently sensed colors. All the noematic moments that we, in the natural attitude, see contained in the object and as related to it, are constituted by means of the immanent data of sensation, and by virtue of the consciousness that, as it were, animates them. In this regard we speak of apprehension as of transcendent

[30] Translator: "*reell*." Whereas "*real*" for Husserl designates the type of existence or "reality" peculiar to transcendent things, "*reell*" for Husserl depicts what is actual, *wirklich*, without it sharing the ontological status of a real transcendent entity, *res*. Accordingly, *reell* concerns the intimate immanence of consciousness. What is "*irreal*" from the perspective of the "*real*" can also be "*irreell*" from the perspective of the "*reell*"; this would bear on what Husserl understood generally as "noema" or "sense."

apperception: It characterizes consciousness' accomplishment which is to bestow on the mere immanent contents of sensible data, on the so-called data of sensation or hyletic data, the function of exhibiting something objectively "transcendent." It is dangerous here to speak of represented and representing, of interpreting data of sensation, or to speak of a function that outwardly signifies through this "interpreting."[31] Adumbrating, exhibiting in data of sensation, is totally different from an interpretation through signs[32].

"Immanent" objectlike formations, accordingly, are for their part not given to consciousness through apperception. In their case, "being given to consciousness in the original" and "being," "*percipi*" and "*esse*" converge. And indeed for every Now. However, they are to a large extent bearers of apperceptive [18] functions, at which time something non-immanent is exhibited in and through them. Now the *esse* (for transcendent objects) is in principle distinguished from the *percipi*. In every Now of external perception we do have an original consciousness, but genuine perceiving in this Now, that is, that feature in genuine perceiving that is primordial-impressional (and not simply retentional consciousness of the past phases of the perceptual object) is a conscious-having of what is being adumbrated *originaliter*.[33] This is not a pure and simple having of the object in which conscious-having and being coincide; rather, it is a mediate consciousness, provided that only one apperception is had immediately, a store of sense-data referring to kinaesthetic data, and an apperceptive apprehension through which an exhibiting appearance is constituted; in and through it, we are conscious of the transcendent object as adumbrating or exhibiting *originaliter*. Time and again we have the following situation in the process of ongoing perceiving in every Now: In principle, the external object is never purely and simply had in its original ipseity. It appears in principle

[31] *"durch dieses 'Deuten' hinausdeutenden Funktion"*

[32] *signitives Deuten*

[33] Perception is original consciousness of an individual object, of a temporal object, and for every Now we have in perception its primordial-impression in which the object in the Now, in its momentary point of originality, is originally grasped. But it must be shown that original adumbration necessarily goes hand in hand with apprentation.

only through apperceptive exhibition and in ever new exhibitions; as the latter progress, they bring something new to the original exhibition from its empty horizons.

Yet, it is more important for our ends to recognize as inconceivable that something like a spatial object, which gets its original sense genuinely by means of external perception as adumbrating perception, would be given through immanent perception, be it human or superhuman intellect. But from this it follows as inconceivable that a spatial object and everything like it (for instance, an object of the world in the natural sense), could be exhibited in a discrete, self-contained manner from one point of time to the next, along with its entire ensemble of features (as completely determined) that make up its temporal content in this Now. In this respect we also speak of adequate givenness as opposed to inadequate givenness. To express this theologically and in a drastic manner, worse service cannot be rendered God [19] than conceding him the ability to make an odd number even and to transform every absurdity into truth. Inadequate modes of givenness belong essentially to the spatial structure of things; any other way of givenness is simply absurd. We can never think the given object without empty horizons in any phase of perception and, what amounts to the same thing, without apperceptive adumbration. With adumbration there is simultaneously a pointing beyond what is exhibiting itself in a genuine sense. Genuine exhibition is itself, again, not a pure and simple possession on the model of immanence with its *esse* = *percipi*; instead, it is a partially fulfilled intention that contains unfulfilled indications that point beyond. The originality of exhibiting the transcendent thing in the flesh necessarily implies that the object as sense has the originality of apperceptive fulfillment and that this harbors inseparably a mixture of actually fulfilling and not yet filled moments of sense. This is the case whether they be moments of sense only prefigured according to the general structure, and apart from that open indeterminate and possible moments, or whether they be moments already distinguished by being specially prefigured. This is why the talk of inadequation as a haphazard lack that a higher intellect could overcome is an unsuitable way of speaking, indeed totally preposterous.

We can formulate a principle here that will become much clearer in our future analyses. Whenever we speak of objects, no matter what category of objects they may be, the sense of this manner of speaking about objects originally stems from perceptions as lived-experiences originally constituting sense, and therefore an objectlike formation. But the constitution of an object as sense is an accomplishment of consciousness that is in principle unique for every basic type of object. Perception does not consist in staring blankly at something lodged in consciousness, inserted there by some strange wonder as if something were first there and then consciousness would somehow embrace it. Rather, for every imaginable ego-subject, every objectlike existence with a specific content of sense is an accomplishment of consciousness. It is an accomplishment that must be new for every novel object. Every basic type of object in principle requires a different intentional structure. An object that *is*, but is not and in principle could not be an object of a consciousness, is pure non-sense. [20]

Every possible object of a possible consciousness is however also an object for a possible originarily giving consciousness; and this we call, at least for individual objects, "perception." It is absurd to demand of a material object a perception that has the general structure of an immanent perception, and conversely, to demand of an immanent object a perception that has the structure of external perception. Both sense-giving and sense require one another essentially—and this concerns the essential typicality of their correlative structures.

In this way it is the nature of originally transcendent sense-giving, which external perception carries out, that the accomplishment of this original sense-giving is never finished as one expanse of perception progresses to another and so forth in whatever manner the process of perception may advance. This accomplishment does not simply consist in bringing to intuition something new in a fixed pregiven sense, as if the sense would already be prefigured in a finished manner from the very beginning; rather in the process of perceiving, the sense itself is continually cultivated and is genuinely so in steady transformation, constantly leaving open the possibility of new transformations.

Let us note here that in the sense of concordantly and synthetically progressing perception, we can always distinguish between an unceasingly changing sense and an identical sense running through the changing sense. Every phase of perception has its sense insofar as it has the object given in the How of the determination of the original exhibition and in the How of the horizon. This sense is flowing; it is a new sense in every phase. But the unity of the substrate *x*, which holds sway in a steady coinciding, and which is determined ever more richly—this unity of the object itself, that is, everything that the process of perception and all further possible perceptual processes determine in it and would determine in it—this unity runs through this flowing sense, through all the modes, "object in the How of determination." In this way, an idea that lies in infinity belongs to every external perception, the idea of the completely determined object, of the object that would be determined through and through, known through and through, where every one of its determinations would be purified of all indeterminacy, and where the full determination itself would be devoid of any *plus ultra* with respect to what is still to be determined, what is still remaining open.

I spoke of an idea lying in infinity, that is, of an unattainable [21] idea. For, the essential structure of perception itself excludes a perception (as a self-contained process of courses of appearance, continually passing into one another) that would furnish absolute knowledge of the object; it excludes such a knowledge in which the tension would collapse between the object in the How of determination (which is changing and relative, remaining incomplete), and the object itself. For evidently, the possibility of a *plus ultra* is in principle never ruled out. It is thus the idea of the absolute self of the object and of its absolute and complete determination, or as we can also put it, of its absolute individual essence. In relation to this infinite idea which is to be seen, but which as such is not realizable, every perceptual object in the epistemic process is a flowing approximation. We always have the external object in the flesh (we see, grasp, seize it), and yet it is always at an infinite distance mentally. What we do grasp of it pretends to be its essence; and it is it too, but it remains so only in an incomplete approximation, an approximation that grasps something of it, but in doing so it also constantly grasps into an emptiness that cries out for fulfillment. What is constantly familiar is constantly unfamiliar, and from the very beginning all knowledge seems to be hopeless. To be sure, I said "seems." And we do not wish to commit ourselves here straight away to a hasty skepticism.

(Of course, the situation is entirely different with immanent objects. Perception constitutes them and appropriates them in their absoluteness. They are not constituted by constant sense modification in the sense of an approximation; only insofar as they become in a future are they laden with protentions and protentional indeterminacies. But what has been constituted as present in the Now is an absolute self that does not have any unfamiliar sides.)

We have rejected a hasty skepticism. At all events, we should have initially made the following distinction in this regard. Given that an object is perceived and that we progressively come to know it in the perceptual process, we had to distinguish [a] the particular empty horizon that is prefigured by the process running its course and that is attached to the momentary perceptual phase [22]
with its prefiguring, and [b] a horizon of empty possibilities without this prefiguring. Prefiguring means that an empty intuition is there that provides its general framework of sense. It belongs to the essence of such a prefiguring intention that when pursuing a suitable, appropriate direction of perception this would have to occur: [either] the process of determining more closely, which is a fulfilling process, or as we shall address later as a counterpart, disappointment, annulment of sense, and crossing out. There are also, however, partial horizons without such a firm prefiguring. In other words, aside from definite prefigured possibilities, there are counter-possibilities for which there are no support and which remain constantly open.

Speaking purely in terms of the sense-giving process of perception itself, we can say, for example, that when something like an illuminated appearance, a shooting star and the like flashes in my visual field, e.g., while gazing at a star-studded sky, it is a completely empty possibility that is not prefigured in the sense, but is left open by it. So, if we confine ourselves to the positive sense-giving process of perception together with its positive prefigurings, the question we pose is both understandable and obvious: whether no enduring and ultimately abiding self of the object is even attainable in going from the non-intuitable empty prefiguring to the fulfilling process of determining it more closely; put differently, whether not only newer and newer objectlike features can enter into the horizon of perception, but whether, in the process of determining more closely, even these features already grasped imply a further determinability, *in infinitum*, hence themselves continually and constantly maintaining the character of the unfamiliar *x* that can never gain final determinacy. Is then perception an "exchange" that can in principle never be "cashed in" or "realized" by new, similar exchanges, whose realization leads again to exchange and likewise *in infinitum*? The fulfillment of an intention is carried out by being exhibited in the flesh, to be sure, with empty inner horizons. But is there nothing at all in what has already become exhibited in the flesh that would bring with it a definitiveness so that in fact we are left stuck in an ostensibly empty business of exchange?

We feel that it cannot be so, and in fact looking more deeply into the structure of the series of perception, we come up against the peculiarity that is summoned to solve the difficulty initially for praxis and its intuitive sensible world. Also in the case of [23] incomplete fulfillment, that is, in the case of fulfillment laden with indications, it is the nature of genuine appearances as fulfillments of prefigured intentions to point ahead to ideal limits as goals of fulfillment that would be reached by continuous series of fulfillment. But that does not happen right away for the entire object, but rather for the features that have already come to actual intuition in each case. In view of what is genuinely exhibited in the appearance, every appearance belongs systematically to some type of series of appearances to be realized in kinaesthetic freedom in which at least some moment of the shapes would achieve its optimal givenness, and therefore its true self.

The phantom[34] as a sensibly qualified bodily surface functions as a basic frame for the object of perception. The bodily surface can exhibit itself in continually diverse appearances, and likewise every partial aspect that comes into relief. For each of them we have distant appearances and near appearances. And again, inside each of these spheres we have more favorable and less favorable appearances, and in ordered series we arrive at optima. In this sense, the distant appearance of a thing and a manifold of distant appearances already point back to near appearances in which the form given at first glance[35] and its fullness appear at best in the total overview. This [optimal] form itself given at first glance, which we have for instance when looking at a house from a well chosen standpoint, gives a framework for sketching in further optimal determinations that <would be brought about by> drawing closer, where only single parts would be given, but then, optimally. The thing itself in its saturated fullness is an idea located in a sense belonging to consciousness and in the manner of its intentional structures; and it is, as it were, a s<ystem> of all optima that would be won by sketching in the optimal frameworks. Thematic interest that lives itself out in perceptions is guided by practical interests in our scientific life. And that thematic interest comes to a rest when certain optimal appearances, in which the thing shows so much of its ultimate self as this practical interest demands, are won for the respective interest. Or rather, the thematic interest as practical interest prefigures a relative self: What suffices as practical counts as the self. Thus the house itself and in its true being, and specifically [24]
with respect to its pure bodily thingly nature, is quickly given optimally, i.e., experienced as complete for that person who regards it as a buyer or a seller. For the physicist and the chemist, such ways of experience would seem completely superficial and miles away from its true being.

I can only say in a word that all such highly ramified intentional analyses, which are difficult in themselves, belong for their part to

[34] Translator: The "phantom" for Husserl is the "schema" of the concrete material object, that is, examined without regard to a possible nexus of causality.

[35] *oberflächliche Gestalt*

a universal genesis of consciousness, and here especially in the genesis of the consciousness of a transcendent reality. If the theme of constitutive analyses is to make understandable how perception brings about its sense-giving and how the object is constituted through all empty intending as always only exhibiting optimal appearance-sense in a relative manner, and to make this understandable from perception's unique intentional constitution according to intimately inherent components of lived-experience itself, according to the intentional noema and sense, then it is the theme of genetic analyses to make understandable how, in the development proper to the structure of every stream of consciousness, which is at the same time the development of the ego—how those intricate intentional systems develop, through which finally an external world can appear to consciousness and to the ego.

<DIVISION 1: MODALIZATION> [25]

<Chapter 1: THE MODE OF NEGATION>

<§5. Disappointment as an Occurrence that Runs Counter to the Synthesis of Fulfillment>

We must now expand our insights in a new direction. Up until now we have investigated the continuous, unitary courses of perception in which the unity of an object is maintained concordantly. This took place by virtue of the unanimity of a coinciding that bestows fulfillment upon the intentions, namely, the intentions that are aroused as perception progresses. The process was a constant process of expanding knowledge. This expansion proceeds through discrete syntheses of perception in such a way that a thing, already fairly familiar through a previous perception, occasionally gets perceived again under a simultaneous remembering of previous perceptions, that is, in straightforward recognition. As we can easily see, the new acquisition of knowledge continues the previous acquisition with respect to new sides. But all of this concerns syntheses of fulfillment, that is, of concordance.

There is, however, an occurrence that runs counter to fulfillment, namely, disappointment; there is an occurrence that runs counter to determining more closely, namely, determining otherwise. Instead of the acquired knowledge being preserved and enriched further, it can be placed in question, annulled. In short, there is something like the difference between the modalized consciousness of being in distinction to the originally non-modalized consciousness of being, and we are now in a position of gaining deeper insights into the structure of the modalities of being and their constitution, and noetically speaking, insights into the structure of perceptual belief and its modifications such as "doubt," "supposition," "negation," etc.

As we know, the intentional systems occurring in the normal case of perception that we have described (noetically speaking, the [26] apprehensions apperceiving the respective sensation-complexes) have the character of actual or potential expectations. That is, if in perceiving I instigate a kinaesthetic series, for instance, a certain head movement, the appearances will run their course in a motivated succession such that they accord to my expectation. Thus, in the normal case of perception, all fulfillment progresses as the fulfillment of expectations. These are systematized expectations, systems of rays of expectations which, in being fulfilled, also become enriched; that is, the empty sense becomes richer in sense, fitting into the way in which the sense was prefigured.

But every expectation can also be disappointed, and disappointment essentially presupposes partial fulfillment; without a certain measure of unity maintaining itself in the progression of perceptions, the unity of the intentional lived-experience would crumble. Yet despite the unity of the perceptual process occurring with this abiding, unitary content of sense, a break does indeed take place, and the lived-experience of "otherwise" springs forth.

There is also a lived-experience of "otherwise" without a break, a disappointment of a regular style, which by virtue of its regularity can be anticipated and which thus can even be prefigured in the empty horizon. In other words, there is a steady consciousness of alteration whose phenomenological analysis is fundamental for [understanding] the constitution of a change. Change is a continuous process of becoming otherwise; however, this becoming otherwise maintains unity, namely, a unity of the object remaining concordantly the same as the substratum of its continuous alterations in and through which it becomes otherwise, and in and through which it becomes otherwise time and again.

Let us now already assume a unitary object, be it unchanged or changed, that first abides "concordantly" in the continuity of the original experience, "getting known" better and better. But then all of a sudden, and contrary to all expectation, green rather than red

shows up on the back side that is now becoming visible; instead of the ball-shape indicated by the front side, an indentation or something angular appears, etc. Prior to the ensuing perception of the back sides, perception in its living flow was intentionally prefigured toward red and ball-shaped; perception's referential indicators were determinately directed toward red and ball-shaped. And rather than being fulfilled in this sense, and thereby being ratified, the intentional prefigurings and referential indicators became disappointed. The general framework of sense is retained [27]
and fulfilled, and only at this point, only after we have these intentions, does "something else" occur: a conflict between the intentions still living, and the contents of sense being newly instituted intuitively along with their more or less full intentions. We have a system of continual concordance once again insofar as the insertion of this new framework into the old one restores concordance. But in a partial system we have a superimposing group of intentions that exist in the relationship of disappointment with those upon which they superimpose. After we saw the green and the indentation, and after they lasted concordantly during the course of corresponding appearances, the entire perceptual sense gets altered, and not merely the sense in the current expanse of perception; rather, from it the alteration of sense radiates back to the preceding perception and all its previous appearances. They are reinterpreted in their very sense as "green" and "indented." Naturally, this does not take place in explicit acts; but if we were to go back actively, we would necessarily find the altered interpretation explicitly and consciously, that is, the continual concordance that has been produced. But layered beneath this is something that does not accord with it, and actually what does not accord pertains to the entire series that has flowed-off insofar as we are still conscious of the old apprehension in memory. But it especially comes to life at that place where "green" and "indented" emerged. Occurring here is not only the phenomenon of conflict involving both contrary determinations, ball-shaped and indented, red and green; rather, the "it is not ball-shaped and not red," the empty red-intention, is "annulled," negated by the superimposing "green," that is, by the full superimposing green-perception; and with it the substratum itself, the thing itself, which

in the original perceptual series bore the sense determination "red" at the corresponding place of its shape, is in this respect crossed out and at the same time reinterpreted: It is "otherwise."

<§6. Partial Fulfillment—Conflict through Unexpected Sense Data—Restored Concordance>[36] [28]

Our[37] considerations had taken a new turn in the last lecture. The study of the structure of perceptions with respect to their intentional accomplishments enabled us to gain deeper insights into the essence of modes of being and into the way in which they are intentionally constituted. In the normal case of perception, the perceived object gives itself as being in a straightforward manner, as existing actuality. But that "being" can be transformed into "dubitable" or "questionable," into "possible," into "supposed"; and then "non-being" can also occur here, and in contrast to this, the emphatic "it really is," the "it is indeed so." Correlatively, (i.e., in a noetic regard), one speaks of a believing inherent in perceiving; from time to time we already speak here of judging, that is, of judicative perception. In the case of normal perception—what is usually and straightforwardly meant by "perception"—even if the object is believed to exist with certainty, this belief can pass over into doubt, into taking-something-to-be-possible, into rejecting, and again into an affirming active acceptance. What one so hotly debated under the rubric of the theory of judgment in the newer logical movement since Mill, Brentano, and Sigwart is at its core nothing other than the phenomenological clarification of the essence and logical function of the certainty of being and modalities of being. Here as everywhere, the phenomenological method alone brought to light the problems of pure consciousness and their genuine sense. That is, it concerns understanding how consciousness necessarily equips sense with modalities of being in every sense-giving that it carries out, and it concerns understanding which feature of

[36] Editor: For §§6-8 see Appendix 1: <Descriptions of the Phenomenon of Conflict without Regard to Position-Taking> pp. 425ff.

[37] Editor: Beginning of a new lecture.

constituting consciousness is to be made responsible for this accomplishment. Here the source of really radical clarifications is perception; and for reasons that will become clearer below, transcendent perception is privileged where these clarifications are concerned. What we have said holds even though the specific concept of judgment, the one that dominates the [inner] logic of theory does not even occur yet in the framework of mere [29] perception. Still, the modalities occur precisely here, and it is no coincidence that perception and judgment have these modalities in common. From there we will be able to show that the modes of belief necessarily play their role in all modes of consciousness. Moreover, we must gain clarity so that we can surmount the confusion that blinded such a brilliant researcher as Brentano concerning the questions of belief and judgment, and on the other hand, so that we can understand the constant role of modalities in logic. Let this suffice as an indication of what is to come.

Our analyses up to now have illustrated that every phase of perception presents itself as a system of rays of actual and potential intentions of expectation. During the continual elapse of phases—and during the normal case of perception, during the so-called perception that occurs usually and straightforwardly—there is a continual process of inciting actualizations, then further, there is the continuous fulfillment of expectations whereby fulfillment is always a process of determining more closely. But we also have now the occurrence of disappointment as a possibility that runs counter to the fulfillment of expectations. In order for a unity of an intentional process to be maintained, however, a certain measure of thoroughgoing fulfillment must be presupposed under all circumstances. From a correlative direction this means that a certain unity of sense must be maintained throughout the course of changing appearances. Only in this way do we have the constancy of *a single* consciousness, a unitary intentionality spanning all phases during the course of lived-experience with its appearances.

What happens now if a disappointment occurs in the [perceptual] process rather than a fulfillment, regardless of whether a changing or unchanging object had been constituted in it perceptually? So, for example, we see a uniform, well-rounded red ball; the flow of perception has flowed-off for some time, and has done so in such a way that this apprehension has been fulfilled concordantly. But now as it progresses a part of the back side that was non-visible gradually appears, and contrary to the way in which the sense was originally prefigured as "uniformly red, uniformly well-rounded ball-shaped," there occurs the consciousness of "otherwise," disappointing the expectation: "Not red, but green, not ball-shaped, but indented." Such is its sense [30] now. A general framework of sense has been maintained in thoroughgoing fulfillment; only a part of the anticipating intention is affected, the part belonging precisely to that place on the surface in question, and the corresponding sense-part gets the character of "not so, rather otherwise." Here a conflict occurs between the still living intentions, and—emerging in newly instituted originality—the contents of sense and the contents of belief, together with the horizons proper to them.

But there is not only a conflict. By being presented in the flesh, the newly constituted sense throws its opponent from the saddle, as it were. By covering it over with the fullness of its presentation in the flesh as the sense that is now demanded, it overpowers the former, which was only an empty anticipation. The new sense "green" in its primordially impressional force of fulfillment is a certainty that has a primordial force which overpowers the certainty of the anticipation of "being red." We are now conscious of it *as* overpowered; it bears the character of "null." In addition, the "green" is on the other hand integrated into the old framework. The "being green and indented" that occurs in the primordial impression, and in the entire aspect of the thing from the side in question, advances; in accordance with its sense, and as long as we remain on this one level, it advances the preceding series of appearance of which we are still retentionally conscious in a concordant procession.

<§7. Retroactive Crossing Out in the Retentional Sphere and the Transformation of the Previous Perceptual Sense>[38]

But surely, a certain duplication in the content of sense essentially belongs to the entire phenomenal state of affairs. Just as the unexpected New and "Otherwise" covers over and voids the prefigured sense "red and ball-shaped" that was prefigured in the train of perception up to this point, so too does something take place retroactively in a corresponding manner for the entire preceding series. That is, the perceptual sense is changed not merely in the momentary primordially impressional expanse of perception. The noematic transformation radiates back in the form of a retroactive crossing out in the retentional sphere, transforming its accomplishment of sense that stemmed from the previous perceptions. The previous apperception that was harmonized with [31] the consistently unfolding "red" and uniform "round" is "reinterpreted" *implicite* as "green" and "indented."

This retroactive crossing out and "reinterpretation" essentially means that if we were to bring the retentional elements (i.e., the series of appearance of which we are still freshly conscious, but which have become completely obscure) to intuitive givenness in an explicit remembering, we would notice the following situation in memory: We would find in all the horizons of these retentional components not only the previous prefiguring in the previous structures of expectation and fulfillment, just as this prefiguring was originally motivated at that time, but we would find superimposed upon it the corresponding transformed prefiguring that now points continually to "green" and "indented." But it does it in such a way as to characterize the conflicting moments of the old prefiguring as void. However, insofar as these moments of sense are mere moments of a unitary sense organized in a tight uniformity, the entire sense of the series of appearance is altered modally, and this sense is at the same time duplicated. For we are still conscious of the previous sense, but as "painted over," and where the corresponding moments are concerned, crossed out.

[38] Translator: Paragraph title modified.

Accordingly, here we are studying what the phenomenon of "otherwise," of "annulment," of nullity, or of negation originally looks like. We recognize as basic and essential that the superimposition of a new sense over a sense that is already constituted takes place through repression, just like correlatively in the noetic direction, there is a formation of a second apprehension, a second apperception that is not juxtaposed to a first one, but lies over it and contends with it. Belief clashes with belief, the belief of one content of sense and one mode of intuition with a belief of a different content in its mode of intuition. The conflict consists in the peculiar "annulment" of an anticipating intention, the annulment of an expectation through a full primordial impression for which disappointment is only another expression. And specifically, it is an annulment that concerns an isolated component, while the concordance of fulfillment advances where the remaining components are concerned. The objectlike moment "red" and its anticipated "being" is immediately affected by the annulment, and is that which primarily bears the character of "not." It is only now as a result of this that the thing itself as the substratum of the presumed red is crossed out in the process of belief: The thing "intended" as being red over-all is not; this same thing is rather green at such and such a spot. After the original, [32] simple, and normal perception is transformed by being crossed out, we have once more a perception that is like the normal perception insofar as the transformation of sense, which goes hand in hand with being crossed out, produces a perception of a unitary and thoroughgoing concordant sense, and we constantly find the fulfillment of intentions as perception progresses: With the substitution of "green" and "indented" everything is now in agreement. Yet, there is assuredly a difference insofar as the system of the old perceptual apprehension is also retained in retention for consciousness, and this old perceptual apprehension is partially imbued with the new one. We are still conscious of this old one, but with the character of being annulled. The previous normal sense of consciousness is crossed out in the manner stated above, and the new sense is imposed upon it. We can also say that the old sense is declared invalid, and another sense is interposed as valid. These are only different expressions for the negation and substitution of a new fulfilling sense for the intended one.

Let us draw the most important results from this. First, an original negation here essentially presupposes the normal, original constitution of the object, which we have described above as normal perception. The constitution of the object must be there in order for it to be modified originally. Negation is a modification of consciousness that shows up as such in accordance with its own essence. Secondly, the original constitution of a perceptual object is carried out in intentions (where external perception is concerned, in apperceptive apprehensions); these intentions, according to their essence, can undergo a modification at any time through the disappointment of protentional, expectational belief. This modification takes place together with the superimposition of countervailing intentions arising here essentially. But it happens in such a way that some of the intentions are not only somehow affected by those intentions opposing them, but rather as affected in a special way, they are affected such that through this process, these intentions alter precisely their entire intentional accomplishment. Concretely put, and drawing on our example: The green that emerges as opposing the intention toward "red" does not alter anything in the intention toward "red" insofar as we remain conscious of it as the intention toward "red." The character of consciousness of an "annulled," an "invalid" <intention> [33]
emerges now, and accordingly red has the modal character of "void."

In contrast, what is newly perceived has the character of "being valid," even though what is newly perceived disappoints the intention. Likewise, we can say that in such a contrast, every normal perception is a consciousness of validity—every perception in which an occurrence such as disappointment or similar occurrences have not yet arisen. But if we compare the unaltered consciousness, on the one hand, with the consciousness that is altered by being crossed out, on the other hand, and if we make this comparison in view of the content of sense, then we will see that while the intention is indeed transformed, the objective sense itself remains identical. The objective sense still remains the same after being crossed out *precisely as* a crossed out sense; thus, the content of sense and its modality of being are distinguished: On the one hand, it has the mode of straightforward, uncontested concordance, on the other, it has the mode of being contested and being crossed out.

<Chapter 2: THE MODE OF DOUBT>

<§8. Conflict between Two Superimposed Perceptual Apprehensions Having the Same Hyletic Stock>[39]

Let us now consider still another affiliated, possible type of occurrence that exhibits the mode of transition to a negating annulment, but that can also occur as a lasting condition. I mean the phenomenon of doubt that can be resolved, be it in the form of negation or also in the form of affirmation, in the former case [i.e., negation] like in the previous example that we already cited concerning an illusion being unmasked in consciousness: What is initially seen as a human being becomes dubious and is finally disclosed as a mere wax figure. Or conversely, doubt resolves itself in the affirmative form: Yes, it is indeed a human being. During the doubt of whether it is an actual person or a wax figure, two perceptual apprehensions obviously overlap. One of them [34] lives in the normally flowing perception with which we began; we see a human being there for a time, concordantly and uncontested like other things in our surroundings; they were normal intentions, partly fulfilled, partly unfulfilled, being fulfilled normally in the continual succession of the perceptual processes, without any kind of conflict, without any kind of break. And then what follows is not a clean break in the form of a decisive disappointment, not a break such that a perceptual appearance of a normal intentional type clashes with a component of expectation that was aroused, and crossing it out with its fullness, covers it over and annuls it. Rather, in our present example we have the situation that all of a sudden the full concrete content of the genuine appearance

[39] Editor: For §§8-11 see Appendix 2: <Sense and the Modality of Being in Perception and Remembering> pp. 431ff.

(alongside the original empty horizon and the original prefiguring) gains a second content, one that overlaps the other: The visual appearance, the spatial shape filled with color, was previously endowed with a halo of intentions of apprehension that gave the sense "human lived-body" and "human as such." And now the sense "dressed up wax figure" is superimposed upon this. Nothing has changed with respect to what is actually seen; indeed, they even have more in common; they both share apperceived clothes, hair, and so forth. But in one case it is flesh and blood, in the other, wax.

If we go back to the ultimate structures, we can also say that one and the same stock of hyletic data is the common support for two overlapping apprehensions. Neither one of them is crossed out during the period of doubt; they stand here in a mutual struggle; each one has, so to speak, its own force, each one is motivated, demanded, as it were, by the previous perceptual situation and its intentional content.[40] But demand opposes demand, one contests the other and is wronged in the same way by the other. There remains in doubt an undecided struggle. Since an objectlike formation is only constituted by the empty horizons together with the common, genuinely intuited core, we accordingly have a bifurcation of original, normal perception (which only constituted [35] one sense in concordance), into a duplicity, as it were, into the form of a doubled perception. We have two interpenetrating perceptions by virtue of the common core-content. But actually, this expression does not really fit. For its conflict also means a certain mutual repression: If the one apprehension overpowers the common intuitive core, if it is actualized, we will see a human being for instance. But the second apprehension, which is directed toward the wax figure, has not become nothing; it is suppressed and put out of commission. Then, for instance, the apprehension "wax figure" obtrudes, and accordingly we now see the wax figure; but now the apprehension "human being" is no longer functional, but is suppressed.

---

[40] The choice of these expressions is not arbitrary; it shows that all apprehensions consist in tendencies, motivated in their interconnections, and that they have their "force" in this motivation. See below (pp. 81-2) where the propensity to believe is first introduced for doubt. Correlatively we would have to speak of a *inclinatio ex*.

This holds, however, not only for the momentary situation of perception, for the Now-phase. For we recognize even here the essential retroactive efficacy of the conflict on the process of lived-experience that has flowed-off: We recognize even in this lived-experiencing the crumbling of the singular-sense consciousness into a many-sense consciousness; that is, the process of becoming bifurcated, with its apperceptive overlapping, advances into retentional consciousness. If we explicitly presentify the expanse of perception preceding the doubt, then it will no longer be there as having its singular sense like any other memory; instead, it has taken on the same doubling; the apperception of wax figure is everywhere superimposed upon the apperception of human being. But of no less importance—indeed, it is of the utmost importance—is the fact that the doubling is not really a doubling of perceptions, even though the fundamental character of perception, the consciousness of something being presented in the flesh, exists in both cases. If the apperception of human being suddenly changes into the apperception of wax figure, then the human being will stand there first in its presentation in the flesh, and then a wax figure. But in truth neither of them are there like the human being was prior to the onset of doubt. Evidently, the mode of consciousness has altered, although the objective sense and its modes of appearance, now as before, has the mode of being presented in the flesh. In fact, we have not yet thoroughly taken into account the essentially altered mode of belief or mode of being. The way in which we are conscious of what appears in the flesh is different. Instead of it being given to consciousness precisely as being there in a straightforward manner, like in normal, univocal perception, i.e., in perception running its course concordantly, it is now given to us as questionable, as dubious, as [36] contentious: It is contested by another givenness, a givenness in the flesh, a givenness of another <apprehension> permeating it and in conflict with it.

We can also express this in the following manner: Consciousness, which gives its object in the flesh (originally), does not only have the mode of presentation in the flesh, which distinguishes it from presentifying consciousness and empty consciousness (both of which do not present the same sense in the flesh); it also has a variable mode of being or a variable mode of validity. Original, normal perception has the primordial mode, "being valid *simpliciter*"; this is what we call straightforward, naive certainty. The appearing object is there in uncontested and unbroken certainty. What is uncontested points to possible contestations, or even to breaks, precisely to those we have just described, and by becoming bifurcated, they undergo a modification in their mode of validity. In doubt, both presentations in the flesh contending with one another have the same mode of validity, "questionable," and each presentation that is questionable is precisely in dispute and contested by the other.

We already see here that what has been demonstrated for perception as the consciousness of something being presented in the flesh must be carried over to memory. For, modalization also occurs in memory by virtue of a radiating back into retention, and therefore into remembering that makes [what is retended] explicit. Naturally, we only have in mind here the segments of the past for the same object that continues to endure as being present in the flesh. While normal memory (by virtue of the fact that it is the reproduction of a normal perception) gives the reproduced element in the normal mode of validity, of certainty as certainly existing, the memory that is encumbered by discrepancy due to that radiating back yields the altered mode of validity, "questionable," questionable whether it was this or that, whether it was a human being or a wax figure.[41]

<§9. Resolving Doubt through the Transition to a Ratifying Certainty or Negation>

The possibility of deciding, of resolving, and the possibility of their potentially active forms belong to the very essence of doubt. In contrast, doubt itself means indecisiveness, consciousness is [37] indecisive consciousness. In the sphere of perception a decision is necessarily carried out such that as we progress to new appearances (for instance in the free orchestration of

[41] For the completion of our descriptions concerning the essence of the situation of doubt, see pp. 81-2.

corresponding kinaesthesės running their course), a fitting fullness that suits an expectation is integrated into one of those empty horizons engaged in a mutual dispute. This is the most original form of decision. Under the given intentional situation, modified or completely new sense-data that arise demand precisely [those] apprehensions that complete the remaining uncontested intentions; they demand the apprehensions to complete the intentions in such a way that the source of the contention is quelled, and what is especially motivating the doubt will be annulled through the force of a primordial impression. Fulfillment through a primordial impression is the force that mows everything down. We move closer to it, we clasp it with our hand, touching it, and the dubious intention of wax we just had gets the priority of certainty. It gets this through the concordant transition to new appearances that do not accord with the apprehension of human being and its unfulfilled horizons, and negates the latter through its fulfilling weight of being presented in the flesh. With respect to the one instance, a negation takes place in this decision; in particular, it takes place with respect to the apprehension of human being that is guiding the original perception and that then becomes modalized as dubious. In the opposite case, an affirmation would have occurred, or what amounts to the same thing, a ratification of the original perception, but which later became dubious. That which appeared in the flesh would have then received the modal validity character of "yes, really."

So, in a certain respect even the ratifying Yes, like the No, is a mode of modifying certain validity and is distinct from the entirely original, entirely unmodified mode of certain validity; the straightforward constitution of the perceptual object is carried out univocally in this mode, and without struggle. But I said "in a certain respect." For speaking of "modalization" is ambiguous. On the one hand, we can mean each transformation of the mode of validity as distinct from the original mode of validity, naive certainty, so to speak, that is not ruptured by discrepancy or doubting. And on the other hand, we can mean a transformation where the validity-mode of certainty is concerned, where it ceases to be certainty. The primordial mode is certainty, but in the form of the most straightforward certainty. Just as an affirming decision occurs by passing through a period of doubt, so too do we have a [38] restoration of certainty; when something turns out to be real[42] "in effect" I become certain of it anew. And yet, consciousness is altered now. Passing through a period of doubt to a decision gives to consciousness precisely the character of resolved consciousness, and it gives to its noematic sense the corresponding character that expresses itself in the "yes," "in effect," "truly so," and in similar turns of phrase.

Here as everywhere it becomes clear to us, and it will become even more clear to us later, that everything that consciousness undergoes through changes and transformations, even after the transformations, remains sedimented in it as its "history," and this is, so to speak, the destiny of consciousness. But since consciousness is what it is as consciousness of something, as a process of giving sense, this means that every such transformation shows up in the sense, and that even where the objective sense is the same, yes even where the mode of appearance is the same, it expresses itself as a modality, as a transformation in this sense.

If we observe consciousness as a whole being at variance with itself, we find constituted uniformly within doubting consciousness the disjunctive "*A* or *B*," in the negation, the "not *A* but rather *B*," and further, in the affirmation, the "not Not-*A*, but *A* indeed." In this way, straightforward "being" of the objective sense is transformed into " doubtful being" or what amounts here to the same thing, "questionable being," and then possibly through a decision it is transformed into "not being" or into "being indeed." In phenomenological considerations above all, in considerations whose intentions are directed toward an ultimate understanding of consciousness and the accomplishments of consciousness, we must persistently orient our regard toward both these sides, toward the noetic, toward the side of the lived-experience, and toward the noematic, toward the side of that which gives itself to consciousness in the life of consciousness, toward the side of sense and its modes that are so diverse. This we must do in the sphere now under consideration.

---

[42] *wirklich*

Already guided by being and modalities of being, and in the focal orientation toward consciousness and toward the ego that carries out this consciousness, we find the original mode of naive perceptual certainty, or if you will, naive perceptual belief. Then we find the modified modes, doubtful uncertainty, negating as the negative decision that crosses out a certainty and annuls it in the form of a positive contrary certainty. We find further the affirming, the renewed becoming-certain, a certainty, however, that is in the form of a ratifying experience. We also speak here of [39] active acceptance, just as in the opposing instance of rejection. We see here that active acceptance is something different from naive certainty, and unlike the latter, presupposes passing through uncertainty as doubting, questioning. Let us note in passing that when we speak of questioning, we are not concerned here with the wishful intention to decide, an intention that co-determines its sense. It does not concern us here, and logically, is not essential to it.

Finally, let me mention some important parallel expressions: "To hold to be true" where any kind of certainty is concerned, and "to hold to be false" where "rejection" is concerned. Correlatively, we have on the side of sense the expressions that we have constantly employed, "being certain," "not being," and so forth, and appropriate to the latter way of speaking, we also have the "true," especially as an expression for the "yes really," and the "false" as an expression for "not being." We want to note that the concepts of true and false occur here as expressions for the modes of being we have characterized. In fact, all analyses of the origins of these concepts must begin at this point. I say begin. For we have not even hinted at how these concepts will develop all the way up to the full concept of truth.

<Chapter 3:
THE MODE OF POSSIBILITY>

<§10. Open Possibilities as the Indeterminate Scope of Intentional Prefiguring>

We still have to address the important group of modalizations with respect to possibility and probability. They fall completely within the scope of uncertainty, where we understand by uncertainty not merely a privation of certainty, which would encompass the case of negation; rather by modalities of uncertainty, we mean those that do not pertain to decision at all. Whenever consciousness has lost the mode of certainty and has passed over into uncertainty we will also speak of possibilities. But also more than that. For in this sphere we encounter several concepts of possibility.

Let us mention first of all the concept of open possibilities with respect to the following context: What is intentionally prefigured in the apperceptive horizon of a perception is not possible, but [40] certain. And yet, possibilities, indeed a whole range of manifold possibilities, are always included in such prefigurings. The prefiguring for the non-visible side, which is given in the perception of a thing from the front side, is as we know an indeterminately general prefiguring. This generality is a noetic trait of consciousness emptily pointing ahead, and correlatively is a [noematic] trait of sense for what is prefigured. Thus, for example, the color of the thing's back side is not prefigured as an entirely determinate color when the thing is still unfamiliar to us, and when we have not yet looked at it from the other <side> with more precision. "A color" is indeed prefigured. But potentially still more than that. If the front side has a pattern, then we will expect the back side to follow this thoroughgoing pattern; if it is a uniform color with a medley of specks, we would also possibly expect specks for the back side, and so forth. But there is still indeterminacy. Pointing ahead has now, like all other intentions in normal perception, the mode of naive certainty; but it has this mode precisely according to what it gives to consciousness and in the manner in which it gives something to consciousness, that is, according to the sense in which it gives something to consciousness. What is certain, therefore, is some kind of color in general, or a "color in general broken up by specks," and so on, i.e., indeterminate generality.

Let us reflect on the consequences of this. Naturally, we use the term generality here only as a provisional aid for an indirect description pointing toward the phenomena themselves. For we do not have in mind here logical concepts, classifying or abstracting generalities. Rather, we have in mind simply this intending into the future peculiar to perception as it is given in perception with its mode of consciousness of indeterminacy. Belonging to the general structure of every empty intention, and likewise to such an indeterminate pointing-ahead, is the possibility of explicating it in the shape of presentifications. We can freely form presentifications that bring to intuition what is non-visible, for instance, by imagining that we are walking around the object. If we do this, then intuitions appear with entirely determinate colors. But we can obviously vary these colors freely within the scope of indeterminacy.

What does that mean? If we are purely directed toward a mere bringing to intuition[43], in other words, toward a quasi-fulfillment of perception through presentified perceptual series, a concrete intuition with determinate color will occasionally set in. But this determinate color has not been prefigured, i.e., it has not been demanded. What is presentified is there as certain, and indeed as *the* back side; but it is so precisely in an indeterminacy-consciousness that does not indicate this determinate color appearing there in a fortuitous manner. If other intuitive presentifications occur with other colors, certainty will be extended just as little to these. For none among them is settled in any particular way in advance. None among them are demanded. [41]

Let us contrast this with the following case. In the instance of an actual fulfillment in the actual progression of perception, the appearance of color fulfilling what is indeterminately prefigured is constituted in itself as certainty. In this case, what takes place with certainty is a specification that determines [more closely] and thus

[43] *Veranschaulichung*

a gradation of knowledge. The new emerging expanse of perception with its content of certainty contributes a concreteness by determining more closely the generality that was indeterminate and prefigured; this concreteness is encompassed by the unity of perceptual certainty and therefore uniformly fulfills what is prefigured, the anticipation. The fulfillment is at the same time an increase in knowledge. (The specks that are determinate.) However, that is not the case with the illustrative presentification; every other color can serve just as well for the color that has just appeared. Presentification is only endowed with the mode of certainty to the extent that it retains its mode of indeterminacy in relation to the coloring, despite the determinate coloring occurring in it. Only in this respect does it distinguish itself from a determinate memory, like the kind [of memory] that we would have of it were we to presentify the back side to ourselves again after having actually perceived it. Accordingly, it is clear that every presentification that is merely an intuitive presentification prior to an actual acquisition of knowledge must have a modalized character of certainty in relation to the quasi-determining content. But this uncertainty has the distinctive feature that in it the color that is given fortuitously is precisely a fortuitous color for which not just anything could arise, but rather some other *color*. In other words, the general indeterminacy has a range of free variability; what falls within it is encompassed *implicite* in a similar way, and yet it is not positively motivated, not positively prefigured. It is a [42]
member of an open range of closer determinations that can be adapted to this framework, but beyond this they are completely uncertain. This makes up the concept of open possibility.

<§11. Enticing Possibilities as the Tendencies to Believe within Doubt>[44]

Motivation prefigures something positively, and yet does so in the mode of uncertainty; what this means can be illustrated even more clearly by contrasting it with another kind of possibility that

[44] Editor: See Appendix 3: <Evidence of Possibilities as Such and Modal Modification *in infinitum*> pp. 436ff.

we will come to know. Let us look back to the phenomenon of doubt. Whenever we speak of doubt, we also speak of propensities to believe. What occurs in the front side that is seen, together with its apprehended sense for the back side, may prefigure something determinate. But it does so ambiguously and not unequivocally. This happens when we become unsure whether what we see is a complete thing or a piece of scenery, for example. That yields a conflict in consciousness that plays itself out here in empty premonitions of what is to come, which was not the case with our example of the wax figure/human being. In this new case, the struggle can take the form of a static state of suspension. But as soon as the ego directs itself toward it and even carries out a presentifying intuition, the struggle passes over to a dynamic play of opposing [senses], to a doubting vacillation. This issues in a propensity to believe for each side. Namely, by initially actualizing for itself the motivations leading to the one side, the ego experiences the concordant demand that is coming from this other side. By exclusively surrendering itself to these motivations, as it were (while what speaks in favor of the other side remains out of action), it experiences a force of attraction, a propensity to turn toward it in certainty. But this also holds likewise when actualizing the opposing intentions. In this way the normal egoic act of perception is modalized into acts we call enticements to believe. From the side of the objective senses, from the side of the objects given to consciousness, we also speak here of enticements to be, which is to say that affection issues from the side of the object, that the object exerts on the ego an enticing demand to be, just as if the object were its hostile partner. The sense itself has the propensity to be.[45]

Now, we likewise call this enticement, possible (regarding it outside of its relation to the ego); but it determines a [43]
fundamentally different concept of possibility than the concept of open possibility that was described previously. The difference between them both becomes completely evident precisely in their contrast.

---

[45] Translator: See the discussion of affection, Division 3.

<§12. The Contrast between Open and Enticing Possibilities>

In principle, an open possibility does not imply a propensity. It does not exert an enticing demand to be; nothing speaks in favor of it; there is no demand directed toward it, even if there were one inhibited by opposing demands. Thus, we do not speak here of enticements at all.

Let us call these new possibilities problematic possibilities or questionable possibilities. We do this because the intention to make a decision arising within doubt between one of the enticing factions of the doubt is called a questioning intention. We speak of questionableness only where enticements and contraposing enticements play off of each other, when something speaks for or against them. The most direct expression for these possibilities, however, is enticing possibilities. It is completely clear that they designate a totally different kind of modalization than the modalization of open possibilities. For modalizing consciousness in each instance has a fundamentally different origin.

We can even characterize open possibility as a modalization of certainty. But this modalizing consists in the fact that an indeterminately general intention, which itself has the mode of certainty, implicitly bears in a certain way a diminution of its certainty with respect to all conceivable specifications. For example, if within an indeterminate generality a speckled color is demanded in certainty, then the fulfillment is restricted insofar as precisely "some kind" of color with "some sort" of shaped specks is demanded. And every specification of this type fulfills this demand in the same way. The specification fulfills the demand; thus, something of a demand also belongs to it. But not only does each one of them exert the same demand; rather, the demand is an implicit one insofar as each contingently occurring specification is essentially grasped in accordance with the indeterminately general demand; it is co-demanded in accordance with it, while as has been shown no actual demand aimed precisely at this specification is directed toward it from the beginning and at present, be it a diminished or even an unrestrictive demand.

This is entirely different where the enticements are concerned, [44]
where each one is intended in its particularity.

It is now clear that we have determined a closed and exactly limited group of modalities from a primordial mode of straightforward naive certainty. We have done this by [recognizing] that they are modalizations by virtue of conflict, namely, by virtue of an originally and straightforwardly certain demand with opposing demands. Problematic consciousness with its problematic possibility belongs to this group. We therefore make a fundamental distinction between the modalities arising from conflict, and the modality of open specification.

We can continue our exposition of problematic possibilities by noting that they and only they appear with a different weight. The enticement is more or less enticing; and that also holds particularly when comparing all potentially diverse problematic possibilities that belong to one and the same conflict and that are bound synthetically through this conflict. For, the conflict, the bifurcation of a consciousness into reciprocal inhibitions, creates a unity; noematically speaking, it is the unity of opposition, of the possibilities bound to one another through it.

<§13. Modes of Certainty As Such in Their Relation to Enticing and Open Possibilities>

It is of no little consequence now to regard a unique group of modes of certainty that have this peculiar trait: Certainty remains certainty. [These modes] concern the differences pertaining to the "purity" or "completeness" of certainty.

Let us imagine the following situation. I believe that it is so; I do not doubt; I am not indecisive; I carry out the unbroken thesis: "It is so." But it can nevertheless transpire that while I am so entirely certain, so "sure," it may also be the case here that there is much that speaks against it being so. Another being (or several like it) stands before me as an enticing possibility.

Such opposing enticements, opposing possibilities, can have differing weight; they exercise a stronger or a weaker pull, but they do not determine me. Determining me in belief is just the one possibility for which I am resolved, for which I have decided [45]
earlier, perhaps in a process of passing through doubt.

We see, moreover, that a concept of "conviction" belongs here. Different witnesses speak and present their testimonies, having different weight. I weigh them and decide for the one witness and his testimony. I reject the other testimonies. Here the weight of the other testimonies can even become null and void. In other words, they lose every bit of weight; in truth they have no weight. And yet, it must also be the case that they do retain a weight (and do not turn out to be downright false). But this one testimony in particular has such an "overpowering weight" that I decide for it and do not "accept" the others, do not "take them up," and in this sense reject them. I take the side of this testimony, I come out in favor of this testimony, and against the others.

Yet I can potentially mark the differing weights without deciding in favor of one of the enticements. I hold it in abeyance like the others. I wait possibly for an "objectively decisive" experience, withhold an opinion, wait for an experience that turns up one of the possibilities as an "indubitable" reality, namely, for one that will negate and annul every one of the other "possibilities" thereby robbing them of their weight. In this sense, one can characterize these groups of modes of certainty as modes of conviction.

Thus, the modes of impure (or incomplete) certainty are modes of certainty that refer to the realm or sphere of what is enticing. Let us ground this impure certainty phenomenologically, and more particularly, in the original field of perception. Then we will see still finer distinctions.

Something entices me as a possibility, something speaks in favor of it; but there are other opposing possibilities, and something speaks in favor of them, too; or rather, this or that speaks "against" the others. Or I am "conscious of" only one possibility. For example, a cloudy sky together with humidity speak in favor of a thunderstorm, but not "for sure."[46] It entices in [46]
this way, and it does so in varying degrees, changing according to the particular circumstances.

[46] What does "being" conscious of mean here? Being in relief. I am "unconscious" of the other possibilities or entirely indeterminate other possibilities; they are not awakened, but inhibition is still there.

It may be the case here (a) that I am conscious of this possibility in and through its enticement and nothing more: I "do not let myself be determined by it." (b) I am inclined to decide for this possibility, I go along with it for awhile, as it were, and let myself be drawn into it; I am ready and willing to follow its pull. Insofar as the enticement as such means an affection on the ego, which from the perspective of the ego corresponds to a being-drawn, a "propensity" lies in the enticement itself. But the fact that I let myself "willingly" be drawn in, that I am about to follow after it, is still something new phenomenologically. However, here this "following" can be inhibited by opposing propensities, or not be "efficacious" at all. (c) This efficacy means that I straightforwardly yield to the inclination, perhaps without inhibition, that I take up its position, that I definitively "decide" for this possibility. I believe, I am "subjectively certain" that there will be a thunderstorm, and fetch my raincoat and my umbrella. We can then speak of presumption or of a presumptuous certainty in a specific sense. This is similar to the way in which we believe the one witness in the clash of testimonies, although the testimonies by the other witnesses are not shown to be simply false since they still have weight; but they have a weight that we no longer accept. It is not merely the case that the one testimony whose enticement is privileged is stronger: We lend it validity, believing in it in our subjective certainty; and this inner Yes means a No! for the other testimonies of the opposing enticements. They do not hold for us, that is, they are not valid for us "subjectively." In itself, in its own phenomenological character, this presumptuous certainty is characterized as an impure certainty. To be sure, the decision is made, but it is gnawed at internally, so to speak, weakened by the opposing possibilities whose weight is still there and still weighs upon us, only that we deny them their validity. That gives presumptuous certainty an inner character distinguishing it clearly from pure certainty. Obviously, this impurity, this murkiness, has its degrees.

We must take note of still another distinction here. When we say that "something speaks in favor of one or several possibilities" we confront an ambiguity that points us to different phenomenological connections.

(1) An enticement refers to leeways[47] of possibilities, and these possibilities are not merely imaginary possibilities. Insofar as this is the case, something "speaks" in favor of them all. [47]

(2) But that is only to say that they are "leeways," and from them the determinately directed expectations, inhibiting one another or uninhibited, (the determinate "signs"), mark out all sorts of things. And this is what we had in mind when we spoke in the more restricted sense of something speaking in favor of possibilities. And we hold firmly to this concept.

Whenever we have certainties that refer to leeways of open possibilities we speak of "empirical, primitive certainties." All types of external perception belong here. Each perception implicates a leeway of specifications at every moment within the certainty of a general prefiguring. Yet nothing speaks in favor of these specifications in their particularity. We can also say that the same thing speaks in favor of all the open possibilities of a leeway, they are all equally possible. This implies that nothing speaks in favor of the one possibility if it speaks against the other.

(a) Certainty is a pure certainty; only one single possibility is just right; "something speaks" in favor only of it and lacks the character of mere enticement. It is a complete certainty, complete precisely in the sense of this purity that has no "opposing motives." The raised hammer will fall!

(b) Certainty is an impure certainty.

But in comparison to the immanent sphere, and to the givenness within the immanent present that is evident in not being able to be crossed out, another opposition also emerges now, namely, the opposition between:

(α) those empirical-primitive certainties that implicate precisely the leeways of other possibilities, even if nothing speaks positively in favor of these possibilities (except for the decisive ones). Non-being here is not ruled out; it is possible, just not motivated.

(β) absolute certainties whose non-being is ruled out, or again, if you will, is absolutely certain. There are here no open opposing possibilities; there are no "leeways."

---

[47] *Spielräume*

But there is still a question here concerning how we place what we have just said in relation to modes of evidence.

I can have leeways, "real possibilities" given in evidence, as in experience. On the other hand, I can have an apodictic exclusion of opposing possibilities, of a conceivable being-otherwise. And accordingly the decision can be "evaluated." (Empirical certainty—apodictic certainty.) Still, I can also be conscious of emptily meant possibilities and opposing possibilities without such evident givenness; and I can decide, endorse a possibility when something speaks in favor of it presumptively, etc. [48]

That is clearly a unique theme and yields its own distinctions.

We have become familiar with modes of certainty, modes of straightforward belief. On the other hand, certainty can be "modalized," which is to say, can cease to be certainty at all: For instance, it can pass over into an enticement, and moreover, into an inclination to follow the enticement, but still without leading to a decision. Accordingly, this would be exactly not a resolve, not certainty, but a modalization of certainty. Likewise, [we have modalization as] a doubting understood as being divided in the vacillating propensity to believe one or the other; and in this indecisiveness, [we have the propensity] to believe the one or the other, striving further to reach a decision, to seek certainty. And likewise [we have a modalization when] calling-into-question, despite certainty; certainty is bracketed, put out of play, etc. Put more precisely, under the general structure of certainty, straightforward belief, we have different specifications, different modes of certainty: Empirical and apodictic certainty. We have still further distinctions within empirical certainty, namely, distinctions of act. Within empirical certainty and certainty in general, and potentially in doubting, distinctions of act can emerge as transformations of the mode of certainty. But there is always certainty! We had come to know impure certainty as a certain decision for an enticement. But there is also a decision that remains within uncertainty.

Let us consider doubting and questioning. Doubting is a doxic mode of comportment experienced as being torn between two or more possibilities; it is a vacillation between different possibilities concerning a judicative intending, namely, between possibilities that are enticements, and whereby judicative intending is precisely not actually passing judgment, not having certainty, but is rather a "problematic judging." I am not certain, but I am inclined to believe that *A* is; something speaks in favor of *A*, and I "would like" to judge it so. We see that this "I am inclined to" can initially [49] mean the same thing as "something speaks in favor of." Thus, both expressions can be construed as merely correlative expressions. On the other hand, we distinguish from this an inclination-toward as an inner complying, as it were, a sort of deciding for, and yet without being decidedly resolved. I am about to follow the counsel, but an inner "contra-diction" inhibits me, a propensity that inclines me no less to believe otherwise. The decision is inhibited. I can inhibit this propensity, and I can explicitly become conscious of the enticement without already setting off internally to make the decision, that is, without "following" the lead, and perhaps only at this point then curbing or inhibiting myself.

Doubting is a wavering in decisiveness, and each member [of the doubt] which is not being decided upon is still a mode of making a decision. But it can also happen that we decide for the most weighty problematic possibility, for the one that is most strongly affective. What takes place here, however, is not making a decision in certainty, but a particular mode of making a decision that is peculiar to the enticement. Then we have presumption as taking-it-to-be-likely.

Where several problematic possibilities are separated and united we have then a consciousness of problematic disjuncts, the consciousness of "it is 'questionable' whether *A* or *B*"—to be sure, without taking questioning in the narrow sense of the word.

The question arising in doubt is a striving after a decision that issues from the comportment of doubt, a striving that is being motivated within this comportment of doubt. Or, it is a striving to reach a certainty that is motivated from an inhibited decision, from a decision that remains incomplete. But is not the propensity itself such a striving like when we have a singular propensity toward an enticement? Is the question: "Is that so?" a striving to overcome the inhibition, and to reach the corresponding resolved certainty? With respect to the genuine sense of a question arising in doubt, namely, as a multifaceted indecisiveness: Is this the striving intention to decide the doubt, to overcome the inhibition here or there, and to arrive at a certainty? Certainty of the fact that *A* is, then, annuls all opposing propensities. And it is not only the case that certainty annuls the propensity for *A* insofar as it is transformed precisely into a certainty, losing its inhibition, by finding completion in certainty; it crosses out the opposing [50] propensities, too, insofar as these cannot also pass over into certainties. To decide for *A* means to reject *B*, *C*, etc., with certainty.

In any case, it is now characteristic of doubt and also of the question arising in doubt that I am not convinced in advance about what is there in certainty; and it is not the case that I have only put this certainty out of play.

There is a type of question, then, that has different foundations, namely, where I am already interiorly resolved that it is for instance *A*, but where I call into question whether *A* is or *B* is, etc. (in other words, without being in a position to want to decide a doubt that has gripped me). But how do I arrive at this? What sense can that have?

Certainty can be incomplete, impure, and I seek a more complete or a completely pure certainty.

In[48] the previous lecture we distinguished between impure and in this sense incomplete certainties, and complete or pure certainties. And we have done this with respect to a particular type of certainty, namely, certainties of transcendent perception. Let us consider a little more closely the modifications of certainty about which we speak.

Such a certainty is impure insofar as it has the mode of making a decision for an enticement, more specifically, a subjectively secure decision for an enticement, even though opposing enticements are there, and against which the ego decides despite their weight; the ego does not accept them, although due to their weight, they "demand" being validated. Naturally, this demand or claim consists here in this weight itself, that is, in the affective force that the enticement exercises on the active ego. By affective force I mean a tendency directed toward the ego, a tendency

[48] Editor: The beginning of a new lecture.

whose reaction is a responsivity[49] on the part of the ego. That is, in yielding to the affection—in other words, by being "motivated"—the ego takes up an endorsing position; it decides actively for what is enticing, and it does so in the mode of subjective certainty.

A "pure" certainty occurs here when the opposing enticements completely lose their weight, when, accordingly, they are entirely [51] crossed out as the experience progresses; they are experienced as straightforward nullities. "What is there" is decided upon from the standpoint of the matter,[50] by itself, and in making a decision, the ego follows the decision on the part of the matter. It does not need to take sides; it does not need on its part to back one of the possibilities. Each one of the other possibilities, as the possible ground for taking a position, is yanked out from under its feet, and the only ground as the ground of a certainty pertaining to the matter is there of its own accord. The ego finds itself standing on it, and then merely establishes itself subjectively on its ground.

We have a simpler case still where we could not be further from speaking about making a decision, because the opposing enticements are lacking from the very start, and in their place are open possibilities. Let us take an example from external experience: Watching the smith, I expect the raised hammer to fall down and bend the iron; watching the glass fall, I expect it to strike the ground and break, and so forth. – Alternative possibilities are there: Some unforeseen effect can intervene, an accidental bump can make the glass fall on the nearby straw mat instead of on the stone floor, etc. Every event as a physical event is surrounded here by a horizon of open possibilities—but they are open; nothing speaks in favor of them in this given moment; the expectations are straightforward certainties that are not inhibited; a modalized expectation does not oppose other like expectations in relation to an enticement.

[49] *antwortende Tätigkeit*
[50] *Sache*

<Chapter 4: PASSIVE AND ACTIVE MODALIZATION>

<§14. Position-Taking of the Ego as the Active Response to the Modal Modifications of Passive *Doxa*>[51]

What strikes us now is the ambiguity in speaking of a decision that come to pass on its own or in the matter itself, namely, as undergoing a decision that just arises, and the deciding position-taking that is carried out on the part of the ego as the ego's reaction. Looking back at our initial introduction to the modalities [52] of being and the modalities of belief, we recognize that everything that was disclosed at that time was initially a modalization taking place purely in the intentionality of perception, potentially as completely passive; and in any case it had to be taken purely that way at first. It is clear now that we must distinguish between the passivity and the activity of the ego: (1) modal modifications of passive *doxa*, of passive intentions of expectation, their inhibitions passively accruing to them, and the like; (2) responsively taking a position that is peculiar to making a decision, doing so actively, as proceeding from the ego.

Furthermore, it is also clear that the concepts of belief and of modalities of belief themselves undergo a modification with this distinction. For now we have to separate out the essentially different processes and occurrences of passivity and activity according to their constitutive accomplishments. Thus, we have:

(1) Initially in passivity the syntheses of concordance or discordance, of uninhibited intentions that are fulfilled freely, or inhibited intentions that are crossed out, etc. And correlatively, we have in the noema modes of being that maintain the identical objective sense, possibly in connection with an opposing sense.

(2) The active position-takings of the ego, the active decisions, convictions, letting oneself be convinced, and taking sides, etc., and finally the activity of conviction in the broadest sense (where we no longer speak strictly of testifying for or against). These

---

[51] Editor: For §§ 14 and 15 see Appendix 4: <Levels of Decision. Receptivity and Spontaneity> pp. 439ff.

activities also have their noematic correlates. We must note here that it is not a matter of merely making passive intentionality patent; it is not a matter of becoming merely cognitively aware in perceiving, a mere living through the enticement that is taking place in an attentive turning-toward, that is, a matter of merely becoming consciously attentive to enticements, nullities, and the like. Rather, the ego passes its judgment in its own position-taking, *it makes* a decision for or against, and so forth. One may well say that here lies the specific source for what we normally mean or can mean by judging. "Conviction" expresses more: Issuing from the passive perceptual situation, letting oneself be determined such that one has a judicative position and then has a judicative determination. Thus, we also understand why in practice, judging and conviction become equivalent expressions. [53] We will see shortly that this position-taking or this group of position-takings that occur here are completely non-independent from the standpoint of intentionality, namely, insofar as they presuppose the occurrences of passive *doxa*. Let us note in advance that these position-takings, this validation and its transformations, further, are not to be confounded with other modes of comportment by the ego that belong to the sphere of judgment, especially not with active explication, colligation, comparison, differentiation, and the like—all of the operations to which we are indebted for the logical forms of different state-of-affairs.[52] In all these actions, judging is always only a process of conferring or denying validity that stems from the ego.

The ego does not always take a position judicatively in this strict sense. When it simply perceives, when it is merely aware, apprehending what is there and what, of itself, is presented in experience by itself, there is no motive for taking a position provided that nothing else is present. There must be opposing motives in play, open or not, effecting a particular consciousness; disjunctive possibilities in a tension of opposites must be at hand. Judging is always deciding this or that, and is thus deciding-for or deciding-against, active acceptance or dismissal, rejection. But that must not be conflated with the modes of being themselves:

[52] Translator: See Part 3.

with straightforward "being," with the "null," and again the "not null" already appearing in the objective sense by merely becoming patent, with the "it is too so" resulting from being twice crossed out. In all these modalities the ego of itself does not need to take a position actively, even though it can also be motivated to take such a position by them.

The noetic Yes and No, however, arise from taking a position specifically as judging. As with every mode of consciousness, we have a noematic correlate. Here, of course, this correlate is the noematic "valid" or "invalid" arising in the objective sense; it occurs in the objective sense with the character of being declared valid or invalid by the ego. Judging in the specific sense is thus the ego's act of *posito*, of positing, in its possible dual shape: In the shape of the ego's assenting decision or in the shape of dismissal, rejection. We must still consider whether this means that the positing itself has a dual "quality" in the sense of traditional logic. [54]

We can at least say this much initially: Where a judging arises in the primordial sphere of a motivating perception, two opposing position-takings become possible, and as the case may be, actual. Both, however, are completely non-independent insofar as they have their motivation founded in what goes on in perception itself, in perception's proper and potentially purely passive course. Perception has its own intentionality that as yet does not harbor anything of the active comportment of the ego and of its constitutive accomplishment. For the intentionality of perception is rather presupposed in order for the ego to have something for which or against which it can decide. By virtue of the unity of this motivational situation, that is, by virtue of its unity arising from bifurcation, both opposing position-takings are intimately related. For example, where two possibilities were in conflict with one another, the decision for one possibility is accompanied, as a correlate, by the decision against the correlative possibility, potentially, if not actually.

If we look more closely at how the motivation bearing on the ego functions, and how the ego reacts to it with an active affirmative or negative response, then we would have to say the following: The motivational foundation for the decision as the ego's firm positing-as-valid, or again for the negative decision, is thus the restoration of perceptual concordance. The split occurring in a conflict, in which perceptual apprehensions are reciprocally repressed, returns to an unbroken unity.

The ego is affected by all of this. As the ego, it is itself at variance with itself in its own way; it becomes torn and finally becomes united. It was inclined to endorse the one apprehension, that is, to carry out the apprehension's tendencies of expectation above all, to let them become active expectations issuing from the ego-center. But it still finds itself inhibited once again; it is drawn toward the opposing tendencies of expectation and is inclined toward the opposing apprehension. If perceptual concordance is restored, a single perception in the shape of a normal perception running its course, then the inner conflict of the ego with itself is resolved. The ego can no longer be swayed in this way or that; for [55]
the annulled apprehension along with its annulled intentional tendencies cannot be carried out, and this is especially the case with its expectations that are directed ahead in a living manner, but crossed out. But the ego does not merely have as its operational field the free horizon of expectation and the intentionality that is now concordantly established. It actively takes up this position, appropriates what is concordantly given as being *simpliciter*. "Active acceptance" is what carries out a peculiar appropriation, determination,[53] thereby establishing this being as valid for me from now on and abidingly.

An important moment emerges here as characteristic of judicative decision-making. It is no longer a question of making something present, of merely making the intentionality of perception patent; rather, it is a question of appropriation through which the active, strivingly active ego appropriates to itself an acquisition, that is, an abiding knowledge. But it does this in a conscious manner. For what thereby accrues to the ego as being valid for it, having the character of being valid for it from now on, that is, continually, and remaining abidingly valid—this belongs, as we have said, to the essence of declaring-something-as-valid, to the essence of the so-called active acceptance that the ego carries out. In other words, it has a validity extending into an open, egoic

[53] *Festlegung*

temporal horizon of conscious life. Perhaps it is not too much to express it in the following way: When I posit something as valid in an affirmative and judicative manner, I mean by this that it is settled for me from now on, as established for the future, and in particular, as being in this way or that. If we were to stand already within the sphere of expressive, predicative judging and in the sphere of communication, then the accomplishment of judgment would be articulated most acutely with the phrase, "I ascertain," or also "I assert that." But we must observe—and this belongs to the essence of judgment—that we do not already find the communicative relation in judgment's first originality; as a rule the communicative relation is presented along with the expression, the assertion[54].

What happens now to the opposing apprehension that has been negated? Naturally, it is still held in retention; the ego was previously drawn to it, and perhaps already inclined toward it in a preliminary fashion. Indeed, it could be that precisely this apprehension was previously concordant in the form of a normal perception, and that it was carried out by the ego by regarding it as a presumptively existing thing. Thus, there are affective motives that also orient or reorient the regard in this direction. But here the [56]
ego responds by dismissing it, by declaring it invalid. The latter obviously turns either against a previous declaration of validity, or against a propensity to such a declaration; that is, it already turns against taking a position and its final result, establishing it.

But now it becomes clear that the affirmative or negative position-taking does not simply exhibit two interchangeable "qualities," like red and blue in the sphere of colors, and that consequently speaking of "quality" is not appropriate here at all. The negating act carried out by the ego is a process of striking down validity; this expression already suggests the secondary intentional character of negation.[55]

---

[54] *Behauptung*

[55] Translator: The process of striking down validity (*Außer-Geltung-Setzen*) should not be conflated with the process of suspending validity (*Außer-Aktion-Setzen*) peculiar to the epoche. For the latter neither affirms nor negates validity; it merely puts it out of play, makes no judicative use of it. See *Ideas I*, esp. 59-61.

Yet something of the highest importance surfaces here concerning every logical concept of judgment. Namely, we have characterized deciding-in-favor-of by taking-possession-of, appropriation as being henceforth valid, as settled for me from now on. Deciding-against means that such a validity, which was somehow expected of us and possibly taken up by us earlier, is rejected—just as we find something analogous with other acts, for instance, when I reject a decision because a motivational situation has altered or when I resist an inclination of the will.

But the "invalid," which I can make valid once more by a slight shift in attitude, corresponds to negation as deciding-against; that is, by judging affirmatively, I can establish it. The "no" or the "null" then enters into the content of what is established. Accordingly, one can also take the concept of judgment in such a way that it deals exclusively with the activity of establishing being, and that it encompasses nullity as a moment of content, so to speak, as existent non-being. In fact, logic and science reduce everything to judgments that establish [something], and with good reason. No matter how much it is possible to deny, there is nothing of denial in theoretical statements; rather, at one time they establish that something is so, at another they establish that it is not so. Accordingly, a judgment that knows only one "quality," establishing something as valid, is our privileged concept of judgment. Naturally, this does not change the fact that making a decision itself does not have a single modality, but develops in opposing modalities: even if cognitive interest, which logic serves, is interested exclusively in establishing [something], in making assertions, and even if every dismissing negation is [57] ultimately thought of as reduced to the position of a *negativum*, and then perhaps still further to positions with an exclusively positive content.

Still, these considerations need to be filled out considerably. Firm active acceptance and resolved rejection are not the only modalities of taking a position in belief that the ego carries out. Rather, it is clear that a subjective-active comportment of the ego—precisely what we characterize in the genuine sense by the term "doubt," or the expression "I doubt whether it is this or that"—also corresponds to what we called, already in perception itself and in its passive course, a split perception, "perception modalized as a doubtful perception." Indeed, I already mentioned previously that the ego itself can be at variance with itself, even though this takes place on the motivational ground of that split running through the intentionality of perception. I am now at variance with myself, I am torn to the extent that I am inclined to believe now this, now that. This inclination-toward means in general, and in active doubting specifically, more than a mere affective pull of enticing possibilities. They entice me as being, which is to say in general, that I go along with the one, and then with the other already in the manner of a deciding for; I confer on it something of validity, though to be sure, [validity is conferred] in a such way that [it] can be inhibited again and again.

This "going along with" on the part of the ego is motivated by the weight of the possibilities themselves. A judicative tendency, which I actively follow for a stretch, issues from these possibilities as enticing. This means that I carry out something like a momentary decision in favor of it. But I am stopped in my tracks by the opposing possibilities making an affective claim on me. Even this wants to be heard, as it were, and makes me inclined to believe. The inhibition here is not a mere privation, but a mode of the phenomenon of an inhibited decision, precisely of a decision being arrested along the way. I do go along for a stretch in carrying out a decision, as it were, only I do not go all the way to a firm decision of belief. Likewise, in such motivational situations, the decisions on the part of the ego, decisions that dismiss the other possibilities and that go against them, are then inhibited negative decisions.

The phenomenon of taking the side of a possibility and internally closing oneself off to the others would belong here [58] especially. In this case, a genuine decision already comes about, a subjective certainty and establishment, assertion; but it is an impure decision, one that is gnawed at so to speak, a decision not made with good logical conscience; this is unlike the case where the motivation for precisely this firm resolve issues from the matter itself as a concordantly constituted experience.

Also belonging here is the phenomenon of presumption, of taking-to-be-likely. If I consider possibilities having different weight, then the strongest weight will possibly motivate me to make a decision for it, it will motivate a kind of privileged active acceptance, which does not therefore mean establishing it or affirmatively asserting it as being *simpliciter*. To be sure, when one possibility has an overpowering weight, or when what continues to speak in favor of it on the side of the matter[56] takes on an overpowering weight from subjective motives, I can pass over even further to a certainty, albeit to an impure conviction. But then one no longer speaks of "presuming" or "taking to be likely." The negative correlate of this is naturally taking-to-be-unlikely, whereby a kind of rejection is expressed, but not a straightforward negation.

<§15. Questioning as a Multi-Layered Striving toward Overcoming Modalization through a Judicative Decision>

What happens, finally, with questioning, questioning that is so inextricably intertwined with doubting? Does it also belong, like doubting, to the rank of these judicative modalities? Within the passive sphere and, in particular, within the intuition that is split in the intentional conflict, the *disjunctivum* corresponds at the same time to the process of doubting and questioning that can be motivated here. In and through their rivalry, *A*, *B*, and *C* are apprehended and united in the unity of the conflict. We can express this in no other way than by saying: We are conscious that "either *A* or either *B* or either *C* is"; and we find just this in expressing the active question and the active doubt, namely, as the content of the question or the content of the doubt. This is to say: I question, I doubt whether *A*, etc.

Stated in our previous manner of speaking: A unitary field of problematic possibilities in the passive sphere precedes questioning and doubting. Naturally, there are at least two of these [59]
possibilities. But it can also be the case here that merely one of these contesting possibilities comes consciously to the fore, or as we also said, becomes patent; meanwhile the others go unnoticed in the background in the manner of empty presentations that are

[56] *das sachlich für sie Sprechende*

not carried out thematically. Every egoic act has its theme that can be a singular theme or a unitary manifold of thematic singularities, which then make up the entire theme in their unity. Obviously, either the theme of a question (like of a doubt) is a problematic singularity whose disjunctive opposing terms remain then unthematic (like when I merely ask: Is that a wax figure?), or it is the entire problematic disjunction (like in the question: Is that a wax figure or a human being?).

What now is characteristic of questioning as an activity that is obviously peculiar to the ego? The passive disjunctive tension of problematic possibilities (of doubt in the passive sense) motivates an active doubting, a mode of comportment that displaces the ego in an act-schism. This essentially and immediately implies an uneasiness and an original drive to get beyond it, to come back to the normal condition of unanimity. There arises a striving toward a firm decision, that is, one that is ultimately uninhibited and pure. Questioning already gives rise to this striving. It frequently happens that the established concordance, and through this the inner unity of the ego with itself that is aimed at, can be lost once again. This recurring experience can provoke a further motivation, namely, it can awaken a drive to overcome this renewed uneasy insecurity. Unlike the other instances, in this case the matter does not rest with striving toward a judicative deciding and toward an appropriation and establishment of the passed judgment; rather, the striving is directed toward a definitive, secure judgment, that is, toward a judgment that the ego can justifiably ground and with regard to which the ego can be subjectively certain of not lapsing again into embarrassing modalizations. This multi-layered striving is expressed in the following two kinds of questioning.

(1) Straightforward questioning in general is the process of striving, from a modal modification (or if you will, arising from a schism and inhibition), in order to arrive at a firm judicative decision. Questioning has its intentional correlate in the question, [60] just as judging has its correlate in the judgment. Perhaps it is clearer when I say that the ego-*actus* of judging as the process of passing a judgment is to be distinguished, naturally, from the *judgment* that is passed in the judging. In the linguistic sphere, the statement as that which is stated corresponds to the stating; what is

written is there as a proposition, what is verbally expressed is there as what is asserted, as what is established. Likewise, we have the stated question as opposed to the activity of questioning.

The proper sense of questioning is manifest in and through the process of responding or in the response. For with the response there is a tension-releasing fulfillment of striving; satisfaction sets in. Corresponding to the various possible responses are [1] the diverse modes and levels in which satisfaction can occur: as relative and yet already as satisfaction, or as full and ultimate, and [2] the different directions in which the questioning intention can go. For example: Is *A*?. The response runs: Yes, *A* is! Or, no, *A* is not. Thus, it has both firm modes of judgment as possible responses.[57]

Since the questioning endeavor is fulfilled, answered in the corresponding judgments, it is clear that the experience of the forms of judgment, forms of judgment that in a parallel manner fit the sense-content of the questions, implies that the questioner already consciously anticipates these possible forms of response, and that they already occur in the articulation of the questions themselves as the contents of the question. Every possible judgment is conceivable as the content of a question; naturally, it is not yet an actual judgment in the question, but a prospective judgment, a merely intended[58] (neutral) judgment, which as the content of the question, points to Yes and No.

Even doubting within developed consciousness is a doubting comportment, an inhibition and being divided when taking a position in relation to prospective,[59] possible judgments.

If the question has several components and is posed as a complete disjunction, then it may read, for instance: Is *A* or is *B*?. Thus, it shows disjunctively the corresponding prospective judgments. When the question has two components it can also read here: Is *A* not or is *B*?, and so forth. The responses turn out accordingly; they are directed toward the prospective, possible [61]
judgments as contents of the question which are taken into

[57] "<Further> Is *A* not?" "Yes, *A* is not." "No, it is not not *A*."

[58] *vorstellig*

[59] *vorstellige*

consideration according to the terms of the disjunction. Wasn't Rome victorious, or was it Carthage? Yes, Rome was victorious, but not Carthage.

Yet, there are still other parallel responses throughout insofar as the response is a decision, affirmative or negative, but not every decision must have the mode of a firm certainty. Even taking-it-to-be-likely is taking a position that makes a decision, even though it may not be satisfying in a definitive manner. Yet, in a way, it already resolves the indecisiveness insofar as the ego, in taking it to be likely, has endorsed one possibility by believing. In fact, we can also respond to the question, Is *A*? with, Yes, it is likely. No, it is unlikely.

As we might expect, then, further mitigated responses are still possible. This is the case insofar as every mode of judgment that still has something of a decision in it, and accordingly, every form of decision that is arrested, can serve as a response as well. For example: Is *A* or *B*? Response: "I am inclined to believe that *A* is." To be sure, this is usually preceded by: "I don't know," or "I haven't made up my mind," "I'm not sure." This shows that the practical intention of questioning is actually directed toward a "knowing," toward a decision in a special sense, toward a judgment in the strict sense. But it is nonetheless a response, even if incompletely satisfying. On the other hand, it would not be a response at all if we were to say, for instance: *A* is fascinating. So, response in the genuine sense of this word is a judicative decision, taken very broadly. Let me retract this statement. For in a certain sense, to say "I don't know," or even "I'm not sure," is also a response to a question. This obviously concerns communicative interaction in which I merely inform the other with my response that I cannot comply to his wish, that I have no response at all for his question. And in fact one can also reply in such cases with the phrase: "I have no response."

Still, our preceding analyses have not yet expressly taken a position regarding the extent to which questions themselves belong to the modalities of judgment. Following our analyses, this does not require a long exposition. Of course, questioning belongs to the sphere of judgment and knowledge, indeed, it belongs to [62]
them inseparably; and it belongs inextricably and necessarily to

logic as the science of knowing and of the known, more precisely, as the science of cognitive reason and its formations. But it does this only because the judicative life, even the rational judicative life, is a medium for a peculiar wishing, striving, willing, acting, whose goals are precisely judgments, and judgments of a special form. All reason is at the same time practical reason, and this also holds for logical reason. Of course, we will still have to distinguish valuing, wishing, willing, acting (which aim at judgments and truths through judging), from judging itself (which is itself not a valuing, wishing, willing). That is, questioning is a mode of comportment that is related in practice to judging. I find myself disagreeably frustrated when I pose a question and do not reach a decision; this may also frustrate me in other decisions relating to my practical life. Accordingly, I wish for a decision.

Yet, questioning is not merely a static state of wishing, but a directedness striving toward a judicative decision that already belongs to the sphere of volition. Only later when we see practical ways of actually bringing about the judicative decision does it become a resolved willing and acting. To be sure, the normal concept of a question is an interrogation directed toward another person, and possibly toward myself in turning back to myself, an interrogation proceeding from myself toward myself. Here, communication with others does not fall within the scope of our reflections, just as predicative questions in their relation to predicative judgments do not. But we can also leave out of consideration turning-toward-oneself that makes oneself the terminus of communication, just like others are made the terminus of communication (for the ego can in fact interact with itself). Thus, primitive questioning is a practical striving toward a judicative decision, and is further a habitual practical attitude that can be possibly effective for a long time, being always on the verge of passing over to corresponding volitions, endeavors, actions, of trying out methods for solving problems, etc.

(2) Having also given a breakdown of questioning *implicite* in what we have just said above, which we are not in a position to carry out further here, we must now consider the levels of questioning alluded to above. First, it is apparent that questioning [63]
can find its firm reply through a firm assertion with which we

seemingly reach a definitive position, and that then, in spite of this, we can renew the questioning. For example, we ask: "Is *A* true?" The answer runs: "Yes, *A* is true." But we ask once more: "Is *A* really true?" And we do this without ever doubting. This may happen in our perceptual sphere, and can be clarified in the following manner: A perception at variance with itself has passed over to a concordant perception that harbors the decision; it has passed over to a concordant perception according to the sense of one of the apprehensions. But all the same, there is always the open possibility that the further course of perception will not ratify affiliated anticipations, and thereby the validity of the apprehension's sense. The need may thus arise to secure it further and to justify the perceptual judgment, to ratify and strengthen it. This can take place by drawing closer, putting perception freely into action according to prefigured possibilities in order to realize and to see then whether it is actually true. Accordingly, the new question is a question permeating the open possibilities of the horizon and relating a justifying question, i.e., a question directed toward actual, true being to the anticipatory intentions. Through the ratification, then, what is already judged as existing is endowed with the new character, "truly and really so," such that we could also characterize this question as a question concerning truth and reality. Naturally, the interplay can be repeated here, indeed the actual and the true are not completely definitive, since new horizons can open up. What we have presented here suffices to bring to light the difference between straightforward questioning and questioning concerning justification or truth that are adjoined as higher levels to straightforward questioning.

The investigations of our previous lectures have presented a portion of the phenomenology of judicative acts in the higher sense, although our necessary analyses of the lower levels were not yet brought to completion. This was due to the fact that the fundamental theory of judgment initially leads to *doxa* and doxic modalities that belong to the passivity of intuition itself. Here it was quite necessary to disclose immediately its contrast with the [64]
higher level of judging that takes place as specific egoic decision making. Otherwise one would have formed the view that the theory of perceptual belief, and likewise the modes of judgment

occurring in passive intuition of every other kind, would already make up a full theory of judgment. But it is important to keep this in mind from the very beginning, and not as an empty generality: that the cognitive life, the life of *logos*, indeed like life in general runs its course in a fundamental stratification. (1) Passivity and receptivity. We can include receptivity in this first level, namely, as that primordial function of the active ego that merely consists in making patent, regarding and attentively grasping what is constituted in passivity itself as formations of its own intentionality. (2) That spontaneous activity of the ego (the activity of *intellectus agens*) that puts into play the peculiar accomplishments of the ego, as was the case with judicative decisions.

intention is fulfilled in a synthesis of the intended object and the corresponding object itself: We can also say that we are making an initial study of the nature of evidence. Making a presentation evident to us is indeed bringing it to originally fulfilling confirmation. Thus, it is not a question of an arbitrary synthesis of identification; rather, it concerns a synthesis of a presentation that is not self-giving with a presentation that is self-giving.

Naturally, we take at first the mode of certainty and positionality as a basis for these presentations. From the outset we see that the important distinction between empty and full or [67] intuitive presentations, a distinction with which we are familiar, especially becomes an issue for the syntheses of confirmation. To be sure, we know that even a perception, in particular, an external, transcendent perception, can occur in syntheses of fulfillment—and not only as a perception confirming an intention; rather, it can even occur as a mere intention that becomes fulfilled in new perceptions. This happens, for example, when we perceive a tree from the front, and wanting to know it better, draw nearer to it and now perceive it in new perceptions; by determining the tree more closely, we also have a fulfilling confirmation. Meanwhile, every external perception harbors its inner and outer horizons, regardless the extent to which perception has the character of self-giving; this is to say, it is a consciousness that simultaneously points beyond its own content. In its fullness it simultaneously points into an emptiness that would only now convey a new perception. The self-givenness of a spatial thing is the self-givenness of a perspectival appearing object that is given as the same in the fulfilling synthesis of appearances intertwining and devolving upon one another. But it is the same object that itself appears now this time in one way, now another time in another way, appearing in other perspectives, always pointing from a perspective to ever new perspectives in which the same object that is exhibited is continually determined more closely, and yet is never determined definitively. For we always expect appearances of newly opened, empty horizons. Thus, where there is no horizon, where there are no empty intentions, there is likewise no [synthesis of] fulfillment. A datum that is given in immanent perception, i.e., that is adequately given in each Now does not therefore admit of any further confirmation with respect to this Now. Still, it does occur as a fulfillment insofar as the preceding perceptual phase already points to what is to come. This fulfillment is a fulfillment of an anticipation and is a definitive, absolute fulfillment, or evidence.

Accordingly, it may now seem that the unity of a synthesis of fulfillment (of a confirming one) would be characterized by the fact that an empty consciousness (be it a consciousness that is standing completely empty for itself, or a consciousness incompletely saturated by intuition) would be synthetically united with an appropriate intuition, whereby what is emptily intended[63] and what is intuited coincide in the consciousness of the same [68] [object], that is, coincide in the identity of the objective sense. One would like to think, however, that fulfillment is certainly a bringing to intuition: confirming an intending, that is, meaning an object but not having the object itself intuitively, or having it intuitively, but still meaning beyond what is already intuitively given, and now passing over to the intuition of what is not yet given. But we will see that this characterization would not work, for not every process of bringing to intuition, that is, not every fulfillment is confirming.

It is of fundamental importance to distinguish between the different possible syntheses pertaining here to intuitions and empty presentations, and to characterize them in more detail. The possible syntheses are determined according to their phenomenological character by the types of intuitions and empty presentations founding them. Conversely, one can become aware for the first time of the different sorts of founding presentations in the different operations of closely related presentations within the synthesis, and in the different character that the synthesis assumes in these cases. Without distinguishing between different possible syntheses, one might easily overlook distinctions within intuitions and within empty presentations that could occur here.

---

[63] *Vorstelliges*

<§17. Description of the Possible Types of Intuition>

Let us pursue this matter more deeply by proceeding initially from the general distinction between intuitive presentation and empty presentation. The intuitive presentation has for its part different modes. Perception is the primordial mode of intuitiveness (as always, understood as doxic positionality). It is to be contrasted with the mode of presentification, which, upon closer inspection has various forms, too. By studying intuitive remembering we have learned that a remembering in itself manifests itself as a presentification of a perception, thus, that it is not structured as simply as a perception. It is a present lived-experience that is not itself a perception; instead, it presentifies a perception in the temporal mode of a past perception, and precisely thereby it presentifies its previous perceptual object as having-been. Every other kind of presentification has a similar structure.[64] Thus, there are intuitive presentations of something [69]
present that are surely not perceptions of that present something, but rather are presentifications of it: for example, when we make intuitively present the back side of a thing that is more or less familiar from a previous perception, or when we make intuitively present the co-presence of other things, like when we intuitively presentify Berthold's Fountain.[65] Here we do not merely <present> it as the fountain seen yesterday in its mere pastness, but <present> it as now and as actual, just like the intuitions that we

[64] Translator: Empathy [*Einfühlung*] as a mode of presentification does not make present a previous or futural perceptual object because the other or the alien can in principle never be given "originally." This is certainly different from a remembering that literally re-presents its object, or from a futural presentification that anticipates a futural object as present, or finally from a co-presentification in which something that is not given in the original at present, but can in principle be present (cf. below 367, 373f., 377f.). Accordingly, rather than translate *Vergegenwärtigung* as re-presentation, which would be well-suited for temporal acts, I prefer the common neologism, presentification, since it includes the full complement of acts ranging from imagination to empathy.

[65] The original Berthold's Fountain to which Husserl refers was located in the center of Freiburg at the junction of Kaiser-Joseph-Straße and Salzstraße. It was erected in 1807 for Bertold III. Herzog von Zähringen and was destroyed in 1944. In 1965 a new fountain was erected for the Dukes of Zähringen, the founders of Freiburg. (This is not to be confused with the fountain built in Freiburg to commemorate the Franscian Friar Berthold Schwarz, which was erected in 1852 and is still standing on the Rathausplatz.)

have here and now of the outside entry ways and the vestibule, etc. To be sure, the memory of the past plays its role here—indeed the vestibule initially appears in the current intuition as a remembering—but what is past extends unaltered into the future in the manner of an object for consciousness. This future proceeds from the reproduced past and does so in such a way that this future is at the same time co-present, relative to our current perceptual present to which these things here in our current perceptual field belong.

Furthermore, we also have intuitive presentifications of what is to come from the future, that is, intuitive expectations. Proper to being subsequent, to what takes place tomorrow, being futural as being-in-anticipation or being expected by us is what we fore-see[66] as a futural duration, for example, the enduring-character of this lecture-hall, of the university, of the street, of the city, etc. Accordingly, we have a consciousness of something futural in an intuitive presentation. Obviously, expectations are not always like this, merely extending the perceptual moment continuously into the future. Something unknown, something singular never yet experienced can also be fore-seen, like an event that is indeed expected, but yet is singularly new, an event that is accordingly expected as completely determinate, as in the case of periodic repetition, or an event that is expected as more or less indeterminate, which is more often the case.

We had focused on a tremendous, new theme in the previous lecture.[67] On the level of passivity, it concerned the tremendous problem of making evident or of confirmation, as well as the closely related problems of mere ratification and corroboration on the level of passivity. The problem of evidence led us back to the distinctive syntheses of coinciding that forms identities, namely to [70]
such syntheses in which intuitions and empty presentations (or intuitions and intuitions) are synthetically united, but whereby empty presentations and their fulfillment once again play an essential role. This happens insofar the intuition, on the one hand, brings empty horizon-intentions into play and the intuition, on the

[66] *voranschauen*

[67] Editor: The beginning of a new lecture.

other, provides the appropriate fullness for these empty horizon-intentions.

A logic that would leave obscure the accomplishment of making evident within logic itself [would] remain hopelessly unclear. But should one not forgo this central problem, then the primary task becomes elucidating the founding level of the passive syntheses of "verification" lying at the basis of all active verification. To do this, however, one must gain deeper insights into the structures of the intuitions and empty presentations that may be functional here. The universal significance that we have repeatedly emphasized, a significance that all these types of consciousness possess for the entirety of a transcendental life as a whole-consciousness, leads us to analyses that are not at all a mere special problem of logic, no matter how important the latter may be. We will be led to insights into the most universal lawful regularities of essences, to the most universal lawful regularities of structure concerning the unity of transcendental inner life, but also to the most universal lawful regularities of genesis.

In the last lecture, we proceeded from a descriptive consideration of the types of intuition that can function in the syntheses of confirmation. They were either perceptions or presentifications; the presentifications were either memories of the past, like when a past lived-experience is presented, or memories of the present, like intuitive presentations of a co-present, for instance, the antechamber of this room, or the co-presence of alien psychic life that is given perceptually along with the alien lived-body; or finally, they were memories of the future, intuitive presentations of an expected future.

Indeed, it occurred to us there that in perception we are nevertheless "horizonally" co-conscious of the past and future. But we are conscious of them emptily, even though they can be exposed subsequently and in an intuitive manner. Likewise in the [71] case of remembering: In every remembering not only is there a past, which can be traced back by remembering, and a future, but there is also a relation to a current present, to the future by perceiving, and therefore to its current future. Finally, even expectation is not isolated and without relation to the current present and to the past of something expected. In all of this we find internal structural intertwinings. We will soon see that it will not suffice to juxtapose as types, perceptions, memories of the past, memories of the present, and memories of the future,[68] and to describe them in an entirely general manner according to the noematic character of their objectlike formations. Or that we cannot be satisfied with a general phenomenological impression and with the evident differences between all types. Only when we understand them in their structural interrelatedness can we also understand how they function in synthetic interrelatedness, including here, as well, how they can function as confirming or confirmed.

This holds not only for types of intuitions, but similarly for the other side, for the side of empty presentations.

<§18. Description of the Possible Types of Empty Presentation>

There are empty presentations of all possible objects in all subjective modes of inner givenness; in other words, corresponding to every mode of intuition is a possible mode of empty presentation. That we relate corresponding empty presentations and intuitions to the same [object] means that through a synthesis they achieve a coinciding with respect to the object.

In fact, we would not be able to speak at all of empty presentations and to attribute to them the character of having a relation to an object if it did not belong essentially to each empty presentation that it admit, so to speak, of a disclosing, of a clarification, of a manifestation of its objectlike character, i.e., that it could enter into a synthesis with a corresponding intuition. By bringing to intuition what is emptily meant there, the synthesis allows us to realize for the first time that something is presented there in an empty manner. We can say without hesitation that non-intuitive presentations are only called presentations in an inauthentic sense; genuinely speaking they do not actually present [72]

[68] Translator: *Wahrnehmungen, Rückerinnerungen, Miterinnerungen und Vorerinnerungen*. See Supplements to the Main Text: "First Version of Main Text Part 2," p. 377, where Husserl explains his motive for using such formulations.

anything to us, an objective sense is not constituted in them; nothing is built up in them as the being of this or that content through actual intentional structures such that we could continuously acquire knowledge of it. What is genuinely presented to us is what is intuited in an originary manner: What is constituted perceptually in its ipseity and in its features, in its different facets, etc., comes to our originary knowledge. But then this is also true of what is quasi-perceived, what is intuitively presented to us in a presentifying modality of intuition, what takes place before our inner eyes piece by piece, reproductively or in the intuitive anticipation of what is to come. Nothing takes place genuinely in the empty presentation, an objective sense is not genuinely constituted. And yet we do say that it presents this or that, i.e., that I am conscious of this or that. In this case, however, we can always place this "something" given to consciousness next to an intuition of it; in the synthesis, we gain an evidence-consciousness, a consciousness that exactly the same [object] that was meant in an empty manner is there in intuition in a genuine way, as the same [object] actually presented.[69]

Naturally, corresponding to every intuition is an empty presentation insofar as an intuition never disappears without a trace after it has elapsed. We are "still" conscious of what it had intuited, now in a non-intuitive way; to be sure, in the end it fades away in a general, undifferentiated emptiness. Every *such* empty presentation is a retention, and the fact that it necessarily connects up with past intentions characterizes a fundamental law of passive genesis. As we know, this law extends further than what is expressed here to the extent that it already plays an unflagging role in the inner becoming of every intuition itself. This is certainly the first aspect of the fundamental lawfulness of the constitution of original time-consciousness: that every lived-experience, speaking most basically, every Now-phase that arises in a primordially impressional manner is continually modified in retention by essential necessity, and this [retentionally modified Now-phase] continues further no less in this manner. What holds for the phases

[69] One must avoid confusing relatively unclear reproductive intuitions and empty presentations, empty intentions.

holds then for the expanses, for the concrete lived-experiences themselves. If nothing new occurs in a primordially impressional manner, then the intuition as such is past, which is to say, it has [73] entirely passed over into the living retention. The objects found in such a retention are disclosed through a process of bringing [them] to intuition; this process of bringing to intuition is connected to the retention according to an essentially lawful possibility, but not a necessity. In other words, they are disclosed in the synthetic transition to an appropriate intuition in the consciousness of them: This is obviously a synthesis of confirmation.

But now it must be said that not all empty presentations have one and the same nature and function; in particular, not all of them have the same nature as retentions do, namely, a nature that we find in the primordial genesis of the temporal stream whereby as retentions [these empty presentations] are connected to any kind of intuition, to any kind of presentation (thus even to any kind of retention); and as we have shown, [this holds] not only *in concreto*, but also in the structural integrity of every intuition itself, which (like every lived-experience in general) can *only* be in the process of time-constituting becoming. With respect to the doctrine of this primordial genesis, we have not only had to speak of retentions, but also of protentions. In our analysis of perception, which was in this regard an analysis of temporal modes of givenness, we have already observed and touched upon the essentially new role of protentions over against the role of retentions. The rubric, protention, designates the second aspect of genetic primordial lawfulness that strictly governs the life of consciousness as the time-constituting unitary stream. Just as a retentional horizon of the past is invariably connected to each impressional present, a protentional horizon of the future is no less invariably connected to an impressional present. Just as one can disclose the retentional horizon, so too can one expose the protentional horizon. Just as the past is first clearly exposed as such through intuitive remembering, namely, as just-having-been, so too is the constitutive accomplishment of protention exposed as the just-about-to-arrive, as becoming originally conscious of the future.

All of this is familiar to us. But when we make the following question the focus of our attention we arrive at something new: whether the two empty presentations as empty presentations are essentially homogeneous, and whether they bring about differently characterized constitutive accomplishments (past – future), for instance, only through a differently regulated order of functioning or merely through an inner complexity. On the other hand, the [74] expressions already tell us something, expressions that we had to choose, in differentiating them, by being intuitively immersed in both situations. In spite of its pure passivity, we spoke of protention as an expectation, and with the colorful image of the present meeting the future with open arms. Accordingly, we already speak this way in pure passivity, which is to say, even prior to [actively] grasping and viewing the perceptual object. We did not use such expressions, and we could not use such expressions with respect to retention. In this connection, there is a difference in the way in which retention and protention function in mindful perception, when we take note [of something] and grasp it. We are mindfully directed, purely and simply, toward the present object, toward the ever new Now that emerges as fulfilling the expectation; and in and through it, it is directed further toward the approaching object. Mindful perceiving follows the protentional continuity. The directedness-ahead, which already lies in passive perception itself, becomes patent in the mindful perceiving. On the other hand, there is however not a directedness in the retentional continuity; there is not a directedness that would follow the trail of pasts being pushed back further and further. One may object here that we can surely also cast a backward turning glance back toward the pasts. While this may be true, it soon becomes clear that there is a tremendous difference in the two cases, and that we must clearly differentiate between the direction of the egoic regard, and the direction in perception itself that already takes place prior to the apprehending regard. In one instance, the egoic regard follows the direction in perception itself, in the other, it does not.

In order to clarify all this it will do us well initially to go beyond protentions as intentions of expectation, and to draw upon other empty presentations that are structurally related to them, and that are at the same time different from all mere retentions. We have in mind making co-present,[70] memories of the present as forms of intuitive presentations, alongside memories of the past and memories of the future. Let us note throughout that these intuitive memories do not, for instance, come first in genesis; instead the corresponding empty presentations are essentially earlier. Thus, the corresponding intuitions first arise through motivations of awakening, already within perception (where empty retention and protention necessarily link up with the primordial impression, and likewise where a concrete empty retention, or even possibly an empty expectation of the future for a [75] new perception, link up with the entire course of perception).

If we now consider the genetically more original modes of making co-present, then at issue, e.g., for every perceptual object, are its entire horizons that are constitutive of it, horizons that belong immediately to it.

In the previous lecture[71] we pointed to the entire compass of empty presentations that share a common structure with protentions and anticipatory presentations. Belonging here are all the horizon-intentions that have been awakened in the concrete connection of possible intuition, for example, when on occasion we become conscious of some elements of an environing-thing-world (that is not perceived) as co-present through particular empty presentations. We recognize this peculiar feature with respect to all such presentations: that they exist with other presentations in a synthetic nexus of a special kind, namely, in a synthetic nexus that lies entirely outside of the genre of identifying syntheses or syntheses of coinciding. In our example of the awakened co-presence of the antechamber, the empty presentation of this antechamber does not occur in an isolated way; rather, it arises in connection with the perceptual presentation in which we survey the seen room with our gaze knocking on the door, as it were. The connection of this perceptual presentation with the empty presentation is a "synthetic" one, which is to say, a unity of consciousness is produced that carries out a new constitutive

[70] *Mitgegenwärtigungen*

[71] Editor: The beginning of a new lecture.

accomplishment, whereby the both objectlike formations receive special characters of unity noematically. More precisely, the perceptual presentation, what appears perceptually in this way or that, points to the emptily presented object as something that consequently belongs to it. A directed ray arises in perception and goes clean through the empty presentation to what is presented in it. From a genetic perspective, we also say with reason that the perception has awakened the presentation, but awakening means precisely both the emergence of a synthesis of directedness in which the one presentation is "directed toward" and in which the opposing presentation is accordingly directed in itself, or in which [76]
the one presentation is characterized as *terminus a quo*, the other as *terminus ad quem*.

Let us now consider such syntheses of consciousness in a wholly general manner, syntheses that have essentially the same character. They are not syntheses that the ego has actively instituted; rather they are syntheses that are produced in pure passivity and that nevertheless can then be produced when the particular lived-experiences entering into connection with one another have arisen from the activity of the ego. If, from the very beginning, we remain focused most simply on the realm that already has our exclusive interest now, the realm of passive presentations as the material for passively emerging syntheses, then we will be concerned generally speaking with such syntheses in which a presentation points beyond itself to another presentation. The latter thereby gains a new inner character that it otherwise could not have. It is the character of the specific "intention," that is, of teleological directedness, of being-intended, of meantness, or correlatively speaking, the presenting is not merely a general presenting consciousness of its object, but rather, is in itself directed toward its object.

There is a danger in this description only insofar as it is not a matter of those very common meanings of the words, "to mean," "to be directed toward," "to intend," which refer to the ego and its acts, whereby the ego, and in a totally different sense, is the radiating point of directedness, of a directedness toward the object. For want of terms at our disposal, we avail ourselves of the apposition, "passive," passive intention. And from here on we will speak only of passively intending presentations. At the outset we also want to name the synthesis in which this intention arises: associative synthesis. We will hold in abeyance for the time being what this association has to do with that association peculiar to the naturalistic psychologist and to empirical psychology; likewise, we will postpone for the time being any further foray into the general doctrine of association as one of the most important of all and completely universal functioning shapes of passive genesis. What we have expounded upon suffices to delimit sharply a class of empty presentations as "intending," as specifically directed toward their objects in the manner of an intention. This is to say that they stand in an associative synthesis, and that in it they have gained their oriented structure from counter-presentations that [77] function as awakening—whether or not we look, incidentally, at this synthetic nexus.

Now we have already said that not all presentations intend an object, and, coming back to our special theme, we now point to the fact that all retentions arising originally in time-consciousness are of this sort, completely unlike all protentions. Indeed, even retentions, those emerging originally, synthetically cohere with one another and with the primordial impression, but this synthesis proper to original time-consciousness is not a synthesis of association; retentions do not arise through an associative awakening directed backward from the impression, and thus, they do not have in themselves a directedness radiating out from there toward the emptily presented past. Therefore, it was not phenomenologically correct when Brentano characterized the regulated connection of retentions to impressions as an original association. One could only speak like this if one employed the word "association" in a completely superficial and frivolous manner for just any connection of presentations to presentations, without regard to what kind they are and how they originally emerge. Association is only at work in the protentional path of original time-constitution, and also functioning there as awakening[72], as we know, is the continual retentional path. Accordingly, traversing the protentional path we have directed presentations, intending presentations in perception (in particular, in pure passivity), namely, expectations.

I said that retentions, as they arise in their originality, have no intentional character. This does not rule out that in certain circumstances and in their own way they can assume this intentional character later. Thus, we should not allow ourselves to get confused by the fact that we do encounter directed retentions occasionally, like in those instances where the ego directs its egoic regard toward something retained.[73] For it holds generally that what is presented, toward which the ego directs its regard—what is perceived, what is remembered, even what is retained—must already in itself be intentional, that is, must already have in its passive content a directedness toward its object. Now, how does a retention get this oriented structure? By a subsequent association, of course. In the normal case of perception, for example, in the [78] perceptual course of a melody, a tonal phrase that has just sounded reminds us of one of the previous tonal phrases that we still hold retentionally in consciousness; it points back to it. Thus, an associative awakening proceeds from the present toward a retentional past that had already arisen originally prior to this association and is fading away. Naturally, the retention in question has now taken on an oriented structure. Likewise, out of the realm of forgetfulness that has apparently become nothing, out of the distant horizon into which all retentions finally sink, one of the stale, non-living retentions that is no longer in relief can once again be awakened as it were; in this case, it initially takes on and must take on the shape of an empty retention that is in relief. The awakening results from some kind of present presentation through association. Thus, from the very beginning every stale retention arising in this manner has the character of a passive intention. The remembered past is remembered by means of a pointing back, and this is analogous to pointing-ahead to what is to come, which is

[72] Translator: Strictly speaking, even though retention can sketch a protentional path, a retention cannot function as "awakening"; rather, it is more proper to speak of a retention being "awakened."

[73] Translator: i.e., a remembering.

peculiar to every memory of the future, expectation. On the other hand, having stated this, we must however maintain the distinction between mere retentions prior to association, and the intentional shape of retentions corresponding to it.

<Chapter 2: PASSIVE AND ACTIVE INTENTIONS AND THE FORMS OF THEIR CONFIRMATION AND VERIFICATION>[74]

<§19. Picturing, Clarifying, and Confirmation in the Syntheses of Bringing to Intuition>

On those frequent occasions when we spoke about intention and fulfillment in phenomenology (and in fact the syntheses of fulfillment do play an enormous role), we certainly did have something particular constantly in mind, but we could not manage to clarify the particular trait of consciousness that the word "intuition" and correlatively the word "fulfillment" characterizes in a general and pure manner. Now, is the concept of intention that we have previously obtained sufficient to attain the particular [79] concept of fulfillment as confirmation? Or does our concept only initially provide a framework for a further distinction that discloses, through a difference, the concept that is actually at issue?

It would be best to proceed by initially considering the possibilities of bringing to intuition, that is, of their verification for empty presentations, and in particular for the intending empty presentations as we have described them up to now. If we consider the types of presentations called protentions, then from the very outset they are presentations directed toward an object by virtue of genetic syntheses in which they emerge according to a primordial lawful manner. And this holds with respect to every expectation or making co-present. Here we find essentially two different modes of bringing to intuition, that is, two different kinds of syntheses of

[74] Translator: Chapter heading modified.

coinciding of the empty intention with an appropriate intuition giving intuitive fullness to the empty intention:

(1) The mode of bringing to intuition which is merely clarifying, merely disclosive, as for example when we picture an expectation that is initially empty, when we picture how it will arrive. And likewise we have the mere "picturing" of a co-present. The generality of expectation is always relatively determinate or indeterminate. By virtue of the generality of expectation, this process of bringing to intuition is actually disclosive only with respect to those components that were, so to speak, foreseen within what was prefigured, that is, [it is] disclosive as actually clarifying the intended objective sense. Only such components have the character of an affiliated filler; only they achieve an actual coinciding in the synthesis. But since intuition must yield a completely pictured image,[75] a concrete one, there is a remnant left out of the coinciding. This remnant is characterized in a purely phenomenological manner as mere filling.

(2) There is an entirely different mode of bringing to intuition, i.e., there is an entirely different synthesis of coinciding between intention and a suitable intuition, namely, the specific fulfillment of intention. Where expectation is concerned, this means that it enters into a synthesis with an appropriate perception; the merely expected object is identified with the actually arriving object, as fulfilling the expectation. Obviously, this is entirely different from merely picturing; it is a genuine confirmation. Thus, merely filling the emptiness is not yet the fulfillment of the intention.

What occurs here beyond what is prefigured, beyond what is determinately expected, we characterize not merely as filling, but rather as determining more closely. As such, the latter has the [80] character of fulfillment. What first comes on the scene as coinciding with the prefigured element is a primary fulfillment. But the overabundance that intuition supplies is also a fulfilling, a secondary one, insofar as it is given as belonging to the object itself which is intended there and is now given in intuition as itself, precisely fulfilling the intention.

---

[75] *ausgemaltes Bild*

Let us compare this with the process of bringing retentions to intuition. We realize that we are not concerned here with all retentions. Those retentions taking place *originally* do remain non-intuitive and sink into the undifferentiated general horizon of forgetfulness that has, as it were, become lifeless—provided that an associative awakening has not taken place. Thus, only directed retentions, namely, retentions that have become intentions by such an awakening are at issue for a synthesis of bringing to intuition.

If we consider the existing possibilities here, we will realize immediately that the process of bringing to intuition as a clarifying process, and the process of bringing to intuition as a confirming one, are not sharply distinguished here, as is the case with protentions. That is, they are not such that the first would be characterized as merely picturing, but then not confirming, and the other as confirming, but then not as merely "picturing." The matter is different here. If an intending retention passes over to an appropriate intuition under synthetic coinciding, then this intuition is indeed an intuitive remembering. Doubtlessly, we can say that through an intuitive remembering we become intuitively clear about what was only emptily intended precisely in the empty retention. Surely the synthesis then accomplishes a sense-clarifying process of bringing to intuition. It discloses the intended objective sense, but an objective sense that is initially presented in an empty manner. But we can and must say no less that this synthesis simultaneously has the character of a confirming synthesis. The empty retention—which as reawakened should actually already be characterized as a remembering, though as a non-intuitive one—is confirmed by a coinciding with appropriate intuitive remembering. Insofar as it is an memory-intention it is fulfilled in the specific sen remembering-intuition the object *itself* is placed befor [81] memory-intention, and it is before us as the very memory-intention had merely intended.

Hence, clarifying and fulfilling confirma separated here; here every synthesis of bri accomplish both *a priori*. In contrast to respect to protentions, there is no memories of the past are concern implies a process of bringing to intuition whose accomplishment is merely a pre-view[76] in which we picture for ourselves what probably will be, or as it must be called in our case, what probably has been. An anticipatory image of a matter is not the matter itself, that is, a merely anticipatory seeing is not an intuition of the self [of the object].

It is not as if a remembering could not also be a "picturing," but what we ascertain here is that it cannot merely be a picturing; rather it is simultaneously and necessarily self-giving and thus fulfilling-confirming. This will become completely clear in what follows (let us keep clearly in mind what we have already expounded upon): Every empty retention, like an empty presentation in general, presents in the way of an indeterminate generality. That is, it has a "prefiguring" for its content, a prefiguring that leaves open and indeterminate what goes beyond it, as is shown precisely through the process of disclosure. Remembering, which alone can accomplish a bringing to intuition here, now brings *the* past directly to intuition as [the past] itself, [namely,] a past that was intended in empty memory; accordingly, as we said, it accomplishes a fulfilling confirmation in the synthesis. What is prefigured initially is fulfilled, and this takes place in all circumstances. But the clear memorial image in its concrete fullness offers more than what the prefiguring determinately intended-ahead.

What happens now to this surplus? We obviously have two possible scenarios here for any of the moments accentuated in this filling surplus. Either the moments are characterized in and for hemselves as belonging to what is remembered itself, to the past elf; and if this is the case, they will have the character of rmining more closely what is remembered, the past, and will elves belong to what is fulfilling. Or they are not rized in this way; and if this is the case, they will have the of mere filler, or as we can also put it, they will have the a mere picturing. A non-intuitive memory of a person met some time ago is fulfilled, for example, by an nbering. Now, if we scrutinize the "memorial-image" more precisely, we will notice for instance that the shape, the reproduced full beard, the glasses, and so forth, have an actual [82] memorial character to be sure, but not the color of the beard, the color of the eyes, etc. What is intuitive there is filling, picturing.

We should note, however, that we merely want to treat occurrences that are purely passive. The word "picturing," as useful as it proves to be for us, normally points to an active comportment of the ego. Where the prefiguring is no longer any help to it, the ego would at least like to get an image, and then pursuing this get various possibilities, various suitable images, possibly waiting for a more complete remembering to flash forth later through associative awakening and waiting for one image or another, according to its content, to take on the memorial character still lacking, the memorial character of determining more closely.

Naturally, we are not speaking here of such activities. Accordingly, mere picturing designates for us the merely distinctive character of filler that does not fulfill, as opposed to the actually fulfilling filler that is given in and for itself as belonging to the directly reproduced past: It is given as past in the mode of "itself."

We can say in general that the processes of determining more closely and picturing are actually always intermingled where remembering is concerned, and that a purely fulfilling confirmation is an ideal limit-case, so to speak, a *limes*, while nevertheless in the opposite direction—and that is a main point of our considerations—there is precisely no *limes*. That is, empty remembering does not admit of a bringing to intuition that thoroughly pictures and that would not be confirming. Now, insofar as it is actually a normal, constant process of picturing bit by bit, it is still an unfulfilled intention. Thus, only the ideal limit-case of an absolutely complete remembering is excluded here.

Every remembering is also an unfulfilled intention in othe respects. On the one hand, it is unfulfilled insofar as it wavers clarity, i.e., is only relatively clear. Obviously, a reproduced im is never absolutely clear, which points once again to an i Unclarity is always intuitively unfulfilled emptiness. Moreon is an unfulfilled intention in another respect, namel the remembering intuitively reproduces only a fragment ned [83] intended past, for instance, only the beginning of an melody. In this respect, then, it is entirely analogous to the case of an expectation being fulfilled; namely, insofar as an appropriate perception occurs, the expectation is fulfilled. Insofar as perception has not yet spoken, it is still an unfulfilled expectation, an unfulfilled intention.

<§20. Intention Toward Fulfillment is the Intention Toward Self-Giving>[77]

The intention is directed toward its object; it does not want to be a merely empty intending toward it; it wants to go to the object itself—to the object itself, that is, to an intuition that gives the object itself, to an intuition that is in itself the consciousness of having a self. But the intention does not really just want to drive at coming into contact with the intended object in a corresponding intuition of the self, merely to find, for instance, that what was prefigured has been brought to fulfillment; even then it is still unsatisfied and strives onward from one closer determination to another, again and again. The indeterminate generality peculiar to what is prefigured in every abiding emptiness is always only a form for the fulfillment intended in an affiliated objective sense, a fulfillment which is to be accomplished in the form of a new, closer and closer determination.

What else does this entire manner of speaking of "driving at" mean than that there is still something more that belongs to the opposition, intention/fulfillment, and to the idea of confirmation than what had been expressly mentioned up until now. It has not only been shown that a directedness stemming from an awakening belongs to all actual intentions and thus to an actually fulfilling, onfirming synthesis. Rather, now this comes to the fore as longing to a confirming intention: This directedness is dentious, and as a tendency, as a striving, it is from the very nning "driving at" a satisfaction. This satisfaction is only le in a special kind of synthesis that brings to intuition, in a is that brings the presented object to self-givenness. And it n such a way that the satisfaction is merely a relative one

77 See Appendix 7: <Belief and Intention> p. 449ff.

and leaves a remainder of dissatisfaction so long as the intuition still contains indeterminacy or mere filling. In general, confirmation has a merely relative character, or rather, confirmation can only become relativized precisely because of this essential relation to a tendentious intention; this is the case even [84] though at any moment there is already, in a good sense, a fulfillment in it. It is not simply an issue of a synthetic unity of presentations for itself, but also of a synthetic unity which concerns the striving that traverses the presentations. Corresponding to this is the double meaning of fulfillment, whose one side is equally expressed by satisfaction, as that relaxation of a striving in which the sought after goal is reached as such.

We spoke of *terminus a quo* and *terminus ad quem* with respect to association as the synthesis between what is awakening and what is awakened by it, between the directedness-toward and what is taken up in the directedness. This way of speaking now takes on a new sense, and a more authentic one. For now it concerns an actual taking-aim. The intention is that lived-experience that is a merely aiming-at, a having-in-sight; its fulfillment lies in the lived-experience of being-at-the-goal-itself. And this is not a seeing that pictures, but rather, a self-giving seeing. It is only that the intention remains an intention so long as something is still lacking with respect to having the goal-self, that is, with respect to what is to be reached. It is also clear that every lived-experience, which is in need of fulfillment, can be characterized as an intention even though, on the other hand, it can exist at the same time in relation to another intention as fulfillment, as confirming lived-experience. The latter, insofar as it is self-giving, the former, insofar as it is incompletely self-giving, i.e., it still contains facets that are possibly unsatisfied.

If we now regard the ego as comporting itself merely receptively, and if we do this within a genetic analysis, then we are obliged to observe that an affection precedes the receptive action. A presentation occurring in the background, a directed one, affects the ego, which is to say, a tendency heads toward the ego; the ego reacts by turning toward, the presentation assumes the shape of a grasping in which the egoic regard is directed toward the object. This yields the most strict, normal concept of intending[78], more precisely, of doxic intending, of intending being, extending far beyond the sphere of simple presenting. But there is more here if the directedness should have the shape of that intention that terminates in confirmation.[79] The presentation now has the form of a striving issuing from the ego, of an *intentio* that [85] intends the true self: As such *intentio* is initially characterized by everything that we have described; it is then a presenting endeavor that wants to realize itself in the continuous acquisition of knowledge, in a fulfilling grasping of the self that is constantly in the process of determining more closely, that is, not just in a mere grasping of the self in general, but rather, being interfused with an endeavor to foray into the moments of the object and to see to what extent they are not yet intuitively realized as grasping the self, in order to bring them to this realization. Through all of this, the ego is the center as the wakeful radiating center of active striving.

But perhaps one will have to say that at root the character of the specific intending and even of this intention can already be without the participation of the ego. Accordingly, when the expression "intentional lived-experience" is employed for any type of consciousness (and this is still very often explicated by saying that every consciousness is a meaning-something, but also often as a mere intending that has not reached the goal of truth), it is only correct to do so when it is correctly understood in accordance with our analyses. One cannot say without further ado that inherent in every consciousness is a striving, an intending toward its object to possess the self [of the object], namely, that there is already a directedness. This has been shown with decisive clarity where original retentions are concerned. Still, it is certainly a general, essential possibility that every consciousness, no matter through what kind of motivation (we mentioned associative awakening in the passive sphere), takes on an orientation, and in this connection takes on the intending directedness toward the very self of a being. At most, there is the possibility that every consciousness become a

[78] *Meinung*

[79] Translator: Husserl should either speak here of egoic verification [*Bewährung*] and not confirmation [*Bewahrheitung*]; or he should articulate a passive striving that is not yet egoically directed. See above, footnote 62.

consciousness that intends being, that it become a striving after knowledge, an endeavoring meaning that is satiated in a synthesis of confirmation. Striving after knowledge in the very broadest sense is precisely a striving directed toward being itself, and already belonging here is the most primitive presenting that has the structure of aiming-at. To be sure, we must not take this striving as an isolated intending, but rather, if we want to have the normal concept of knowing in the logical sense, we must conceive of the epistemic striving in connection with a habitual consistency that does not break off and is not arrested in a single act, and does not break off when the knower sleeps. So much for a preliminary exposition.

Retention and protention, which we have contrasted in our [86] analyses, are brought closer together in certain respects through our radical clarification of the concept and phenomenon of intention. By virtue of its origin, protention, unlike retention, is essentially a meaning-intention. More precisely, protention is to be interpreted accordingly as a "fore-," namely, as an *intentio* directed into the future, as an intending and a striving that is directed forward—whether the ego itself is an ego that intends-ahead, and whether the striving is the striving of an ego, or whether it concerns an egoless tendency that is precisely a presenting tending-toward what is arriving from the future. We also call protention an anticipatory, expectant meaning. In positionality, and specifically in non-modalized positionality, we have a certain belief in this or that presentational content; it does not believe within the present what is given impressionally in the present, but rather anticipates in belief, as an intending into the future concerning what a futural perception itself will indicate.

Let us compare with this a retention that has taken on the form of an intending, for instance, that has the shape of a memory of the past (even if non-intuitive), and has the shape of a tendentiously directed memory of the past. Thus, in it is the intention toward the past that is made available in an intuitive remembering in the mode of self-givenness. It is clear that even this intention is intending-ahead in a certain respect, although the "fore" or "ahead" does not have the temporal sense of future. The present lived-experience of the empty memory means, intends beyond itself in an anticipatory manner, only now it intends something that lies in the past and that would come to fulfilling givenness in the memorial-intuition that gives the past itself. Generally, and without further ado, we see that every intention whatsoever is anticipatory, and this feature is due precisely to the striving that, as such, is directed toward something that can only first be achieved through a realization. Intention anticipates either the effective realization of the self in the future and likewise the effective realization of the self of something that is co-present through perception, or the realization of the self of what is past through remembering.

Yet, we find ourselves in a paradox here. The past has surely long since been realized and can no longer be realized—this does not make any sense. On the other hand, when the protentional striving or tending is directed toward what is futural, it is of course [87]
not actually a striving whose fulfillment realizes or makes actual what is futural. Such a realization can take place now as deliberate, as active, or it can take place now non-deliberately in passivity, like when we pull our hand back involuntarily because of an insect bite (while being perhaps quite occupied with something else entirely). With this, a spatial course of events is realized in the perceptual environing-world.

But all this becomes clear when we focus strictly upon the phenomenological situation itself. In particular, it is worth considering what kind of a striving functions in this intending, and what kind of sought after goal this intending has.

§21 Epistemic Striving and Striving after the Effective Realization of the Presented Object[80]

Naturally, we are not speaking here of a wishing, a desiring, or a willing with the expectation that the intended element should be actual, should become actual, or should have been actual. Thus, even with respect to confirming fulfillment we are not speaking of a corresponding satisfaction of a wish or of a willing that occurs in

[80] Translator: Paragraph heading modified. Given here is Husserl's complete heading for this paragraph.

the joy of something that has become a reality with certainty: in the joy that it actually is, that it has become actual, etc., while previously we had felt within us a sorrow at the lack of this certainty, at a doubt in being, or at the certainty of its non-being. As we said, we are not speaking of this at all. For example, the continual protentional intention in connection with a perception actually taking place harbors, at every moment, the anticipatory certainty of what is to come, the certainty of its futural occurrence (and it does this completely without our help). While listening to a melody as it runs-off, we are certain at every moment that further tonal rhythms will arrive and fit in, and when we hear a melody that is long familiar we are even certain of its completely determined contents. For its part, new strivings may be triggered, fulfilled, or disappointed depending upon whether we like the sounds or not. But this does not have anything to do with the anticipatory intention as such. The aesthetic pleasure or displeasure does not imply a fulfillment or a disappointment for the expectation *qua* protention within the merely tonal perception. Since we are certain from the outset that something is going to arrive in the living perception, there is no room at all for a wishing that it should be, or even for a willing that it should become by [88] realizing it in our actions.

On the other hand, we cannot deviate from our descriptions that we have obtained phenomenologically [which show] that every process of meaning that intends and wants to know is precisely a tendency-toward, a striving. We must only note that it is a differently directed striving, and that on occasion manifold strivings may very well become intertwined without disturbing each other. If the striving that intends and wants to know is not directed toward being, in the expectation toward what is to come, or in the memory toward a past being, then it is a directedness toward the experiential seeing of the self of what is taken for being, or rather it is a directedness toward the experiencing itself so that what is already meant from the outset as being with certainty is or will be actual in the subjective mode of experiencing the self. Specifically, the anticipatory intention bears on a situation in which what is meant in the anticipation[81] as certain will come in the future, as it were, in grasping the self[82]; that is, it is directed toward the actual object in the mode of being presented in the flesh. Furthermore, the memorial intention is directed toward a situation where what is anticipatorily certain as past will become a past that is self-given. Here, then, the aim bears on the modified mode of being presented in the flesh that is provided by the intuitive remembering as a modal modification of perception.

The unique and new acquisition of the synthesis of confirmation is, *first*, the coinciding that results at the end of the synthetic transition, a coinciding of the presentation functioning as an intention with the corresponding experience of the self, whereby the emptiness gets its fullness of the self. In the coinciding, which is given to consciousness as a resultant one, the meant object is identified as a sense that now shows the dual mode of the unfulfilled self and the full self as united; and it does this specifically with the character of "verified intending," namely, with the character of saturation resulting from this process. *Second*, however, what was already uniquely characterized in this lower level has in addition its peculiar character that stems from the higher level, the level pertaining to striving. The striving becomes satisfied, what is experienced in the mode of the self is characterized as an achieved goal that, as achieved, i.e., as resulting from the synthesis, *eo ipso* has the character of the resultant coinciding of an anticipatory intending and an intending in the mode of grasping the self; it therefore has the character of a double-layered fulfillment.

Displacements in striving are not only essentially possible here, [89]
but *de facto* occur quite often. The result-character in the synthesis, the character of "true," can become the sought after goal, and likewise the verifying synthesis as the process of accomplishing [the result] can become the sought after goal, and finally, even the sense of well-being that occurs as a result of the tension endemic to striving being relaxed, as a result of a striving

81 *Vorgriff*

82 *Selbstgriff*

being satisfied, can become the conscious motive and sought after goal, i.e., the willed goal.

Our analyses have made clear the essential distinction between doxic intending in the lowest sphere of presentation, that is, presenting meaning, a striving prevailing in it, and every other striving and being satisfied that is possible on the basis of presentations that are similar with respect to content. In every presenting we have the presentational content as the material of belief and the belief itself; that is, in the full sense of presenting, we have the sense-material and the modal moment "being with certainty," or its modifications. If the presenting has the normal form of an intending intention in certainty, then it can also be the case that there ensues a valuing determined by the particular moments of the presentational content and a striving motivated by valuing. We are interested in the being and the modes of being peculiar to the presented matter: We wish that it were, we find it a practical possibility for us, and then we strive, possibly through our actions, to realize it. If such a striving aims at the actuality of the matter as such, which is of value to us by virtue of these or those features, if it is fulfilled in the production of the certainty of its being, and in a superior manner in grasping the self through experience, it nevertheless does not drive at grasping the self to verify it, but rather, to realize its value, to fulfill the valuing intentions, which for their part are precisely founded in the presentations and the effective realization of their presented objects.

We are occupied with the analysis of the synthesis of fulfillment that corresponds on the level of passivity to that higher synthesis that we call in the level of thought verification, in the normal sense of the term (we spoke directly of confirmation in the passive sphere). This analysis led us to the strict concept of a passive intention as a special shape that a consciousness of something must assume in order for it to be able to function in [90] syntheses of fulfillment. We distinguished two things under the rubric of intention:

First, a consciousness of something does not necessarily need, in itself, to have the distinguishing trait of a directedness toward this something, toward its objectlike formation. This was seen in the case of retentions that originally link up with every perception. Only when an associative ray, when an "awakening" radiates in this consciousness from elsewhere, from another presentation, and directs this consciousness toward its object, does it have in itself precisely a directedness. A different orientation can go hand in hand with this, namely, the ego's attentive orientation. But this is not at issue here. Such awakenings have the character of tendencies, and hence also gradations of intensity; like forces, they can become stronger and potentially weaker. This can be seen, for example, in the intensification that perceptions impart to the periodic terms in retention where something is repeatedly perceived, an intensification that animates them and at the same time secures them as intentional objects. Let us note here that only through these and similar syntheses can the element, which has passed over into a retention together with the new perceptual objects, actually be an object, that is, an intentional object. In perceiving a series of repetitions, but also a melody for instance, the newly appearing objects within pure passivity are not the only objects of the intention (although they are this) toward which the primary directedness aims. To the extent that they retain their force, the objects that have faded away in retention are not given to consciousness in the same way that something new [is given to consciousness when it] arises, and the previous object fades off as something that does not have any connection with the new and does not concern it. The unity of a successive multiplicity, of a diversely articulated process, presupposes that forces issue from something new, or again, that from something new, syntheses interweave as they move back; through this, a unity of intention is instituted, a unity of intention that transports back (or "awakens") the objects fading away in the form of the living intention and uniformly binds this latter intention with the intention of the new object.

Second, as a further moment of the specific *intentio,* we emphasize the tendency toward a terminating in the self, that is, the tendency of the presentations already directed toward self-giving insofar as they are not already self-givennesses. The self- [91] giving presentations, intuitions, which give the presented object in the mode of self, are necessarily intending presentations, namely, directed toward this self [of the object]. But insofar as this directedness terminates precisely in the goal of this orientation itself, the intention is fulfilled.

On the other hand, concretely self-giving presentations are always at the same time intending (in the special sense of intending, namely, as pointing toward fulfillment) insofar as they contain anticipatory intentions, namely, protentional intentions that become fulfilled in the continual progression of self-giving. In other words, concrete perception and self-giving in general are only possible as a continual synthesis of fulfillment. Thus a presentation can only be fulfilled when it is either a straightforward anticipatory intention or when it is intertwined with an anticipatory intention in its complex structure.

Let me note the following here: I have repeatedly wavered, and I do not feel entirely confident about the question [concerning] how the merely tendentious directedness, for instance, of an empty anticipatory presentation toward its objectlike formation (that is, what makes it the "intending" of its objectlike formation), would make it to the further tendency toward fulfillment in the corresponding self (that is, to the tendency of verification). When I gave my lectures a few years ago on transcendental logic,[83] I conceived of both tendencies as basically one and the same, distinguished only by the kinds of effects they have. Namely, the orientation toward the emptily intended[84] object is that mode of presentational tendency toward its object in which the presenting consciousness is not yet intentional. Only as intentional is the tendency precisely a goal-directed tendency which, relaxed and unfrustrated, and operating intentionally, leads to fulfillment, to the goal itself. This goal is thus already its goal in the mode of a frustrated tendency, but precisely in the mode of not being reached. Perhaps this is entirely correct if all involvement of ego-activity is actually left out of play.

If it is a matter of empty memories then the intentional tendency prevailing in them would accordingly be characterized as a

[83] Translator: Husserl is writing this version of his lecture in 1925/26 and is undoubtedly referring to his earlier version from 1920/21 (Cf. "Translator's Introduction, Section 1.)

[84] *leer vorstellige*

reproductive tendency. Associative awakening would thus also be a tendency toward a corresponding reproduction, that is, toward the transition to an intuitive remembering that brings the past as [92] such to self-givenness. Potentially, this tendency can be at work in an uninhibited manner: The empty awakened memory continually passes over into a memorial intuition and, as it were, in a completely awakening manner. The precise analogy for this would be, then, the continual efficacy of anticipations, protentional tendencies as the process of perceiving progresses. Where the intention remains inhibited, and where a fulfillment only occurs subsequently, there the synthesis of fulfillment becomes a discrete one. But naturally—and we must never forget this—all the talk of verification applying to such fulfillments in mere passivity is inaccurate. Indeed, we are not speaking here at all of an active striving and accomplishing that is directed toward a true being which normatively regulates[85] or measures its intending according to what is given as true in the self-giving, although, however, it surely does concern the presuppositions, and in a certain manner, the analogies of passivity, without which that activity could not function.

<§22. The Different Relationships of Intention and the Intended Self. Secondary Verification>[86]

Viewed precisely, our previous considerations have still not yielded any ultimate and pure results. We have clearly discerned that intuitions are essentially distinguished according to whether they are self-giving or not self-giving, and thus that a synthesis is characterized quite differently according to whether it passes over into a self-giving intuition or into a non-self-giving intuition, and then as we also put it, merely pictures. Now, however, what unifies such a synthesis where this synthesis is conceived concordantly is not as a mere accord, as coinciding in the objective sense, but rather, as an accord of the *intentio* (in our sharply defined sense) prevailing in the synthesis in question and

---

[85] *normieren*

[86] Editor: See Appendix 27 to §45.

in its components. The intention running through the empty expectant consciousness is fulfilled in the transition to self-giving, to perception; it is a satisfied intention, so to speak, in the self-giving intuition, being characterized here as a confirmed intention. And likewise in all similar cases. Even when we have a synthetic [93] unity of a perception with a perception, i.e., an original self-giving with an original self-giving (as the process of perceiving progresses), we have the rays of intending aroused by the kinaestheses in the appearances, and these intendings into the future, anticipations that are aroused, become satisfied as it were in the continuously emerging new appearance. But if we keep the phenomena sharply in view, then we discern a distinction here between the intentional ray and the intuition into which the intentional ray enters, and then in general a distinction between the particular consciousness-of (even if it is an empty consciousness) and the *intentio*, which traverses this consciousness as a goal directedness or an aiming. For example, the subsequent retention in strict passivity is an empty consciousness into which (as we have shown) the associative intentions issuing from the living present can sink, intentions radiating-back into what has just been. If, then, reproductive awakening ensues from the emergence of a corresponding remembering, then this intuition has taken over into itself the intending ray, and the latter has here the mode of the fulfillment terminating in the self-given past.

With respect to the presentation of what is expected, the presentation that pictures, we will likewise distinguish the intuition itself (which functions as picturing) from the intention that enters into it; but here the intention is not fulfilled and does not terminate in its sense that has become intuitively given. This objective sense has here just the character of what the intended object is, the character of clarifying, of bringing to intuition, picturing; but as unfulfilled, the intention still goes clean through this "image." It is only when the corresponding experience takes place, that is, [the experience of] self-giving, that the intention is satisfied, that it takes on the character of a goal directed intention, of a seeing into the *terminus ad quem*.

But now we must take one step further. Just as we have distinguished between empty consciousness and intention with respect to the empty retention, so too do we make a distinction with respect to the empty anticipation, and this is significant. When perceiving, I move my eyes this way and that, the intentional rays radiate out from these determinately running kinaestheses. But they are not merely rays that radiate out, which by themselves are nothing, for they are only orientations in presentations; empty presentations are awakened, and running throughout them are the anticipatory rays directed toward the [94] emptily presented objective sense. The emptiness is the potentiality of what is actualized in corresponding intuitions and syntheses of disclosure.

But now we see that corresponding exactly to the fundamental distinction between intuitions as self-giving and non-self-giving is the distinction between the corresponding empty presentations, retention and protention (and to be sure, purely as empty presentations, irrespective of the intending rays belonging to them). Then we see, however, that the empty consciousness of retention is of a fundamentally different kind than the empty consciousness of the empty memory of the future. The uncovering of the one leads to self-giving; to a certain extent, it already has the self in advance, it already has it potentially within itself. But this is not so with the empty expectation. What lies in it as potentiality is nothing other than that intuition which pictures, which does not have any self given within itself. We can express it in this way: Just as remembering brings the past self to givenness more or less clearly, and in the ideal limit-case in a completely clear manner, so does the empty retention have this self in a completely unclear, empty manner within itself; the retention has the self as an originally held, safe-guarded self, as still conscious of the self, and still with a hold on it. When the originally giving impression is over, the self is not lost, despite its emptiness. On the other hand, just as the intuitive expectation only offers a pre-presentation, a pre-view of what is to come instead of the matter itself, so too is the empty consciousness of what is to come an empty pre-presentation of what is to come, but not an empty presentation of the self.

We can also say that whether the expectation is empty or intuitive, it is an anticipatory presentation, only that by the expression "anticipation" we simultaneously mean to suggest that the *intentio* is directed toward what is anticipated.

Elucidating this situation has enabled us to understand that a possible level of ratification lies prior to verifying the self, for example, that presentations radiating-back into the past, that is, into the realm of memories already have a certain character of fulfillment, a ratified one, as soon as the past attains even an empty awakening. Illustrated more precisely, when we have anticipating presentations that refer to our past, as we so often do [95] when for instance we hold an old letter in our hands that we had ourselves written, we initially have an empty pensive presentation, intentionally referring to our own past, but which is at first entirely indeterminate and general. Now, the moment a determinate memory flashes forth we are conscious of a ratified fulfillment, and even before a reproductive remembering occurs. To be sure, that is not a genuine and ultimate fulfillment; it first requires an actualization whereby the past itself is actually and genuinely there for the first time. Still, it is entirely different from the anticipating consciousness that precisely cannot fulfill, even when it becomes intuitive.[87]

There would be still more to study here. In a certain way, anticipating presentations are also involved in the functioning of fulfillment; they are constantly involved, as it were. This was indicated by the fact that every perception in itself, every experience in itself, draws in its wake such presentations as horizons, and more precisely, as intentionally awakened horizons. The moment a conflict between the anticipating intentions breaks out here, the self-giving is also inhibited. The self-giving of the whole, what makes up for example the entire perception as a perception, is not only determined by what in it is genuinely a perception; rather, it is also determined by the anticipatory intentions that are concordantly harmonizing and also sense-giving. Where the realm of anticipation in general is concerned, one realizes that there is something like a one-sided and a

[87] Ratifying the empty empathies [*Einfühlungen*] <through> one another in the development of expression. On the other hand: How does it stand with intuitive empathy? And viewed precisely, how does it stand with fulfillment in the development of expression?

reciprocal corroboration, a ratification in a certain respect, even though no actual verification can take place here. And in contrast to this, abrogation, namely, inhibition, doubt, annulment: We already see that we run up against modalization as related back to the specific element in intention.

Our analyses still need to be completed somewhat. Let me say a few words about the opposition within the realm of positional intuitions, about the opposition between self-giving and non-self-giving, an opposition that has become increasingly significant for us. We have seen that not all intuitions can function as fulfilling-confirming; we have seen that it is a prerogative of certain [96]
intuitions to be self-giving, among which we counted perceptions and memories, while on the other hand there are intuitions like the intuitive anticipations or presentifications that are similar to them such as presentifications of an unfamiliar co-present, and so forth. Here we introduce the term "experiencing intuitions" and note that with it the Humeian concept of impression (in its broader sense) is clarified. That perceptions are self-giving is quite familiar to you and should not give you any difficulties. Phenomenologically, "self-giving" means here that every perception within itself is not only, in general, a consciousness of its object, but that it gives its object to consciousness in a distinctive manner. Perception is that mode of consciousness that sees and has its object itself in the flesh. To put it negatively, the object is not given like a mere sign or a likeness, it is not grasped mediately as if the object were merely indicated by signs[88] or appearing in a reproduced copy, etc. Rather, it is given as itself just like it is meant, and it stands there in person, so to speak.

You could have reservations in allowing rememberings to count as self-giving, however. But as you think more deeply about this mode of consciousness, you will see for yourselves that we also have to speak about self-giving here, only in a modified way; this self-giving shares a common structure with other self-givings, elucidating similar functionings within confirmations. Remembering does not offer the present in the flesh, but in a good sense, a "past in the flesh." For that surely means nothing else than

---

[88] *angezeigter*

a direct return to the past as having-been perceived for us, and as possessing the self of this having-been as such in an original re-execution, as memory. We can also say that perception is characterized as an original acquisition of the object, and remembering as originally going back "in thought"[89] to what is already acquired, as originally having it once again at our disposal.

Referring back to our earlier exposition, let us also note that *both* functions are mutually related in knowledge, and that this relation first makes knowledge possible. The analogy of acquisition and the acquisition being at our disposal once again (which is basically more than an analogy) rightly indicates this. An acquisition that we cannot have at our disposal again is not an [97] acquisition. An object that has been grasped in the flesh can only be there as existent for the subject (having abiding validity in its knowledge and cognition) and can only be an actual object in its surroundings when it is "the identical" to which the subject can return in iterable rememberings. Or [when it is] "the identical" that the subject can mean through rememberings as the same, time and again, or that it can possibly verify in new perceptions, but also as what it can recognize, "the identical" that it has remembered at this time, has perceived at another—which, once more, presupposes remembering.

In contrast to these are the non-self-giving intuitions. Using the expression, "picturing," like we did with respect to intuitive anticipations could tempt us here to oppose such intuitions to self-giving intuitions, regarding the former as mere acts of phantasy. But such a facile characterization of these kinds of intuitions will not do; they are in fact in need of a much deeper analysis. Mere phantasies are not positional lived-experiences, whereas intuitive expectations are precisely expectations; something is believed in and through them, posited as being in the future. The futural thing that is intuitively contrived is thereby not the futural thing itself; it does not appear any differently then a pre-view or projected image through which runs the intention as unfulfilled, so to speak, being directed ahead merely expectantly toward the corresponding self.

---

89 "*geistiges*"

But the structure of the function of such a projected image and of its genesis requires further analysis.

It is clear without further ado that the fundamental distinctions of intentions requiring fulfillment correspond to the fundamental distinctions within self-giving intuitions as such, which are capable of being fulfilling confirmations: since, obviously, not every intention is capable of being fulfilled by just any kind of self-giving in any manner. For example, an expectation can only be fulfilled by a perception, and an intention that is directed back can only be fulfilled as remembering.

Up to this point we have focused exclusively on the syntheses of fulfillment and have not considered the opposing occurrences affiliated with them, namely, disappointment, the synthesis in which a different self appears in place of the intended self, appearing likewise in the mode of the self, whereby, so to speak, [98]
the negative of a coinciding results. The intended self overlaps with a different self in the transition to intuition. It does not achieve the unity of an identifying coinciding, but rather the unity of being otherwise whereby what was intended gets crossed out. Obviously, we return to the doctrine of passive modalization; but through the deeper clarification of the concepts of intention and fulfillment it opens up new insights and a deeper understanding of its essence. Above all it is important to connect up a phenomenological consideration of association with the doctrine of intention. This will open up the possibility of a genetic elucidation of modalities, initially in the sphere of immanently given matters, and then in the sphere of transcendent, externally given matters.

Going back further and further enables a gradual elucidation of how the unity of the life of consciousness itself is constituted as the field of being on another level of being within the immanence of the life of consciousness, how this unity is constituted in the syntheses of fulfillment and in the syntheses of the concordance of other doxic intentions that are further intertwined with the fulfillment; moreover, it enables an elucidation of how things existent in themselves are constituted as a higher level of being in the unity of this life, and the highest level of all, the universe, an entire universe of objective being, of our objective world in its open infinity. But all of this is constituted in concordant syntheses, even though there are always discordances, semblances of appearance, illusions. In a certain respect, the untrue, the non-being is already eliminated in passivity. A thoroughgoing consciousness of one and the same world comes into being through revisions and corrections in the form of consciousness's restoration of the disrupted concordance. I will shortly give a idea of the great problems and the extremely extensive investigations that they demand by confronting the primitive concept of "verification" that we have obtained with the traditional logical principle of contradiction and of the excluded middle. Through this, we will make clear the very special concept of being and the special concept proper to it, normative regulation[90] which, quite understandably, is presupposed by this principle and thus by logic.

But before I pass over to this, I must first still speak about those secondary verifications within the primitive level I just mentioned, the secondary verifications that exist alongside the verification [99] through the self that we treated as the lowest level of the phenomena of evidence. We are considering the realm of intendings, of doxic intentions. Normally, straightforward belief always characterizes intending such that our clarification of intention obviously presents a fundamental portion of the theory of judgment. Intendings can accord with intendings, even where they are not fulfilled. Through the arrival of intendings that accord with a given intending, the latter can get the character of a ratifying intention, and of a stronger and stronger ratifying one without having to speak of an original confirmation or of a process of making evident. Likewise, empty intendings can contend with other intendings and can give rise to an irresolvable doubt, then the conflict can be resolved in the form of reawakened certainties that negate one of the horns of the dilemma and ratify the other. But all of this takes place in the realm of mere empty intentions of belief, while earlier we had studied the parallel occurrences of possible modalization in the originality of perception. For example, if we hold an old letter in our hands it will refer to a person in an indeterminate generality, but we do not know which

[90] *Normierung*

person it is. We recognize the handwriting as familiar, and several people come to mind in our memories, but we are not sure who it is. In reading the first lines of the letter a determinate memory arises concerning the occasion we first received it, but it is a memory that is in no way intuitive, and as we continue to read the letter, the person is immediately specified, the decision is there, ratified.

It is important to note that such ratifying and corroborating functions, which are in accord in manifold ways, actually play a role everywhere and even in self-givings. They are involved in every self-giving, in every perception, and in an entirely originary manner. Their accord belongs to the functions of self-giving fulfillment, just as their discordance can awaken modalizations in them. The "horizons" of perceptions are another name for empty intentions (intentions in our clarified and precise sense) that are integrally cohesive and that are actualized in the progression of perception in and through different orientations. All of these intentions must accord with one another should the perception be normal, that is, continuously self-giving. If they come into conflict with one another then the belief proper to self-giving is inhibited [100] in its fulfillment. For example, as the process of external perceiving progresses optically, the thing is not only intended optically. Intentions of other sense spheres are continually co-awakened and must continually accord in the unity of a synthesis with the genuine impressional ones of the optical sphere. They must do this because they are co-constitutive of the objective sense. However, the moment something does not accord in the tactile sphere, even though it is not impressionally realized, fulfillment of belief is inhibited; the unity of perception as it accords with all intentions in forming the unity of a whole intention is inhibited. But here we must actually go beyond the individual perception and go to the encompassing nexus of external perception in general which makes up a unitary consciousness of the environing-world and an original consciousness in a universal synthesis of all empty intentions, even of the empty intentions co-determining sense. We can see a thing, and everything in the thing jibes; in themselves the tactile features that are co-indicated would assuredly conform. But we look through a stereoscope for example, our hand resting atop the table; through this and also through the continuous preceding perceptual nexus as a whole, we know ourselves to be in a small room, even though the stereoscopic object is a waterfall located on larger isolated rock cliffs. Something does not jibe, and the perception is now broken, modalized; it takes on the character of an illusion, and in this case it simultaneously takes on the character of an impressional likeness of a Swiss waterfall. Now running through the likeness is the intention directed further toward a fulfilling self-givenness that is to be produced some other way.

Thus, we see that in each and every case it is a matter of intentional systems that go together to form the unity of a synthesis and a whole intention arising through it. These unities must be thoroughly in accord so that on the whole the particular intention can have the character of unbroken belief. Modalization belongs to the intention and is carried out in intentions; it is however not carried out in an isolated manner in elementary intentions, but rather in the intentions that are integrated concretely into the synthetic intentions as a whole. Special forms of this, then, are the self-giving intentions which, by constituting a self, are able to fulfill and to confirm non-self-giving intentions.

<Chapter 3: THE PROBLEM OF DEFINITIVENESS IN EXPERIENCE> [101]

<§23. The Problematic Character of a Verification that is Possible for All Intentions and Its Consequence for Belief in Experience>

That we have a consciousness of our own life as a life endlessly streaming along; that we continually have an experiencing consciousness in this life, but in connection to this in the widest parameters, an emptily presenting consciousness of an environing-world—this is the accomplishment of unity out of manifold, multifariously changing intentions, intuitive and non-intuitive intentions that are nonetheless concordant with one another:

intentions that in their particularity coalesce to form concrete syntheses again and again. But these complex syntheses cannot remain isolated. All particular syntheses, through which things in perception, in memory, etc., are given, are surrounded by a general milieu of empty intentions being ever newly awakened; and they do not float there in an isolated manner, but rather, are themselves synthetically intertwined with one another. For us the universal synthesis of harmonizing intentional syntheses corresponds to "the" world, and belonging to it is a universal belief-certainty.

Yet as we already mentioned, there are breaks here and there, discordances; many a partial belief is crossed out and becomes a disbelief, many a doubt arises and remains unsolved for a time, and so forth. But ultimately, proper to every disbelief is a positive belief of a new materially relevant sense, to every doubt a materially relevant solution; and now if the world gets an altered sense through many particular changes, there is a unity of synthesis in spite of such alterations running through the successive sequence of universal intendings of world—it is one and the same world, an enduring world, only, as we say, corrected in its particular details, which is to say, freed from "false apprehensions"; it is in itself the same world. All of this seems very simple, and yet it is full of marvelous enigmas and gives rise to profound considerations.

Let us reflect on the following. We have made clear that doxic life on the level of pure passivity takes on the shape of the passive intention again and again, of a directedness, which, as a tendency [102] operating in an uninhibited manner, passes over to self-givennesses. Thus, running through passive life are ever newly interweaving syntheses of fulfillment. A continual striving after intuition that realizes the meant self, a continual—we couldn't help thinking of the term—confirmation. The fulfilling self as that at which the intention was driving, has indeed the character of truth and henceforth abiding validity for the subject. This seems to be exactly what active cognition, and in the highest level, predicative and theoretical cognition accomplishes. Even it is initially a mere, only active process of intending, even it is a tendentious believing, only now it is a believing that is actively striving, set on fulfillment, becoming operative in evidence. But what is evidence other than seeing the self of what is meant, that is, the fulfilling effective realization in possessing the self; what is it other than the synthesis of coinciding that forms an identity, a coinciding of a merely anticipatory intending with the fulfilling self? Showing-as-true, showing the correctness of the intending takes place precisely through this. It is directed toward the self that is realized in consciousness; accordingly, an evidence that verifies is thus nothing other than the consciousness of the *adaequatio rei et intellectus* carried out in the originality of possessing the self.

No matter how elucidating and even in a certain respect how indubitable our presentation may be, the accomplishment of a genuine confirmation, which strives toward knowledge in the strict sense, surely cannot be exhausted by what activity and passivity have in common here.

Does evidence, does the immediately seen *adaequatio* already yield truth in the full sense? Truth is surely definitiveness. But possessing the self, experience, can come into conflict with another experience; modalizations can occur. Can this not go on *in infinitum* such that definitiveness is never reached? And if there should be such a definitiveness, how would we know it? All knowledge as the grounding of judgment begins with experience. Doubtlessly, it is an initial phase of grounding. But is grounding already accomplished with mere experience? An analysis of experiencing passivity is surely fundamental for the elucidation of this initial phase. But we cannot expect to have come very far with this so soon. From what we have already said, it has already become questionable if behind every judgment there is a definitive truth: because of the relativity of experience. Doubt receives a new [103]
sense if we initially recognize experience as being definitively correct. Does every question have to be able to have a response? A definitive response, and in advance? Put differently, in our view, every judgment has its norm in a truth that is valid in itself, whether we know it now or not and are ever able to reach it or not. Normative regulation, measuring up to the norm, is carried out subjectively in evident verification, since to our mind the true that regulates in a normative manner,[91] precisely as the seen self,

[91] *das normierende Wahre*

comes into our grasp immediately precisely in the experiencing, self-giving intuition.

But are there not many profound difficulties to be found in this in-itself, difficulties that lie in yet a different direction than the one of which we have already spoken? For example, must every judgment that is directed toward the future have a definitive truth or falsity? Must it be decided in advance, only that we do not know how it is decided? Let us follow the matter in [the sphere of] passivity and discern what is missing here. Only from this [sphere] can one grasp the most fundamental of all the shortcomings in the groundings of traditional logic, a shortcoming that concerns the ultimate principle of all logical norms in the most astounding way, namely, the principle of contradiction and of the excluded middle.

To be sure, we think of this principle above all when we speak of logical norms; all knowledge must be logically shaped, a truthful logical knowledge is referred back to the principle of contradiction.

But when we articulate and think through such principles, we notice first of all that we have not yet acquired concepts like truth and falsity here in the full sense, and that speaking of the norm, of a norm of correctness and incorrectness had an incomplete sense. In immanent time-consciousness we have the stream of givennesses in lived-experience, givennesses that are strung together temporally with their anticipations which have the character of an anticipatory believing that is directed-ahead. A spatio-temporal world is given in the stream that is contained in immanent time-consciousness, in the stream of transcendent experiences,[92] the intuitive and non-intuitive. And constantly referring to this spatio-temporal world are manifold lived-experiences of a transcendent believing that is in need of fulfillment. In both respects, belief is not only directed toward the [104] present, but also toward the anticipated future and toward the memorial past; manifold memory-beliefs and expectant-beliefs emerge that can be verified or rejected.

What happens now with the verifiability or refutability of experiential belief in all these respects? What is to be meant by the

[92] *Erfahrungserlebnisse*

axiom we tested that every such belief is either positively or negatively verifiable? To be sure, it does not just mean that the mere possibility of one or the other belongs to the essence of such a believing, and that when the one possibility is taken up as realized, the other would thereby be annulled. It is certainly evident from the principle of contradiction that positive and negative verification, fulfillment and disappointment, are mutually exclusive. But if we want to say that every belief is verifiable in the sense that it is valid or invalid in the usual sense, as it is in the view of the traditional principle of the excluded middle, then quite a bit more is being said here.

Let us get clear about this by drawing a parallel to mathematical judgment, to a judicative believing that bears on what is mathematical. Either it is valid, it is verifiable, or it is not valid, it is verifiable in a negative manner. This certainly implies that whether we ever will or are even able to carry out a verification or not, even without thinking about whether it may ever become a decision of the positive or negative sort, it is surely decided in itself whether the judgment is verifiable or whether it is refutable, already in advance and thus for all actual and possible consciousness in the future. Only we do not know in advance how it is decided; we only first know this in the actual, intuitive verification as the current decision. It is determined in advance, as it were, how the die is cast, whether on the positive or the negative side. If we actually and positively confirm the judgment, then we will know that it was already established beforehand that only a positive confirmation could take place and that the opposite was excluded (and for every conceivable ego).

Let us now turn to the spheres of our external experience as they had been constituted in passivity, and so far as we are able to understand them from there up to this point. We ask with respect to these spheres: Is it an actually intuitable, essential law to be drawn from the structure of the intentionality of experience that every belief, no matter how it arises in the stream of consciousness [105]
and its motivations, is decided in advance according to the possibilities of verification and refutation? How can this "in advance" be understood? Certainly, if a fulfillment occurs, then belief is decided as valid; a prehension of the self has emerged

from mere anticipating, the anticipation has been ratified. But as long as the verification has not taken place, both of the open possibilities do exist. Must it be determined in itself and in advance what alone can occur there, if it is ever to be decided at all and regardless of whether a decision takes place or not? To elucidate the structure of truth or validity is to elucidate this "in itself," and perhaps there are fundamental differences there. In fact, truths of the mathematical kind and other essential truths are fundamentally different from truths like experiential truths. This in-itself is divided according to the correlations: As correctness in itself it belongs to believing, as truth in itself, truth in the strict sense of the word, it belongs to sense or rather to a proposition. The object in itself corresponds to the truth in itself. Now the in-itself belongs to the object.

Thus, we have brought the problem of the empirical in-itself clearly into relief by drawing a parallel to the mathematical in-itself where it is easiest to detect the peculiar trait of validity in itself. And we were able to understand that we are in no way in the position of fashioning in an original manner the axioms of the sort that we previously tested relating to the empirical in-itself. This concerns no less the immanent sphere despite its privilege through the evidence of the ego. In fact, if we conceive of a consciousness that has sounds, colors, and similar hyletic data given in passivity in immanent time-consciousness, and these data are being constituted in the process of becoming, then it is not clear in the least how it should be decided *in itself*, in advance, whether precisely this new tone, or whether a new tone at all should follow a previous tone. And even if an anticipatory belief in a new determinate tone were motivated by the preceding immanent experience, we cannot tell in advance that it should be decided in itself, whether it will actually occur, or instead will fail to appear, or will take place in an entirely arbitrary and altered manner.

What happens with respect to transcendence, that is, the spatio-material world, at least when we conceive of it constituted purely [106]
passively in a consciousness? Naturally, proper to the constitution of a spatio-material environing-world is not only a superabundance of prefigurings of inner horizons for every thing that is actually experienced, but also of outer horizons—which are

reciprocally interwoven with one another, and ultimately all things of experience are connected in the unity of an environing-world with a unitary outer horizon—and thus a superabundance of prefigurings for the path of further possible experiences. But there are precisely prefigurings, there is motivated experiential belief, superabundantly corroborated and ratified through innumerable accordances; but in the final analysis is it not possible for the further experience with its ever new self-givings to continue as it will? Contrary to each and every expectation, contrary to all the overwhelming preconvictions and probabilities? Can it not continue such that everything becomes a confused muddle, such that the entire perceptual world-order is destroyed, such that this world as the unity of experience is no longer even maintained, such that it becomes unstable for consciousness, such that all sense-data lose their apperceptive apprehensions, which themselves only actually grasp appearances in concordant believing? But we have held that there would be the world in itself, and every experiential belief would be valid in itself, would be in itself true and false.

If it is a matter of the future, for example, then it is decided, even where I lack a decision. To our mind, every belief directed toward the future has its truth or its falsity prefigured in advance, once and for all.

However, if we remain in the framework of pure consciousness and consider the immanent and transcendent given matters that are constituted within it in passivity, then as I said, what we have shown has not yet accounted for the clarification of the idea of that in-itself. Where the immanent given matters are concerned, and especially the sense-data, every Now brings with it new ones. But in spite of all aroused anticipations, it cannot be foreseen why it must necessarily be decided which data will occur in the future. And this also holds no differently where the transcendently constituted spatio-temporal world is concerned.

Perhaps some clarification is still required here. This world is given to us originally through external perceptions. Generally [107] speaking, they cohere with one another in continual concordance, and they are likewise intertwined concordantly with self-giving rememberings that potentially span gaps like those of sleep. To be sure, occasionally discordances do also occur. We speak of illusions, experiential belief being ruptured, passing over into doubt; but in the progression of experience, which never undergoes breaks in each and every respect, a thoroughgoing concordance is indeed restored through the changes in meaning and the crossings out just described; that is, running through our consciousness is a sustained unity of world-certainty that is produced again and again over against the disturbances. "The" one world is constantly there, only it is determined more closely and occasionally determined somewhat differently.

The first problem arises here, however: Must it then remain just as it was up to this point, according to the testimonies of our memory? Must an external experience be continually adjoined to another external experience in this way? Can it not be that an external experience is the last one, while consciousness endures? An external experience is assuredly a complex structure of consciousness that emerges in the nexus of consciousness as naturally motivated. Still, must the motivations proceed in such a way that a perception has to be connected up with another perception? We have kinaesthetic courses with which the appearances of things are connected in accordance with associative motivation: thus, certain exhibiting sense-data (in the case of the visual appearance, that is, visual data) along with their apprehensions. By motivation we mean that certain data and their protentional horizons are demanded as co-emerging along with the emergence of other data in our lived-experience. But such associative demands can be annulled in the course of present experiencing. The series of sensation must actually arrive in a certain way in order for the apprehension of a thing to be experienced according to the kinaesthetically aroused pre-demands, and in order for the consciousness of an existing thing to be maintained. If the sense-data were suddenly to begin appearing in a muddle, if our visual field were suddenly to be filled with a confused muddle of colors, the kinaesthetic motivations would [108]
lose their force. What was formerly linked up to the kinaesthetic courses in an expectant manner would no longer be able to occur in the otherwise firmly regulated manner in anticipatory believing, and it would therefore be an end to the play of external

perceptions. Their emergence means precisely a regulated functioning and a continuous further development of cultivated motivations, and this essentially depends upon the actual course of sensations. Yet this is always conceivable as an entirely different course, and as a completely unregulated one. That it is not an unregulated course, but is such a course that makes a continual perceptual flow possible, that is simply a fact. However, if we inquire into the truth of this fact and more precisely, if we ask why what was previously so must still be or will be so, this truth will obviously not be such that it could be decided by recourse to one of those passive confirmations of which we were alone able to speak.

Secondly, even if we do presuppose the truth of this fact, and thus assume that for us, that for the particular experiencing pure ego, external experiences will continually be adjoined in its stream of consciousness and will also issue in concordance every time, this would only be to say that for this ego the unity of a true world will be continually maintained in ratifying judicative intendings. But this is not to say that this world, beyond our present experience of it, is a world determined in advance, determined in itself, such that the decision of true or false would be univocally prefigured for every believing that is directed toward any temporal situation, or for a corresponding believing that is produced hypothetically.

This is illustrated most simply by referring to the difference between the world-view of that part of humanity that is influenced by modern natural science, and alternately, the world-view of the rest of humanity. This world is constantly and self-evidently there for all human beings, and they believe that it will also continue to endure. In their conscious lives they live into a world-future. But by far most human beings do believe that what will come depends to a large extent upon accidents that cannot be ascertained, or that gods decide the world's course according to a momentary whim. Only roughly is there an order that can be foreseen, according to which one can be practically directed, but only roughly. A [109]
conviction was indeed forged quite late in a causality that lawfully and absolutely determines each and every thing occurring in the world; and the sense of this conviction is none other than precisely

this: that each and every temporal being (and in the natural attitude this means all beings in the world) is determined in itself, determined as truths in themselves. From the very beginning, nothing is open in order to have to wait and see first how the Fates of destiny will decide.

Our question rested in this consideration, namely, whether we already attain definitiveness (in the mode of experiential ratification) should an intending into the future actually be ratified by experience. Here, then, the other, last mentioned difficulty comes into play, and in a way that is very touchy where external experience is concerned. Does not external experience evidently lead *eo ipso* into infinity? Each experience is still itself an open intention; it has dimensions that are unfulfilled. Must, indeed, can a synthetically progressing acquisition of knowledge come to an end?

<§24. Development of the Problem of the In-itself for the Immanent Sphere>[93]

Let us turn back to the immanent sphere. Let us directly pose the question with respect to immanence whether self-givenness here cannot be something that is entirely relative, that does not contain any definitive self at all or that does not have this definitive self supporting it as a persistent norm. Can it not be that every self-giving is to be rendered invalid through a conflict with other self-givings, and these again in conflict with others, and so on *in infinitum*? More explicitly, can it not be the case, when any kind of presentation is verified by a corresponding self-giving, that the latter is forthwith annulled through negation, whereby now what is presented would also be given as not actual; that then, however, the self-giving functioning as norm is also negated once again, and so the actual and the non-actual are always only something momentary, something belonging arbitrarily to the process of fulfillment? Or when we take any kind of presentation, is it decided in itself that corresponding to its meant being, to what

[93] Editor: See Appendix 8: <The Apodicticity of Remembering> pp. 451ff. Translator: See also Division 4.

is given in it in the mode of belief-certainty, is a definitive being of the self as true and as incapable of being crossed out for all time?

To be sure, at first we see that immanently constituted being in [110] its living present is not only self-giving as being, but that this being is essentially incapable of being crossed out. The moment we assert that it is not, like we can do at any time, we see that apodictically this assertion is annulled with respect to what is given. Here the indubitable, the indefeasible validity is clear. But what good is it, since its validity is only momentary? What is immanent flows-off and is gone. But where we speak of a true self and of a presentation that is verified definitively, there we reach beyond the momentary consciousness through rememberings in which we repeatedly come back to the same presentation and to its same meant object; and in which, on the other hand, we can repeatedly secure for ourselves and potentially do secure for ourselves the verified self as an identical self, and one that is not capable of being crossed out. Surely, we do have the momentary lived-experiencing (e.g., of an immanent sense-datum that we see in its present becoming) in a certainty that is not capable of being crossed out. But the being that we grasp there is only meant as being in itself when we not only take it as a momentary datum in the mode of the present, but also as the identical *dabile* that could be given in arbitrarily iterated rememberings—that is, when we take it as a temporal datum, for instance, as the tonal datum in its temporality, a temporality that is identically one as opposed to possible orientations like those that are given through variable rememberings.[94]

We see that the temporal form is the form of objects which, as objects, pretend to have their in-itself. All talk of objects thus leads back to remembering. Thus, this does not only hold for immanent objects. Even when we consider a noema, even when we consider that which is momentarily present as sense in the mode of the present and make an objective assertion about it, we

[94] I am not entirely satisfied with this. The object is surely constituted from the very beginning as something temporal, and the momentary phase is an abstraction that we must first construct. The incapability of being crossed out [*Undurchstreichbarkeit*] peculiar to the moment is thus not primary.

grasp it as such a [noema or present object] that can be presentified to us and identified in iterated memories, namely, with this reproductively presentified character, "momentary present." We learn from this consideration that the question [111] concerning how an objectlike formation, an objectlike formation in itself, is constituted, how it can show itself as such originally, leads overall and from the very start initially to the problem of the constitution of an in-itself of remembering, thus, to the question concerning how remembering can be justified and to what extent it can become a source for definitiveness. We must first of all become clear about it.

Indeed, seen more precisely, we are still lacking an intermediate term for a completely systematic exposition. The living present which is structured immanently is as we said not capable of being crossed out so long as it unfolds in constitution; doubt is not possible here. This also concerns, therefore, the span of the living retention proper to it. To make ourselves explicitly clear: Every progressing retention that continues to exist in a living fading-away cannot be modalized. But once again we must add that we still cannot identify an in-itself here. I can keep a hold on a sound fading away attentively, yes even hold onto it more tightly. Here is something of the most primitive activity. But it can also be the case that I am not even attentively aware of it, that I do not actively turn toward it and toward the series of sounds, and yet it exercises a special allure. By virtue of an associative awakening, it gets the character of an intention. In both cases, and in an entirely spontaneous[95] fashion this provocation (this affection) is possibly discharged in a remembering that not only arises in general, but arises as the fulfillment of the intention.

Note that this remembering is something essentially different from a retention, and is not for instance a mere reanimation of it in the sense of an increase in the level of clarity. A clear retention whose essence we grasp in the retentional levels lying most closely to the primordial impression always remains a retention. Every retention is what it is and has its intentional mode only at the place of the streaming perceiving at which it stands. But

[95] *unwillkürlich*

remembering is a kind of re-perception, that is, it is not a perception to be sure, but a being constituted anew, a beginning anew with the primordial-Now, and a retentional fading-away, but precisely in the mode of reproduction. Thus, in remembering all retentional levels emerge "once again," reproductively modified. If such a remembering now occurs in connection with a retention harboring an allure that is just past, it necessarily occurs in a [112] coinciding of sense and in a coinciding of being with the retention. The same sound that has just sunk back and just faded away appears once more on the scene, and I live through its being once more. This can be repeated; I either deliberately or involuntarily reproduce the tone or an entire tonal phrase once more, even an entire melody. What happens here to the incapability of being crossed out peculiar to remembering? Is it to be rejected out of hand? Are we speaking in epistemology from time immemorial about the general possibility of deception in remembering? And does this not hold as well for the immanent sphere?

<§25. Rememberings as the Source for an In-Itself of Objects>[96]

We will obviously have to make distinctions here between close memories and distant memories, (1) between rememberings that are awakened through the retention that is still primordially living, still articulated in itself and found in constitutive flux, and (2) between rememberings that reach into the distant horizon of retention, like with those of an entire piece of music.

1) Rememberings as Arisen from the Awakening of Primordially Living Retention

Where the first are concerned we will say: For what has just past and is still sinking back, which remembering intuitively grasps once again, we have what is absolutely incapable of being crossed out—and we have this even if remembering is repeated,

[96] Translator: See Appendix 9: Both Variations <of Modes of Givenness> pp. 474ff, and Appendix 10: <Possession of the Self and Concealment in Remembering: Reproduction and Retention> pp. 475ff.

whereby the second now creates its evidence from the first and no longer from the retention that has, in the meantime, completely faded away. In spite of all that, we certainly do not grasp the self and the identity of the self in such a repeated coinciding of the self without incompleteness and degrees of completeness. For we know indeed that remembering can essentially waver in its clarity, and can even be intermittent. The different moments of content are more or less veiled, as it were, as if by a fog of unclarity. And yet, [113] it is not one of those concealments in the usual sense, namely, of objects by other objects. The fog of unclarity does not blacken out objects, it is not a real[97] fog. And yet it does conceal, renders the self-giving incomplete. And yet, what is just past as past is absolutely secure, it is incapable of being crossed out, indubitable, and it is absolutely secure with respect to everything that is given from it according to quality, intensity, timbre. It is itself there passing through the fog of unclarity, in all relative unclarity, but just not in an entirely obvious manner, just not as realized ultimately. Accordingly, something is lacking with respect to this incapability of being crossed out. Essentially belonging to this situation is the necessary coinciding that forms identity, the coinciding of given matters throughout the alteration of different levels of clarity, and a certain enhancement in the direction toward an ultimate, most genuine self, the self that is completely evident in its appearance[98] but which is only an idea to be discerned intuitively, an ideal *limes*.

But what is peculiar here is that it does not first require something like approaching this ideal limit in order first of all to confirm a less clear reproduction. In this connection with the living present, it has its original justification in itself, continually. And "original justification" means that it contains a self that is inviolable, although it only stands in a gradation of degrees against a *limes* which in its very nature would alone completely exhibit the "self." The less clear remembering is less saturated, the clearer one is more saturated, it is a "more intensive" self-givenness, but if it is an intuitive remembering at all, it will give just one self and

---

97 Translator: reading *realer* for *reeller*.

98 *das der vollen Augenscheinlichkeit*

not give any other self, or any of its moments.[99] However, empty remembering is not actually a remembering, but an awakening or an affective allure of a retentional sedimentation that is emerging as prominent from the immersion of memory.[100] In a certain sense there are also gradual differences of proximity and distance here.

One must say then that we still have another gradation, namely, [114] with respect to reproductions that reach into the outermost horizon of immersion, even with respect to those reproductions that draw near to it. Namely: Self-givennesses arise here that are indeed actual self-givennesses and are incontestable in this connection, while it is left gradually undetermined how far the actual self-givenness reaches, and what can yet actually be ascribed to it with respect to determinate moments.

2)[101] <Rememberings of a Submerged Past of Consciousness>

The systematic path leads then further to rememberings that do not have their retentional connection to the immediate realm of the present, but rather, that revive a distant, long submerged past of consciousness. We speak here of distant memories as opposed to close memories. Even here, for distant memories, I hold that every remembering has its original justification, and this means that we are to understand essentially that corresponding to every remembering, even to this group, is a necessary idea, the idea of a self that is incapable of being crossed out. My guiding thought here is the following: An intuitive distant memory, when it is not one that fleetingly flashes forth, but is a steady one and is

---

[99] During the course of our lectures the following was added: The justification of close memory does not yield any elucidation of the possibility of the knowledge of an immanent object as being in itself. For we are still bound here to the chain of rememberings, rememberings which adhere to a living retention, which had a departure from it, and were carried by its self-giving evidence. Only when we have first justified the distant memory do we have the possibility of recognizing at any time an immanent temporal object as existing.

[100] *Gedächtnis*

Editor: See Appendix 9: Both Variations <of Modes of Givenness:> (1) of Proximity and Distance within Clarity, (2) of Obscurity as Veiledness, Nebulousness, and Appendix 10: <Possession of the Self and Concealment in Remembering: Reproduction and Retention>.

[101] Translator: Husserl's enumeration.

synthetically iterable and identifiable, has with respect to its objectlike formation essentially only one possible way of passing over into doubt and then turning out to be null, namely, as a confused muddle of rememberings. Thus, becoming discordant, inhibition, and annulment of the belief that is initially unbroken in the self-given past, necessarily leads to the phenomena of bifurcation in which the distant memory in question splits into several distant memories. And it does so in such a way that the unitary objectlike formation of an undivided memory is disclosed [115] as the fusion of single objects, single features and events that belong to the separated memories and are self-given there with partially different objectlike determinations. In the same way it could now happen that every one of the splintered memories lose their character of unbroken concordance, and undergo a crossing out by splitting into yet other memories that are in themselves concordant.

But on the one hand it is still the case that the content of every memory that is characterized as false is only false with respect to the unity of the combined whole, but it remains correct with respect to its parts. What is crossed out is always the whole that has arisen through a commingling, but the elements that issued in the commingling remain self-given, only they belong to different nexuses. On the other hand, however, this process of splitting cannot continue *in infinitum*; it is a muddling up of discrete matters, and thus it must come to an end. Indeed, it suffices that what appears in a memory, essentially, cannot as remembered be completely empty, that its self-giving cannot be an empty title, but rather that it has its source in actual self-givings such that we are necessarily referred back to the idea of a chain of pure self-givennesses that are no longer capable of being crossed out, but are only identifiable with respect to their content and repeatable in complete identity and concordance. Naturally, even here we have degrees of clarity for every portion of genuine self-givenness, and in this respect the idea of the most complete self-givenness as *limes*. Thus even this type of saturation <has> differences of evidence. In both relations we are certainly referred to the active ego and its free activity in which it is guided, precisely, by the experience that memory can turn out to be deceptive, and that, in particular, clouds of unclarity can conceal the comminglings. Accordingly, the ego strives to check its memories thoroughly, to clarify them deliberately, to investigate the intentional nexuses with respect to memories' parts, to disclose the illusion by analyzing, and thus to advance to the true self.

But still necessary for our further understanding is the elucidation of the origins of error in passivity, and in particular of [116]
error in its most original shape of commingling. This problem leads us to a radical portion of the analysis of passive consciousness, namely, as genetic analysis: to the phenomenology of association.

<DIVISION 3: ASSOCIATION>

<Chapter 1. PRIMORDIAL PHENOMENA AND FORMS OF ORDER WITHIN PASSIVE SYNTHESIS>

<§26. The Position and the Delimitation of Themes Concerning a Phenomenological Doctrine of Association>[102]

The rubric "association" characterizes for us a form and a lawful regularity of immanent genesis that constantly belongs to consciousness in general; but it does not characterize, as it does for psychologists, a form of objective, psychophysical causality; it does not characterize a regulated manner in which the emergence of reproductions, of rememberings, is causally determined in human and animal psychic life. For we are working within the framework of the phenomenological reduction in which all objective reality and objective causality is "bracketed." What is there for us is not the world taken as reality with its psychophysical[103] beings and its causalities, but only the phenomena of them, thing-phenomena, human-phenomena, etc., in their intentionality. In this framework of pure consciousness, we find the streaming present of consciousness, we find constituted in every case a perceptual reality constituted as in the flesh. But pasts can also enter into present consciousness through remembering. Put more precisely, in the unity of a consciousness that is streaming in the present, we find concrete[104] perceptions with their retentional components, as well as concrete retentions—all of that

---

[102] Editor: See Appendix 11: <The Concept of Associative Causality> pp. 477ff.

[103] *animalischen*

[104] Translator: See below footnote 120.

in the flow of retention fading away into the distant horizon of retention. But in addition to this, emergent rememberings as well. [118] Between the noematic components of something present and something remembered we find a phenomenologically peculiar connective trait that can be expressed in the following way: Something present recalls something past. Likewise, a second remembering can occur while a remembering runs-off; the second remembering can occur along with the first one in a nexus that is characterized noematically by the fact that the first recalled event recalls the second recalled event. A perceptual consciousness, that is, a consciousness that is constituted originarily can accordingly be characterized as a consciousness that awakens, awakening a reproductive consciousness, and this consciousness can function as awakening in its turn as fetching a past of consciousness, as it were.

It follows from the exposition of our previous lectures that association is a possible theme of purely phenomenological research. For instance, there is still something that remains of the Modern, customary psychology of association after the phenomenological reduction; if one goes back to its immediate experiential material, the phenomenological reduction will initially yield a core of phenomenological facts that also remain within the pure inner attitude and that will shape the preliminary point of departure for further research. If we pursue this more deeply, we will realize that the path is cleared from here toward a universal theory of the genesis of a pure subjectivity, and in particular, initially in relation to its lower level of pure passivity. Phenomenological eidetic analyses of consciousness constituting a temporal objectlike formation already led to the beginnings of a lawful regularity of genesis prevailing in subjective life. We see very quickly that the phenomenology of association is, so to speak, a higher continuation of the doctrine of original time-constitution. Through association, the constitutive accomplishment is extended to all levels of apperception. That the specific intentions arise through it is already clear to us. Actually, Kant already saw that in phenomenological contexts, which we first come across in the natural departure from objective-psychological observations under the rubric of association, not mere accidental facts, but rather, an absolutely necessary lawful regularity is manifest without which a subjectivity could not be. But his [119] brilliant doctrine of the transcendental necessity of association is not supported by a phenomenological eidetic analysis. It does not attempt to show what is actually at issue under the rubric of association with respect to elementary facts and essential laws, and thereby making comprehensible the genetic unitary structure of pure subjective life. On the other hand, I do not want to say that the development of phenomenology has already progressed so far that it would have neatly solved the genetic problems existing here. But it is far enough along to be able to specify these problems and to sketch the method for their solution.

A first group of pure phenomena and nexuses to which the traditional doctrine of association leads us concerns facts of actual and possible reproduction, or more clearly, of actual and possible rememberings. When we practice the phenomenological reduction, they are initially given as transcendental facts. Accordingly, this is still prior to eidetic, essential insight that would intuitively obtain essential necessity and essential laws. Right in the midst of these facts are the phenomena that interest us: the splitting of rememberings into rememberings. These rememberings have been muddled, as we say, such that the memorial images of separate pasts have blended to form a unity of an illusory image. This problem of the fusion of rememberings leads us a step further, then, to the problem concerning the extent to which mere phantasies lead back, through intentional analyses, to rememberings, that is, the extent to which they are products of the fusion of rememberings with respect to their intuitive content.

The doctrine of the genesis of reproductions and of their formations is the doctrine of association in the first and more genuine sense. But inseparably connected to this, or rather, grounded upon this is a higher level of association and doctrine of association, namely, a doctrine of the genesis of expectations, and closely related to it, the genesis of apperceptions to which belong the horizons of actual and possible expectations. All in all, it concerns the genesis of the phenomena of expectation, that is, of those specific intentions that are anticipatory. We could also call this association inductive association. For it is a matter here of the [120] founding level proper to passivity, the founding level of all the active-logical processes that are treated under the rubric of inductive proofs.

Taking them in order, let us now first of all observe reproductive association, naturally, as a purely phenomenological occurrence. We can <call to mind> here the Aristotelian distinction between what is in itself first and what is for us first, namely, what is first from the standpoint of explanatory knowledge (knowledge that makes comprehensible). The associative facts and essential insights of the reproductive sphere that are accessible to us as first (for essential reasons and not accidentally) concern reproductively awakening intentional lived-experiences and awakened reproductions in which we are conscious or become conscious of objects as objects for themselves. It does not initially occur to us that this is something special; therefore, wherever we speak of a consciousness, of an intentional lived-experience, we think without further ado of a consciousness of something offering itself as something for itself, a consciousness of something prominent, existing in a singular manner.

But it is precisely the analysis of associative phenomena that draws our attention to the fact that consciousness must not necessarily be a consciousness of a single object for itself, and accordingly, we touch on a new problem here: how a consciousness of something particular and how a consciousness of explicit particulars becomes possible as a consciousness of a multiplicity and a consciousness of wholeness; namely, a comparative analysis also shows the opposing possibility of many [elements], indeed, a multiplicity being continually fused into a unity within one consciousness, *implicite*, such that consciousness is not a consciousness of a multiplicity, a consciousness that becomes aware of separated particulars in a unitary and yet separate manner. The following example can clarify roughly what that "*implicite*" should mean. A white square, which is in itself completely and homogeneously white without any specks and so forth, becomes prominent as a single square, and many like squares as a multiplicity of particular ones. But however much every square is given as a unity, and given as a unity that is undifferentiated in itself, it is indeed our view that each one can be divided in a number of ways; to our mind each one is in itself a continuum of white whose phases are just not prominent for themselves. Naturally, that is not an arbitrary interpretation but [121] one that has a phenomenological basis.

Let us now set aside the problem of prominent particulars and their implicit multiplicities and only point to the fact that for the time being we can only catch sight of associations, and only of direct ones, by having particular objects given to us phenomenologically or by having in consciousness closely consolidated multiplicities forming unities for themselves or by having articulated wholes, in short, if we have unitary, prominent objectlike formations which, as such unities, recall other unities as past ones; naturally, recalling them as past for us. Remaining within the phenomenological reduction, the associative relation concerns exclusively the *given*[105] objects *as such* in their respective noematic mode, that is, correlatively it concerns the corresponding modes of consciousness. We are not saying that we see this "recalling something" everywhere, but only that we see it in certain cases, that is, that we have grasped it in originary prehension. For example, if winding down a path we catch sight of a cirque, we are reminded of another cirque, one that emerges reproductively. We can note that the reproduced one is not only altogether reproduced, and is not only an altogether reproduced, similar object, but that there exists a certain relation between them both beyond the mere relation of similarity. Something present recalls something reproductively presentified, which is to say, there is a tendency that is directed from the former to the latter and a tendency that is fulfilled by intuitive reproduction. It follows from this that we, as attentive egos, look from this to that by being referred from the one to the other; and we can also say: The one points to the other—even though there is still not an actual relation of indication by signs and designation. Further, the phenomenon gives itself as a genesis, with the one term as awakening, the other as awakened. The reproduction of the latter gives itself as aroused through the awakening.

[105] *bewußten*

Within the purely phenomenological framework, then, there are further distinctions falling between immediate and mediate association, and we observe that both are always intertwined with one another. First, we frequently and easily observe that an *a* recalls a *b* and then this again recalls *c*, and thus that we are not immediately reminded of *c*, but rather, precisely on the way through *b*. But associations can also run their course without being [122] noticed. Just as we fail to notice so many different things that are in our field of consciousness, so too, we fail to notice the connections of association. As usual, we also want to become aware here of our past consciousness and its content by looking back retrospectively, that is, in the manner of reflection. Our interest skims right over the terms when a certain and often very mediate final term especially appeals to us, and in this way, this final term dawns on us like a sudden thought for itself; while the entire associative nexus runs it course in consciousness, it is not noticed in any special manner. For instance, the thought of a magnificent seascape occurs to us during a talk. If we reflect upon how it came to us, we will find for instance that a turn of phrase immediately reminded us of something similar that was uttered during a conversation last summer at the sea. The beautiful image of the seascape, however, completely monopolized our interest.

If we think of intuitive examples in this way, we will then find the similarity of something awakened with something that is immediately awakening as proper to immediate association, as proper to immediate awakening.

At any rate, we find that the awakening does not often lead to an intuitive memory, but instead to an empty presentation. This empty presentation is directed in a determinate manner and as such has a determinate (just not always efficacious) tendency to turn into a remembering, and then through immediate association, to turn necessarily into a memory of something that is similar. We notice, however, that this "something similar" does not remain isolated; in a certain way, the entire past-consciousness is co-awakened; it is from this that the particular thing, which is especially awakened and reproduced, becomes prominent. It is then that something else, another particular thing can still be preferred in a special way in the manner of awakening in the framework of this entire past, and thereby be predestined to a possible intuitive reproduction. The memory of a painting by Titian transports me into the Uffuzzi Gallery; but only special traces from that present become especially awakened and appear: certain other paintings hanging there, or more prosaically, the image of the yawning museum attendant, etc.

Certainly, we can quickly skip over this entire past present insofar as the more effective awakening of this past goes back into [123]
another, deeper past, or following the outline of the future, pursues the incidents that followed there.

These are of course familiar occurrences; we can all find them in ourselves and take note of them. That every memory is open to the question concerning how we arrived at it, a question that arises even in everyday discourse already shows that it concerns matters of fact that must have arisen in everyone's experience. For phenomenology, it forms a point of departure for a treatment and description in the reduction, and then above all, for the method of eidetic research, of research into essential necessities. Proceeding from examples we consider, for instance, whether immediate association as such is possible, is conceivable, if we were to give up the relation of similarity between what awakens and what is awakened, or again, whether the awakening of something through association, of something whose content is alien, is possible in a way that is different from the transmission occurring by means of an awakening through similarity.

One sees then that essential laws surely prevail here. Every awakening goes from an impressional present or a present that is already non-intuitively or intuitively reproduced toward another reproduced present. This relationship, or as we can say forthwith, this synthesis presupposes a "bridging term," something similar; from here the bridge arches across as a special synthesis by means of similarity. Transmitted in this way, a present enters into a universal synthesis with another past present, correlatively a full consciousness of the present enters into a universal synthesis with another submerged consciousness of the present, a synthesis which serves as the framework for special syntheses of awakening and for special reproductions.

This gives us a rough and ready beginning; and now it is a question of first making comprehensible in a more precise manner how certain awakenings come about, namely, how a similarity among a variety of similarities becomes privileged to build a bridge, and how each present can ultimately enter into a relation with all pasts, how—extending beyond the living retention—it can enter into a relation with the entire realm of things forgotten. It is obvious that only through this can we completely solve the problem concerning how the pure ego is able to become conscious of the fact that it has behind itself an endless field of past lived-experiences as its own, a unity of past life in the form of time, as a [124] life that is in principle everywhere accessible to it through rememberings, or, what amounts to the same thing, is capable of being reawakened in the core of its being.

But could subjectivity in truth have its own past, could we speak meaningfully of this "having" if in principle every possibility of remembering were lacking, and further, if the genetic conditions of this possibility (the conditions of possible awakening) that are to be disclosed through essential insight remain unfulfilled? We realize, then, that it really concerns nothing else than clarifying the fundamental problem, the basic, essential conditions of the possibility of a subjectivity itself. What must belong to it so that a subjectivity can have the essential sense without which it could not be subjectivity, [namely,] the sense of an existing[106] subjectivity being for itself, and precisely thereby of a subjectivity constituting itself as being for itself? Certainly, a complete phenomenology of reproductive awakening concerns and exhausts this problem only with respect to the one side, namely, with respect to the constitution of one's own past, or rather, the constitution of the self-having-been in endless immanent time. But we will see that the supplementary part, the other half of the problem, is the realm of the phenomenology of inductive, anticipatory association. Here we will make clear the essential conditions of the possibility of a subjectivity that can know itself as identically one, having its inherent endless future life. Can know: through the possibility that is demanded here of a

[106] *seienden*

verification that is not to be grounded initially by the self-giving of something futural in the future current life, but rather through a novel verification [that takes place] constantly in the determination of anticipations that are to be made possible and that are already being prefigured in an indubitable manner in the particular present.

The problems that we have formulated are not entirely new to us; they are only new formulations of the problem of validity-in-itself that had set this entire Division of lectures in motion.

<§27. The Presuppositions of Associative Synthesis. The Syntheses of Original Time-Consciousness>[107] [125]

Let us now turn back to association, and let us consider how this new field is to be taken into a deep mining phenomenological investigation, and naturally into a systematic investigation that begins with what is elementary, which every association presupposes. We do not need to seek the very first beginning. Clearly, what is presupposed is the synthesis that is continually accomplished in original time-consciousness. In the concretely full, streaming living present[108] we have present, past, and future already united in a certain mode of givenness. But this manner in which subjectivity becomes conscious of its past and future life along with its inherent intentional contents is an incomplete one. The aforementioned manner would be meaningless for the ego if there were no awakening, for the retentions are empty and even sink into the undifferentiated retentional background. Our consciousness of the protentional future is especially empty. On the other hand, there would be no progress without this beginning. In the ABCs of the constitution of all objectivity given to consciousness and of subjectivity as existing for itself, here is the "A." It consists, as we might say, in a universal, formal

[107] Editor: See §29 and Appendix 12: <Note on the Fundamental Founding of the Doctrine of Original Time-Consciousness> p. 479, Appendix 13: <Primordial Present and Retentions> p. 479ff., Appendix 14: <The Accomplishment of the Association of Simultaneity> p. 482ff., and Appendix 15: <Unitary Consciousness and Its Correlate: the Identical Objects> pp. 486ff.

[108] *Lebensgegenwart*

framework, in a synthetically constituted form in which all other possible syntheses must participate.

Still many other types of syntheses are transcendental in the special sense, as apodictically necessary for the genesis of a subjectivity (which is indeed only conceivable in genesis). As we said, these syntheses run their course together with the synthesis constituting the temporal form of all objects, and thus must co-relate to the temporal content, the temporally formed content of the object. In his nearly overwhelming genius (overwhelming precisely because he did not have at his disposal the phenomenological problematic and method), Kant has already sketched out an initial system of transcendental syntheses in the transcendental deduction [126]
of the first edition of the *Critique*. But unfortunately, he only had in mind there the higher lying problem of the constitution of a spatio-worldly object[109], of an object that transcends consciousness. Thus, his question is only this: What kinds of syntheses must be carried out subjectively in order for things of nature to be able to appear, and thus a nature in general. But lying deeper and essentially preceding this is the problem of the inner, the purely immanent objectlike formation and the constitution, as it were, of the inner-world, that is, precisely the constitution of the subject's stream of lived-experience as being for itself, as the field of all being proper to it as its very own. Since the spatial world is constituted through consciousness, since it can only be there for us as existing and can only be conceived at all by virtue of certain syntheses carried out in immanence, it is clear that the constitutive problems of the world presuppose the doctrine of the necessary, most general structures and the synthetic shapes of immanence that are possible in general. Hence, we are to seek here in immanence what are in principle the most general syntheses, especially, as we said, the syntheses concerning content that extend beyond the transcendental synthesis of time, and which as such, according to their general character, are discernible as transcendentally necessary. Our task, therefore, will be to seek precisely these syntheses.[110]

---

[109] *Gegenständlichkeit*

[110] Editor: See Appendix 15: <Unitary Consciousness and its Correlate: the Identical Object>.

It is extremely important to note that under the rubric of time-constituting syntheses, we consider not only the syntheses of the temporal object's retention and protention that belong on the whole and for itself to some kind of temporal object, but also that the concrete, full living present is a unity arising from a synthesis encompassing it. And further that in the continuous process of streaming from living-moment to living-moment, a synthesis, particularly a higher level synthesis is carried out once more. What is constituted universally through these syntheses is known under the rubric of coexistence and succession of all immanent objects in relation to one another. In every living-moment various kinds of temporal objectlike formations are constituted, each one has a momentary Now together with the momentary horizon of retentional having-been and the momentary protentional horizon. [127] This momentary structure is continually in the process of streaming and is synthetically united, precisely thereby constituting this temporal object as identical with itself, e.g., a lasting sound. But still other objects can be constituted in the same concretion[111] of life, now and again in the same living-moment and in its continuous process of streaming; each object can be constituted through a parallel constitutive structure, for example, another sound, a color, etc. Simultaneity is necessarily constituted here; the temporality of one such datum would not be constituted for itself and yet have nothing to do with the temporality of another. A single Now is constituted that brings the Now of the one datum into a unity of identity with the Now of the other. And likewise, the entire formal structure of the subjective temporal modes for both data, and thus for all data constituted from the same Now, is in a coinciding that forms identity—the diverse primordial impressions are connected to a primordial impression, inseparably; a primordial impression streams as one such that all special impressions must run off at an absolutely identical pace. This provides for the fact that there are not many times corresponding to many objects; rather, the proposition goes: There is only one time in which all temporal courses of objects run their course.

[111] Translator: See footnote 120.

Accordingly, corresponding to every Now is a universal synthesis. Through this synthesis, a universal concrete present is constituted, a present into which all particulars that are set off from one another are integrated. Further, the fact that the Now streams in and through temporal orientations implies at the same time another universal synthesis in constituting life whereby we are conscious of the presents coursing as a sequential unity. What is given to consciousness originally as existing simultaneously and sequentially is thus constituted from an originally synthetic unity as existing simultaneously and successively.

This is the most general and the most primary synthesis that necessarily connects all particular objects of which we become conscious originally in passivity as being, no matter what their content may be and however else they may be constituted as unitary objects with respect to content. We had directed our attention to what necessarily gives temporal unity to all distinguished and distinguishable objects: being with-one-another.[112] But naturally, the synthesis of time-consciousness also contains (and already as a presupposition for possible coexistences [128] and succession) that synthesis in which one object is constituted as identically one or (what amounts to the same thing) as enduringly one in streaming manifolds.

If, now, time-consciousness is the primordial place of the constitution of the unity of identity or of an objectlike formation, and then of the forms of connection of coexistence and succession of all objectlike formations being given to consciousness, then we are still only talking about that consciousness which produces a general form. Mere form is admittedly an abstraction, and thus from the very beginning the analysis of the intentionality of time-consciousness and its accomplishment is an analysis that works on [the level of] abstractions. It grasps, it is only interested in the necessary temporal form of all singular objects and pluralities of objects, or rather, correlatively it is only interested in the form of manifolds that constitute the temporal object. An object is something enduring, as constituted this way and that in consciousness. But it is something enduring with respect to its

[112] *miteinander*

content, and that this is the case, that it has this content for this object, that content for that object, all this is constituted in consciousness itself, or otherwise we would not even be conscious of it. The extent to which several objects are differentiated with respect to content, or the extent to which an object is divided or divisible with respect to content, is the extent to which we have (or rather, the extent to which there emerge for us) particular objects that are with one another in relationships of coexistence and succession. But what gives unity to the particular object with respect to content, what makes up the differences between each of them with respect to content (and specifically for consciousness and from its own constitutive accomplishment), what makes division possible and the relation between parts in consciousness, and so forth—the analysis of time alone cannot tell us, for it abstracts precisely from content. Thus, it does not give us any idea of the necessary synthetic structures of the streaming present and of the unitary stream of the presents—which in some way concerns the particularity of content.

<§28. Syntheses of Homogeneity in the Unity of a Streaming Present>[113]

Let us remain in the continual synthetic unity of a streaming present; let us initially not draw upon any of the functions of [129] remembering that we should only take into account later regarding their genesis and new accomplishment, and let us make just as little use of the functions of bringing the future to intuition,[114] the expectations that spring ahead beyond continual protention. We even leave out of play all types of phantasy, all types of conceptual acts, valuing and willing activities, without prejudicing the case concerning their indispensability or dispensability for subjectivity. We presuppose that objects are constituted, immanent objects already prominent, singularities or unitarily closed groups

[113] Editor: See Appendix 16: <On the Connection of Similarity> pp. 491ff., Appendix 17: <Sensible Connection of Similarity. Sensible Uniformity and Eidos> pp. 495ff., Appendix 18: <Association and Synthesis> pp. 505ff., and Appendix 19: On the Phenomenology of Association> pp. 512ff.

[114] *Vorveranschaulichung*

or wholes, as articulated in explicit parts. Standing as we are here before the questions concerning association we must obviously forge a beginning with this in order to set our eyes on new primordial syntheses. But we can only set our eyes on something, grasp something directly where we have something prominent for itself.

The most general connections of prominent objects that are determined with respect to content are obviously similarity or uniformity and non-similarity; or let us say more concisely: connections of homogeneity and connections of heterogeneity. Connection as constituted by consciousness is an accomplishment of syntheses of consciousness, and here we encounter something new, consciousness' synthesis of homogeneity. Where real objects are concerned, there are surely good reasons for contending that mere similarity does not create any kind of connection, any real connection between them. That two people are similar to one another, for example, with respect to their noses, does not produce any real bond between them. However, we are speaking about immanent data, for example, about concrete color-data in the unity of a streaming present, thus, that are given to consciousness in immanent coexistence under some longer constituting duration. But these necessarily have a unity through consciousness, a unity of kinship, as similar to one another or uniform with one another: Several discrete color-data in the visual field are grouped together; they are especially united by virtue of their similarity; [and they are united in] different ways as well. This kinship has its degrees and according to them it unites them now more strongly, now more weakly. The most complete kinship or similarity is uniformity, that is, it makes the strongest connection of homogeneity. What is given there as coexistent in one [130]
consciousness with regard to the particulars that are uniform does not remain for itself indifferent to the other ones; rather, it is united into a special group that is now a multiplicity as unity; in the broadest sense of the word it is already a whole that has bound the particulars together purely through kinship. But similarity accomplishes the same thing, only to a lesser degree, remaining behind uniformity.

Let us regard statically the coexistence of two or more objects that are related with respect to content; although their unity through homogeneity already emerges in this case, and although the differences in the intimacy of this connection emerges if we regard a variety of such cases together, the deeper characteristic among the combinations will be disclosed in a kinetic regard, as it were, in a comparative transition from one component of the relation to the other. The new uniform characteristic is given in such a transition as a "repetition" of the same thing. This is not so with something that is merely similar, though in a certain respect a sameness does come to the fore in the relation of similarity. When we compare the components of the relation, we find that a kind of overlapping of one consciousness over the other takes place; in spite of the modification that it undergoes, the one consciousness is preserved through the transition as a consciousness of the same first object and coincides with the second consciousness, with the consciousness of the second object, and in the case of uniformity, are congruent. That is to say, in spite of the consciousness of duality that is maintained, an identity-consciousness results, the consciousness of a same content, a same what-content. The repeated uniform characteristic is completely the same with respect to content, completely congruent. In comparing matters with similar characteristics we find two things that stand out, (i) the synthetic coinciding in a commonality, that is, in a sameness, and yet (ii) the synthetic conflict of particular matters of this commonality that repress one another reciprocally in the process of overlapping. Repressing means that the one conceals the other, that the concealed element tends toward unconcealment, then breaking through conceals the previously unconcealed element, etc. Thus, when a red square overlaps with a blue one. Visually there is similarity, but there is not uniformity. Red and blue clash.

We can also observe this: Where there has been a coinciding through overlapping, the coinciding without conflict (the coinciding of uniformity) does not yield at that moment anything more of duality, of splitting in two with respect to content. A [131] fusion with respect to content is carried out in the synthesis of the two respective consciousnesses, specifically, the fusion forming a singularity of community with regard to content. But while there is also something of fusion here in the overlapping of similarity, there is not a pure fusion and formation of unity; rather, there is unity as presupposition, as the ground of a concealment, and thereby of repression and eruption. Accordingly, we will say: What is present statically as the uniform connection of discrete contents (or as the merely similar connection of discrete contents) is itself already a mode of the two respective syntheses as coinciding. It is coinciding *par distance*. And obviously, an affective tendency simultaneously belongs to it; for something that is prominent for itself functions affectively. And a connection of something prominent stemming from homogeneity exercises a unitary and augmented affective force or tendency upon the ego, whether it gives into it or not; and this affection relaxes, becomes fulfilled in a turning toward, which when completely at work simultaneously produces the kinetic form of a transition from one term to another and the kinetic form of an overlapping, thus producing the results we described. All of this takes place, however, prior to the occurrences of the higher lying activities of cognitively fixing the common element as something concretely general or as a generic generality proper to a higher level.

What we have expounded upon also holds where successions of discrete objects have been constituted in the unity of the streaming living present, and in particular, as successions of uniform or similar objects or processes. Thus, pure sounds follow one another, and among them for instance a series of repetitions of uniform sounds, where we say that it is "the same sound" being repeated.

When[115] we use overlapping as a way of disclosing what is present in the phenomena of uniformity and similarity of discrete data, we see, as the previous lecture showed, that a fusion in distance is at work in every synthesis through homogeneity; in the case of mere similarity this fusion is inhibited by a distinguishing, a contrasting. If we employ the same method of disclosure, we notice further that these types of connections through homogeneity can be connected differently by bridging terms, thus forming different groups of homogeneity that have the single terms in common. For [132] example, a red triangle is in a unity with other differently colored

[115] Editor: Beginning of a new lecture.

triangles, and if we pass through them and compare them explicitly in pairs, this will mean that they are similar to one another as triangles, indeed, they will be completely uniform. The same red triangle, however, can form a uniform group with other figures that are not uniform but that are all red—uniformity with respect to red, difference or (what amounts to the same thing) similarity merely varying with respect to figure. For ultimately, all figures are similar on the one hand, all colors are similar on the other. Thus, formally speaking, we distinguish "the uniform element with respect to *a*" from "the uniform element with respect to *b*," and likewise for everything similar. We notice that already prior to making a comparison, but by virtue of such intertwining syntheses of homogeneity, red distinguishes itself from figure, that the two respective moments in the red triangle come into relief as moments and therefore exercise an affective force for itself.

Let us look more closely at this situation by simplifying our examples. Within a streaming present, the distant affinity of the red triangle with another red triangle grounds fusion, a kindred unification forming a pair. The fusion or unification of the same triangle with another triangle yields a pair again: Both pairs are now linked by a binding term, namely, by a term of identity. We realize in this case that each respective pair has also moved into a relation of homogeneity. And in fact, speaking in an entirely general manner: Even groups of similarity, even homogeneous pairs as groups become united purely through homogeneity. Two coexistent pairs that are paired by being red coincide as pairs; they form a pair of pairs such that at the same time the corresponding terms coincide in a red-congruity. If, instead, we take pairs that have one term that is not uniform, more precisely, pairs that have one term that is similar in distance, there will be a less integral unity of pairs produced between the pairs as such. They are not completely unified as pairs; they do in fact form a pair of pairs, and the terms that correspond to one another by being paired as red for themselves are, for example, more intimately connected with each other; but because there is a difference with respect to the other terms, both pairs diverge as if each one were going its own way. Now, if we allow the one term to become identical as before, we will then have two relationships diverging from it that [133] will nevertheless still have something of a materially relevant community, but in relief. Precisely by this, moments come into relief through the bridging term, namely, as different moments, as moments grounding the two similarities. Obviously, here lies the beginning of the internal particularization and division within passivity, and thus the presupposition for explicating the internal features and parts within activity, and ultimately within judicative activity.

New relationships, i.e., forms of unity also come to the fore here in the framework of homogeneity—the relationships of the object and of the inner, dependent feature, and of the object as a whole and as a part in the special sense of a part that is possible for itself. Certainly, they only come to the fore in the strict sense through active explication and a corresponding identification; but it becomes clear that the syntheses in question are being prepared in passivity already in the lower level. An example of the process of being divided into parts within passivity would be, for instance, a unified succession of a short tone and a long lasting tone. Here a sustained piece of the longer lasting tone, which does not really come into relief that sharply, is already distinguished from the rest through coinciding.

Other special syntheses arising from homogeneity that have just as much fundamental importance concern the gradations of similarity leading up to uniformity, according to which the pairings and connections of groups are themselves fused more or less intimately, more tightly or more loosely. Namely, the stronger similarity among terms also determines the intimacy with which the pairs themselves are fused together into a unity, into the unity of a group and into the unity of a group of groups.

<§29. Primordial Forms of Order. Supplementing the Previous: The Phenomenon of Contrast>[116]

Here, however, we encounter the problem of the unity of order and other fundamental problems in connection with this, the

[116] Editor: See Appendix 19: <On the Phenomenology of Association>.

problem of continuity as continuous order in undivided fusion. What are the primordial phenomena here? [134]

Up until now we had presupposed, as primordial phenomena, the unity of a prominent object, the multiplicities of prominent objects as the mere multiplicities of coexistence and succession, but then also multiplicities as connected multiplicities with respect to content, that is, connected through homogeneity. Now, what characterizes the order that is obviously a phenomenon in the sphere of homogeneity, and the disorder opposing it? One primordial phenomenon is the disorder, for instance, in the form of a bunch of spots in a visual field that is otherwise uniform. As uniform spots, they can also be thought of as an order in the visual field; we may be able to give them an order *idealiter*, or we may be able to parcel them out in any ordered manner, so that we would see order and then would have order given as a phenomenon of passivity. Thus, even this is initially a primordial phenomenon.

Primordial phenomena especially include the phenomenon of uniformity and the phenomenon of a series of gradation: [in the first instance] a unity of similarities, of sheer color-data, namely, in coexistence, but [in the second] a graduating from red to blue as the unity of an ordered sequence. Or a series of triangles as an order of size consistently getting larger. We can take as a parallel primordial phenomenon, however, a temporal ordering of just such data that follow one another and are unified according to an order through similarity, through uniformity, or through a gradation according to color or according to the size of the shape. On both sides we can describe the concrete unity of order as already being at hand: Proper to it is not merely a general formation of a pair through similarity, but a special similarity that we call here gradation. Then the connection of pairs with one another such that the graduated term, the final term of a pair, is the beginning term of the next pair, the term as a point of departure for a new gradation, etc. Let us note that coinciding in a gradation still has a special intimacy, even though it is not a uniformity. The lesser one is not simply repeated in that which is graduated, and yet it is the same, but still more than that. Obviously, a special binding force in the ascending concatenation, however, is inherent in the increase, in the gradation; for every pair of data that is relatively graduated is itself again an increased gradation in contrast to a previous pair of data that is relatively decreased. Just as a single graduated pair more strongly exercises an affection than just any similar pair, so too is a concatenation of graduated pairs, in the unity of a gradation that is being augmented in the [135] repetition, a stronger affective unity than a different unity of similar pairs.

Indeed, now the question arises: What brings about the concatenation, the process of joining together? How do concatenations (as opposed to mere collections) come about in mere passivity, as opposed to mere collections in which everything forms a bond with everything else, singularities and connections, in a disordered manner?

We are led here above all to the primordial phenomenon of succession, that is, to a primordially ordering accomplished by time-constituting consciousness. What is already prominently accentuated in it going from one present to the next and is constituted in the train of presents as a lasting unity of identity is linked together temporally; that is to say, the temporal relations are from the very beginning and by essential necessity linked together in being constituted. If we pay attention purely to the formal aspect of time, we can even discern here relationships of gradation that are built up as the most original relationship within necessary concatenation. All temporal relations among temporal objectlike formations are constituted in subjective modes of orientation, and thus there is, in consciousness, the concatenation of past element with what is more past, then with what is still more past, but whereby a unity of identification runs clean through the flux of this relativity, a unity that constantly produces a concatenation of order between the identical, successive data themselves. Now, if this primordial order has produced a necessary concatenation between data, however they are determined with respect to content, then the gradations determined with respect to content can now link up as such in a temporal sequence and thus form unitary series of consistent gradation. Likewise unitary series can also arise with pure uniformity and similarity with respect to content. We should recognize that the source of succession must give the series a higher force than they would possess if they were not to operate from this source. For succession introduces a new moment, to be sure, a necessary moment of uniformity; it introduces the uniformity of the sequence as sequence, and if they are enduring uniformly, then possibly in addition to this, it introduces the duration of single data of sequence as a uniform duration.

But what happens now in the coexistence that is still not ordered in itself by the order of succession? Let us consider the visual field and the special data that occur in it. They have a unity [136]
of homogeneity, but that is not yet order. Nevertheless, there are orders here, and as in time, independent from the content that is ordered there. In the visual field, taken in a purely immanent manner, there are possible series that, admittedly, we can characterize in language only by borrowing from the language of the perceptions of [transcendent] objects. Thus, an order of coexistence of random specks of color or sharply delimited figures in the order of right-left and in the order of above-below, or in a certain path going right and then up, etc. [This holds] likewise in any other possible direction of being ordered.

We realize that within the visual field there are multifarious orders that are related to one another, relations that are not clearly understood prior to analysis. We see that there is also a field of ordered loci here that prefigure concatenation in coexistence just like there is an original field of order with varying temporal loci that prefigures the field of succession. Even here form and content are distinguished in a new way: [a] the form of positional[117] order, and [b] as content, the thing ordered (as we will see) what stands at this or that visual locality[118] occupying it.

But surely there is a great difference here: Succession is a singular, a "linear" concatenation that is always uniform. In the visual field, however, we do not always have all data in a concatenation, in an identical linear order; rather, various series of concatenation can be formed, and these various series of concatenation can be formed simultaneously, originally, such that

[117] *örtliche*
[118] *Lokalität*

many lines are contained within the field as systems of local positions; they appear filled out, now with this content of the object, and another time with a different content of the object, making order possible for it in advance, and all of these linear local systems go together to form this one field-form, just as a detailed analysis teaches us. Nevertheless, similar to the form of succession, we have a prefigured form of order and in it, now these and now those contents as ordered, forming a unity.

But a problem immediately arises here: Insofar as every order necessarily becomes an allure to progress through local laws of order if the order comes into relief as a unity through a special affinity with respect to content, do not the local lines of order that are distinguished in experience bring along their layers of temporal apperception? But progressing through it also [137]
immediately fashions a temporal order which, however, can also be freely carried out in an inverse order by identifying the ordered data.

What holds for the visual field holds for the tactile field—for all authentic fields as such that are unities of locality. Not for the auditory field; it is not an authentic field. For here we lack every possibility of ordering the coexistent.

Let us now pass over at once to new primordial phenomena, still keeping with phenomena of unique syntheses. We had considered similarity, uniformity, and gradation, connection through uniformity and connection through gradation but we had only done this in a rough manner. We also encountered sequences and naturally the intertwining of sequences as well which form an encompassing sequence, and likewise the intertwining of orders of coexistence. All of that for *prominent* data.

But if we consider one such datum, then features can become prominent within the datum itself, and among these features, the feature of inner evenness, like an evenly colored white square, and again the feature of an inner non-evenness, of an inner self-dissociation of indeterminate specks, which pass over into one another and flow into one another in an undifferentiated manner. Or also, and especially striking, a continuous process of gradation and thereby flowing, continually in the process of gradation, for instance, in the sense of something becoming-redder-and-redder, or in the sense of a colorful ribbon whose colors lead into one another as in a prism. This problem immediately comes to the fore, namely, of relating these and similar occurrences of inner continuity (each one of which initially makes up an undifferentiated global character of a content) back to the phenomena of discrete matters that are prominent.

How far have we come in our previous lecture?[119] Let's take a look back for a moment. In order to sketch a natural way toward a phenomenology of genesis and, in passivity, toward a phenomenology of association, we considered the structure of a living immanent present which is itself indeed the most universal genetic phenomenon. We find in every such present essentially a hyletic core; a unified multiplicity of sensible data (visual data, [138] sound data, etc.)—unified in the most loose manner—is essentially and constantly constituted in simultaneity and living succession. In this respect, we take impressional consciousness to reach as far as the retention that is still living reaches. Multiple data that are prominent for themselves must emerge in wakeful conscious life; we have now described what gives unity to such sensibly impressional data in the living present, to all data as a whole, and to singular groups, and to entire sense-fields for themselves. We found special unifications arising from homogeneity within the all-encompassing forms of coexistence and succession. Everything in a present that is prominent and at the same time homogeneous is connected. Accordingly, every sense-field is a unitary field for itself: Everything visual is connected through visual homogeneity, everything tactile through tactile homogeneity, everything acoustic through acoustic homogeneity, etc. We speak of unitary sense-fields in the broadest sense. They are heterogeneous, and accordingly are only united by the temporality of the living present. In addition to this formal unicity, every field has in itself a materially relevant unanimity, a unanimity with respect to content, which is precisely the unanimity of homogeneity with respect to content.

If we consider a field for itself and disregard the general homogeneity, we will have special connections arising from

[119] Editor: The beginning of a new lecture.

homogeneity, especially multiplicities that become prominent in a unity, like a group of red figures and a group of blue figures. The greater the "similarity," the homogeneity, the narrower is such a unity of groups, the more it is a unity of prominence.

I still have to add an important supplement here, for I had unfortunately forgotten the other day to discuss beforehand the phenomenon of contrast as a primordial phenomenon. Everything that is in relief for itself within a field comes into relief from something precisely in this field. This yields a specific, fundamental concept of differentiation. We could also contrapose concretion and discretion, whereby we now understand concretion in a more or less literal sense.[120] Uniform things and things that are very similar grow together as it were. Here, things that are very similar are so similar that they do not break apart in the separating conflict that creates distance when coinciding occurs, but rather blend, even if not purely.

We still have another especially useful expression for the relationship involving the unity of a prominent datum and that datum from which it is set in relief: contrast; but it is not used only [139]
for extreme cases, like the contrast between a loud bang and a soft background noise or tonal background. Both concrete fusion and contrast belong to a homogeneous multiplicity of groups: Each term of the multiplicity is a term for itself through contrast, but they are not opposed to each other; indeed, they are especially united with one another by a fusion without contrast, for example, red specks on a white ground. Alternately, inherent in every contrast that remains a phenomenon of homogeneity, there is

---

[120] Translator: The German term "*Konkretion*," like the English term, "*concretion*" comes from the Latin "*concretus*," meaning, growing together. It derives its meaning from the roots, *crescere* "to grow" and *con* "with or together." In this case, what is abstract is not what is less specific or less particular, but what is not joined together, not connected. In fact, Husserl uses the German term, "*Konkreszenz*," in place of "*Konkretion*" on occasion and in order to emphasize this "more or less literal sense" of *Konkretion*. Although I will render *Konkretion* as "concretion," and will use the English term "concrete" for *konkret*, one should not forget its resonance with its root sense as "concrescence" and "concrescent." To maintain this connection, moreover, I translate the German term "*Diskretion*" as "discretion" and its cognates as "discrete." Likewise, these terms have the sense of not being joined together, not being connected, and hence are to be understood as "discrescent."

something of fusion, there is something that unites the concrete data homogeneously and at the same time disturbs concretion by rupturing its continuity. To be sure, in order to understand this we already need the more detailed expositions of the previous lecture. In any case, any deeper clarification presupposes a clarification of primordial phenomena: prominence under contrast, on the other hand, fusion, concretion of prominent data in similarity without distance. Since data that are prominent for themselves become united in a discontinuous fashion, fusion here is a fusion at-a-distance; we will soon get to know fusion at-close-proximity.

We then turned to the forms of order of prominent data and of multiplicity of data, and we distinguished the universal order of succession from the special forms of coexistence proper to particular sense-fields, if not to all fields. The universal form of coexistence that is accomplished by time-constitution itself is not a form of order. For that we need special forms such as visual locality or the local order of tactile data proper to the tactile field.

The universality of successive forms of order means that all coexistences together form a single order of succession in every living, streaming field of present. If we take any sense-field, that is, a field of coexistent homogeneity, then its stock of data will be a stock of concretely existing data: concrete, not only with respect to momentary coexistence, which cannot be anything for itself, but rather also with respect to succession. Something constituted as an existing datum and as prominent for itself is constituted as enduring, possibly beginning now, lasting awhile, and ceasing. Before any closer analysis, we discern here the distinction pertaining to primordial phenomena, the distinction between the longer or shorter duration of such a temporal datum that is [140]
concretely self-contained, just like the phenomenon of something being in a shorter or longer "state" of duration, enduring, becoming. Likewise within becoming as a primordial phenomenon, we can distinguish something that has already become in the shape of invariability from something that has become in the shape of variability. Everything that has already become, in one mode or another, has its firm place of order in the universal form of succession. This is a form of unity, namely, a form of order into whose movement everything that is still in the

process of becoming is simultaneously integrated insofar as it has become. Distinguished from the concrete groups that have formed a special unity in each coexistence are the groups as concretions of succession that are being formed in a special intertwining in succession, (e.g., a configuration of black colored specks in the visual field, on the other hand, a sequence of light signals or a sequence of sounds). Both groups as concretions will simultaneously prevail here because this or that particular datum within one field will begin anew and something else will cease, thus belonging within the living past and to its materially relevant fusions. In both of these inseparable relations, concretion is only possible as fusion in the form of order, that is, as the fusion of something that is ordered temporally.

We already know (and a closer analysis will immediately show), that this order of all discrete matters is in itself a continual one which, in its continuity, runs through everything that is prominent for itself as a unity. In other words, every prominent datum is not merely juxtaposed with other data in the living relations of succession. Rather, it has in itself an inner synthetic structure and in particular is in itself a continuity of sequence. This inner continuity is the foundation of a continual fusion with respect to content, fusion at-close-proximity. The duration of a content, of a color in the visual field, of a sound in the acoustic field—whether they are data giving themselves now as unvaried or as variable—does not have the character of an unanalysable quality; rather, as a phenomenological analysis immediately shows, it has the feature of continuing, of stretching from phase to phase; in what has become, it is the finished temporal extension, expansion. In the living process of temporal ex-tension, the contents that are extended in it normally fluctuate; delimited partial domains come into relief more or less sharply; but [141] whatever becomes prominent in it is necessarily seen in the form of an internal succession, conveyed by a flowing that is unexplicated and undifferentiated.

Now, we easily see that in this inner continuity of the temporal extension of a temporally extended content, the material content is not there in a juxtaposed manner, but rather, that the materially relevant unity is only conceivable as continually ordered, as temporally extended. A concrete unity, the unity of an immanent datum, is only conceivable as the continuity of content in and by virtue of the continuity of an extension as the continuity of a duration. I said, "in and by virtue of." For it also becomes clear that what provides the material content precisely with a materially relevant, inner continuity and thus unity, is in the first place rooted in the most original continuity of temporal extension. All continuity with respect to content (e.g., the continuity with respect to the content of a sound from a violin) is the unity of a continual fusion passing from phase to phase; but the content can only meld together continually in the continual process of becoming in the order of time. Even what we call a continuous graduation according to quality or intensity in coexistence, e.g., in the visual field, is only imaginable as a steady continuity of similarity in temporal continuity. Only in temporal continuity is there an intimately inherent unity in which two things that are similar to one another can become a "real" datum, a datum existing for itself.

But certainly, other essential laws hold here. We must not think that temporal continuity can be filled with just any content: We cannot mix color data together with sound data to form a unity of an immanent temporal datum; homogeneity in continuity belongs to a self-contained datum. But we are bound even further. A certain materially relevant continuity must be held with respect to content. The unity of the tone is dissolvable *idealiter* into a tonal phases. These phases have unity through successive fusion in accordance with temporal continuity; this fusion can only be made possible as unitary in the flux of continual temporal becoming if the data meld together in a materially relevant manner without distance, continuously from phase to phase. Thus there can be no break at any point, that is, a distance in content occurring abruptly. To be sure, there are various possibilities here. Continuity need not prevail in all moments with respect to their content. For [142] example, if the continuity in the tonal quality is maintained, e.g., as a continuous uniformity with respect to the quality *c*, then a break in intensity can take place, a sudden change from loud to soft. Every such contrast makes divisions, is a breaking up into sections. In addition, every section is then in itself a unity. But a sound cannot consist of sheer discontinuities; it can only be one sound with particular discontinuities. And actually then it is already a multiplicity. On the other hand, however, it is indeed a unity for itself if for its part, in its being divided, it sets itself off in prominence from something else by means of an encompassing contrast, for instance, from a general background of sound.

The unity of the impressional present can offer diverse successive unities and multiplicities as unities of continuity that, in the unity of a continual, temporal fusion under contrast, have turned into concrete special unities within passivity in the manner we have described. Thus, there are a variety of successive data that are respectively real bonds.

<§30. Individuation in Succession and Coexistence>[121]

But the successive form of the field of the present is a unique one. Something that is concretely simultaneous, and likewise, something that is simultaneous as an abstract phase has its identical temporal duration, or rather, its identical temporal locus. Fundamental here is the distinction between temporal uniformity and temporal identity. For example, several successive sounds can all have a uniform length of duration. But the uniform element in this duration is the uniformity of the temporal shape, but it is not the identity of the temporal expanse, the expanse of an order of temporal loci.

We are standing here before the problem of individuation, indeed, before the problem of the most original individuation within immanent time-consciousness. Every immanent temporal object has its temporal shape, that is, its temporal length: if, in a special sense, it is a "singular" and not a multiplicity. If it is a "multiple" object like a melody, then not only does the duration of the melody as a whole, as its temporal length, belong to the temporal shape, but also the special ways of filling out and not filling out this temporal length, i.e., not only the temporal lengths of the single sounds, but the pauses, too. In the repetition of the melody, this temporal shape is completely uniform. However, an [143]

---

[121] Editor: See Appendix 20: Time as the Form of Individuality and Subjective Transformation, pp 519f.

individually new temporal object is thereby constituted—new, but completely uniform. The temporal shape (like every quality that makes the temporal shape concrete) belongs to the temporal object as its "essential" character. We distinguish from the temporal shape the place of this temporal shape in time, or better, this temporal shape itself in its individuality through which it is an element of universal time itself, as a local system built up out of single temporal loci. In the same measure that we repeat a sound, say, in full uniformity, we will have a full uniformity not only with respect to content, but also with respect to duration. Every point of time here corresponds to a point of time there. But the points of time themselves as temporal loci in the unique temporal order are irreplaceably different; they are in principle incapable of being repeated. All the objects take part in this unrepeatability. Each one is itself and singularly unique as the object of its system of temporal loci that belongs to the universal system of temporal loci, to the system of singularly unique time. But the temporal shape and the fullness of the temporal shape are repeatable and yield, through comparison and through induction, the general essence that is individuated here.

We need to carry out similar expositions with respect to locality. Locality is what orders the coexistences peculiar to a homogeneous region (of course, not necessarily each one of them), and it is locality that can individuate something uniform and simultaneous within a homogeneous region. Just like every successive unity (formed from fusion and prominence) has its temporal shape and individual location, so too does every unity of coexistence of a local field belonging here have a field-shape that is potentially repeatable in uniformity within the same field, and it has its process of individuation through the individual location that is unique and cannot be repeated. The elementary analysis as the analysis of phases of a local field, as the analysis of phases of a continuity of coexistence, leads again to the fact that it (e.g., like the visual field) has the form of a continually ordered local system analogous to time. Only it is a two-dimensional local system that in itself accordingly allows more diverse and more intricate local shapes. Repeated here is the set of laws (and everything else that belongs to it) that the qualitative filling-out of local continuity must continually obey for a leap in quality not to be able to take place at each point of the phase of the local order. Only that the unity running along a local continuity is not a unity of continual [144] becoming, which would be precisely succession.

We must note well that the temporal and local locus as that which individuates a concrete temporal objectlike formation in succession and coexistence is not something like a distinguishing trait or even something like a hallmark, an index whereby every individual thing is distinguished from every possible thing like it. The traditional discourse of individual difference in contrast to specific differences might easily suggest this. The latter refer to something specific, to something that is generally uniform with respect to content according to temporal shape or size, and according to qualification; something specific, something that is generally uniform as such, is to be grasped generally and conceptually by induction as a concrete essence, divided according to *genus* and *species*.

In any case, something that is originally constituted in consciousness as an object, that is, such that the object is grasped as it is itself *originaliter* in consciousness, is constituted in essential necessity within original time-consciousness as continually identical and enduringly identifiable—and therefore also as identifiable beyond the sphere of the living present by means of the concatenation of remembering. The ever new continual, synthetic identification that has a varying content is necessarily maintained in the continual constitution of immanence, and then within the constitution of immanence, the constitution of transcendence in an unmixed manner; it thereby necessarily forms the system of successively ordered coexistences. The uniqueness of the temporal locus is accordingly nothing other than the correlate of the form of the respective process of identification that is carried out in its unique nexus, that of constituting life in which this object is constituted as this object and is only identifiable as this object precisely because consciousness can go back to its original process of constituting, can reawaken it, and can find it as this same process of constituting, as this object that is recognizable at any time.

On the other hand, insofar as the constituted object and the reawakened object, and many other objects besides that have been respectively constituted, exist in relations of homogeneity and are connected according to uniformity and similarity in syntheses of homogeneity while they have become prominent through contrast—insofar as this is the case, interconnections of comparability will then arise, common and distinct features become prominent, and thus, the possibility of logically conceptual predication. Accordingly, every object necessarily not only has its *toion*, its comparable or its specific essence, whereby [145] it can become linked with other objects generally and conceptually according to genus and species; as the presupposition of all such syntheses of homogeneity and comparisons grounded in them, it also already has its individuality, its thisness constituted beforehand. That is, it is the same, constantly recognizable, and as such it belongs to the determinate nexus of originally constituting life. Even this life has its unity of identity in each phase, and is constituted in original time-consciousness. No matter how much this may look like it leads to a precarious *regressus in infinitum*, I believe that a reflective analysis can overcome this difficulty perfectly well.

With this crisp differentiation (in which modes of constituting the object as an identical object are necessarily intertwined with the constitution of the temporal form of order and with the formation of syntheses of homogeneity made possible by the constitution of temporal forms of order), it is clear why identity, on the one hand, and (predicatively) uniformity and diversity, [on the other,] are so closely related and yet remain distinct in principle. Correlatively, syntheses of coinciding are distinguished as syntheses of coinciding that form identities and syntheses of coinciding [that form] non-identities.

<§31. Problems of a Phenomenology of Sense-Fields>[122]

From here we could proceed further toward a systematic phenomenology of the universal temporal field and of sense-fields, toward a typification of occurrences grounded in the essence of these fields, a typification that is carried out from the perspective of eidetic analyses. For the fields as systems of order, and with respect to the mere form of these fields, at issue is the formation of essential concepts and axioms that lead to a grounding of a geometry and a topology of these fields: figure, line, point, distance, segment, direction, size, straight line, etc. On the other hand, at issue is the typicality of possible qualified structures, and temporally, the typicality of the forms of becoming peculiar to variability and invariability, to overlapping, to rivalry (rivalry of [146]
the visual fields), to covering-over, etc.

Now would be the place to allude to problems in the phenomenology of sense-fields that have localization: [1] the possible types of the disintegration of the unity of a sense-field or [2] the possible types of concrete occurrences. [a] Occurrences under the rubric of concrete independent contents that could subsist for themselves, which is another way of saying that the sense-field can be conceived as being contracted to those contents. [b] But also occurrences of the typicality of possible change that would be a change of shape and a change in quality corresponding to the distinction of shape and coloring of the shape as its qualification. Where changes in shape are concerned there are distinct types of an expanding extension whereby the shape continuously expands to encompass the new pieces; these elements now become prominent or they do not (in accordance with their qualifications), or they become diminished. There is especially a contraction of shape that produces both the intuitive *limes* forms, line and point, from the typicality of a plane, whereby every line for its part also admits of a kind of contraction transforming it into a point. The point is a punctual plane if it approaches the type, pure point as a *limes*, while still admitting of

[122] Editor: With respect to §§ 31-36 see Appendix 19: <On a Phenomenology of Association> pp. 512ff.

a kind of contraction. Likewise the line that is a plane which has a linear form. The *limes*-shapes still have a sense if we remain within sensibility and do not import any notions of which we are not in possession, like a mathematical *limes* formation. These *limes*-forms within sensibility function like those of quality, for instance, pure red, pure white, which are also not mathematical *limes*. We evidently arrive at the same *limes* in another way, namely, by division for which we let one part (in a sharp contrast) change from one qualification into another. A border, indeed, a line results from the leap in quality that occurs by dividing one plane into two planes, for instance, when both parts having been evenly colored, a color completely uniform in itself, are set off in relief from one another through sharply contrasting colors. A "gradual" transition through a gradually changing qualitative graduation yields a border only if the transition first goes very slowly and then proceeds very quickly, and then very slowly again. Separate bands with a velocity have a linear form, and [147]
should they become a pure line, we would have to modify the velocity so that through a sudden increase in velocity, it would finally spring over into a line.

Certain formally similar concepts need to be formed for the continuous temporal order and for each local order as well, especially for the visual one. Thus, the concept of point, of segment with segment-length and segment-orientation, the concept of series as "straight" series. With respect to its form, time is a mono-dimensional continual "straight" series ("homogeneous"); likewise, the visual field with respect to form is a two dimensional manifold that is to be grasped as a continuous double series (a series of series).

In abstraction from any qualification (which is to be thought of now as freely variable), two points within the visual field form a punctual pair that is connected by means of fusion. Every such unity of a pair can be uniform and non-uniform from two different perspectives: according to size (the size of the distance) and orientation. The unity of a pair regarded according to orientation forms a connection that is not uniform whereby two correlative relations are determined: $a>b$ and $b<a$, with the axiom: if $a>b$, etc. A manifold of points is prefigured through two points, a manifold whose entire distances lie "in the same direction." All the points of this point-manifold form a mono-dimensional, continuous manifold that is called a straight line. Time is a straight line. Departing from each point there are many straight lines in the visual manifold; every new, second point that has been added to the first determines such a new manifold.

Accordingly, the following axioms are valid: If a is in direct vicinity to b, and b is in direct vicinity to c, then a is in the direct vicinity to c. With respect to orientations: If $a>b>c$ then $a>c$. And every segment has a length, segments are comparable according to length (size). For each segment there is the same segment, etc. in each direction. All of these are axioms that are exactly expressed for the *limes* and which hold approximately for the approximations to the *limes*.

Let us bear in mind here that the fields are themselves limited and that the expansion *idealiter* of the fields into infinity is a kind of idealization that one can entertain, but which, however, does not imply any necessity. But however the *limes* may arise here, it [148] would not make any sense to infer that straight lines meet (somewhere in the far distance). Of course, the homogeneity of a field is an idealization. For sharply defined lines, and even lines that approach being sharply defined, do not run through the actual visual field everywhere in the same way, etc. Still, constructing an idealization does sketch the field of possibilities.

Problems of a phenomenology of sense-fields (to be carried out systematically): A theme that is missing includes the overlappings that arise spontaneously in passivity under fusion through affinity, or under conflict. Affinity especially is to be defined as unifying prominent elements: undifferentiated fusion as a counterpart to contrast. For example, an overlapping occurs when an "image" is suddenly superseded by another at some point in the visual field, or when maintaining the shape, the coloring suddenly changes over into another.

An important phenomenon that belongs here is the rivalry of the visual fields and potentially also the phenomenon of the concealment of a visual content through another, like when one eclipses another. Objective apperceptions should surely not be included here, and bracketing them takes some effort.

<Chapter 2: THE PHENOMENON OF AFFECTION>

<§32. Affection as Effecting an Allure on the Ego. Contrast as its Fundamental Condition>

We must not get embroiled in a phenomenology of the sense-fields here. It must suffice to have pointed to it. There is still another direction of research bringing us much closer to questions of association, a direction characterized under the familiar rubric of affection, and even though we are familiar with it, we especially want to focus on it now. By affection we understand the allure given to consciousness,[123] the peculiar pull that an object given to consciousness exercises on the ego; it is a pull that is relaxed when the ego turns toward it attentively, and progresses from here, striving toward self-giving intuition, disclosing more and more of [149] the self of the object, thus, striving toward an acquisition of knowledge, toward a more precise view of the object.

Consciousness constitutes partly explicit objects, that is, prominent and actually affecting, and partly implicit (parts and moments) that have not come or have not yet come into relief, but which, however, insofar as they can be brought into relief under "favorable conditions," are still taken into account under the standpoint of affection. In this respect we must distinguish between the actual affection and the tendency toward affection, the potentiality of affection that is not empty, but that is rooted materially in essential conditions. Sensible data (and thus data in general) send, as it were, affective rays of force toward the ego pole, but in their weakness do not reach the ego pole, they do not actually become for it an allure that awakens.

Affection presupposes prominence above all else, and within the impressional sphere it presupposes that prominence which we could find in it alone, namely, when as before we left out of play every consideration of the distant horizon of forgetfulness and likewise, of course, the realm of rememberings. Thus, prominence was for us a prominence that takes place through fusion under

[123] *bewußtseinsmäßiger Reiz*

contrast with respect to content. In a certain way, affection is now a function of contrast, although not of contrast alone. Insofar as the most original affection is to be seen as the affection generated in the impressional present, contrast is then to be characterized as the most original condition of affection. Connected to the gradation of contrast is a gradation of affection, but also already a tendency toward affection. The same contrast can, for instance, actually exercise an allure on the ego, another time it can be that the ego will not be reached by the affective tendency. Extremes of contrast are so strong, make such a forcefully efficacious prominence that they drown out, so to speak, all competing contrasts. Thus, there is something like a possible competition and a kind of concealment of active tendencies by especially strong ones. For example, particular colored figures becoming quite prominent affect us; affecting us at the same time are noises like the sound of a passing car, the notes of a song, prominent odors, etc. All of this takes place at the same time, and insofar as we turn [150] to it alone, listening to it, the song wins out. But the rest still exercises an allure. But when a violent blast breaks in, like the blast of an explosion, it drowns out not only the affective particularities of the acoustic field, but also the particularities of all other fields. What otherwise spoke to us, no matter how little we paid attention to it, can no longer make it through to us.

If we rule out such extremes, then diverse objects can be prominent for themselves through contrasts insofar as each one fulfills the essential conditions of unity from a specific affinity of concrescing fusion and contrast. But for this not all of them need to become prominent for the ego, to say nothing of actually forming groups, configurations, which as such are actually there for the ego—that is, which as such syntheses actually affect the ego.

What gives a single prominent datum the priority of affection? Yet in its interconnection, the single datum is dependent upon the others for its affective force, as these are dependent upon it. We stand in a relativism of affective tendencies, and the question is, what kind of laws and ultimately essential laws can prevail here?

In our considerations of the lowest genetic level we formulate the problem in an abstract manner that is necessary for a systematic [phenomenology of] genesis: We do so as if the world of the ego were only the impressional present and as if transcending apperceptions arising from further reaching subjective lawful regularities did not play any role at all, as if there were no modes of knowledge acquired in the life of the world, aesthetic and practical interests, values, and the like. Thus, we consider functions of affectivity that are founded purely in the impressional sphere. Accordingly, we may only take from the sphere of the heart[124] some feelings[125] that are co-original with the sensible data, and say: On the one hand, the emergent affection is functionally co-dependent upon the relative size of the contrast, on the other hand, also upon privileged sensible feelings like a passionate desire founded by a prominence in its unity. We may even allow originally instinctive, drive related preferences. Naturally, these would be unique themes of investigation whereby a suitable experimentation (not an inductive-objectively oriented one) would be quite possible: It would have the task of producing [151] favorable conditions for the production of pure cases of the kind in question.

<§33. Laws of the Propagation of Affection>[126]

But let us pay more attention to another direction. Suffice it to say that in the relativism of affective tendencies something, at some point, has necessarily become affective as such. Now, are there not laws concerning the propagation of this first affection? Where the object is concerned, we can also characterize affection as the awakening of an intention directed toward it [i.e., the object]. Put differently, are there laws of propagation of intentional awakening? The most privileged case here is where affection results in attentiveness, grasping, the acquisition of knowledge, explication. Then this lawful regularity would of itself pass over into the lawful regularity of awakening or again would lead the attentiveness further, or which is to say, would lead

[124] *Gemütssphäre*

[125] *Gefühle*

[126] Editor: See Appendix 21: Sensible, Multi-Radiating Affection. Sensible Group—Genuine Collective Objectlike Formation, pp. 520ff.

thematic interest further, and would even possibly lead the graspings and acquisitions of knowledge further. Naturally, introducing the language of awakening already alludes to the fact that we are dealing here with something that is so closely related to associations in the common sense that we could already speak of associations in a broader sense, of primordial associations, where there is not yet any question of reproduction. Now, there is a law here saying that every original awakening in propagation, that is, in the associative transference of awakening to new data, is bound through homogeneity. Original association is carried out in our sphere of hyletic sensibility exclusively within each one of the sense-fields for itself. In other words, every sense-field forms for itself a unique, self-contained realm of affective tendencies, capable of forming organizing unities by means of association.

How does the lawful functionality of this formation of unity look now? We spoke of the structure of the entire impressional field and, running parallel to this within the form of time, we spoke of its division into the impressional special-fields, the visual field, etc. As a whole, and with respect to its special prominent characteristics and special groups, each impressional special-field is a unity of temporal becoming, coming into being and passing [152]
away, whereby relatively enduring coexistences are formed as groups, into which possibly new terms enter, excluding others. On the other hand, in this process of becoming, integrally cohesive unities for themselves, unities of temporal sequences that are homogeneous and connected with respect to content are formed for each sense-field: melodies, successions of color, and the like. But when is what we view in abstraction from the questions of affection, like something existing and becoming, which arises purely through concrescence and contrast—when is that there as actually affective on the ego? When does a group, a configuration of the visual field or tactile field, when does an acoustical series like a melody actually become prominent for me?

Where successions are concerned, it would seem that only when they have been formed affectively, only when the affection beginning with one or more points is propagated as an actual affection under the conditions of concrescence and of contrast, and potentially under the conditions of affect[127]—only then does a new formation of unity first come about.

Accordingly, in our analyses of possible immanent objectlike formations and, precisely speaking, of possible hyletic objectlike formations, we would have disclosed the temporal and local form as essential necessities (as well, the mere significance of the *conditions* of the possibility of hyletic formations of unity as the essential necessities of filling these forms with respect to content), while the formation of unity itself, the actual formation of hyletic groups and particular data existing for themselves would still depend upon the remaining factor of affection that is not taken into account.

Naturally, what we said about consciousness, namely, that essential functions of affective forces are disclosed here, can be applied immediately to the elementary components, to continuities through which objects of sense are structured with an order. If a datum that has just begun is awakened, the awakening will continue to proceed further; the awakening accompanies the datum that is continually unfolding, but will lose the datum in this unfolding, for example, through a diminution of intensity, through an effacement of contours and the like with respect to affectively significant conditions, so that in the case of a datum that has an unfavorable beginning and that would otherwise not come into relief, the continual transference of affective force will exercise a steady awakening upon what is less favorable, consequently working against the affective diminution, halting it. In its [153]
transition to *pianissimo*, the beginning loud tone carries the tone in affective force to the softest *piano* that would otherwise remain unnoticeable.

The forms of order obviously have a special affective significance. The universal, successive order [has a special affective significance] insofar as everything is constituted as in the process of becoming, and affection primarily follows the constitutive process of becoming. But not, for instance, as if succession as such would be a principle of association. Association does not function as the association of succession

[127] *Gemütsbedingunen*

from one field to another. What is essential is not succession, but rather a nexus that is akin in its own order.

In the final analysis, do not the essential, lawful regularities of the immanent formation of unity that we have described—those of the formation of particular, self-contained objects, wholes, groups, configurations—do they not express the mere conditions of the possibility of such unities, while the actual emergence of these unities itself is dependent upon affection and association? Let us consider the following in relation to this question. Everything that we expounded upon earlier concerning the formation of unity through the coalescing of matters that are kindred and coming into relief through contrast had to be tied to the instances of unities and connected multiplicities. These unities could only be there for us either as given directly in the scope of attentiveness, or through the fact that we grabbed hold of unities subsequently by reaching back into the past horizons of a sphere of attentiveness; these unities were given to us without and prior to our becoming attentive, and yet, they had at least affected us to some degree. Thus, affection (and obviously the awakening transference of affection as well, hence, association) played its role overall.

Accordingly, the question arises: In their regulated dependency of those essential conditions of the formation of unity, but also codetermined by novel essential laws, do affection and association not first make possible the constitution of objects that exist for themselves? Are there not regulated inhibiting, weakening counter potencies which, by not letting affection arise any longer, also make the emergence of self-subsistent unities impossible, unities in other words that would not emerge at all without affection? These questions are difficult to answer; and they are especially [154]
difficult if we wish to make our way from the sphere of the living present into the sphere of forgetfulness and to comprehend reproductive awakening, as will be necessary to do later. I do not need to say that the entirety of these observations that we are undertaking can also be given the famed title of the "unconscious." Thus, our considerations concern a phenomenology of the so-called unconscious.

In order to shed some phenomenological light in this darkness, let us begin with clear examples that will enable us to catch direct sight of awakening in its propagation and the motivation determining it. We assume that something prominent is already constituted, it may even already be affective. It is evident that not every affection can have arisen through the awakening of another affection. Every instance of affection through an isolated extreme contrast, like an explosion, illustrates this essential possibility. Certainly, we do not need such extremes. While taking an evening stroll on the Loretto Heights a string of lights in the Rhine valley suddenly flashes in our horizon; it immediately becomes prominent affectively and unitarily without, incidentally, the allure having therefore to lead to an attentive turning toward. That in one stroke the string of lights is affective as a whole is obviously due to the pre-affective lawful regularities of the formation of unity; because of them possibly other groups of lights in the visual field will also simultaneously be there affectively as prominent special unities, and this *ceteris paribus*. Indeed, the straight string of lights can have an affective priority, that is, can have the affective priority of a stronger affective allure, of a stronger prominence for us. In these examples, we are conscious of an articulated unity (however much it, as articulated, has within itself special prominences that are materially relevant) in an unarticulated affection. On the other hand, we are not conscious of them through an awakening from the outside, from the part of other unities. We are quite comfortable in characterizing such cases of unawakened affection as awakenings if we characterize them as the zero-point of awakening, similar to the way in which the arithmetician counts zero, the negation of number, among numbers.

Let us now revise our example and suit it to the case where a propagation of affection takes place, that is, where an awakening radiates from an affection and is directed outwards. The string of lights is already in the process of awakening, even if it only be in a [155] zero-awakening. One of the lights suddenly changes its coloring from white to red with sufficient intensity. Or we only alter its intensity; it becomes especially luminescent. It now becomes especially affective for itself, but at the same time this accentuation obviously benefits the entire string, which in other respects remains affectively unarticulated. We will have to say that a new affection has ensued and that an awakening ray issues from it (or that with respect to both string segments a double ray of awakening issues from the light situated in the middle); this ray becomes unified with the awakening force of the string that is already affectively at work, forming a unity of an awakening that is stronger. Obviously, we will also be able to say that a type of propagation of affection takes place with respect to every articulated whole that was directly affective as a whole: namely, the propagation of affection to its members. Insofar as this is the case we did not need to alter the original example. We could have said at the outset that the awakening of the entire string is distributed among the members from the very beginning, apportioning to them mediately a special affection, doing it within the entire affection of the whole. Also needing to be studied here is a certain opposition of fusion and contrast insofar as when the radiating single light obtrudes too strongly, it can also inhibit the spread [of affection] to the accompanying interconnected lights.

Here is a different example from the sphere of successive objects that are in the process of forming a unity: A melody sounds without exercising any considerable affective force, or if this should even be possible, without exercising any affective allure on us at all. We are occupied with something else, and the melody does not affect us for instance as a "disturbance." Now comes an especially mellifluous sound, a phrase that especially arouses sensible pleasure or even displeasure. This particularity does not merely become affective for itself in a living manner; rather, the entire melody is accentuated in one stroke to the extent that it is still living within the field of the present; thus, the affection radiates back into the retentional phases; it is initially at work by accentuating [the retentional phases] in a unitary manner, and at the same time it has an effect on the special matters that are prominent, on the particular sounds, fostering special affections. With this, the motivational causality is completely and immediately evident. The particularity of the sound has made me attentive. And through this I became attentive to the entire [156]
melody, and, understandably, the particularities thus became alive to me.

We also see from these examples an important distinction regarding the objects' mode of givenness for the subject of consciousness. Sometimes the givenness of the wholes, their affective prominence and therefore the possibility of grasping them, precedes the parts, sometimes the givenness of parts the whole. That depends upon the varying conditions of affection and fundamentally upon the conditions determining the materially relevant unity in a materially relevant manner. Insofar as this is the case the distinction can be relativized. If the lights in the string of lights were to radiate in a temporal succession, then the string as a whole in its givenness would naturally follow upon the givenness of the particular parts. But in the final analysis, we do arrive at original singularities, that is, at objects that must be given beforehand as a whole in all circumstances and for essential reasons in order for their parts to be given.

Let us now attempt to tease apart the elementary relations of affection. In the living present the primordially impressional emergence has *ceteris paribus* a stronger affective tendency than what is already retentional. For that very reason, affection has a unitary tendency toward the future where the orientation of the propagation is concerned, intentionality is predominantly oriented toward the future. Naturally, this should not mean that the progressive fading away of retentions is merely a weakening of affection; rather, it is the nature of retentional modification not to alter the affinities and contrasts with respect to content in the same way that a materially relevant modification takes place within clarity; rather it produces a new dimension of blurred distinctions, a growing cloudiness, a murkiness that essentially decreases the affective force. However, this does not preclude the fact that an awakening emanating from the more luminescent sphere of the affection that is still in full force works upon this penumbra, thus, a backward working affection in contrast to the affection that is constantly making its effect forward. If, for example, an acoustical object is awakened right when the entire field of acoustical data had been dormant, as it were, then the neighboring past is co-awakened insofar as it has an acoustical affinity to what was especially awakened; as one can easily see, it points to a [157] respective affective past-horizon. Every object structured in a completely undivided continuity would belong here, like a completely even tone, and similarly in the visual field, an even color. This does not preclude parts from subsequently coming to the fore precisely in and through an awakening—for example, when something that becomes a particular part by being set off through internal differentiation is placed next to uniform elements in similarity, or in another kind of uniformity, and when in overlapping or mere coinciding at-a-distance, this particular part carries out a kind of internal division of something continuous right within the consciousness of the absence of internal distinctions within the impressional sphere.

In our examples we had our eye on awakenings that are maintained in the line of enduring coexistence, like when brighter rays of a light are transferred to the string of lights, or that are maintained in the line of successions that are sinking back into the successive series: like the awakening radiating back from the mellifluous tone toward the series of tones given to consciousness in a living retention.

Obviously, however, the associative awakening also goes forward in the direction of the future. Even the new tones of the melody, the entire futural series that is intertwined along with the new becoming profits from the first awakening. Tonal formations, which perhaps would not have fulfilled the conditions of affection, become noticeable now, indeed they even become the objects of thematic interest. Thus starting from the place of awakening, the awakening also expands in a forward direction, following the path of the fusion of objects.

In the other example, since we are already attentive to this one row of lights, we would likewise be especially prone now to become attentive to a newly radiating row of lights, even a relatively weak one, and so forth.

We obviously have something analogous here in primordial association to the two different kinds of association in the customary sense that we distinguished earlier: associations turned toward the past (and as we will add, potentially reaching into the present by means of reproduction) and associations as awakening anticipations.

The analysis of these kinds of examples and observations of them carried out within an eidetic attitude show without further ado that within every living present (and restricted initially to the sense-data that are being unified within it), affections are constantly at work beyond themselves; we always find affective [158] awakenings, that is, associations. The principles of the constitution of hyletic objectlike formations, namely, as connected coexistences and successions, the principles of local systems that constantly function to individualize and their fillings, the principles according to which contrasts and inner fusions (concrescence) take place, are constantly operative. They constantly form essential conditions of affection and the transference of affection as awakening.

If we thus presuppose the structure of the present according to primordial impression, retention, and protention, then with regard to it there are, on the one hand, associations, awakenings that radiate into the co-present and the retentional pasts, and on the other hand, associations that head in the direction of the future. Heretofore we have substantiated only the former with examples, for they are, as we see, the more original. In other words, the distinction we made at the outset in our initial, rough consideration of association in the customary sense of the term (namely, between the backward turning and first association, on the other hand, the forward turning, anticipatory, secondary association), also holds here when we deal with primordial modes of association with respect to the living sphere of presence, and when the concept of association is expanded.

Let us therefore investigate more closely now the immediate awakenings that are directed nearby and that are turned backward. In accordance with the lawful regularities of the structure of the concretely living sphere of the present, the entire primordial impression belonging to it can only have the form of an articulated coexistence, articulated according to the sense-domains and according to the data for themselves, objects for themselves that are connected and separated within the sense-domains. So far as succession is concerned, it has the form of articulated succession such that homogenous orders and nexuses of succession must be formed, that is, for every sense-sphere in itself. Every primordial-impressionally constituted object belongs within an order that is homogenous and successive with it.

<§34. The Problem of the Relationship between Affection and the Formation of Unity>[128] [159]

Yet, in accordance with what we have already suggested in our previous lecture the essential conditions of affection can now mean two things. Either it means this: If we presuppose this objectlike structure of the field of living presence, then all affection is, or again, all emergent awakenings of one objectlike formation by another are subsequently bound to this regulated structure. And in particular, they are bound in such a way that all associative awakening can only accompany the regulated forms of connection according to which multiple unified objects of a higher level arise from single objects. Our examples belong here. Objects were already given to us, and we left it open whether they are already there for us through affection; our only object of consideration was the progression of affection. We saw this progression, i.e., association, fundamentally bound by the kind of higher object formation that arises through affinity and contrast. For this reason, association could not pass over from the objects of one sense-domain into those of another, and had to follow persistently the unifying affinity of prominent objects within a specific domain. It also had to be determined through the contrasts of such objects by their "background" objects.

[Or] secondly, one could also understand the essential conditions differently according to the other interpretation of the constitution of the typical structure of objects peculiar to every living present, an interpretation that was suggested previously in our last lecture. One can tentatively say that affinity, continuity, contrast are relations that need not yet be viewed as an actual fusion-in-itself, as actually producing a unity in and through prominence. Only what keeps to certain forms of such relations can actually be fused. But perhaps we must distinguish here between (a) unconditionally necessary fusions that are carried out in a fixed lawfully regulated manner under all circumstances, namely, in such a way that we would have to consider that even if

[128] Editor: See Appendix 21: Sensible Multi-radiating Affection. Sensible Group—Genuine Collective Objectlike Formation, pp. 520ff.

affection were to be everywhere functional, no matter to what degree, it could not exercise any special accomplishment on the formation of unity; and (b) fusions, formations of unity that are [160] owing first to affection. Let us see if such a theory is tenable.

Among the first, the unconditionally necessary fusions, we would obviously have to rank those fusions that constitutively bring about the fixed form of the field of living presence: the temporal form and the form of the local field, but also what should manifest itself as the firm form of unity of another sense-field. This would mean that just as the original process of streaming within immanence cannot at all be a special accomplishment of an affection, but rather is an original continuous fusion that necessarily takes place, and yet has the conditions of its structural possibility in order for the necessary continuity to be precisely fulfilled continuity; this must likewise be the case with respect to the streaming *hyle*. That is, hyletic fusion must be carried out in the fixed necessity of temporal constitution, a hyletic fusion arising from original continuity as successive fusion, and again without any accomplishment from affection. Furthermore, there must likewise be able to be unities of coexistence; and the most original unity is the unity arising from coexistent continuity: Every sense-field for itself is such a unity, constituted without any affection.

Now, however, every sense-field also has special unities that are constituted—here we are approaching the realm of affection. The wakeful life of the ego is such that the ego is explicitly affected, affected by special unities that are, precisely through this affection, given to, graspable or grasped by the ego. But does affection not also presuppose here the work that is accomplished, to wit, the special fusion that necessarily ensues when the formal and the material conditions are fulfilled? In fact, if we examine these conditions, then we must certainly admit that continuity can be a more or less complete continuity, a multi-layered or a single-layered continuity; this is a matter concerning the contents themselves and does not as yet concern affection at all, initially. Thus, for example, that on the underlying basis of the local continuity of positions, as the fixed form of the visual field, the colors occupying it continually extend and thereby continually fuse as the coloring of the extension; that the colors do this in a continuous uniform fusion or qualitative graduation in and through the continuity of intensity; and again, that such a continuity with respect to content is only maintained for a time, that it makes a [161] leap at a certain point and produces a contrast—all that is a matter of the fixed formation of unity. It is evident that affection has nothing to do with it.

What happens now when we pass over to the higher level of objectification, that is, when we take into consideration the varying homogeneous syntheses of multiplicities, spatial and temporal configurations? When we carry out overlappings there, and then speak of a coinciding and fusion *par distance* independently of an actual overlapping, when we speak of a kind of attraction of prominent, uniform, and very similar <contents>, when we speak of a kind of synthesis at-a-distance, then it already looks like an associative awakening at-a-distance and suggests that the latter is responsible for the emergence of a fusion at-a-distance here. Naturally, affection does not make the relationships that are grounding in the particularity of the contents.

But perhaps it is so that initially every concrete, particular constituted thing, everything that is constituted in relief for itself and not constituted first through affection, necessarily exercises an affection to some degree or other. This degree would then depend upon the particular conditions. But then it can be the case that among simultaneous special objects, things that are similar to one another emerge, and by simultaneously affecting the ego, take on a peculiar unity: Initially [it is] a unitary point of prominence for the ego, but later it becomes a materially relevant connection to a higher articulated unity. On this interpretation, it is first through the synthetic unity of affection that the synthesis of fusion at-a-distance, of the unity arising from affinity, would come about.

One could attempt to interpret this situation likewise in the case of succession and its configurations, like melodies, and so forth.

However, upon a closer inspection of the possible interpretations we find what we have just uttered to be untenable. The division that is made here between the constitution of objects of a higher level as opposed to the objects that are constituted in an original singularity is unjustified because it is incomprehensible that fusion should first be generated through the unity of affection.

(in which little or nothing is affective with respect to the objective sense) into another empty retention in which more is affective now, that is, more "emerges" from out of the "fog," potentially in an especially salient manner, and is grasped. In fact, that is the [175] first form of a disclosive awakening.

If we inquire into how such awakening takes place, the answer will run: Just as within coexistence an intuition becomes unified with an intuition, and initially within primordially impressional coexistence, an impression becomes unified with an impression, becomes unified not only in a continuous local juxtaposition, but also through a synthesis of coinciding in distance that takes place essentially, and just as in this case coinciding and affective communicating are inseparably bound together, so too when we pass over to succession; in this case, what has finally become non-intuitable achieves a synthesis with what is actually intuitive, and especially with what emerges in the new impressional vivacity. Affective communication would mean that every contribution of affective force by any "member" of something connected in distance through homogeneity and prominence augments the force of all its "comrades." Moreover, according to our principle interpretation this means further: Only by virtue of affective force does connection come about at all, and the affective force is *eo ipso* there so long as intuition takes place. Within the streaming living sphere of the present there cannot be individualized intuitions. The extent to which the conditions of near-connection and distant-connection proper to the intuitions' "content," to their objective sense, are fulfilled is the extent to which the interconnection comes about; only that depending upon the affective relief, there will exist differences of salience, differences of affective intensity of the prominence of these or those particular objects and of these or those particular interconnections.

Let us now turn to succession. We naturally assume, and it is indeed actually quite evident in a certain way, that distant syntheses as well as near syntheses can also take place in succession. Without this there would not be a world for us, and above all, as the basic feature of a possible world of experience, there would be no world given originally in experience, not only with respect to coexistence, but also with respect to the successive order. However, in original experience one thing constantly follows another, there are processes, and not only those that continually stretch back, but also those in which temporally distant objects and processes are involved in a unitary manner, e.g., the successions of hammer blows or the sequence of tones as in melodies, etc.

But would the like be possible if these syntheses were bound to [176] the extraordinarily narrow sphere of successive and actual intuition? It is clear that the awakening of a concealed element already plays its role here in order to make possible any expanse of successive syntheses, even if not exactly an arbitrarily large expanse. When the second hammer blow rings out, how does the synthesis come about between it and the first one which has in the meantime become intuitively empty or even completely empty? How can the synthesis be the synthesis of the succession of uniform blows, how can it arise before us in a unified manner as a uniform pair? The contents that are genuinely intuitable there are not uniform. But as the retention changes, precisely the identical objective sense has been maintained, and the uniform sense-content, only once in the mode of the impression and the other time in the mode of emptiness, grounds the synthesis. But the emptiness is "clouded over," has more or less lost the efficacy of the distinguishing feature with respect to content. So, how then can we become conscious of uniformity? The answer naturally runs: No matter how incompletely affective the one objective sense is, it is still there as the first, obscured hammer blow, and can thus still enter in a community of sense with the new hammer blow. From the affective force of this hammer blow and as a consequence of this community, an affective awakening now goes back into senses that are uniform. The affective awakening does not bring the uniform sense to intuition for instance, but does indeed effect an un-covering. That is to say, now it has its own affective force—the reanimated past hammer blow, just like the new hammer blow. The condition is thus fulfilled so that uniformity as such can now actually be produced synthetically. That is, as a unique, as an affectively prominent synthesis of uniformity.

Let us now imagine hammer blows progressing even further. We experience the chain of hammer blows extending relatively far back, given in consciousness in an actually living manner, and actually constituted in a synthetic manner; we experience it as a chain only by virtue of an awakening running back. Finally, however, the chain is lost in the zero-horizon, the efficacious concatenation presupposes an affective awakening that still has an ample positive force. The force of intuition is limited, and the diminished forces of the derived intuitions are limited, and it finally becomes nil in the radiating back. –The concatenation that is already formed, the pair, the triplet, etc., are new objects and [177] sink back as objects of a higher order, etc. Thus retroactive awakening concerns them as such.

<§37. Retroactive Awakening of the Empty Presentations in the Distant Sphere>[138]

If the final remainder of affectivity is lost as the retention becomes obscure, and if the process itself has lost the final remainder of vivacity of the streaming process, what is essential is of course preserved. Earlier I thought that this retentional streaming and the constitution of the past would continue to go on incessantly even within complete obscurity.[139] But now it seems to me that one can dispense with this hypothesis. The process itself ceases. Thus we have the following, never ending typicality: The primordially instituting process of ever new sense-objects goes on and on ineluctably at the head of the living present; it is bound to the lawful form of the entire objectlike structure; the primordially instituting process begins again and again with impressional momentary data that immediately crystallize as coexistent objectlike formations and are systematically articulated; these formations are continually modified in retention while ordered articulations of succession are simultaneously constituted. But this retentional modification leads further and further into the one nil.

---

[138] Editor: See Appendix 24: <Effect and Cause of Awakening> pp. 531ff.

[139] Editor: See "Consciousness and Sense—Sense and Noema," pp. 591ff.

What does this nil mean? It is the constant reservoir of objects that have achieved living institution in the process of the living present. In it, they are tucked away from the ego, but quite at its disposal. Heretofore their being consisted in nothing other than the process of their being experienced originally, that is, in their being-given-to-consciousness *originaliter* and remaining-held-in-consciousness in this process of a streaming, synthetic coinciding that forms identity. Where the retentional line of identity is concerned, this being and even this remaining-held-in-consciousness does not cease when the process has reached its end in the zero-point. The constituted object, the identical element, is no longer constitutively vivacious; thus, it is also no longer affectively vivacious, but the sense is still implicitly there in a "dead" shape; it is only without streaming life. It is not efficacious for new constituting—*notabene* for new, originally experiential constituting. How it can become efficacious and even [178] constitutively efficacious in a new shape is the problem of association. Initially, however, we want to say that every accomplishment of the living present, that is, every accomplishment of sense or of the object becomes sedimented in the realm of the dead, or rather, dormant horizonal sphere, precisely in the manner of a fixed order of sedimentation: While at the head, the living process receives new, original life, at the feet, everything that is, as it were, in the final acquisition of the retentional synthesis, becomes steadily sedimented.

One is only acquainted with sleep by waking up; so too here, and in an entirely original manner. Waking up sedimented sense can initially mean that it will become affective once more. And we do not need to rehearse how this is possible. As we have already mentioned, it is all the same whether we conceive empty presentations (empty intentions) that are still living as being awakened or ones that are already fast asleep. The motives must lie in the living present where perhaps the most efficacious of such motives (which we were not in a position to take into consideration) are "interests" in the broad, customary sense, original or already acquired valuations of the heart[140], instinctive or even higher drives, etc.

Certainly, there is an unmistakable distinction between the awakening of the entirely submerged sphere, the distant sphere, and the awakening of empty intentions (or of their contents) that are still in wakeful life or that are standing on the edge of life. In the latter case the thing awakened is once again integrated into the originally living constitutive nexus—as has become clear with the example of the expansion of the near sphere through repetitive affection (the hammer blows). But the distant sphere, with its sedimentations of all previous accomplishments of previous living presents, yields an awakened sense by rousing any empty presentation that is related to what is in the distance; it yields a sense that stands outside of continual, living connection with the sense being constituted and coming to life in the moment of awakening. This is very significant. All intuitions and presentations, which are originally at home in the framework of the living present, that is, which belong to the continual unity of its syntheses and its downward directed movement of identity, are integrally related in the unity of this continuity. We speak without [179] further ado of integrally cohesive intuitions and of other kinds of presentations. The awakening of the submerged past, however, yields presentations that are not integrally cohesive with those of the living present. Accordingly, these syntheses in which these presentations can enter, by virtue of their affective force, together with the presentations that have arisen originally and that are capable of being constituted originally, stand in contrast to those syntheses of presentations that have arisen originally with each other. These syntheses have a fundamentally different character. We will have to pursue this further.

But first: Awakening is possible because the constituted sense is actually implied in background-consciousness, in the non-living form that is called here unconsciousness. Here awakening is also the production of an affective communication and therefore the production of a relevant synthesis, of an objectlike connection which, as connection, is actually produced like a simple object,

[140] *Gemüt*

affective for itself. Once something is constituted in the manner of an object, it can connect with anything else that is already constituted as an object; in this instance, precisely two kinds of conditions are to be fulfilled, on the one hand, the conditions of content, of both senses of the objects, on the other hand, the conditions belonging to the consciousness of them. Only something that is grasped explicitly, that is, something that exercises an affection, can become connected in consciousness. Certainly, should the connection of both objects itself be given as a connection, that is, be given intuitively, so too must these objects be self-given. There is a further, important special case where the connection of objects should yield a unity of objects in the special sense of a connected whole understood most broadly. For this, the intuitive modes of consciousness of these objects must be integrally cohesive, fit together into the universal unity of a constitution encompassing them.

Awakening the obscure distance is initially an empty awakening. Where the content is concerned, the awakening is made possible in accordance with the principle that makes a unification with respect to content and affective communication possible in every present: The fundamental condition is namely the "similarity" of contrasting discrete matters, and everything belonging to it. One color can awaken a concealed color, a pronounced sound, a sound that has become masked. A rhythm can also awaken another rhythm, e.g., the rhythm of knocking [180] blows can awaken a similar rhythm of signal lights. (Insofar as this is the case, awakening can certainly transgress the sense-fields.) Through our analyses of the necessary structure of a living present, the vague principle of association of similarity and of contrast takes on a unequaled richer and deeper sense. In this structure lie all associative preconditions of content. The first synthesis, which is made possible through the affective communication gained by the transference of affective force, is of course precisely the synthesis of the similarity that has become currently given to consciousness, a similarity between the awakening and something presented in an empty manner, the awakened, this similarity in the essential noematic mode of the "recalling another."

If the matter rested here, precious little would be accomplished. Although the awakened element obviously lies in the direction prefigured by the course of the constituted past that has come to life as the past-directedness and, as it were, as the objectlike formation in the most distant past, what we naturally count as the past would still be quite far from being constituted. A genuine object to be grasped and disclosed as identically the same again and again is first constituted with the help of remembering, as we have already ascertained long ago. Even the past and the synthesis in the course of pasts, which are carried out in the small circle of the living present, does not yet yield the genuine constitution of the object, but rather, only a fundamental portion of its constitution.[141]

<§38. The Transition of Awakened Empty Presentations in Rememberings>

In the last lecture,[142] we stood before a new level and accomplishment of association. The first level, which we addressed under the rubric of primordial association, was that systematic or systematizing affective awakening that makes possible the objectlike structure of the living present, all kinds of original syntheses proper to the formation of unity of manifolds.

The second level, the one we just treated, was the awakening that radiates back, which illuminates once more the darkened [181] empty presentations, bringing the contents of sense implicit in them affectively into relief. Belonging here is the especially important case of awakening the presentations of the zero-sphere.

The third level is the level of the transition of such awakened empty presentations in reproductive intuitions, and that means here, rememberings.

One may well promulgate the principle with certainty that belonging in general to empty presentations that have undergone

---

[141] Translator: In the last two instances, I have rendered *Gegenständlichkeit* as object. See "Translator's Introduction," Section 3.

[142] Editor: Beginning of a new lecture.

an awakening is the tendency to transmute them into self-giving intuitions, which at any rate goes by way of rememberings.

In any case, the law holds that rememberings can only arise through the awakening of empty presentations. Accordingly, they can only appear as consequences of empty presentations, which for their part have arisen from an awakening in order to enter into the living present; rememberings that have arisen in the most immediate way only appear through the awakening of those empty presentations that arise in a fixed necessity of the structural elapse of a living present in itself. Thanks to the establishment of this origin the entire, essential lawful regularity of reproductive association is prefigured. It becomes entirely dependent upon the fulfillment of those lawful systems that make possible the association on the previous level, that is, make possible the awakening of empty presentations, whereby we already come back then, naturally, to the sense-structure of a living present and to the affective conditions proper to it generally. I hardly need to say that through this, the traditional laws of the association of similarity, of contiguity, of contrast in their obvious, but vague and entirely incomprehensible truth obtain an intelligible and precise sense through essential insight. However, the precision lies in the systematic analyses and affiliated laws.

Let us now examine the situation according to what is most essential. Let us take a distant awakening leading back in one blow from the present into a deeply lying layer of sediment in the zero-sphere, that is, into a submerged, distant past. There is now an empty presentation that makes us conscious of a past singular datum or of a singular connected nexus. Naturally, it is awakened [182]
by a uniform or a very similar nexus in our present. Thus, a similar synthesis is necessarily given with it, a synthesis comparable to what must have been formed in the original present itself in a primordially associative manner, and therefore also with the noematic character of all association. The awakened element has itself the noematic character of something to-be-awakened through the awakening element (the character of recalling something).

Further: Just like, awakening and connection progress[143] in accordance with the standard of affective force within the primordial sphere of all associations, and are determined ever anew in it through the influxes of affective forces, so too [do they progress] with the arrival of the awakened distance. A firmly constituted objectlike nexus, only having fallen asleep, is awakened at a location, at the location of a present that has passed by. Note well that it is a matter here precisely of the objectlike formations, of the inner connections and prominences, the connections and orders that had actually just been constituted and are now forever enclosed in the zero-sphere. Naturally, the awakening can now progress, living on the affective force of what was first-awakened, potentially augmented through the stimulation of affective interests[144] etc. that arise in the progression. When awakening progresses, only individualized, especially strong sense-moments of the distant present in question will come into relief in the emptiness, just like rough contours in a dimly illuminated fog. An even more favorable case is when awakening passes over into remembering. Of course, this transition is carried out as the synthesis of identity; the synthesis of identity is the accomplishment of intuitive remembering, the accomplishment of the re-constitution of the objectlike formation, but in the mode of coming-back-again to something familiar, in the mode of being-presentified-again, not genuinely experiencing, but rather as if one would experience. Now the past present is reproduced in the vivacity of the noetic-noematic flux with all accomplishments—with all accomplishments of remembering, which in the ideal case, are completely intuitive, while in truth remembering wavers in clarity and distinctness, thus, mixed with empty moments, a middle stage between pure, complete intuition and empty presentation.

[143] Translator: reading *fortschreiten* for *fortschreitet*.

[144] *Gemütsinteressen*

<§39. The Difference Between Continuous and Discontinuous Awakening> [183]

The greater richness of affective efficacy is advantageous to the progression of awakening. But it is clear that awakening is not tied to only one reproductive present; rather, it can pass over continuously or in leaps into the other layers of sedimentation.

The direction toward higher situated layers, that is, the direction forward into the relative future is prefigured in a regulative manner and *a priori* for the possible continuous progression, while only the progression of awakening that takes place in leaps is possible for the transition to the underlying layers. This is to be understood in the following manner: The reproductive tendency is a tendency toward the re-constitution of the objectlike formation under consideration. Thus, just as the original constitution as such is now a process directed forward into the future, so too is the re-constitution. The object is not in the present as something fixed and dead; rather, it becomes in the living constitutive nexus in which, moreover, it is not isolated; instead, in the intertwining of the becoming and being-transformed, it is a term in the process of becoming, the term of an objectlike-whole being shaped progressively.

The moment the tendency toward re-constitution starts to be realized, we stand in the process of becoming which is just what the tendentious memory strives to re-produce. Accordingly, here is the field of continual awakening. On the other hand, every awakening radiating back is naturally discontinuous, whether it goes near-by or into the distance. If the reproductive process goes tendentiously forward, then the impoverishment and submersion into the old zero-sphere is repeated in the process reproductively; the discrete awakening that radiates back can acquire something once again from the zero-sphere in repeated reproductive tendencies. The awakening can jump from one sedimented layer to another, which can be now higher, now deeper; it can jump in leaps and without any determinate order. In this way, manifold possibilities develop for rememberings that follow upon one another and that are entirely and immediately disjointed. Each one is characterized as a beginning expanse of streaming re-constitution, but one that is sinking once more into powerlessness. All these rememberings are nevertheless referred back to the [184] continuously integrally cohesive system of sediments existing in concealment—a coherent nexus that would however only be reproductively realized if we would continuously reproduce, if we *could* continuously reproduce our entire life from the very beginning and in one stroke.

The emergence of rememberings yields a peculiar enrichment of that living, originally constituting present that we conceived heretofore in a necessary abstraction, purely as a process of immanent hyletic experience—of original experience. Now joining the particular intuitions of this living experience, which are connected in thoroughgoing internal continuity, that is, particular intuitions that cannot be isolated at all, are rememberings like tattered rags, rememberings arising from an entire previous experience that is reanimated. United in a certain way with the current present through association, they do not have any direct experiential connection with it, they arise in a narrow sense, disjoined. Different rememberings are also without an internal coherence between each other, where they emerge in leaps through discontinuous awakening. This difference between the interconnections arising from a constitutive continuity and the interconnections arising through a mere distant-awakening is highly significant. On the one hand, we understand from this that in the continuity of original experience the phenomenon of modalization cannot emerge—original experience is a nexus of thoroughgoing concordance in absolute necessity. Only first with remembering, in more general terms, with the phenomenon of reproduction does there arise the possibility of connecting discordances, the possibility of modalizations. Before we examine this more closely, we want to bring the principle contours of the theory of association quickly to a close.

<Chapter 4: THE PHENOMENON OF EXPECTATION>

<§40. Motivational Causality Peculiar to Expectations>

We now turn our attention toward a new fundamental direction concerning the lawful regularity of association. It will already have occurred to you that I have, so to speak, kept dead silent [185] about the long familiar phenomenon of protention and of expectation in general—so essential to the structure of every original experience, and thus, to the sphere of the living present. All formation of unity and every nexus in experience is rooted in homogeneity, formal and material. Or, as we can also say, in uniformity or similarity. Homogeneity effects the institution of the unity of experience only in impression and retention, specifically, in the temporal form of the present and past in the constitutional form: enduring coexistence and the sequence of something enduring. The future, however, which as a more or less prefigured horizon belongs to the constituted objectlike formation, arises from the continuous and discrete protention according to the laws of the formation of expectation. The future does not fashion the unities of experience in the original sense, it presupposes them.

Yet viewed more precisely, we must recognize that even already in the hyletic primordial sphere of immanence, expectation is not only possible as the expectation of something futural. We have already alluded to the fact on various occasions that there is also an expectation within the impressional present, as is evidently the case with respect to the phenomenon of lacking a term which is otherwise part of configurations that exist in the synthesis of similarity. Something similar recalls something else that is similar, but it also allows something similar to be expected in coexistence, as in succession.

Obviously, all the analyses that we have undertaken are already presupposed for the elucidation of the phenomena of expectation, be they tied to the synthetic formations of coexistences or successions. The structural formation of a living hyletic sphere of the present occurs in the process of impressions and retentions under the essential conditions of fusion and segregation, connection at a distance proper to this process. This structural formation is presupposed in order for expectation to be able to make its own marks in this structure, and then in order for a fulfilled or annulled expectation to be able to occur as a future phenomenon.[145] It is quite evident that with expectation we [186] simultaneously move into the most original headwaters of modalization where genesis is concerned.

Memory—taken in the broadest sense, thus initially the most original retention—already precedes protention in what is elementary. If any kind of *a*, a sound for instance, is in the steady process of melting down new impressional phases, if it is, in other words, a course of continual connection corresponding to certain essential conditions in this original process of becoming, then a futural horizon, that is, an expectational horizon is immediately there along with it; a progressive process of becoming analogous to the previous becoming is expected according to the previous becoming, expected according to the same continual style of the course: If the invariability is of this or that qualitative and shaped content, then it is the invariability of a uniform content, and if it is variable, then the variability will likewise be prefigured, but according to the same style of variability as before. Thus, it is a matter of a necessary motivation: In accordance with what is given to consciousness retentionally, there is "to be expected" something new on its way having a uniform style. Obviously, the expectational presentation is characterized as a presentation of a new kind, specifically, as a presentation of a second level, as the replica of the more original presentation of the past. With respect to the object, what is expected is naturally not like what is perceived, that is, characterized as being in the flesh and as present; likewise, it is not characterized as what is remembered, as just-now, and in the subsequent expectation, what has previously been; but rather, it is characterized as in accordance with what has been, with its primordial image, as what is anticipated or as a projected image or model of being prior to its actual being.

[145] The abstraction of a purely passive world of sensation must still be circumscribed more precisely. To be considered are kinaestheses as originally "free," "subjective" courses. Editor: See Appendix 25: <Kinaestheses and Potential Expectations>.

If we go beyond the momentary singularities that are immediately at work, existing at the head of the constitutive process, over to the successive nexuses constituted in an articulated manner, then every past constituted in a unified manner, that is, every succession flowing-off in a unitary manner, is projected into the future as an expectation, even if it has already become emptily presented. Accordingly, we have in the progression of experience on the one hand an ever new coexistence formed out of original experience itself, and on the other hand we have it interwoven with a structure of expectation, with a layer of anticipation, which is partly fulfilled by the experiencing, partly in conflict with the experiencing and [187] annulled. Naturally, we likewise have a progressive structure of expectation in the temporal series, anticipating the futural experiences on the basis of the coexistence that is already given, and anticipating the futural sequences of coexistences on the basis of the entire expanding sequence of coexistences that has heretofore run-off.

Let us examine the situation in the sphere of the living present more closely. The occurrence of something futural is expected through its similarity to what has occurred in the past, like already happens in the most primitive case of a steady protention. We must take as a basis unified data connected through community. If p is connected to q in the order of succession and p' uniform with p immediately occurs in the sequence, in other words, as a repetition, then (in accordance with what we said earlier) it recalls the p that is submerged in retention. The latter gets a shot of affective force and is passed on further to q. In this connection, the q' following the p' that has just occurred will also be expected in original essential necessity. This associative expectation obviously presupposes association as an awakening reference-back of "memory." Let us now advance one step further.

This also holds for the distant sphere provided that it has been awakened precisely in an associative manner. The distant p can be complex in any manner, an enduring complex, a coexistence that is potentially constituted successively by the approach of new terms to form a self-contained and then enduring connection, such that after it has become a self-contained configuration given to consciousness, *q* then ensues. We then say, under the enduring "Circumstances" *C=q* joined with *p*. If we assume that the Circumstances *C* would be constituted in a unitary manner in a previous situation of consciousness of the distant past and then a *q* had ensued; and if we assume in addition to this that now in the currently present new situation of consciousness, the similar Circumstances *C'* would have been (implicitly) repeated, then in the event that the previous *C* and their *q* have been awakened, the occurrence of *q'* will now also be necessarily motivated as arriving. If the awakening has become a distinct, clear remembering, then the occurrence of *q'* will be given in original evidence as something to be expected, and in particular, motivated by the memorial *q* in the previous situation *C*. Here we can [188] directly see motivational causality as a necessity; we can say in evidence: I expect *q'* here because I have experienced *q* under similar circumstances, and this "because-thus" is given in evidence.[146] Correlatively: I infer "inductively" in complete evidence the present, similar arrival from what has arrived under previous, similar circumstances. Like every inference, this too has necessity and yields in essential generalization an evident law of inference. Only that here the open possibility that something else will indeed occur belongs to the evidence of the motivation of something arriving as something to be expected.

<§41. Strengthening and Inhibiting Expectational Belief. The Function of Expectation for the Formation of Configurations>

It is further evident that the anticipatory belief of expectation has a differentiation of force, that is, a gradation, and that this force grows with the number of inductive "instances," that is, with the frequency of what has occurred under similar circumstances. It is likewise clear in the alternate direction that when under the Circumstances *C*, now *q*, now *r*, now *s* has occurred, exclusively, the motivated tendencies of expectation will be inhibited. It is likewise clear that the increase and the inhibition of intensity do

[146] Translator: The "because-thus" structure of motivation was also treated in *Ideas II*, especially, 238 ff., 241 ff., 244 f.

not merely emerge in a contingent and arbitrary manner, but rather can be motivated in a necessary and evident manner.

If for example earlier in the circumstantial situation *C*, *a b c* have occurred, and in the current similar situation *C'*, *a'* has occurred, then according to what we already said, *b'* and then *c'* are naturally motivated as arriving. If now, however, *b'* has actually occurred, then obviously *c'* is doubly motivated, since here the law gets applied once more. Thus, the occurrence of the expected *b'* does not only ratify the expectation, which was already awakened by the occurrence of *a'* and was also mediately awakened for the arrival of *c'*; it also strengthens this expectation.[147] If the motivational situation is clear, that is to say, if the awakened, previous situation of consciousness and what has occurred in it has resulted in a clear remembering, then the [189] motivations and their augmentation are also given in evidence, that is, the confirmations and the corroborations in their graduation are also given in evidence along with every actual occurrence of a term of the series concerned.

Let us note, moreover, that when in the place of an expected *a* only a part, α occurs, the supplementary part β is now "missing," it is "lacking." Indeed, proper to expectation is a certain coinciding between the present and the associatively awakened remembered past, the retentional past and potentially the past that is distant from us; and the surplus of fulfillment as the surplus of a coinciding, the surplus of too much and too little, becomes prominent within the coinciding; what has not occurred stands there in the consciousness of "lack."

Likewise, with the other types. If *a* has repeatedly emerged under certain circumstances or as a final term in a regular sequence, and if the corresponding expectation is there through associative awakening, then the absence, the non-occurrence becomes salient, the expectation is disappointed; the present temporal field, that is, the sense-field is filled out, but filled out "otherwise." And it does so of course in conflict with the content

---

147
a *b* *c*
a' *b'* *c'*
b' *c'*

of the current sense-field that is actually realized. In the further progression of the genesis of consciousness such an absence acts as a diminution of the force of expectation that has been gathered through repetition. In the clear reproduction and reference-back of the present situation to the previous situations, the positively and negatively motivating instances become evident, each newly emergent having-been motivates with uniform force, each one speaks with uniform force in favor of a reoccurrence, each absence likewise speaks against it, and this "speaking in favor of" and "speaking against" is given here in evidence. But in the empty sphere of retention the forces—and with them also the forces of expectation, blind like any drive—are [either] compounded [or] inhibited; in any event, the typicality and lawful regularity of expectation is, we see, thoroughly dependent upon the typicality and lawful regularity of reproductive association and thereby is also mediated by the typicality and lawful regularity of original association in the livingly streaming sphere of the present.

If one configuration has formed a prominent, specially bound unity through the affective forces governing in it, then the force will radiate out from it to accentuate objects that will fulfill the [190] conditions for forming a uniform configuration—initially within the same living present. The part "demands" the whole—something uniform awakens something else that is uniform, which is not yet at all constituted as a unity explicitly for itself; and it does not demand the whole by a pure and simple awakening, but rather by a co-connected "expectation," by the demand as coexisting as co-belonging to the unity. Even the force of this apperceptive expectation increases with the number of "instances"—or with habit, which amounts to the same thing. At the same time, a more forceful formation of unity occurs here—a habitual one—as the regularity of the expectation being ratified; on the other hand, [we experience] the disappointment that accompanies becoming conscious of the lack of the habitual [connection], and [we experience] the disappointment that accompanies becoming conscious of something other than the habitual [object expected].

With the number of instances, I said. Reflecting on it more precisely, what does that mean? If a plurality of data emerge in the same present as it continues to develop, data that can go together with the configuration in question, then the entire configurations in question will be awakened by the awakening that reaches back and that issues from the particularities. And these configurations radiating ahead in a protentional-expectational manner will awaken the projected image or model of this configuration, allowing it to be expected, and through this the coalescence of the configuration will simultaneously be favored once more as fulfillment. In this way, the anticipation is at work "apperceptively," it is co-productive in the configuration of the coexisting objects.

Naturally it is exactly the same with respect to the apperceptive efficacy of the distant pasts with respect to the influence that they exert upon the formation of new configurations by awakening their already configured unities: here continually according to "uniformity," which can be "great" similarity. The past configurations that have been awakened overlap with the present, and are capable of being generated as similar in this present, in this way they become generated and are then coincident with the old ones, they are "familiar"—not individually, but rather according to their type.

The unity of apperception that has become, the configuration in this motivation, is surrounded by configurations that are awakened and that are similar to it, coinciding with it in obscurity. At the same time, the nexus of the terms of the configuration is not merely a nexus with respect to content, like when the nexus has become a fusion, forming a special unity beyond the particularities by virtue of the affective circumstances, but rather apart from [191]
affective motivation, which governs selectively in the configuration. (Various figures could have been formed out of the chaos, and could still be formed, but the path of awakening that was privileged favored the path of special unifications of a lower and higher level up to the entire configuration in question.) A unity of integral togetherness has become a unity in which the terms, the connections of subordinate levels (the parts), refer to one another as to be expected reciprocally in being together, as being demanded reciprocally and fulfilling this demand. The integral togetherness—the unity formed from habitual

connectedness (and by virtue of diverse experience)—the force of unity, increases with the frequency. Finally, the unity is surrounded by an open and clear horizon of something long familiar and of something that has been experienced this way time and again, and will also be found this way at once in the new instance with "empirical certainty," and it presents itself as such a whole without further ado.[148]

[148] The formation of a type.

<DIVISION 4: THE IN-ITSELF OF THE STREAM OF CONSCIOUSNESS> [192]

<Chapter 1: ILLUSION IN THE REALM OF REMEMBERING>

<§42. Overlapping, Fusion, and Conflict of Rememberings of Different Pasts>

The question concerning the intensification and inhibition of associative awakenings naturally demands a fundamental investigation. The living force of awakening, which radiates out from the impressional present, flows over into the retentional sedimentations of memory in accordance with the principle of similarity: The prominent intentional objects, the singularities and connected complexes of the intuitive present, link up with similar objectlike formations that are implicitly constituted in memorial sedimentations; they radiate toward them *via* the awakening force. But actual prominence and then further, awakening, reproductive renewal, is only of benefit to few. The awakening forces can reciprocally combine with one another, but can also inhibit one another, and it remains a question upon what the actual revival and reproduction can depend when the departing force has a uniform level of intensity. A uniform object with a uniform force of prominence necessarily has behind itself another consciousness that has run-off, that is, another horizon to be revived; and something that gets remembered at one time through the associative force of the respective object, and at another time, that does not get remembered, depends upon its formation. It remains a question how far essential insight reaches into this dark sphere of retention. It is otherwise in the realm of already reproductively developed memories and of the expectations that are awakened through them. There are obviously many kinds of clear essential laws here whose systematic delineation would be an important task.

I would now like to say something that relates to the problem I posed earlier and not yet solved: how memories can become discordant, how a muddle of different pasts is possible, and how the appearance[149] of a unitary past, which later turns up through splitting as illusion, as mixture [is possible]. [193]

Remembering—more or less empty, or again, more or less intuitable, right to the *limes* of ideal reproduction—is according to our presentation nothing other than the phenomenon of awakening the past objectlike formation that was constituted in the original living present and has sunken back into retention, and has finally become completely void of force. Thus, just as this process of wakeful consciousness, of the wakeful constitution of consciousness and of the sinking into quiescent sleep is a process that never breaks off, and therefore the sediments of the unconscious element are continuously layered upon one another, so too is the potentiality of awakening one that continues to infinity. And it is evident then that belonging ideally to every remembering is a possible continuity of awakening, a continuity of possible rememberings that will lead to the living present in which we stand right now. Every single remembering and every such continuous series must, it seems, be concordant. How should an "error," an "illusion" have sense here? Certainly, it is merely a reproduction of something already there in the unconscious, and this is itself something that has arisen from original constitution. Since this original constitution fashions sense originally, it certainly cannot falsify its sense. What is that supposed to mean? But remembering as mere reproduction is not in a position to do that either. Remembering would then indeed be productive in a peculiar sense, in a way that is incomprehensible at first glance.

If we have not demonstrated a source of sense at all for such concepts as illusion, error, nullity in the sphere we have considered up to now, this would seem to be a fault in the theory. For we already know that remembering can be deceptive even in the immanent sphere. Accordingly, we must first discover the

149 *Anschein*

source of this deception and therefore the origin of this concept of deception.

If we ask how discordance is possible, if we thereby ask how an overlapping and interpenetration of memories, which belong to [194] diverse pasts, is possible genetically, then we must point to the following: Memories emerge as awakenings of components of the subsoil of memory. The latter contains ordered sedimentations—layered in a fixed order—of all particular retentions, of all presents that have been constituted. If now several retentions occupying diverse positions in this order are awakened together by virtue of what they have in common, awakened from an awakening present or from an already reproduced memory, then these retentions will never be able to be remembered at the same time in a completely intuitive manner. It is an essential law in general that the original temporal field can only be filled out once in a completely intuitive manner. That is to say that even the perceptual present does not allow one remembering to become completely intuitive at the same time as and as long as it, the perception, is actually completely intuitive. The moment we put ourselves back into the past and in this way actually see intuitively, we are removed from the perceptual present, or we experience a peculiar competition. At one time we have the full intuitive present, then a clear memorial image breaks forth, then the one shines through the other, which however is damaging with respect to the intuitability of both. The same holds to a still higher degree where two reproductions are in question from the very beginning. Each one constitutes a fulfilled temporal field in the stable, steady formal system of the living Now and of what has just-been, and this formal system can only function constitutively once, can only be filled out one time, namely, in completely intuitive, i.e., actual constitution. What is given in the unity of a full intuition is given as a present with a fresh past belonging to it. In general, the living present as a formal system of original constitution is only given originally once in a lived-experiential manner.

Now, if both memories have a uniform affective force from the very beginning, a contest will ensue. The least fortification of affection, then, will give a preference to one of the reproductions, and depending upon the circumstances, will lead *mutatis mutandis* to its reanimation and then to the break through of its continued development and to a transition to the other reproduction. If there is a bridging term of similarity between both reproductions, that is, [195] an association, then both intuitions that undergo a transition from one to the other coincide in a congruent manner according to this term; they are gradually fused through this term in the transition, and fused without distance if they are completely uniform. Let us take the case of reproductive association in genesis, which indeed implies an order in the emergence. If a reproduction brings into play another reproduction through its bridging term *a*, then the awakening *a* must coincide with the awakened *a'*, and in such a manner that initially the awakening one covers over the awakened one by virtue of the force of the completely intuitive memorial situation which forms the point of departure; that is, it does not allow it to push through properly to an intuition. For the increase of force that is apportioned to it will, reaching back, also become advantageous to the force of the awakening *a*. But where the similarity is very great, in particular, where an identity of sense is constituted (which the similar terms carry within themselves, for instance, as appearances of the same [sense]), there exists a tendency toward complete fusion. Through the similar terms that are coinciding with one another and becoming fused with one another, both reproductions stand in a relation of overlapping with respect to their temporal fields. What extends beyond the coinciding similar terms, coinciding in the sense of a congruence, also exists in a certain negative coinciding, in a relation of conflict, but in a certain relation of correspondence—namely, corresponding to the generally uniform structure of both temporal fields and the fields of coexistence proper to them, like the sense-fields.

Every[150] remembering constitutes in an intuitive manner a fulfilled temporal field as a reproductive "present" that is in the process of becoming constituted; it is a "present" of a fixed formal system to which all alteration with respect to content is bound: inalterably a Now and a continuum of just-having-beens. While a remembering seizes hold, so to speak, of this formal system and

[150] Editor: Beginning of a new lecture.

fills it intuitively, no other remembering can be there, namely, as an actually intuitive remembering. But it can be awakened, and only through the victory of the other remembering that is taking place intuitively, [can it be] suppressed. Two intuitions find themselves in competition such that each one has approximately [196] the same overpowering force, and in contrast to all the other affective stirrings. Then the least priority of reproductive force, which will be of benefit to the one, or rather, to its preliminary stage (the awakened retention), will allow it to break through to an actual intuition whereby the opposing intuition becomes forestalled and suppressed. This can lead to a change, a transition from the one intuition to the other. A certain coinciding of the sense-contents can already occur here. A coinciding according to commonalties takes place essentially with each transition of consciousness; and, after all, there can never be a lack of such commonalties; for at least the temporal form is something held in common, even possibly where something spatial is remembered, something extended in a local field, the local shape.

But in general, that is a mere overlapping of images, a concealment that in itself does not produce any fusing, any unifying synthesis that is tied to the conditions of content. However, the matter is otherwise if an associative connection exists between both reproductions, noematically speaking, if the one thing reproduced reminds us of the other. The similar bridging terms form precisely a bridge that allows the one reproduction not only to flow over to the other in the transition, and proceeding from the awakened bridging term of the other side, to the contiguous bridging terms. For in this case the common bridging terms coincide in a special way. They fuse, each one with a different intimacy according to its similarity; they are congruent most intimately when they are uniform or even when they reach a congruence of identity through an identity of sense constituted in them.

Let us consider this case of reproductive association in its genetic development; an order of becoming is sketched out here. The one reproduction R, the associating one, arouses the other reproduction R' through its bridging term a. The bridging term a initially awakens the other term a', and in such a way that a, being congruent with *a'*, not only covers it, but fuses with it by virtue of the materially relevant similarity; and the covering over in question here only means that the *a'* does not achieve its own intuition, that in its own being it is in the mode of repression, and yet according to a common sense in which *a* is maintained, it has entered into this common sense.

The awakening that spreads out from the *a'* now tends to bring the *R'* progressively to intuition (tracing the connective lines of [197] contiguity), and intuition can set in gradually, in sections or as a whole: But it does this in conflict with the intuition *R*.

We note here that even the non-congruent regions of *R* and *R'* stand in a certain coinciding. We could well characterize the incongruence of the fields of *R* and *R'* extending beyond *a* or *a'* as a negative coinciding or as a covering over. In a certain, systematic way they correspondingly refer to one another: such that the one correspondent is in conflict with the other, such that the intuition of the one suppresses that of the other. Thus for both especially, the forms held in common correspond to one another: The temporal form, "original present" of *R*, is in a relationship of correspondence with the temporal form of *R'* according to this or that Now, and according to every level of what has been, such that through the common form the correspondents cover over each other in an incompatible manner; in the form, Now, the reproduced element can be given only once, and the other reproduction awakened in this form must meanwhile remain repressed. This also holds for the forms of coexistence that run through the stream of lived-experience under constant formal coinciding: for the realms of sensation, the forms of sense-fields proper to them, the analogy of the spatial form as the form of the optical field with its localities and of the spatial form of the tactile field. But also, on the other hand, with respect to the constantly constituted transcendent thing-world, the spatial form, and specifically in its constant mode of orientation. These forms produce a connection insofar as, when *R* is intuitable, the *R'* that is awakened through it is not only awakened in general with is sense-content. Rather, the original form of temporal orientation belonging invariably to the sense-content of *R'*, the field-forms belonging to the realms of sensation of the *R'*—all the like has a necessary and original affinity to the corresponding forms in the *R* that has become intuitable, coinciding intentionally with it. But what is formed in the concretion of the *R* excludes what is formed in the concretion *R'*. They can only become intuitable one after the other.[151]

<§43. The Possibilities of a Repressed Memory Breaking through to Intuition. The Disclosure of Illusion through the Transition to Higher Levels of Clarity> [198]

If, now, the affective force of *R* and the affective force of its inner affective connections is depleted to such an extent that the corresponding forces of *R'* become preponderate, then various possibilities are given.

1) Either (and this is the most obvious one) the terms in the connection emerge out of the dark; the reproduction spreads out from here, moving along the connections, and is raised to increasing clarity, and finally, the entire past present comes completely to life. (Every element that becomes vivacious represses the corresponding element of *R*, corresponding to the formal systems of succession and coexistence.) This implies that the entire affiliated temporal field stands there concretely filled out, but also that as the process of bringing to intuition progresses, as it radiates back to the bridging term, the latter breaks away from its coinciding and fusion with its complementary term. Now it comes to an intuition for itself for the first time. We then completely live, as it were, a portion of the past present. The associating situation, the current present or the first memorial situation *R* from which the awakening departed, is entirely submerged.

2) But there is also another possibility. By virtue of their own affective force, the interrelated elements of the stimulating intuition *R* stand their ground. On the other hand, by breaking through, the elements of the associative situation triumph. Both

[151] *Nota*: Earlier, I introduced protention as the title for anticipatory intentions. In this case I should strongly emphasize that the empty intentions which are found in memories are not anticipatory protentions; this also follows from what was said earlier. They are not anticipations of perceptions, but rather of memories. This requires more consideration.

can prevail without linking up, whereby each one of these *disiecta membra* bears its own protentional horizon of the repressed supplementary elements for consciousness. The fixed forms as frameworks are then spottily filled out with the contents of R and R'. This is similar and in fact essentially related to what happens in the competition within the visual fields—and for good reasons. Here we have parallel phenomena: the alternating victory of an optical image that is in itself unitary and concordant. Then the phenomenon of the piebald muddle, which also gives itself as a muddle. And finally, the phenomenon that we now find with [199] reproductions as well, the phenomenon of a combined image. In fact, rememberings can link up to form a connected remembering that has joined the elements of different memories into an intuitively concordant image. What makes the fusion into an illusory image is the force of apperception: If the components *a* and *b* have frequently been connected in a present and have been connected to form the unity of an objectlike formation, like they occur together here in an intuitive manner in the irruption of memories overlapping with one another (and in the irruption of the forms of temporal fields becoming momentarily fused through overlapping), then the components *a* and *b* will here, too, be taken as connected in the form of the unity of such an objectlike formation. But this unitary perception and the unitary intuitive image is an illusion. That is to say, to put it at first in a suggestively rough and provisional manner: The components *a* and *b* are, now as before, elements of intentional wholes, they have repressed supplementary elements that protest from the subsoil against the demands directed in the illusory image from *a* to *b* and vice versa, and they protest above all against their reciprocal fulfillments, although the protests are too weak, not loud enough to lead to a clear doubt and to a negation. But the illusion will be disclosed as illusion as soon as the affective force of the suppressed one becomes livelier and even now wins out, as soon as it leads to a reproductive development and to an unfolding, and now both the situations and their segregated temporal fields diverge intuitively, and each one developed to completeness and clarity. The illusory image will show itself in these transitions precisely as a conglomeration and fusion of different memorial images that are concordant in themselves, and the suppressed element within the fusion and its force of belief will come to life behind the fusion.

Let us expound upon these thoughts in a more precise manner and bring them to a conclusion so that what is essential will come to the fore. We spoke of protests that would be too weak to be able to break through and become "audible"; we spoke of the suppressed protests possibly coming-to-life—what is implied in this line of speaking? How are we to understand the unity of a memorial image and the corresponding concordance of the deceptive memorial belief that must abandon its unitariness in the [200] alterations of such affective transformations and that must let itself be discredited as illusion? How do we understand this discrediting itself, this claim to legitimacy and rejection of it? Obviously the gradation of clarity plays a great role here. We pose the question: Would it be conceivable that a memory, a unity of reproductive intuition in which a unitary objectlike formation is intuited according to coexistence and succession, would ever become dubious and negated if it were to remain presented in constant graduating clarity as identically the same intuitable givenness? Would it be conceivable that a reproductive intuition of ideally complete clarity could become negated by just any affective transformation, by just any synthesis with other rememberings? To pose such a question means to answer in the negative. A remembering that offers an "illusory image," an image that is to be parceled out to different rememberings, cannot be a completely clear one. All mixtures take place in the mode of unclarity. Only in unclarity can motivations, can expectations, potential coherent nexuses, their reciprocal fortifications and inhibitions, lose their efficacy, entire layers of the like can become, as it were, dimmed down; they lose their affective relief, fade away into the background. Every disclosure of an illusion is carried out in the transition to higher levels of clarity, just as every disclosure of a reproduced intuition through continual confirmation takes place in the progressive clarification of what is less clear.

<Chapter 2: THE TRUE BEING OF THE SYSTEM OF THE IMMANENT PAST >

<§44. Confirmation of Self-Givennesses by Expanding into the Outer Horizon, on the One hand, and by Approximating the Idea of Absolute Clarity, on the Other>

In[152] our summary of the previous lecture with respect to the disclosure of a deception or an illusion in the sphere of [201] remembering, we said that the illusion as illusion is disclosed in principle only in the transition to higher levels of clarity. The illusion that is not disclosed is only possible in a self-giving consciousness of a relatively lower level of clarity—principles which by the way hold quite generally for deceptions and for the disclosure of deceptions—something that we will expound upon further here. Every confirmation is a process of bringing something concealed to light, a process of bringing it to the clarity of self-giving. If for their part self-givings admit of further confirmations and the free ego actively demands them and aspires toward them, then as we know this implies that a self-giving also has, generally, its concealments, and that what remains within it in an empty or veiled manner can be brought to the light of clarity, namely, in the form of new self-givings. We know further that verification can take place positively and negatively, essentially, in two mutually related types, the concordantly confirming and the disconfirming. Even these processes can take place in a medium of unclarity, of potential emptiness, of mere intending. For their part, these processes are self-givings for the true and the not-true, that is, for verified being or invalidated being, the nullity, only insofar as the self-givings take place "in the clear light of day" for the moments of the object as their self-givings related to the moments of the object in an accomplishing manner—thus, insofar as the synthesis of fulfillment actually imparts the fullness of the self to the emptiness; and likewise for the negation, insofar as the conglomeration [having arisen] through overlapping, the

[152] Editor: Beginning of a new lecture.

suppression of the suppressed (i.e., of one party struggling for the intuitable validity) has become visible. Therefore, a progression to newer and newer, richer and richer clarity, namely, to the clarity of self-giving belongs to the essence of all processes of confirmation, processes that in this way confirm in an evident manner. But in this case not only does the progression to further clarity play a role in general, that is, the mere progression of self-giving to self-giving; also playing a role is the progression to higher levels of clarity. This is especially essential, we believe, for the disclosure of an illusion.

Let us take this up more clearly. Confirmations of empty presentations lead to self-giving presentations which themselves demand confirmation once more. But now, two things belong to the confirmation of a self giving: First, the synthetic apposition of new self-giving intuitions with respect to those components of [202] self-giving that are merely appresentational and not really presentational, that is, not actually already self-giving for the objectlike formation in question. Elements or sides of an object, of an object that is self-given in an intuition, can remain outside of genuine self-giving (for instance, with each external perception, say, of a house, the non-visible sides and parts of the house), we distinguish something that is genuinely perceived, genuinely presented, from something that is merely co-perceived, but genuinely speaking which is as yet only intended in an empty manner. We also speak of an empty outer horizon. And precisely with respect to remembering. A situation or a process extending further on in time, like a symphony, appears intuitively, but upon closer inspection only elements of the situation, small segments of the symphony, are genuinely self-given, even though we do "mean" or "intend" the whole. Thus even here we have a merely appresented outer horizon.

Secondly, like external perception remembering, too, has an "inner horizon," namely, with respect to the element already presented, with respect to the element that has already achieved self-giving. Even this genuine self-giving makes still another confirmation possible. Hence, required for the confirmation of the concrete, full self-giving is not only the foray into the outer horizon and its fulfillment through appropriate genuine self-givings, but all these self-givings demand confirmation once again, namely, progressive clarification. "Progressing": Here is the place of the gradations of clarity. All self-giving, specifically, all self-giving that is already limited to actual presentation, stands under laws of the gradation of "clarity" by which we understand in the most general sense a gradation of self-giving whose ideal *limes* is an absolute or "pure" self-giving; absolute clarity is only another word for this. This "pure" is to be understood, for instance, like the "pure red" in the series of entirely different gradations, those of the nuances of red. In both, there is a consciousness of graduation that is continually progressing in the process of self-givings, a consciousness in which the idea in question is self-given as an evidently motivated anticipation, as a degree that is never self-given, and yet is self-given in evidence as an anticipatorily motivated *limes*, self-given precisely as "idea." [203] The further this graduating process goes on, the clearer and purer the self-giving is with respect to the *limes*. All self-givings are full of such gradations, and, I said, all self-givings are potentially operative in such gradations; and by this we also mean that it concerns a universal law of consciousness in general. For consciousness not only exists, but consciousness exists in the form of the self-giving, in the form of a constant original self-giving *qua* perception, and in the form of a self-giving in the form of reproductive memory.

Examining this more closely, we will see however that immanent perception is only a perceptual self-giving with reference to possible immanent rememberings. Certainly, it originally constitutes a self [of the object], but a self that is identical and identifiable for the ego only by virtue of the manifold possible rememberings, which for their part are thoroughly subject to the law of gradation of all self-giving. A self, an object, we already said earlier, is only there with reference to the active ego, "available" for it as something that remains at its disposal, something identifiable again and again. And precisely for this reason, we already speak of a constituted self in the passive sphere because the conditions for something being freely at our disposal are already prefigured there. They are prefigured there for the immanent constitution of inner perception through corresponding rememberings, which, even if to a restricted extent, are a realm of freedom, of free generation, of the free graduation of clarity, like the realm of free expansion in the form of the fulfillment of the empty intentions in a persistent continuation in the series of newer and newer rememberings. An immanent self that has been primordially instituted is a constant possible telos for the active ego on the basis of possible rememberings; or rather it is a member of a realm of being in itself that is a universally possible telos for this ego. The path to it however is the path of attaining it, the path that is called confirmation, positive and negative. According to what has been presented, it is a dual path: the path of expanding self-giving through the broadening of rememberings, and the path of the self-givings that have already been won, self-givings approaching their *limes*, approaching the idea of absolute clarity.

<§45. The Primordial Transcendence of the Past of Consciousness and the Idea of its Complete Self-Giving>[153] [204]

Consider this significant point: All self-givings that are called upon to bring out the self of the object (here in the sphere of memory, but likewise in all spheres) as a true and identifiable self that is available, are in a certain sense "transcendently" giving intentional lived-experiences; this concerns all self-givings that enter into a possible active identification. Lived-experiences of remembering arise in the immanence of the primordial present, but what they presentify once again, the past, is transcendent to the lived-experience and to the entire stock of things constituted originally in the present. In this way, our own entire past of consciousness is given to us in a transcendent manner in the respective present, an entire past of consciousness that can be expanded any way one likes through new rememberings. Thus, every self-giving here has its own transcendent [self], and every

[153] Editor: See Appendix 26: <Repetition and Essential Identity of Rememberings> pp. 536ff., and Appendix 27: <Two Fundamental Concepts of Evidence: Self-Giving as Such and Pure Self-Giving> pp. 538ff.

self is transcendent in an original and a good sense—this certainly presents us with a curious paradox.

The stream of consciousness and its immanent time is the first transcendent in the manner of a primordial source, namely, it is the transcendent self, which in the immanence of the originally streaming present is primordially instituted, and then through rememberings issues in a self-giving and a verification of the self that are freely at our disposal in just this present. The stream of consciousness lives with streaming, and simultaneously becomes objectlike, objective for its ego; the stream of consciousness becomes an object as the transcendent self that comes to incomplete and approximate self-giving in rememberings and in syntheses of remembering of a particular present. For the ego it corresponds to the idea of a true self, to the idea of the true past of consciousness, as the idea of complete self-giving. According to what we said previously in a general manner, this idea is two-fold: The one concerns the orientation toward clarity and its *limes*, the other toward expansion, insofar as the entire self of the stream of consciousness is in question. Obviously, and in a curious way, this idea is itself essentially a streaming one because here the object, namely consciousness, is precisely a stream that carries out within [205] itself ever new primordial institutions. In just a bit we will want to address what holds for the primordial transcendence of remembering, that is, what holds for the primordial self, for the primordial objectlike formation that is constituted in the streaming of the living present as the existing stream of consciousness in the universal form of immanent time—that this also holds for transcendence in the customary sense, for the objective world, and correlatively for the realm of external perceptions. This transcendence of the spatial world is a transcendence of a second order; in relation to it, the stream of consciousness is called an immanent object.

But before I go further in this direction, let us complete our analysis that we have undertaken with good reason in the primordial sphere of remembering, and in particular with respect to the gradations of clarity. At the basis of every remembering is a retention that was prominent from the very beginning or has come into relief through a subsequent associative awakening. Its

awakening entails being singled out from the memorial background. Through remembering it has become a revived self-giving, which as a reconstitution essentially runs continually toward the present at least for a time, but then for self-evident reasons is able to follow in both directions in leaps. Essentially, every portion of self-giving is more or less clear, and this gradation of clarity is *eo ipso* the gradation of the fulfillment of the underlying corresponding moments of the empty retention, the gradation of bringing the self of retention's empty sense to intuition. We also call this gradation a gradation of disclosure because essentially the intuitability not only has an awakening force in general, but an increasing awakening force; it increases according to the level of clarity. Correspondingly, along with the increase of clarity there are more and more differentiations of sense that become distinguished and that come to the fore in self-givenness. Thus, the inner horizon is disclosed more and more when we ascend through repeated rememberings of the same [inner horizon] from an unclear self-giving to clear ones. And even where no new differences are discernible, there is, in the graduation, a more intense fulfillment and approximation to the true self, to the anticipatory ideal *limes* of complete clarity, to the *limes* of the idea of absolute fulfillment. Naturally, running [206]
through every such series of graduation, which is carried out in the unity of a consciousness, is a synthesis of coinciding that forms identities, namely, through coinciding according to the same sense and according to the same graduating self, or rather, a graduation according to the content of the self, according to the content of truth. We can also say: We are conscious of what appears in the mode of the self as something determined closer and closer, determined ever more completely (here we do not mean it is determined predicatively).

Accordingly, the degree of concealment is also to be characterized as a degree of relative indeterminacy, of relative poverty, emptiness, which runs through every self-giving. Self-giving is always an incomplete fullness, that is, fullness that is attenuated by emptiness. It has an inner horizon of emptiness in the form of attenuation. In those syntheses of the graduation of clarity we are also conscious of a deficiency that occurs in the transition because of the coinciding that overlaps. In the *limes* we would thus have the idea of an unattenuated self, of a self that could no longer be satiated any further, and this is the norm of all approximation, the norm of the true having-been; it is the true self of every having-been that is intended.

But in place of pure series of gradation, through which runs the concordance of ratifying belief, there can also be syntheses of discordance; it can happen that in the graduation of clarity, a part of what is self-given gets progressively fulfilled and continues to progress in the sense of an approximation, but that the accompanying process of the expansion of self-giving, which radiates out from this place, will lead, with the clarification, to clear and clearer and clearer supplementary elements that clash with the corresponding elements of the entire image given originally in relative unclarity. The image splits into two images, remembering into two rememberings referring to different pasts; each one maintains itself in increasing clarity. In the condition of complete inner clarity, only one remembering can be concordant, and all other completely clear rememberings are either mere repetitions of it or they harmonize in a system of integral togetherness. That is to say that if one follows through with every such remembering, completing it, to the point of fulfilling its empty outer horizons, and if one does so again and again, then this operation will yield one and the same complete self for every clear [207] remembering: The completion of every clear remembering leads to the same temporal nexus, to a single filled temporal nexus of immanent having-beens.

On the other hand, conflict also belongs in a certain way to the sphere of ideally clear remembering; that is to say, with the corresponding conditions of the community of sense, overlapping and a partial coinciding, partial repression can be generated, and the synthesis of conflict can be produced. But if a mixed image is also to be possible in ideal clarity then it is indeed necessarily crossed out here. The easiest way for you to see how something like this is possible is to imagine "in phantasy," as we say, that a house which is presented in memory is displaced onto another street, and to do this in full clarity, but yet in a way that has been crossed out in evidence.

However, if we take normal rememberings, then the gradation of clarity and the reference back to an ideal self belong to their essence, but such that both cases of determining more closely and of determining otherwise, of verifying approximation to the self or of negation are essentially open. But necessarily belonging to the latter case is the fact that parts of self-giving are confirmed through verification, [whereas] crossing out concerns lumping together things that do not belong together.[154]

What we expounded upon earlier did not merely present us with special features for a theory of remembering and for the theory of a norm of true being that essentially rules over the manifold of possible rememberings of an ego. Essentially, and understood from *a priori* laws of genesis, living, streaming consciousness contains a realm of true being that gets continually richer. But in accordance with the primordial institution in constant identity with itself, it becomes an abiding realm of true being, a realm of objectlike formations in themselves that are pregiven, available for the active ego and its active apprehension, identification, verification, and invalidation: But not only available in a current experience, but rather as a lasting and an enduring in-itself; in a certain respect, the actual process of experience is incidental to [208] this in-itself. To be sure, this in-itself of the stream of consciousness is on another level of being by virtue of the fact that its future is not likewise in itself.

We must make the marvelous feature of this situation abundantly clear. Consciousness is not only a streaming original present in which rememberings occasionally show up, and lived-experiences do not merely follow upon one another in the process of streaming through the steady form of this primordial present, and not only are there lived-experiences in an ordered sequence, in a fixed temporal form, lived-experiences that make up a unity of a past stream terminating in an ever new present. Rather, that this is the case is a fact that is available for the ego. It is a truth that it can ascertain as an active ego. The stream of consciousness up to the Now is a true being and it is for the ego, whether it notices <it> or

[154] The important supplementary exposition belongs here (pp. 260ff.).

not.[155] Every past lived-experience has been: in itself. And yet [it is there] for the ego, namely, in such a way that it is something available, something truly existing, something recognizable for the ego. If that were not the case, then we would not be able to speak about a stream of consciousness at all, and we can easily see that if it were not its nature to bear within itself an "in-itself," a true being of consciousness itself (according to the primordial conditions of passivity that make active cognition possible), then in a world that is already constituted objectively-externally, no empirical ego could attribute to another ego a stream of consciousness and everything else we reckon with it. It is therefore the most radical problem of a transcendental logic to understand how—in the streaming egoic life that is the life of consciousness—this life itself can be constituted as a true being; as a true being, I say, that is, as the ideal correlate of possible verification, as an ideal norm which, according to essential laws, lies in consciousness, and more precisely, in the self-givings belonging to it. And this ideal norm is implicitly contained within consciousness, so that later in its freedom the ego can seize hold of it and can make it manifest. In this case, the true self that is to be disclosed in the processes of concordant clarification is essentially not something fortuitous and fleeting, belonging only to this [209]
process as its momentary *limes*. Indeed, two self-givings, two rememberings, which emerge in a synthesis of identity as rememberings of the same past, can essentially turn up as rememberings of something different. But for essential reasons, the connecting consciousness of identity cannot be annulled if both terms are verified in a continual identity with themselves as they continuously approach the *limes*. The synthetic consciousness "*A* is identical to *B*" is essentially transformed if *A* and *B* approach their *limes* in a parallel manner and are verified as true, both of them in a pure self-giving of identity; then the identity is also given as a true self. Accordingly, belonging to every true being, once manifest, is the ideal possibility of repeating a pure self-

[155] In the lecture: Objectivated consciousness as a first universe, as a totality of objects. But I forgot to say that this totality does not only have a mobile endpoint in the Now, but also a beginning—as a problem. I made an allusion to the parallel between the universe as object and its reference back to the universe of consciousness.

giving, and all iterations are capable of being integrated into a synthesis of identification and only of identification, according to which the true is thus the identical and unique, like the corresponding identity itself. Certainly, all that would remain hidden, that is, the ego would be intellectually[156] blind to it if it were only to live in passivity. On the other hand, if it lives in an active manner, it does not have to exercise a cognitive activity; but in this case it is blind to everything "in-itself," to everything that it has not brought to self-active cognition, that it has not brought to a verifying appropriation which is actually carried out in free acts. But in each case, everything that makes the accomplishment of the active ego possible is already prepared in passivity, and it stands under stable essential laws according to which the possibility of this accomplishment can be comprehended.

Thus, the ego has a first, absolutely and necessarily constituted sphere of the in-itself, a first sphere of absolutely secure, true objectivity[157] without which it is altogether inconceivable as ego. We distinguish between [a] the current living ego along with the living present of consciousness with the primordial impression, retentional modifications and the [retentional] horizon's end, and [b] objectivated consciousness, the objectively temporal system of lived-experiences that presents itself in the primordial immanence of that present of consciousness partially, incompletely, unclearly, and potentially demonstrable as false in the orientations: noetically in rememberings, in the momentarily fresh retentions with the limit of the primordial impression. [210]

In originally living consciousness, however, other objectivities[158] also present themselves, namely, through the medium of the current, living lived-experiences of consciousness. These contain then not only the intentionality that leads to the constitution of immanent time and of objectivated immanence, but also a second intentionality in which different and other individual objectivities are constituted. Thus external perceptions, external memories, etc., are lived-experiences that have their objective

[156] *geistig*
[157] *Gegenständlichkeit*
[158] *Objektivitäten*

being in the first sense, insofar as they get their place in immanent time in the form of objectivated subjectivity for itself. On the other hand, they are presentations of things, of animals, of human beings, etc., and what they present in them is something objective in space existing in itself and in objective time existing in itself.

Thus, we have come to understand how consciousness prior to all activity of the ego manages to objectify itself, or how it manages, according to the essential laws of original genesis, not only to have its own past in general, but to be able to gain knowledge of it. Belonging essentially to the genesis of consciousness are the possibilities for series of rememberings to be able to be continually awakened in the form of series of fulfillment for every remembering in which the true past being is constituted as an absolute *limes* of clarity such that it cannot be crossed out. Consciousness not only is and not only becomes; an absolute norm is also inborn in it, so to speak, an absolute norm for each of its rememberings and for the closed universe of rememberings. Certainly, this norm would remain hidden and the ego would be, so to speak, intellectually blind to the true being of this sphere if it were only to live in passivity. But if proper to the essence of an ego and of an egoic life is also the possibility of a free activity, and if the possibility of a free activity extends essentially to rememberings, and if further, proper to it is the fact that it seeks fulfillment for such self-givings and strives after true being, then an attainable truth with respect to the ego's sphere of the past is prefigured for the ego by virtue of the essential structure of the underlying basis of passive life. What is prefigured for the ego is the true being of its own past life as a necessarily valid idea; it lies ready, even though higher thought processes are required to realize them as already prepared and necessarily valid. [211]
All of this holds, as we said, for the past of consciousness.

<Chapter 3: THE PROBLEM OF A TRUE BEING FOR THE FUTURE OF CONSCIOUSNESS>

<§46. Disappointment as an Essential Moment of Expectation>

Our considerations did not have any bearing on the direction of the future. While streaming consciousness constantly projects a protentional horizon ahead of itself, it is not clear that this prefiguring should be so determined and so compelling that we are able to say that a norm of true being as a binding one is secured for the future. What I have experienced[159] actually was, even if I have momentarily forgotten it, and it remains for me a true reality; progressing from rememberings to rememberings, I am able to awaken it once more and to legitimate it in pure immanence: Here, the norm is contained securely within me. But how about for the future?

Expectations can really only be fulfilled through perceptions. Thus, they are also essentially susceptible to disappointment in all circumstances. Perception brings something new; that is its nature. To be sure, it may have a prefiguring that stems from the past of consciousness, something new arrives in accordance with something already familiar, something already constituted as past for me. Potentially, the motivation can be evident to me and it can be of a force that outweighs all counter forces; it may even happen that no counter forces are at work, and further, that none of them can be exhibited in the past. But it is certainly clear that it is perception that first decides, and that something new can be a slap in the face to all expectation. Think only of prefigured sequences of sensations like melodies. The "must" that expectation harbors is not an absolute necessity of being, but rather the necessity of an anticipated being. It seems that we must say: In its life, the ego has a more or less determinately anticipated future ahead of it. Moreover, it may have the freedom to picture intuitive [212] possibilities that in the framework of this prefiguring project a fulfilled future according to the pattern of the past. But in this

[159] *erlebt*

case, only the form of a possible future and the fact that something in general will undoubtedly arrive is *a priori* incapable of being crossed out. The ego living toward the future also naturally experiences what arrives right at the moment when it has become present and when the ego actually perceives it. From this point on the in-itself of the objectlike formation in question is secured; but it is first of all here, in the actual perception, that it is instituted as a reality which is freely identifiable and demonstrable in its self at any time. The unfamiliar is then transformed into something familiar, and in a certain respect every object is a familiar object when it is an existing, true object for the ego, and in its true being, knowable for the ego.

The future is the realm of the unfamiliar, and insofar as this is the case, it is initially not a realm of the in-itself, not a realm of true objectlike formations that are pregiven to the ego in their truth, but rather a realm of indeterminacy that the ego occupies with objectlike formations only insofar as it is sure that determining fulfillment will later constitute an objectlike formation. This is what we must say, so it would seem. And yet an expectation can also constitute an objectlike formation and make the future determinate.

<§47. The Constitution of the Objective World in Its Significance for the Determinate Prefiguring of Futural Consciousness>[160]

How does the immanent stream of consciousness accomplish this? In fact, our inner life actually does accomplish this. Naturally, in the phenomenological reduction we grasp an essential type of consciousness through our factual consciousness, an essential type of a streaming lived-experience in which an objective future is constituted, at least to a large extent, and according to the belief of contemporary science, universally. Thus, not an indeterminate future that must first wait until we actually experience it in order for it to take on the character of a [determinate] object for us, that is, in order for it to be able to [213]

[160] Editor: See Appendix 28: <The Problems of the Definite Determinability of the World> pp. 541ff.

become demonstrable in the manner of an in-itself, in the manner of a true being along prefigured paths of verification, and to be able to become determinable again and again as the same according to a norm. But rather, precisely a determinate future which is determinable in this way, in advance and prior to actual experiencing. I do not ask now whether it belongs to the essence of a consciousness in general, according to immanent necessity, to have to be so shaped, that is, that it must correspond to such an essential type of consciousness. I assert that our given consciousness and its essential type has this marvelous feature, not only to constitute objectively (in the sense of demonstrated absolute necessity) its own temporal being with respect to the past up to the present, but also its future. If we ask, however, how our streaming consciousness brings that about, our answer runs: through the fact that it constitutes within itself an objective world in a transcendent intentionality. A stream of lived-experiences of spatio-material experience runs throughout the stream of the lived-experiences such that all things experienced in the course of the steady sequence of perception are referred to one's own lived-body with which everything else that emerges in the course of lived-experience under the rubric of psychical lived-experiences is then constituted as intertwined in a regulated manner; psychical lived-experiences that are bound to the lived-body are regulated as psycho-physical lived-experiences.

Yet, first of all we have to attend to the main point: Things are constituted in immanent lived-experiences, but are not themselves [constituted] as a lived-experience, neither as a past lived-experience, nor as a futural one. They are essentially constituted as intentional unities, as identifiable unities of actual and possible perception, so that they, extending beyond the current perceptions, are unities only appearing in them. And they are thereby constituted as enduring unities reaching *via* the experienced past and present into a non-experienced past and present, but also *via* the experienced past and present into the future, and reaching into it objectively. The thing is constituted from one side, but it is more than what I see of it; it has sides that are presently non-visible, but that in themselves befit it. Likewise, the thing has, in itself, a future, it is in itself not only with respect to the past and present, regardless of whether the past and present have been genuinely experienced or are now genuinely being experienced; the thing is [214] also what it will be, in itself, no matter how little I may be familiar with it. In my lived-experiences of so-called external perception, such an external existence is constituted with such a sense-giving of the in-itself. Much is indeterminate here, partly in the current perceptual thing or the thing-complex, partly with respect to its surroundings, which are constantly co-constituted as a horizon—and yet everything here is constituted as objective, as in itself; everything is constituted as determinable indeterminacy. A being that can be legitimated, a true being in itself, lies at the basis of all of this; all error, all illusion has its norm in a hidden truth, but a truth that is to be attained.

If we accept this claim as belonging to the sense of nature as the objectivity of actual and possible external perception, then we will now have to consider that this nature is constituted precisely in the immanent sense-giving of these perceptions, that is, of consciousness. The following is accordingly clear: In the natural objective attitude, nature presents itself as a universal nexus of thing-like objects in which the objective past, present, and future is determined in itself. According to this claim, the futural course of nature can be legitimated from the standpoint of every experienced present, ultimately, logically knowable and predictable—it is thus in the objective attitude.

But it is now clear, I say, that by virtue of the constitution of nature as a nature, which universal time fills out objectively, a rule of lived-experiencing and initially of perceiving—a rule encompassing the whole of consciousness and therefore also its future—is prefigured. There are norms of verification, there are originally prefigured ways of possible verification that the active ego in its freedom can discover and survey. That genuine verification is possible at anytime—this is intrinsic to the sense of every objectivity being experienced, whether we understand this objectivity in a constitutive manner or not. Every genuine verification prescribes a determining rule for the course of perceptions, of past and future ones. Just because a thing existed does not mean that we would have to carry out a perception and a progressive legitimation relating to it; but given that it existed, what is prefigured along with it is something that I could have seen or must have seen at some time by appropriately running through my kinaesthetic data; as it is also prefigured that I could see it and the manner in which I could see it in the future, and the manner in which I could continue to legitimate it as that existing- [215] being by the appropriate position or movement of my lived-body. Naturally, this yields the regulation of my perceptual appearances and therefore the regulation of my current or possible perceptual lived-experiences. Certainly, this concerns a narrower sphere at first, a sphere that we first instinctively prefer; it is a sphere in which we do not pay attention to the abnormal functioning of our lived-corporeality (which of course is itself only taken into account as constituted from a phenomenological perspective). However, every abnormality that belongs here as well—a blow to the eyes that modifies our visual images, a burnt hand whereby the tactile appearances break the rule of normality, and the like—even such abnormalities I say only indicate new rules for the interconnections between lived-experiences; they, too, belong in a grand preshaped constitutive nexus; set down in this nexus is transcendent objectivity, the universal objectivity of a world constituted as existing in itself, and therefore a universal regularity encompassing the course of lived-experiences, a regularity that prefigures a firm determination for future consciousness from past consciousness. The objective existence of the world has its correlate in the unity of experiences that are concordant and continually verifying, experiences that can now only run their course in such a way that they continually verify precisely this existence. The possibilities of experience are freely at the disposal of the ego as paths of possible verification; it orchestrates their flow through its kinaestheses.

Thus, we see that the stream of consciousness, in which a nature is constituted in a thoroughgoing manner, has a marvelous inner organization. This lies in the fact that we not only always have external experiences, but that a steady rule is prescribed for all possible experiences, those that are actually transpiring and those that are freely possible. This prefiguring is not a blind and fundamentally senseless prefiguring coming from the outside; rather it is one that is accessible to ego-consciousness in the form of knowledge: just like the objective reality of nature is not a fact that exists for itself and without any relation to the ego, but rather is a fact for the ego, namely, through actual and possible experiences and verifications that are prefigured in it—verifications which in their self-giving of a transcendent world are referred to determinate paths of verifying legitimation of true being and of bringing out its true being. In other words, the spatio-temporal world and the correlative regulation of the stream of [216] consciousness not only exists, but exists precisely for the ego, it is for the ego as a pregivenness, an availability, as a readiness for possibilities of cognitive activity that are to follow. A consciousness that would be entirely without organization is completely inconceivable. For try as we might to imagine a consciousness without regularity, and try as we will to conceive of each new present bringing new data in an entirely unorganized manner, the general structure of consciousness in general would still prefigure certain binding relationships; and in this respect it also prefigures a steady regulated order as we have come to understand, namely, that everything that has emerged in a primordially impressional manner must be held in retention and that the ego's own secure past must thereby be constituted for it.

Yet, an infinitely much richer organization, and an organization that encompasses consciousness as a whole and in advance for the future means, as we now see, the constitution of an objective world. At first, the constitution of a physical nature. But this is only a lower level, only a constitutive sublayer of the entire world pregiven to us. This world is also a psychophysical world. In it we find animals and human-beings given objectively, we find consciousness objectively tied to their objectively experienced natural lived-bodies, objectivated consciousness, so-called psychical lived-experiences that maintain through it an integration into objective space and objective time. Other kinds of consciousness also participate in objectivation, kinds of consciousness that we previously only mentioned in passing and did not consider in more detail: the kinds of consciousness like emotion and willing; and not only functions of passivity, which we have alone pursued up until now, participate in the formation of sense, but also creative activity peculiar to the functions of reason. The world that lies in front of us is not merely psychophysical nature, but also a personal, communal world, and a cultural world with manifold special types of objectivities that are there for the ego who experiences them, constituted in the ego, in the immanence of its stream of consciousness, objectivities that are indices in the ego for regulations of consciousness of higher and higher levels. For every kind of objectivity that is there for us there are kinds of self-giving, and then obviously corresponding to them, affiliated ways of legitimating the corresponding true being which as transcendent being, is always given perspectivally. Thus, [217] ever new rules of a possible course of consciousness are indicated, and modes of preparedness are developed in consciousness itself; but this regulation encompasses each and every consciousness, even emotional-consciousness and willing-consciousness, since this, too, participates in constitutive functions.

<Transitional Methodological Considerations>[161] [218]

<§48. Consciousness as a Storied Structure of Constitutive Accomplishments. The Disciplines of a Systematic Investigation>

Naturally, all of this is of the greatest interest not only from the standpoint that we utilized as a leading clue, according to which the constitution of transcendence is a means of prefiguring futural consciousness and of giving to it as well the steadiness of identifiable elements, objectivity. The great theme of transcendental philosophy is consciousness in general as a storied structure of constitutive accomplishments in which ever new objectivities, objectivities of ever new types, are constituted in ever new levels or layers, in which ever novel self-givings are developed, and belonging to them, ever novel prepared ways of possible legitimation, of possible ideas of true being. All other levels are thereby taken up into the higher ones, but are not lost in

---

[161] Translator: Section heading modified. With the inclusion of the "Active Synthesis" Material as Part 3 of the Main Text, this section can no longer be function as a "concluding consideration" as it did in Hua XI, namely, as the end of the Main Text.

them; rather they are themselves ready at any time for corresponding orientations of interest and demonstrations.

It is necessary to make all of that intelligible through the phenomenological method, that is, in pure consciousness and in a systematic order. The leading thought is this: Nothing can be grasped in a stream of consciousness, or rather, in its ego, without this consciousness having accomplished the corresponding intentional genesis from its material of hyletic components and according to essential laws, that is, according to laws that are purely and simply irrevocable. This is an intentional genesis whose emanation is the respective consciousness of the object, and whose sedimentation is the respective retentional system in which we find the preconditions for the in-itself of this type of intentional objectivity and for its normative regulation. Consciousness is an incessant process of becoming. But it is not a mere succession of lived-experiences, a flux, as one fancies an objective river. Consciousness is an incessant process of becoming as an incessant process of constituting objectivities in an incessant *progressus* of [219]
graduated levels. It is a never ending history. And history is a graduating process of constituting higher and higher formations of sense through which prevails an immanent teleology. And belonging to all sense is a truth and a norm of truth. History in the usual sense of its relation to human culture is only a highest level, and even this we see has its in-itself prefigured.

The path of radical phenomenological investigation must trace the storied structure of constitution, and it must itself seek out and bring this structure to light. Of course, it can do nothing else but take its departure from the objective world as it presents itself immediately and naively; and in it mere physical nature—dead, spiritless nature—becomes distinct as a lower level that is relatively easy to tease apart insofar as all creative accomplishments of the free ego remain out of play with respect to the constitution of nature and insofar as one abstracts from all such accomplishments in this consideration of the world. An obvious task here, one that is both encompassing and in itself already very difficult, is the study of ready-made intentionality within the self-giving of nature in multifarious modes of external perception. Many of our expositions were developed in this direction before the Christmas break. Offered to us in these expositions is, so to speak, a static understanding of the noematic contents of external perception and thus of the ready-made phenomena of external nature, or rather, an understanding of the variety of phenomena that integrally cohere in a regulated manner and that are able to coalesce noematically, and while running their course, bring into view the phenomenal unity of a spatio-temporal infinity and a nature interwoven with causal dependencies. Precisely such analyses open to a multifaceted phenomenology of lived-corporeality and of psychophysical animality.

But the further course of our considerations showed that this is only a beginning. The general theory of belief points us further down the path. The preferred consideration of concordant nexuses of experience must be superseded by a consideration of possible occurrences of modalization, of bifurcation, of negation and then the occurrences of verification. What is to be accomplished in this regard for every kind of objectivation only first became and becomes clear to us in the self-objectivation of consciousness with [220] respect to its past; and at the same time we also became sensitive in this most fundamental and most primitive sphere to the problem of genesis, a problem that we had also tackled there. The same would now have to be accomplished as well for the constitution of nature. This entire network of possible concordant and discordant modes of givenness of nature, of possible verifications and invalidations, must be integrated into the history of constituting consciousness, and it must be made intelligible genetically how, and according to which essential laws, something can be prepared in consciousness, how the latter can arise in consciousness and can make possible these kinds of constitutive nexuses and the normative regulation according to ideas of truth.

It must suffice for us to have at least brought the problematic to light, and at those places where we have carried out our analysis, especially where we have undertaken the work for the most universal and at the same time the most primitive constitution of immanent temporality, to have gained a clear understanding of the style of such investigations. By virtue of the tremendous variety of problems that belong here, these investigations lead to an entire range of transcendental disciplines, to a transcendental science of nature, or rather, to a science of a possible nature in general, we could say to a science of a transcendental physics; taken in its full sense, this would encompass a transcendental science of space and a transcendental science of time; in the same sense they will lead to a transcendental psychophysics and psychology, to a transcendental science of personalities, of individual personalities and personalities of a higher level, thus to a transcendental sociology related to possible personal communities in general. Likewise [they will lead] to a transcendental science of culture as a transcendental science of possible accomplishments of communities in general. They all treat, statically and genetically, the corresponding constitutive problems.

These [constitutive] disciplines stand in an intimate relationship to the *a priori* sciences which (only in part developed in a pure manner) explicate the *a priori* of the pure "essence" of the respective regions of objectivity or their affiliated existential forms. We also call such sciences, ontologies. The essence of a nature in general, its pure idea, naively develops so to speak the [221] ontology of nature, especially the essence of space, pure geometry, that of time, the pure doctrine of time, that of specific *physis*, of physical materiality, pure mechanics, a pure science of the possible causal formations of physical being. The fundamental concepts of empirical physics, fundamental concepts in the most strict sense of concepts of principle, are not any different than the leading concepts of the ontology of nature or of rational physics. These concepts and the axioms that are grounding in them constantly serve the physicist as norms for all possible physical being and for the empirical truths to be established for it. They serve the transcendental philosopher if we assume that ontologies that are developed naively and dogmatically exist as leading clues to transcendental analyses.[162] In theorizing, the physicist places himself on the soil of a nature that is given in experience; he wants to determine it theoretically according to its true being. The rational physicist, generally speaking, the pure geometer and mechanic, the ontologist of nature, places himself on the soil of

[162] *transzendentale Leitfäden*

the givenness of essences of the pure idea of space, of time, of a possible nature in general.

But the transcendental phenomenologist takes nature and a possible nature in general purely as the correlate of the consciousness of it. "Material object" designates for him a type of meant and potentially self-given object that he regards purely in this correlation and in the phenomenological reduction. The ontological fundamental concepts, which in principle explicate the essence of a space, of an objective time, of a materiality, serve him as indices for certain systems of verification; the systematic series of self-giving, which terminate in the true self, contain this self as a *terminus ad quem* that is distinguished noematically. The general knowledge that all true being is constituted noematically as an ideal *terminus* in the processes of self-giving, and that it must essentially be constituted in this way, and that unique lines must correspond to every essential moment in true being in the constitutive nexuses of self-giving, leads to the fact that one will begin the phenomenological investigation into an object-type like physical nature precisely with the [following] consideration: What belongs essentially to something like physical nature?—in order then to see how something of this sort gives itself to consciousness and how it gets legitimated in consciousness in a verifying manner [222] with respect to all its essential aspects, i.e., how these nexuses of legitimation must be shaped according to noesis and noema. This task, which is in no way a trifling one, of the systematic explication of a highest concept for a region of being, for example, of the concept material nature, is however already accomplished in the corresponding ontology or would be accomplished in it were it to be established in a completely scientific manner. Thus, phenomenology and ontology exist in a kind of alliance. The principle difference of method, not merely, but already with respect to the basis of their work consists in the fact that the ontologist takes the idea of nature as a kind of ideal reality, that he situates himself on the soil of this idea in order to investigate it according to its properties; for example, as geometer the ontologist takes the idea of space as given and inquires into which essential properties and ideal shapes belong to it; while the transcendental phenomenologist does not have as his theme an existing ideal space, but rather, the idea of a consciousness in general in which an objectlike formation in the form of spatiality can be given. He does not pass a geometrical judgment, but judgments about all the transcendental possibilities, upon which rest even the possibility of geometrical judgments and geometrically true judgments.

<PART 3: ANALYSES CONCERNING ACTIVE SYNTHESIS: TOWARD A TRANSCENDENTAL, GENETIC LOGIC>[163] [3]

<INTRODUCTION. CIRCUMSCRIBING THE INVESTIGATION INTO THE ACTIVE EGO>

<§49. The Relation between Activity and Passivity>

Let us now turn, then, to the transcendental consideration of the accomplishments of activity. We said repeatedly that a consciousness of the object is actually and genuinely carried out only first in egoic acts; an object—an object as object—is only first there for the active ego. All of the concepts that refer back to the concept of the object: "identical sense," "being" and "modalities of being," "true being" and "verification"—all of these get their genuine character only first within the framework of activity. We must now elucidate what this means, and therefore we must first of all completely elucidate what is actually accomplished on the lower level of passivity as well. It is endemic to the nature of the situation that we can only speak of these lower levels if we already have before us something constituted in activity, ready-made; and if we abstract from activity, then the lower level is at first unavoidably and essentially still indeterminate, so that we can also have the purity of understanding the accomplishment of <the> lower level only with the successive investigation into the higher level. In addition, every accomplishment of activity itself in turn sinks in a regulated manner into passivity, and is sedimented in the accomplishments

[163] Translator: The following pagination to the German text corresponds to Husserliana XXXI.

of original passivity, which once more demands successive processes of purification.

Passivity is what is in itself first because all activity essentially presupposes a foundation of passivity as well as an objectlike formation that is already pre-constituted in it. Thus, this also holds for the spontaneous accomplishments of genuine *logos*. In general, we can say: The investigation into the active accomplishments of [4]
the ego, through which the formations of the genuine *logos* come about, operate in the medium of an attentive turning toward and its derivatives. Turing our attention toward is, as it were, the bridge to activity, or the bridge is the beginning or *mis en scène* of activity, and it is the constant way in which consciousness is carried out for activity to progress: All genuine activity is carried out in the scope of attentiveness.

We are familiar with the fact that there are differences in the mode of attentiveness and that what we call negative attentiveness, or the counter mode of all attentiveness within passivity is called affection. Something can be noticed in a primary fashion; if this is the case, then the ego is attentive <in> a distinctive sense, the ego has turned toward it in a primary sense; but something can also be noticed in a secondary fashion; a single thing or several things in the unity of a single grasping can be called to our attention in a primary fashion or can be noticed in a secondary fashion and, for instance, can still be held onto in a secondary manner after it was primary. The affections proceed to the ego from out of the passivity of the background; they are what are presupposed [for the ego] to turn toward. Carrying out this turning toward, the ego complies with the affection; it directs itself toward what is exercising the affection. Now, before we enter this sphere, we must make the limits of our entire previous transcendental investigations clear. Naturally, we will keep these investigations in mind as we proceed in order to facilitate our task.

<§50. The Constitutive Accomplishments of Affect-Consciousness[164]. Affection of Feeling and Turning Toward in Feeling. Will and Desire>

Up to now we have left out of consideration *affect-consciousness* and its constitutive accomplishments, even though it also already plays its constant role in the passivity of the life of consciousness. At this time we want to say a few words about it so that we can reach a better understanding of the sphere with which we are occupied. It was the *sphere of presentation*[165], a term that is unfortunately extremely ambiguous, a term taken in a certain sense that is now determinative, *the sphere of objectivating* [5] *consciousness in the specific sense*.

We had in the back of our mind some kind of consciousness as the consciousness of something, as the consciousness of an object,[166] and this *something* was the same thing given to consciousness running through varying and possibly very diverse lived-experiences, namely, by virtue of those peculiar syntheses that have a continual or a discrete coinciding of sense. As long as the concordance of this coinciding sufficed, the One, the objective sense, was characterized in the mode of *being*. As this thoroughgoing unity, every constituted thing for itself in the background of consciousness exercises an affection on the ego precisely as one. If a turning toward ensues, this One and this being with its objective sense has emerged from the obscurity of passivity; it has become something grasped; the ego has turned toward it and is possibly occupied with it in manifold ways. By following the affection, a theme is made out of the identical object; I sidle up to it, as it were, in order to get to know it better, to determine it, to know it in its true being.

---

[164] *Gemütsbewußtsein*

[165] *Vorstellungssphäre*.

Why sphere of presentation? Sensuous sphere would be better.

[166] Yet that can only mean that sensuous consciousness underlies all valuing, consciousness that constiutes matter, and in a higher level, consciousness that constitues nature. Or, intitially, relative objectivating.

But now what is constituted within passivity as identical, as *an object*, can lead to a *feeling*[167] already within this passivity; it can be characterized as pleasurable or unpleasurable, as agreeable or disagreeable, varying according to the context in which the respective consciousness of the object occurs, thus, correlatively, according to the noematic horizon in which the object in question noematically occurs. That is to say, the consciousness of the object founds a *novel* consciousness: a layer of consciousness of the intentionality of feeling[168], which exhibits a novel intentionality. Certainly, even this new layer yields an accomplishment with respect to the object, the object that was already constituted in the lower level as the object of such and such sense. Sedimented in it, or rather, in the noema, is a new moment precisely as the character of feeling, for instance, as "pleasurable," or in the case of frustration, as "painfully lacking," and the like. The same thing occurs noematically with respect to what is given to consciousness as being or what is given to consciousness in the corresponding modality of being. It is something that is already there and [6]
discovered with a turning toward directed [to it] in a fitting manner, in its own way, constituted once more in the manner of an object, like consciousness as a whole and all its noematic contents.

But if we hold firmly to the directedness toward the underlying object in relation to the self-contained objectivation through which it is given to consciousness—toward the object to which feelings relate—then it is evident that its objective sense does not undergo any enrichment through the process of objectivation, that nothing at all enters into the objective sense, that it is not determined more closely, like what happens precisely through the fulfillment of its empty horizons or by virtue of its connection to other objects by means of relative predicates.

The object is constituted in and through the objectivating that underlies the intentionality of feeling, and is progressively constituted in a distinctive path of identifications, a path that is trodden by means of this objectivating and that is prefigured according to further possibilities. The intentionality of feeling as

[167] *Gefühl*
[168] *Fühlen*

such has nothing to seek down this path. On the other hand, the entire life of consciousness is certainly a progressive objectivating, too, precisely in this sense: Not only is every consciousness itself constituted as an object, but each one has its accomplishment which is, so to speak, seized by the objectivating, and which is interlaced with further objectivating accomplishments by the objectivating. Objectivating consciousness is irrevocably bound to every consciousness; what it itself brings to consciousness as non-objectivating is at the same time given to consciousness with an objectlike character; in this way, a path of possible identification is instituted. If an objectivating consciousness is already at hand, if an object is, so to speak, already instituted, a feeling consciousness can be built upon it and can enter into a peculiar relation with it, into a relation that only a consciousness relating to another consciousness can enter: namely, in such a way that neither are juxtaposed to one another, but rather in such a way that the one consciousness, the founded one, has its object in that which the other consciousness has instituted as object, and that now this object gets a new character; this new character itself becomes an identifiable one by virtue of this essential feature of every founded consciousness: only to be able to relate to an object through [the relation of] foundation by accomplishing something for it, by precipitating something on its noema, and simultaneously by objectivating the latter as sense.

Owing to this, the sense contents that have arisen from non- [7]
objectivating lived-experiences can be intertwined with them in higher objectivations; objectlike formations like objective values, like, e.g., works of art, economic goods and so forth that are given to consciousness as matters that are laden with objective value predicates. The latter are predicates that obviously arise from the intentionality of feeling. These value determinations are not the arbitrarily varying characters of feeling, they are predicates, that is, elements that are identifiable; but the sources from which the objectivation for these predicates are drawn are the feelings and the contents of them accruing to the matters in question. But in the final analysis, one must distinguish here between the intentionality of feeling itself, and the objectivating—be it passive or, in higher levels, active—the objectivating that objectivates the contents

arising in the intentionality of feeling and that makes use of them in order to constitute new predicate layers with respect to matters that are constituted in other ways.

If we go back to the genetically most primitive series of development, what is carried out in an manner entirely independent from all accomplishments of feeling is a graduated objectivating, beginning with the hyletic data coming into relief, up to things of the senses like visual things, then to intuitive material things, whereby the varying feelings do not even enter into the unity of identity of the respective constituted objects. In this way, nature is constituted as mere physical nature in manifolds of a pure objectivation without the objectivating interlacing of the accomplishments of feeling. A concept of presentation, precisely as mere, pure presentation, is characterized through this kind of pure objectivation (presentation—feeling—will).

One can also point to the following for the purpose of drawing a contrast. Each time we spoke of affection, we had in mind a mode of consciousness being carried out as the consciousness of an objectlike formation, that is, each consciousness was treated here to the extent that each one is either purely objectivating or founded, and then also according to its higher accomplishments seen from the perspective of objectivating, of following the series of intentional identification. Affection was a mode of carrying out objectivation; the constituted object exercises an affection, exercises an allure of gradually varying intensity on the ego as the ego of possible knowledge. On the side of the ego, it is a tendency [8]
to turn toward and to pass over into active objectivation, into the system of determining and verifying accomplishments in paths of coinciding that form identity. Thus, already within passivity, affection is a specific modality of objectivation for the ego. The ego is aroused in a special way here, although it does not yet orient itself actively. We must judge *dynamis* according to entelechy, and for this reason we called the ego here the cognizing ego.

If we pass over now to *feeling*, we also find here a *different mode of execution*, that of *passivity*, which means here the mode of intentionality of feeling without active egoic participation and the corresponding active mode. Again, we have different degrees of affection, different degrees of the tendency toward the corresponding turning toward, of the *turning toward in feeling*. But now when the feeling emerges out of its latency, what is new is *not attentiveness* in the usual sense, but rather a *mode parallel to attentiveness*. And the progression is not an objectivating process, but rather an *intentionality of feeling*, an unfolding of feeling, albeit a feeling that is *active* now, an intentionality of feeling on the part of the ego that is actively referred to the corresponding matter at hand.

The active intentionality of feeling has become *pleasurable or displeasurable activity*, the ego regards the object, namely, the underlying object whose consciousness of it has itself taken on the active form of execution. *The object lies in the scope of attentiveness*. But the role of this objectivating is to serve, the *ego lives in pleasure*, for instance, it regards the object with pleasure. But in this case neither the pleasure nor the character of feeling with respect to the object has for its part become actualized as object, that is, it has not become a theme of identity. Instead, what is required for this is *a new attitude*, that is, a *new active objectivation* that directs precisely *the thematic gaze toward the "pleasurable" and identifies it* and determines it *as object*. While the ego is a cognizing ego when it is attentive and when it is engaged in other *active objectivating* behavior, it is the *feeling ego* when it is *turning toward in feeling*. Feeling, while not a theme, is [9] not a passive feeling at this time, but rather an intentionality of feeling radiating out from the ego. Accordingly, we also distinguish the *objectivating* affection from the *affection of feeling*.

Further, just as objectivating consciousness has its own syntheses, which on the active level are the specific cognitive syntheses, judgments, so too does the consciousness that carries out the intentionality of feeling have its own syntheses and likewise its own modes of modalization as well, modes that accord with modalities of being through the relation of foundation. Belonging here, to give just one example, is the sense of lack that is founded in the modality of non-being, the lack of what would be given to consciousness as gratifying for the feeling ego in the modality of being.[169]

---

[169] Not given in the lecture and probably written in a moment's haste.

Striving, desiring, and shunning are so tightly connected to the intentionality of feeling (which can have manifold possible qualities that always move within the general typicality of positivity and negativity) that one might be tempted to see in them[170] only the modalities of feeling. Desire is related to those affections—of which we have spoken as tendencies of a turning toward, be the tendency of turning toward an objectivating or a feeling one—by the fact that it likewise is tendentious, but yet with an entirely different character. It is a tendency that occurs in the way that both passivity as well as activity are carried out, everywhere an intentionality of feeling in the mode of striving that, unlike the tendency of turning toward, is striving after, or driving away from, shunning, that is, it has its positivity and negativity, like feeling in general. Its fulfillment is a relaxation that results from realizing [the striving, etc.], in the change into the corresponding joy of fulfillment: At root, joy lies in the arrival of what was lacking. On the other hand, the tendency to turn toward is relaxed precisely in the turning toward, in the activation of passive consciousness, of objectivating, feeling, possibly desiring consciousness.

Realizing a desiring is a synthetic consciousness that can still run its course in a special sense actively and passively: namely, voluntarily and involuntarily. Volition is not a mere desire; it belongs in the more general sphere of pure activity. Should one [10]
say that there is no originally latent volition? Should one say, it is effective in passivity only as habitual resolve, that is, only through the fact that formerly it was precisely a current, firm resolve? The concept, "will," has been taken so broadly sometimes that every egoic activity, that is, every active presenting, intentionality of feeling, etc., has been apprehended under its name, i.e., attentiveness has been taken as the activity of the will. The strict and genuine concept of will, however, designates only a special mode of activity which spreads over all other regions of consciousness insofar as all activity can occur in the form of voluntary activity. It seems to me more and more that the will is not a separate mode of consciousness, but rather a special and

---

[170] Translator: Reading *ihnen* for *ihm*

higher form of activity that can appear everywhere under certain essential conditions that lie in presupposed objectivations and intentionality of feelings.[171]

<§51. Playful and Positional Consciousness>

In the last lecture, we contrasted objectivating consciousness and feeling-consciousness only to circumscribe our own direction of investigation. We did not want to go into the special feature of the intentionality of feeling, and in connection with the latter, into the functions of the heart[172] in general and the functions of willing, and we also did not want to go into the way in which the contents arising in these spheres of consciousness undergo that exceedingly significant objectivation through which a spiritual environing-world, a world of culture is constituted over and above mere nature. If such an objectivation has come about, then it has at all events the general feature of such an objectivation, [namely,] we have paths of possible identification, possibilities of determination. We want to get to know what holds in an entirely general manner for an objectivating activity, and what kinds of forms of unique structures generally arise from it.

Certainly, the realm of activity is *eo ipso* a realm of free [11] volitional activity; here the ego—be it fleeting, be it steady—directs its activities toward cognitive goals that are as such naturally of value to it. Insofar as this is the case, feelings, strivings, volitions, play a constant role here. Yet to a large extent we can circumscribe the accomplishments of cognitive activity and make them intelligible without entering into a systematic investigation of all kinds of consciousness, which, to be sure, would be required for a universal, transcendental elucidation of the eidetic interconnections between subjectivity and objectivity.

If we limit the scope of our analyses in such a way, we must, on the other hand, also expand <it> and still take into account the *constitutive accomplishments of phantasy*. These accomplishments

---

[171] All of this is very obscure, and remains behind what was previously said. Much is lacking to make it definitive. In any case, what I have said here is completely unsatisfactory.

[172] *Gemüt*

are a great field of free activity for the objectivating ego, but they already play a role in passivity that is very much in need of clarification. When the term "phantasy" is understood in the way we understand it here, the accomplishments of phantasy have the significance of a curious general modification that encompasses all types of consciousness with all of their noematic structures, a modification that I want to circumscribe in broad strokes in order only later to provide some descriptions of the manner in which phantasy provides the basis for its own formation; and I want to do this by considering the forms of judgment and the categorial concepts proper to them.

We all know the difference between an actual perceiving, that is, having an object given to consciousness in originary givenness, in the consciousness of actual existence, and on the other hand a quasi-perceiving, like we have for example when viewing a painted phantasy-landscape, with phantasy-houses, trees, and the like. The latter are there with the character of originarity, and we also speak freely here of perception, and yet [this] consciousness is something else entirely: The things are there not as an actuality *simpliciter*; rather, they are there as a quasi-actuality, as a phantasy-figment. We also say: as an aesthetic "illusion"[173]. But this is unlike the case of a bifurcated consciousness and the resolution of it, where it is a matter of a consciousness of a deceptive illusion, where two intuitions as it were struggle with belief: [where] the one that inserts itself into the concordant belief-unity [12] of the encompassing experience of the environing-world is the true one, [and stands] in contrast to the one that is crossed out and is not inserted into it. In our example there is no conflict, there is also no resolution on the basis of a conflict, no negation and no affirmation here. The perceptively exhibited phantasy-world is a world for itself and yet not an actual world, [it is a] mere "image." On the side of consciousness we must say here that it is an analogue, a counterpart of a perceiving, experiencing consciousness, but it is not itself a perceiving, an experiencing; it is an experiencing-as-if. But in this case it is not a reproduction. Since the objects are given as being presented in the flesh,

[173] *Schein*

consciousness does not have the character of a presentification in the mode of a memory. But even though they do this, we do not take that existence seriously; constituting consciousness is a playful-consciousness; being, in which the objective sense is characterized, is a playful being.[174]

This is similarly the case in the comparison of a remembering as an experience to a reproductive phantasy which, by the way, could appear completely uniform to a memory where the intuitive content is concerned. Reproductive phantasy is illusory[175], but not illusory the way a deception is, like in the conflict of rememberings that interpenetrate one another; rather, a "phantasy-image" stands before us as mere "image," as play, and the reproductively given image in this case can be imagined as present, or also as past, or also as future: But all that is play, and being as present-being, etc., is itself playful being. This can be seen in the fact that we do not speak here "seriously" of an attestation, but only of a phantasy-play of an attestation. An experience is binding precisely as a serious act that actually constitutes a being, an act that has actual retentions and protentions at work admitting actual fulfillment, or in the opposite case, then, of a necessarily actual rejection. A phantasy-experience is the play of an experience, and in play the progression of concordant verification can be added onto it, but the opposite [can occur] in play as well. In play, everything is possible, so to speak, one thing like the other. In fact, phantasy is a realm of freedom [13] and that means arbitrariness. We can continue to play this way or that[176]. Consciousness is continually a consciousness of the "as if" and as such has the character of this modification through and through, a modification that we call here precisely play: phantasy.

But the same can be said now of each and every [mode of] consciousness. We can contapose a playful analogue to each one of these [modes of] consciousness. Thus, phantasy signifies a general modification that in itself points back to an unmodified consciousness. We call unmodified consciousness positional

[174] Is this equivalence of image consciousness and phantasy tenable then? That is certainly not my view at all.

[175] *vorschwebender Schein*

[176] *das Spiel ... weiter spielen*

consciousness; being and modalities of being, which are apportioned to the object in positional consciousness, are serious ones, are positings. Thus, a phantasy-experiencing does not only correspond to experiencing as positional, intuitive consciousness of individual objects, but also, in play, to actual wishing, a phantasy-wishing, a phantasizing into a wishing. A phantasy-desiring [corresponds] to an actual desiring and willing, etc. The correlates of phantasy are fictions, "phantasy images," the correlates of positionality are actualities, meant or true actualities: actual object, actual deception, actual probabilities, etc. In phantasy, all of this is meant with the term play, non-being and being alike.

On the other hand, even phantasy has its constitutive accomplishment. Namely, the counter-image of a thing, which is being attested to seriously in experience in the corresponding nexuses of concordant fulfillment, is a fiction, a fiction that in play, being concordantly maintained, yields precisely the image of a concordant being, the image of a true existing thing. And thus corresponding to each type of truly existing objectivity is an image typical of it, a concordant, fictitious objectivity. A fictitious objectivity as fictum, an image as image, yields a kind of objectivity, an ideal, noematic one insofar as what the ego has formed in this manner in play can be identified in the repetition of play and in maintaining the constituted sense, and can then be explicated in cognitive acts. Every such formation, which nevertheless first crystallizes into a firm unity in free activity, is called a free possibility, a possibility of phantasy.

Up to now we spoke of phantasies purely in themselves, free [14] phantasies with their playful actualities, and we conceived of them without any tie to the current actuality of the respective ego. But it is also a common occurrence, and an extremely important type of connection between positional and phantasy consciousness, that a positional actuality is refabricated, that, for example, a red house that is given to us in positional experience is reconceived, fabricated as green in conflict with the red that was experienced.

The synthesis is carried out in the form of the on set or the assumption of a unique modalization of original consciousness of being, a modalization that presupposes phantasy. The actual positing of the house forms the basis [of this modalization] and remains in effect, but it is transformed in the following way: "assuming that this house were green." In this case, the new positional accomplishment of assumption is an accomplishment of spontaneous activity. Still, a non-spontaneous phantasy-play can also allow formations to arise that insert themselves into our world of experience and obfuscate it—not of course without repressing matters given in experience. In this insertion, they have a tie to actuality and a piece of solid ground of actuality underlying them, and in tolerating them, we also already have in them a kind of beginning, only that we have not generated them in synthetic acts and have not formed hypotheses. We certainly do not have to go into this. What we have presented suffices to contrast playful and positional consciousness, and to distinguish pure phantasy and phantasy that is imbued with positional consciousness.

<Chapter 1: ACTIVE OBJECTIVATION> [15]

<§52. Cognitive Interest and Striving for Knowledge>[177]

If we now delve into our investigation of active objectivation, the latter, as we know, necessarily refers us back to realms of objects[178] that are already pre-constituted, realms of objects that are contained in the potentiality of the background. They will be, in part, well-known objects that we got to know little by little in previous acts. They have sunken into the background with their structure constituted in activity, and if we are able to take note of them again, we then encounter them with the character of familiarity <in order> to be reconfirmed in the mere renewal of activity as the ones with which we are familiar. In part, they will be objects that are strange to us, objects that nevertheless can have the apperceptive mint of actively constituted objects insofar as the apperception already followed in the background, so to speak, the

[177] Editor: See Appendix 29: <Theme and Theoretical Interest> pp. 548ff.

[178] *Objektitäten*

model of the previous activity. No object can be given to developed consciousness without such a prefiguring. An object that exercises an affection from the background, but that does not yet bear any traits that stem from active accomplishments, is actually a limit-concept for us, an abstraction, but a necessary one, since we see necessarily that whatever exercises an affection must already be a unity of constitutive manifolds, and that where a unity is constituted in this blind way, many kinds of occurrences of disruptive inhibition, and hence preliminary stages of modalization, must already occur as well. But certainly, despite all of its ties through [passive] positionality, so long as the ego has not actively formed its world, we cannot expect the firm path of knowledge given by a teleological relation to guiding ideas, and even the constitution of firm unities of identity, which unities, as genuine objectivities, give to the ego an environing-world and a rule for its further activities. And the ego must continually intervene with ever new formations; it must not allow the objectivities that are already formed to be abandoned in the passive background and, so to speak, allow decay to rule. The [16]
organization of the realms of being, the realms of truth for the ego, especially of an objective world as the environing-world of the ego, is an accomplishment acquired only through its activity and in higher formations through its fully conscious positing of goals and goal oriented activity. What concerns us is understanding the levels of this accomplishment, the originally prefigured system of their typical strides forward, and in this typicality, their necessary strides forward.

Pregiven are not only positional actualities, naturally, meant actualities, but also the play of phantasy, emergent fictions, however they may have arisen, pure or alloyed. We do not consider them initially even though they do play quite an encompassing role in the path of knowledge, for all eidetic knowledge rests on them.

Let us take as a natural point of departure positional pregivennesses, [i.e.,] any kind of individual objects. We assume that they are given in perception, that they are unfamiliar, or more clearly, unrecognized; the active objectivation, which we call cognition in the broadest sense, is not yet exercised with respect to them, the memory of them is not attached to them in the manner of a re-cognizing. And still more restrictively: Such an object emerges, we follow the affection, *we turn toward it, we grasp it.* In a special way, we now live through the continual unity of the objective sense, that is, the continuity of the streaming and varying consciousness, a continuity in which the continually unitary self is constituted. We are continually directed to this self, to the object of experience; we actively carry out the continual consciousness of the experiencing: The consciousness of existence is hereby a *living awareness, a living belief. But in this firm directedness toward the object*, in the continuity of its experiencing, there is an *intention* that *intends beyond* what is given and beyond its momentary mode of givenness toward a progressing *plus ultra.* It is not only a progressive conscious-having in general, but a *striving onward* to a new consciousness. This striving is founded in an *interest* in the enrichment of the self [of the object] that is *eo ipso* being augmented with the grasping, according to its content streaming toward the ego. Interest is a feeling and a positive feeling, but only [17] apparently is this feeling a sense of well-being with respect to the object.[179]

It can be that the object itself also stirs our feeling, that it is of value to us and that we therefore turn to it and dwell upon it. Indeed, it can just as well be that it is not of value to us, and that it awakens our interest precisely by its repugnance. The interest of which we are speaking here is a feeling, but one that is directed in a very peculiar manner. Namely, even if an object motivates our turning toward it through a value that we feel in it, the sense-content of the object is necessarily enriched as soon as we grasp it, in part, by its merely intuitive persistence in perception, in part, by the ensuing awakening of its obscure horizons: for instance through our involuntary eye movements and head movements, and through which ever new appearances of the object make ever new sides of it intuitive. Here, the object given to consciousness is the same, but its sense is enriched in the varying mode of the manner of givenness, and at the same time this process now bears with it a

[179] Theoretical interest. Cf. the more precise exposition on content and object of theoretical interest <see below p. 548f.>.

constant open horizon, related to possibilities and expectations for ever new enrichments of this sort. Connected to this is a distinctive feeling, the joy in this enrichment, and in relation to this horizon of expanding and increasing enrichment, a striving to get "closer and closer" to the object, to appropriate the self [of the object] ever more completely. This striving can of course also take on the form, then, of a genuine volition that becomes the will to knowledge in higher, and then intelligible levels. The interest that we have described is the motive of active objectivation, of "knowledge or cognition" and is therefore called "*cognitive interest*." And accordingly the striving toward knowledge that is also usually meant under the concept of cognitive interest is also a peculiarly directed striving that is not to be confused with a "desiring-the-known-object."

Cognitive interest can be fleeting and secondary insofar as it is [18] obscured by other interests, and where it reigns, <it can> also have from the very beginning the character of a transference of feeling. But the extent to which it is at work as being directed toward knowledge, it fashions pre-conditions for levels of newer and newer cognitive accomplishments that have different forms.

<§53. Explicative Acquisition of Knowledge>

For the sake of simplicity, we want to leave all other kinds of interest entirely out of consideration and operate with the idea of a pure and unbroken cognitive interest by which the ego (which is conceived of as the subject of the acts to be described) is guided and, as it were, is guided as long as we attribute [this] to the ego. We also determine the concept of *theme* now. It designates the *object* as the substrate and center of a *unitary interest*, more precisely, the entire object toward which the ego is constantly directed in the mode of attentiveness, while its cognitive interest is satiated by its richer and richer givenness, but which also strives onward toward newer and newer givenness.

Let us now turn back to the formation of knowledge in perception. We have just thematically grasped a perceptual object and have pursued it for a time through the unity of intuition with interest and in active grasping. As long as nothing else takes place, the object is indeed only "the indeterminate object of empirical intuition," to speak with Kant. But the matter cannot rest here very long. The "examination," as the fitting German word suggests,[180] goes immediately onward, and when we use this word, we immediately think of a "more" as well. The continuity of examining necessarily falls into discrete, distinct steps. The examining that would continually stream forth in a serial fashion would become a mere staring at if it would not take shape in distinct shapes and pass over into a chain of single graspings, of single acts, acts that would form an internally connected unity, a poly-thetic unity that binds the single theses together. Examining the object, the first thing to strike my eyes especially, for instance, is its luster, then its surface color, then its shape, then a portion of [19] the object, for instance, the roof of the house, then for instance the special features of this portion, the color, the shape, etc. (In this case we take the object as a world for itself and do not let the attentiveness spread to the rest of the world, to its neighbors. What would happen here will occupy us later.) Thus, single acts follow one after another. The object, every object, has properties—so runs the self-evident ontological talk. The object is given to consciousness—at least on the level of higher developed consciousness—in such a way that it has (or will immediately gain) its inner prominence, its inner affections, and the examination follows them. In this way it comes to the progressing acquisition of the knowledge of its inner determinations. More precisely: If the object, the house, is called *S*, and these so-called inner determinations, α, β, γ, then the process generated by the cognitive interest in *S* does not simply issue in the sequence: grasping *S*, grasping α, grasping β, etc., as if the graspings, the first one and [then] the other ones, would have nothing to do with one another, as if a change of *themata* were to occur: like when cognitive interest in an object has grown weary, or has been overcome by the interest in a second object, and then again by a third one, where, in these instances, attentiveness is provoked by the corresponding powerful affection. Rather, in the entire *process*

---

[180] Translator: "*Betrachtung*." Husserl is referring here to the antiquated sense of the term as "trachten nach etwas": endeavoring or striving after something.

of single acts, which lead from the grasping of *S* to the graspings of α, β, ..., we get to know *S*. The process is an unfolding examination, a unity of articulated examination. Throughout the entire process the *S* keeps the character of the theme, and by gradually getting a hold of it, moment by moment, part by part, it is precisely "moment," "part," generally speaking, property, determination. In this way it is nothing for itself, but rather, it is something of object *S*, something from it and in it. We get to know it by grasping the properties. In the developmental process, the indeterminate theme *S* becomes the substrate of the emergent properties and they are themselves constituted here as its determinations.

But how is it that the ego, in grasping α, is aware of recognizing *S* in it? Why is the α given to consciousness in a manner that is different from the *S* or from any other *S′* toward which the ego [20] turns right after having turned toward the *S*? What we will think about initially does not suffice: The *S* must be held onto in the transition to the other one rather than simply relinquishing the hold on it, and that means that the cognitive interest in *S′* represses that of *S*, not in the sense that it simultaneously annuls it, destroys it. It only pushes it out of the central place, out of the place of primary attentiveness and examination; it becomes something noticed in a secondary fashion. It is still held onto actively, only that it has the shape of what has retreated due to the pressing nature of the *S′*; it has the shape of what has retreated but is still maintained.

All of this designates significant noematic modifications, but they are just as possible for the cases in which the grasping and examination of a concrete-unitary object passes over to another object as when it passes over from an object to its properties.

Certainly, the designated modifications essentially belong to the process of the explicative acquisition of knowledge, and with this we have thus accomplished the beginning of our description. But what is required now are more differentiated, further descriptions, as is evident from the cases that run counter to these. We realize here that the consciousness of something like a determination of the object and correlatively of the substrate—the subject of determinations in such processes, whose description we now attempt—only now arise at all; we notice that in the articulated activity and in the unity of the encompassing synthesis, which synthesis connects the succession of the acts grasping *S*, α, β, ..., an original formation of sense is constituted with respect to the objective sense of *S* and the emergent α, β, ..., a formation of sense by virtue of which the terms "subject" and "determination" originally gain their significance. The corresponding concepts have their foundation of abstraction in these sense-forms. In other words, we are standing at the place where the first of the so-called logical categories originate; and proceeding from here the primordial sources of all other categories will be opened in a similar way. We will fully clarify what this means.

<§54. The Active Synthesis of Identification. The Most Strict Sense of Objectivation>[181] [21]

Now back to our description, and <we> are guided back to it by the following questions: How is it that the *unity of the theme* S *does not get left behind*, indeed, that *S* remains the dominating theme and not merely the theme that is held onto, namely, when we pass over to the graspings of α, β, ...? How is it, when the latter are grasped in their turn and therefore have become *themata* in a certain respect—how is it that they do not have the same "weight" as *S*, that they are merely *themata* in which the dominating theme *S* is efficacious and the thematic interest in *S* is not decreased in them, but is increasingly fulfilled? How is it that we speak of an explication as the unfolding of *S* in its determinations? How, on the other hand, is it that the *S* is determined as α, as β, etc.? If a thing is initially grasped in undifferentiated generality, and further in its specificity according to shape, color, according to any kind of element coming into relief, then this way of grasping is different from the way in which a thing, then a sound, then a smell, are grasped synthetically in succession. In the first case, the synthesis with respect to the objective senses is a synthesis of continual coinciding that runs

[181] Editor: See Appendix 30: <Determining as Cognitive Act> pp. 549ff.

clear through it, running clean through the sharply differentiated steps of the act. But in the other case the coinciding is lacking. In both cases, and in all possible cases taken together—in which the ego progresses from grasping to grasping in a synthetic operation, in the combining unity of one cognitive interest—in both cases, I say, is a certain intellectual[182] overlapping of all things grasped. An overlapping is carried out with each synthesis, even if entirely different objects are observed in a unitary fashion. The ego functions as continually active in and through the sequence of steps; it is still directed toward the object of the first step with the second step, and in spite of the privileged place that the new one occupies as being something that is grasped in a primary manner, it is now directed toward both together with the new one and through the new one toward the previous one. Both together are actively taken up in the ego, the indivisible ego is in both. The [22] succession of the rays of attentiveness and rays of grasping has become a double ray.

But now there is an essential difference whether a synthesis of coinciding takes place in accord with sense in this synthetic activity, and especially a synthesis of identity for instance, or whether this does not take place. If we pass from a color over to a sound, this is not the case. If we pass from a color over to another color, always in a synthetic manner, we will already have syntheses of coinciding that coincide by overlapping in the manner of uniformity or similarity. But even if we take our case of the thing and the property of the thing, and generally, the case of the object and the property of the object, then an entirely peculiar synthesis of coinciding that forms identity arises. It is not a total identity, like when we pass over synthetically from one presentation to other presentations in the consciousness of the same object, and actively identify it with itself. Instead, it is precisely a peculiar identification, precisely that identification in which an object and its determination coincide, in which the consciousness of the substrate of sense and <of the> determination (property) originally arises.

---

[182] *geistige*

Yet we must add that a modification of the corresponding coinciding must also already ensue passively, already when the passive affection of *S* and the passive affection of α meet in the passive ego, or, even in the case of the wakeful ego, when the *S* is abandoned while the interest falls to one of its properties, so that the latter becomes a unique theme and *S* has relinquished its thematic character. The living *S* that is still fresh, still exercising an affection, coincides with the property that has been drawn from it. But then we do not have a unity of knowledge, then the *S* is not characterized as the substrate of determination for the ego, and the determination itself is not characterized as a determination. Should this be the case, then the identification must be one that is actively carried out, it must be an act running through the thematic unity of both terms, an act that we can describe in the following way: The *S* as theme initially undergoes a general examination that is lacking any determination. An affecting moment α, which is passively "enclosed in *S*," now penetrates to the active ego. But this ego is abidingly interested in *S*; as such it "concentrates" its interest, that is, its *S*-interest in α. The fullness of givenness of the [23]
S is enriched in the grasping; but this takes place because it itself is given to consciousness as *S* only in its particularity. The concentration on the particularity therefore fulfills and enriches the interest in *S*.

In this discursive activity, however, an active synthesis of identification is carried out between *S* and α, and it founds the active fulfillment of the thematic intention. The *S* in this case has passed from the beginning mode of undetermined generality over to the noematically new form of a *S*, a *S* that in the explicit identification with α and in the concentration on α has become the determinate *S*, has become the subject of the determination, α.

If the cognitive movement passes over <from> α to β, then the regard that is concentrated on α is expanded in order to be deepened in the concentration on β. But the acquisition of the first partial knowledge has precisely not been lost. After it is carried out, *S* is from now on the *S* determined as α for all further cognitive steps. With this it has a new mode again. The moment when the active two-fold synthesis *S*-α has come about, we have the *S* in the mode of the original acquisition of its determination by α. But afterwards the acquisition remains as the acquisition in retentional modification. The *S* held onto goes with the new sense formation S_α into the new determinations. The S_α is determined further as β, the $S_{\alpha\beta}$ as γ, etc.

After the current processes of knowledge acquisition, processes of explication, the object is constituted abidingly as the object determined by the respective determinations, even if it has sunk into passivity. It has internally taken up, *as habitual knowledge*, the sense formations constituted originally in the described acts. In the initial view of a later new perception, this view is given to consciousness of course with the empty horizon of acquired knowledge, and every new explication has the character of a repetition and of a reactivation of the association of the "knowledge" that was already acquired. *Even similar objects* are now apprehended according to familiar laws in an analogous sense, and the apprehension has to be verified in the actualization of its cognitive intentions by the corresponding explicative syntheses actually coming to light. The developed consciousness, the consciousness of the subject that has already operated with respect to all types of objects as explicating consciousness, will [24]
hardly be able to have objects given that are not already apprehended in such a "logical structure," that is, that are not already apprehended at least in the empty form of determinability, as the substrates of properties that are prefigured in protentional indeterminacy as chains of properties that can be explicated.

Every object now harbors, in apprehension, an implicit horizon of properties, of familiar and unfamiliar ones. But this implication is entirely different from that of objects that we must conceive as found in a still undeveloped consciousness, as entering into the first original determinations. An essential difference is alluded to by our use of symbols: Whether an *S* already arises in the shape of apprehension, $S_{\alpha\beta\gamma}$, whereby this shape, however, only lies as the empty prefiguring in the protentional horizon, or whether the *S* arises without such a horizon. In the latter case, the activity of explicating *S* yields a first, fully original chain of determination for *S*, in the former case, however, it also yields an original chain, but as filling-out in this entirely different sense as determining more closely an already prefigured horizon and possibly as bringing-to-givenness-once-again a horizon that is already familiar. What is familiar here is the *S* as it is shaped by property-like determinations.

But it is precisely here that we must begin. Let us once again clearly draw our attention to the fact that the determination of *S* through its gradual emergent properties, a determination that is carried out in the continuity of thematic interest within the activity of the ego, is an accomplishment that is in the process of being brought about, being built up here in the mode of originality, that it, then, like everything that is constituted in consciousness, passes over into corresponding non-original modes, since precisely consciousness itself can take on the mode of retentional modification, finally, the mode of property sedimented in the horizon of memory, and from there can take on the mode of the re-emergence.

Let us add that rememberings are of course also possible, namely, those rememberings that give back all such modes reproductively; in this case, however, there is also the possibility that an object *S* is remembered, in relation to which <we> now first grasp certain properties in a determinative manner that we [25] had not grasped before, that is, that are not remembered as determinations of the previous determining processes. But whether we operate now on the basis of perception or reproduction, the difference remains that determinative explications of a *S* have their basic shape of originality in a manner that is entirely analogous to how we speak in a modified way in another context of originality in perception or, in other modes of self-giving. And, in fact, we see that original activity, so far as its unity reaches, possesses the character of the unity of self-giving. What comes about in synthetic activity is once more noetically speaking a consciousness that, as a consciousness, has its What, its objectlike formation, and in the progressive intermeshing, even has an ever newly formed objectlike formation. It is a progressive creation of sense, a progressive constitution of the object, specifically, an original and self-giving one, precisely in the course of actually creative activity. But in the unity that is being constituted in manifold ways, there are several constitutive accomplishments. The one that is for us initially the most important accomplishment we designate as *objectivation in the most strict* sense. "Object" in the complete and genuine sense is identical with itself and is originally constituted as the thematic object for an ego in identifying activity. There are as many fundamental formations of identification that we have for a theme as there are modes and, as it were, sides of objectivation. (But we will soon see that all these modes of objectivation are necessarily related, and are found there ready as possibilities, even if only one of them is carried out.)

<Chapter 2: [26]
THE FUNDAMENTAL STRUCTURES AND FUNDAMENTAL FORMS OF JUDGMENT>

<§55. The Possible Objectivation of Judgment and of its Components. Syntactic Matter, Syntactic Form, and Syntagma>[183]

We have chosen as our point of departure the explication of a *S*. *S* is determined by partial self-identification as α, β, γ, and thereby it remains the one and the same *S* in the unity of this multiform activity, in the progression to newer and newer concentrations under mediating distinctions. Initially [this unfolds] in a continual fashion. But we will soon be able to take a further step whereby even this ipseity can be brought to a higher objectivating form in a synthetic activity.

Just as the *S* can only be given in this process as object, namely, in the form of a substrate, so too can the determination, the property as property (namely, of a *S*), only be given in this process. And further: The *S*, constituted in this form of substrate, is non-independent, as α, constituted in the form of determination, is non-independent; both are constituted in relation to one another. And in fact not only subject and determination (or a plurality of determinations) are constituted, but both together constitute a connecting, harmonizing unity. We have already said previously, and we say it now in a still more precise manner: With every self-contained step that is carried out, the entire synthetic deed of the

[183] Editor: See Appendix 31: Syntactic Formation, pp. 551ff.

ego makes up a consciousness, a self-contained intentionality; corresponding to the entire synthetic deed of the ego is a self-contained noema, a self-contained objective sense with a mode of being. But precisely the constitutive whole as whole—this is what we have not yet observed more closely.

If we call the *objectivating process a process of judging*, here, a process of a unitarily determinative judging, and [if we call] every self-contained determinative process that has the character of an act, a *judicative act*, then corresponding to it correlatively as its [27] What, specifically, as its objectivity, is *the judgment* or also, *the proposition*. "Judgment" is ambiguous, since it designates judicative lived-experience, evident in the predilections of contemporary psychological and logical discourse, but not proposition. But apart from the grammatical resonnances, we also use the term for more general cases. Thus where we require complete clarity, we say judicative proposition (*propositio*). We will also have to distinguish the objective sense contained in it and its mode of being: We speak of *propositional sense* and *propositional mode of being*.[184] But do these ascertainments not come into conflict with our understanding, since we did say that constituted as an "object" is something that is actively identified as theme, and in determinative judgment this is the *S*, the subject of determination, and no longer the constituted property (no matter the degree to which it is constituted in a differentiated manner in the form of the property accruing to the constituted property), and even to a lesser extent is it the judicative proposition (no matter the degree to which it is found noematically as a self-contained and articulated unity)?

How the conflict is remedied is clear. What "is found" in this way, what is in this way co-contained in the framework of our active constitution of an object is an index for essential possibilities for new active objectivations whose theme is thus the property and, on the other hand, the judgment itself. In determinative judging, the *S* is objectlike; in the progression of the determination it is altered noematically, in a certain respect it also

[184] Cf. <332 ff.> concerning proposition distinguished from state-of-affairs. Editor: See §14: The Distinction between State-of-affairs and Judicative Proposition.

takes on sense-structures that are being annexed, sense-structures that are also transformed in the mode of givenness. It is subject in a newer and newer way, and subject in the judgment, "*S* is α, β, γ...." Or actually, the chain of judgments "*S* is α," "$S_{(\alpha)}$ is β," "$S_{(\alpha, \beta)}$ is γ," etc. But the respective judgment itself is not given to consciousness as an object in the judging, it is not itself constituted as an object, but in the appropriate change in attitude it can at all events become a theme and subject of determination in a new judging, and this is likewise the case where the determinations are concerned. What is there for itself can exercise an affection for itself, but it must first be "there." The primary judgment "*S* is α" must first be constituted where *S* is objectlike, then the judgment can itself become objectlike, like when for instance it becomes a subject in the following way: when it contains the "*S*" as subject and the α as predicate. This path of objectivation is prefigured *a priori*. Likewise a determination that is constituted originally in our judging, but is not objectlike in the active sense, as for example the bodily shape, can become a theme in its own right. In grasping it, the *S* does not determine itself now; [rather,] this very *S* gets determined, and now has taken on the form of the subject. And now its previous form of determination does not belong to the sense. We have yet to treat the cases in which even here the path is prefigured. [28]

Let us note that *S* as the subject of judgment has the form of the subject, but during the judicative process, *S* is the object, and not *S* in this [subject] form. For this we need quite another step that grasps its component, the subject of judgment, in a determinative manner on the basis of judging that produces objects.

Now, one could still raise the following objection. There is certainly much that is ready for possible objectivation; but not everything is equal, and especially in judging, to be sure, the subject may be distinguished as the theme of the determination; but the determinative predicate is also distinguished, and in a manner similar to *S*. By the *S* being determined by it, by α, α itself does indeed belong to the theme and is itself in a thematic hold, albeit in service of the *S*. In a certain, general way, it is thus also "objective," and if it is made independent as subject, it has only changed the form of objectivity; it has assumed the distinguished form of subjectivity, the form of substratum.

This is completely correct. Everything that lies in the thematic regard when we judge has in a broadened sense the distinguishing trait of something given to consciousness as an object, in contrast to what does not lie in the thematic hold. And we already recognize that everything given to consciousness as an object in this sense can be given to consciousness in this way in a varying mode, and within this in a possible mode of substrate as well. In this way, in the judgment, "This tree is green" (leaving out of consideration the general conceptual and the verbal expressions), "green" is there "objectively" as predicate; in the judgment, "This green is brighter than that other green of the object over there," "green" is an independent subject, but abstracting from the forms, it is objectively the same, and can also be identified as the same. [29]

We have gained important knowledge from this; we encounter fundamental distinctions now that must be carried over from determinative judgment to all types of judgment. They are distinctions that we gain of course in reflective attitudes, all of which refer back to a normal attitude, that of the execution of judgment itself.

If we judge that *S* is α, is β, is γ, then we now distinguish: (1) the identical *S*, the α, the β, etc., in their objective ipseity, just as they are found and are maintained in the thematic regard (2) from the *S* in the form of the subject, from the α in the form of determination, etc., just as they belong to the content of the judicative proposition. The *S* itself, α, β themselves can be given in different shapes of this type.

If we pass over to other forms of judgment, we will distinguish the form of the subject and that of the object, for example, the tree as subject, and the tree as object, whereby the thematic content of different forms can be the same. And we previously saw the same thing already with respect to the α, β ... as well. Accordingly, two <or rather three> concepts are determined:

(1) The form as the component of the form of judgment or as the form that makes the content adaptable to a judgment, we call the *syntactic form.*

(2) Its content, what is thematic and maintained identically in different syntactic forms, we call the *syntactic matter*[185] or *judicative core*.

(3) What is formed syntactically as the unity of matter and form we call the *syntagma*. The judgment is a *syntactic unity* that is completely self-contained; it has in itself syntactically formed matters, and it can only have syntactically formed matters; and the matters that are formed in this way in particular divisions have their encompassing syntactic unity that contains still further "forms" of judgment: To be sure, for them, the talk of form is less suitable; for the correlate of identification, what is expressed in the "is" and many in other forms proper to it, *do not form* the matters, *S*, α, β etc. *in a similar fashion*.

Traditional logic always spoke of the *termini* of a judgment, [30] without ever bringing it to phenomenological clarity, for example, "Socrates is tall." Seen precisely, these *termini* are not, for instance, subjects, predicates, and the like, but the syntactic cores in the subjects, in the predicates. We indicate here general structures of judgment, but now it is necessary above all to ascertain the general ideality of judgment and <to determine> its relationship to time.

<§56. The Ideality of the Judicative Proposition in the Sense of Omni-Temporality>

In the progressive process of determining, judging essentially builds upon judging; it is not only added on externally, but builds the unity of a new judging, and it correlatively builds a judgment as a constitutive accomplishment out of the already constituted judgment. Just as judging is a process of becoming, so too what is originally pre-constituted as objective—what we call judgment—is a unity of becoming; the becoming is a becoming created on the part of the subject from materials of passivity, of course. The original core of judgment's being, that of constitution, is a being in the mode of created being, that is, a being in the form of temporality.

[185] *Stoff*

Certainly, this temporal being is nothing less than the being of an *individuum* or of the individual event. A judgment is not an *individuum*. The difference that arises between the two signifies *fundamental types in the mode of temporality as the form of the objectlike formation*. Now, when we arrive at vindicating the *a-temporality* of judgments and of the entire realm of cognitive structures that are built out of them or are contained in them, this *a-temporality will itself be brought to light as a distinctive shape of temporality*, a shape that a-temporality fundamentally distinguishes from individual objectlike formations. If we rest provisionally with our presently delimited sphere of determinative judging, and even continue holding firm to intuitive individuals as the point of departure, the difference already appears to us here in a certain way. If the object is an intuitive one, for instance, given [31] in its duration as red, then the judgment that brings this fact to light through the explication of the object, as the judgment carried out now, is constituted in the process of becoming and is referred to the Now, that is, to a certain stretch of time belonging to the judgment itself, a stretch of time that is different from the duration of the object.

If we repeatedly carry out the same judgment in arbitrary rememberings that yield the unitary consciousness of the same object in their concatenation, then each time it will have its new constitutive becoming of self, its new duration, and possibly even the judicative tempo will be a very different one. And yet the judgment as the judicative proposition is one and the same. This is to say that all such judging actions essentially enter into the unity of an encompassing complete identification; there are manifold acts, but identically one propositional sentence in all of them. It achieves original givenness only in a temporal act that has its determinate temporal locus, possibly in several [or] in any number of temporal acts with many [or] even any number of temporal loci. But the proposition itself does not have any binding temporal position; and its process of becoming built up that belongs inseparably to it does not have the individuality of the arbitrary act. The ideality of the proposition as the idea of a synthetic unity of becoming is the identity of something that can emerge in individual acts at each temporal position; at each temporal position in which it emerges, it

emerges necessarily in a temporal manner [as] becoming in a temporal manner, and yet is the same "omni-temporally." Thus, running through the temporal manifold is a supra-temporal unity found in it. *Supra-temporality signifies this omni-temporality.* Found in each of these manifolds is the same unitary structure, and in this way it is found in time, and of course as given to consciousness. If I make a judgment now, then the What of the judging, the judicative proposition, is given to consciousness in the mode of the Now, and yet it is not bound to any temporal position, it is not bound to any judicative action found there, and it is not represented in any action by an individual moment, by an individual particularization. It is itself at each position and becomes itself at the position where a corresponding judicative act is unfolding. But while the individual element has "its" temporal position and its temporal duration, beginning at a position and passing away at a position, and is past, such an ideality has the temporal being of supra-temporality (it is contained in individual acts, in individual temporalities that can emerge at any position, and nevertheless [32]
[remains] identically the same in all of them); it has the being of omni-temporality, which is however a mode of temporality. We will hear later that this holds similarly in a certain way for all ideal objectlike formations. But only similarly. We will hear later that the ideality of the objectlike formations of sense and the objectlike formations of the proposition is not the ideality of the species that is particularized as the *eidos* in individual moments.[186]

<§57. The Forms of Judgment Issuing from Determinative and Complete Identifications>

Given this, we can proceed in two directions. The nearest direction is the one in which we systematically pursue the shapes of judgment that are opened by our point of departure: We will go down this path. But I would like to point out from the very start that all of our discussions are burdened with a limitation that can

[186] The judicative proposition has no restriction? See accompanying page. Appendix 32: <The Ideality of the Objectlike Formations of Sense and the Ideality of the Species> pp. 553f.

still essentially influence the concept of judgment itself. We had an experiencing as our point of departure and, at all events, a positional consciousness. But it is clear from the very start that even phantasy, multiform play-consciousness, <can> bring to us objectlike formations [in their] pregivenness, and that we can freely and actively exercise entirely similar processes of an attentive turning toward, of grasping, of explicating a substrate in terms of its properties in relation to a world of phantasy that is already constituted in the background. But we will have to pose the fundamental question concerning which character these judgments have, judgments that operate on the basis of play, on the basis of the "as if," more precisely, not phantasized judgments, but judgments about phantasy objectlike formations and, in contrast to these, which character the judgments have that operate on the basis of positing, judgments not about actualities posited [in the modality of] "as if," but about actualities posited in all seriousness. We will have to ask to what extent the difference between founding positionality and quasi-positionality, between seriousness and play, approaches the conception of a concept of [33]
judgment in the strict sense. By this we have in mind, of course, the concept of judgment that lies at the basis of traditional logic, which is completely in the dark concerning its phenomenological origins. I would not like to delve into this problem here so that we can initially slake our thirst on the shape, on the systematic development of the forms of judgment, and therefore on the development of the forms of active objectivation. But first I had to point to the problem so that you will keep in mind that our concept of judgment that relates to the ground of positing has its limits precisely with this distinction, and must obtain its final demarcation only by drawing the necessary contrasts.

Let us make one more observation. We started with explicative determinations, with individually intuitive objects. But the idea of the objectlike formation was expanded in and of itself. For we saw that after something has been judged, the judgment itself, this ideal objectlike formation, can just as well become the subject of determination, and then in like fashion its ideal components, etc.

We explicitly recall that no matter what and no matter <how> something is there as ready-made and exercises an affection—that is, even every objectlike formation that arises as the constitutive accomplishment of egoic activity—this "something" can be the subject of determination and can deliver up features in the form of a property; this in turn, then, occasions judgments that themselves can in turn become substrates from which we can in turn draw judgments: an ideal-unending process.

Let us first observe the forms of judgment that are immediately contained in the framework of the determination. The conceptually most simple form of the determinative judgment is "*S* is *p*." Its most universal significance is certain. According to what we have previously said, whatever *S* stands for, no matter what exercises an affection on the cognitive ego, whatever is in the position of being noticed, can become the subject of determination, and the process of determining can obviously be concluded for itself with its first step. If the determinative movement goes on in this direction—which is another possibility—then the form "*S* is *p*, *q*, *r*," etc., arises, and depending upon the particular conclusions, [we will have] the particular forms "*S* is *p*, *q*," "*S* is *p*, *q*, *r*"; *S* is etc.

Indeed, there are different things to be noted here: In the most [34]
simple determinative judging, *S* goes throughout the entire movement of judgment in the unity of a firm thematic hold. It is that which lies at the basis, the substratum, which runs throughout and without regard to the determinations set off in relief.

Where the *p*, *q* is concerned, the *p* is grasped, then *q*, while *p* becomes something secondary, but remains firmly held onto. Let us note now that in this case *q* enters into a special synthetic unity with *p*. By *S* being a theme that is held onto throughout, a determination is continuously being prehended; now the new one is not only prehended for itself, but at the same time grasped along with it. The intellectually prehending hand that already possesses *p*, grasps *q* next and holds onto it along with the other, etc. With this, a separate form occurs for the unification of what is held onto singly, forming an encompassing hold, the collective form, which is linguistically expressed by the "and." For this reason, we must express "*S* is *p* and *q*," etc., in a more precise manner. In this case belonging to every term is a separate synthesis of identity with the *S*; identifying rays, as it were, run from one *S* to *p*, *q*, etc. They are not however only connected in the *S*, but have the collective relation on the side that is determined. The one thematic interest grasps them together in their successive ordering that however is ideal, since the ideality of the proposition does not harbor any individual temporal loci. Only the ordering is constituted.

But still another form occurs to us. If we take the judicative process in the middle of its movement and we allow it to be interrupted, two things are possible according to the kind of thematic interest [in play]. It can be limited to the *p* or to the *p* and *q*; it is not a limitless interest in *S* or it does not maintain it as such. It limits itself. Or the determinative movement breaks off, but the thematic interest, the cognitive intention, is maintained in its limitlessness. In one case, we have the forms, "*S* is α," "*S* is α and β," and further similar forms that get richer with respect to its terms; in the other case, [we have] the forms "*S* is α, etc.," "*S* is α and β, etc."

The curious "etc." appears here [as] a fundamental form in the sphere of judgment. It can be and will be pre-constituted in general, even with limited judgments, but not as taken into the form of judgment itself. Namely, an open horizon for the new properties to be expected will have arisen as the determining process progresses according to the protentional laws, naturally, [35] beyond the succession of the currently constituted determinations. Every articulated[187] intellectual movement progressing in a uniform style brings with it such an open horizon: an open one; *for it is not the next term as a single one that is prefigured, but the progression of the process*, one term and then another, etc. But it is precisely a matter of fixing this style of an open process and the peculiarity of a consciousness constituting it: what is expressed in the "etc.," in the "and more of the like." Obviously, however, this "etc." may or may not enter into the shape of judgment depending upon the thematic interest; thus, it creates differences in the forms of the judgment itself.[188]

We have therefore designated, at root, an infinity of forms. The term "infinity" itself means as much as the "etc.," only that it says more, namely, it continues to give the further *plus ultra*. When we

---

[187] *gegliederte*

[188] This should have already been said in the context of protentional laws.

use the concepts of number here for the purpose of a characterization, we can say: The forms cultivated with and without the coda, "etc.," can have one, two, three terms, and so forth. But certainly, we cannot therefore say *a priori* that any kind of determinate object will yield or can yield infinitely many property-like determinations, or even, in the sense of objective truth (of which we have not yet spoken here), that every object must have infinitely many properties.[189]

New forms arise if we conceive of the determination carried out in such a way that the *S*, after the determination by *p* is carried out, shifts in turn into the primary grasping and then issues in a new determination, but in a thematically coherent manner—this is something that is possible *a priori* at any time. Thus, not merely in the form "*S* is *p*," and "*S* is *q*." Naturally, neither of these judgments have any unity in themselves, and let alone a unity of judgment, even if they be carried out by the same ego, which [36] could indeed also ensue at different times and without any interconnections.

If both are carried out in a present, possibly even by means of remembering one of them, the *S* that is given to consciousness twice in different modes arrives at a passive coinciding without further ado, even if there is a division between them due to an interruption of interest. But if the unity of theoretical interest remains unbroken, then not only will the succession of both judgments, "*S* is *p*," "*S* is *q*," be bridged through a passive coinciding of *S*, but the thematic activity will run throughout this bridging in the *S*, and then we will have a judgment that is built up out of two judgments. Thus, if the identification of the *S* with itself reaches into the unity of thematic activity, we will have two identifying activities with the one identifying activity running though them. "*S* is *p*" and "the same *S* is *q*."

We get contrasting forms when we allow the identifying connection to be carried out on the side of the determination. We judge, "*S* is *p*." We imagine that now a thematically determinative interest in *p* is awakened, and this is determined as α, while the

[189] Editor: See Appendix 33: <Main and Subordinate Determination and the Division into Main and Subordinate Clause>.

interest in *S* remains the secured and overriding interest. We will then have a new formation of form. Above all, the transformation of the form of the determination *p* into a substrate of determination, is for itself already a principle of the formation of forms in judgment. In this regard, lacking a fitting expression, I spoke of nominalization in my *Logical Investigations*, but only in relation to non-independent features. Here I want to use the more general expression, "becoming independent."[190] The *p* becomes independent in becoming the substrate. This is accordingly already a universal formation of form; for we can distinguish forms of judgment in general where there is a determination that has become independent and, in another case, where it has not.

Let us observe an important, special group of forms of judgment that belong here. Granted that we judge "*S* is *p*" and if the becoming independent of *p* leads to the judgment "*P* is α," then, if a unity of thematic interest encompasses both, an identifying synthesis ensues that produces a unity of judgment. [37] Initially, one will begin: "*S* is *p*, and this, namely, this *P* is α." For example, "This thing is red, this red is saturated." But this second proposition must assume the form of a subordinate clause, since the interest belonging to the second judgment is subordinated to that of the first; in other words, since the dominating interest is the determination of *S*, the second proposition in turn gets the form of the attribution. This thing is saturated-red, it is saturated existing red. The closer determination of red co-determines the *S*. The form of a determination arises that is now determinate and as determinate is determinative.

In the present context we have seen different identifications function in connection with determinative, that is, with partial identifications. They were obviously complete identifications. But complete identifications can also determine forms of judgment in a different way, initially, in an entirely independent manner. If this is the case, they will yield peculiar judgments that are on a par with determinative ones. Finally, we could have begun with completely identifying judgments instead of with determinative

[190] *Verselbständigung*

ones. Indeed, the determinative ones seem to me to merit a priority genetically.

What is identified will normally already be determined to some extent, even if minimally. In this case we must distinguish: the continual fusion formative of identity—*S* is recognized and immediately explicated in a progressive manner, for instance, by reappropriating the interest and the previous "conviction"—or distinctions are made and actively identified. We never speak of acquired convictions, of their reanimation, reappropriation, etc. It is nevertheless quite conceivable that a thematically grasped *S*, still prior to being explicated, reaches a unity of identity with an S' that is given in another mode of givenness. We would then have the identifying judgment, "*S* is identical to S'." The form becomes fruitful through its tie to determinations. And here all emergent forms of syntactic terms, in the framework of determinative objectivations, give to us many (ideally prefigured) possibilities to shape more richly the identifying judgment: thus, for example, "*S*, which is α, is identical to S', which is β," or "S_α is identical <to> [38]
S_β," or "S_α, which is β, is identical to S', which is β and γ," and so on, in many regulated combinations that can easily be derived from it. Further, we can once again bring to the unity of judgment several identity-judgments through overarching identifications, for example, "*S* is identical to S', and the latter is identical to S'' as well," etc.

<Chapter 3: THE SYNTACTIC AND THE OBJECT-THEORETICAL DIRECTIONS OF EXAMINATION> [39]

<§58. Sensible and Intellectual Objects. Substrate-Objects and Determination-Objects. Independent and Non-Independent Objects>[191]

In the formations of form that we have carried out up to now, we are thoroughly concerned <with> those things that are ideally iterable; that is, the principle of each particular formation of form can repeatedly be in operation. In this way, the series of property-like determinations, which can be connected to a subject collectively in the form "S is α and β and γ," is ideally infinite; likewise the series of subordinate clauses connected attributively to a subject is ideally infinite. This is likewise the case with the principle of the formation of forms; [this principle says] that a determination can be made into a subject of determination for a new determination. Etc. I refer to this because it can be seen in a like fashion where the futural formations of forms are concerned.[192]

Let us note further that our point of departure from a single concrete individual object as the main substrate was merely an exemplary one, or can be taken as such in retrospect. For no matter what exercises a unitary affection, no matter what enters into the attentive regard, we can make it a substrate, then conceptualize the idea of a substrate as such and now intuitively see the possibility of all such iterable forms of the total and determinative identification that we also called forms of judgment. Accordingly, in the generality of this mode of observation, the concept of substrate can leave open whether or not it is a matter here of substrates that have arisen from the thematic becoming independent of a determination. This having arisen in no way [40]

[191] Editor: See Appendix 34: <Absolute Substrates and Substrates as Determinations that have Become Independent> pp. 556ff.

[192] Cf. <314> concerning what we will treat subsequently. The distinctions of the *termini* of judgment according to the origin in the sphere of judgment itself, etc. Eidetic generalization of the distinctions that have been won.

needs <to> mean something that is merely arbitrary and relative, as we will immediately see.

This holds likewise for the distinction between the substrates that stem from passivity and those that stem from activity. There is nothing prejudiced in the generality of our forms of judgment even for this distinction, and even this distinction has behind it an absolute distinction. Let us now elucidate this distinction here. If the ego pursues an affection, it grasps it and makes what was already constituted passively an explicit object. It exercises receptivity, an actively self-giving intuition, and if it exercises further determination, if it carries out total and partial identifications with respect to the received object, and thus [carries out] other activities that interconnect, then arising now in the sphere of synthetic activity itself will be new objectlike formations, the judicative objectlike formations, states-of-affairs, which themselves can then in turn become substrates of identity and of determination.

But everything that is actively constituted sinks into the background and is transformed into a passivity; emerging from memory, it can similarly exercise an affection like other types of passivity. And yet it continually bears within itself *the stamp of its origin* and its essential peculiarity. It can actually be received only in the explicit execution of the same multiform activity from which it has arisen. Only then is a *grasping of the self* [of the object] possible. The self-giving constitution of such objectlike formations is essentially a synthetic activity. Thus, this must have already taken place so that what was constituted in a multiform fashion can be grasped in a grasping of the self and so that it can become a theme. By contrast, there is an *original passivity*, an *originally passive constitution* of objects, whose activation is *sheer reception*: the sensible objects. All objectlike formations that have arisen from activity—which can only become passive after they themselves, or ones similar to them, had been actively "engendered"—point back for their part to founding objects that were originally constituted in a passive manner. *The intellect* is a name for constitutive accomplishments of objects that the ego has given to itself through the activities of identification. The self-giving is a creative self-giving. *Sensibility* is the contrary, <it is a [41]

name> for constitutive accomplishments without the participation of the active ego; the grasping of such objects is indeed an activity, but it is a mere receiving of a pre-constituted sense, and further explicating, judging already presupposes this sense.

The difference is not a mere relative one, although even intellectual objects, as ideas, can exercise an affection from [the sphere of] passivity and can then be received. Precisely this receptivity is fundamentally different for both; in order to be an actual grasping of the self it needs to be an active-synthetic accomplishment where intellectual objectlike formations are concerned, and where sensible objectlike formations are concerned, this is not the case.

Also belonging here is the distinction between *original substrates* and *original determinations*. What has emerged in the judgment as a determination can become independent thematically, it can become a substrate. Thus, the same substrate can appear now in this form, now in that form. And yet there is an absolute difference even here in the background. There are objects that can occur in original self-givenness in the form of substrate only by having occurred previously as determinations. And, on the other hand, there are objects for which this is not the case. Functioning as determinations is not essential for the latter; we call them *original substrate-objects*. For the other ones, the form of determination is essential, namely, their *a priori* original form; we call these objects *original determination-objects*. A shape, a color, is to be given originally only as the determination of a concrete object. A shape cannot become prominent without the shaped object becoming prominent, and the object is grasped first, even if the interest immediately passes over it, and then the object's color is grasped, which may perhaps immediately usurp the main thematic interest.

So it is with respect to all non-independent moments: Their *concretum* is grasped earlier than they are themselves, and prior to this they occur in the form of determination, in that of the "on something" before they become the substrate. This is otherwise when we have a pile of stones that are strewn about, given in an intuition that brings it into relief, or when we have an articulated whole like a series of markedly different houses: The particular affections of the members here are contained in the unity of the whole affection. But from the very beginning the plurality or the [42] whole (the unity of the street) can be grasped and become the substrate, and any particular thing can become the substrate just as well.

These distinctions about which we have just spoken, the distinctions in judicative objectlike formations of course also signify universal distinctions in judgments that we can also meaningfully characterize as formal distinctions of judgment. But those universal forms of judgment that we have enumerated up to now have a universality and so encompass all such distinctions of judicative objectlike formations, and are insensitive to all such distinctions.

We can also characterize this *dual differentiation of forms of judgment* in the following way: If we imagine objectlike formations as objective sense, as they are always constituted and ready to occupy our thematic interest, then judicative activity can come into play from different directions, and these objectlike formations, together with those that are formed out of them or together with them, become the *termini*, the syntactic matters for syntactic forms whose possible system of forms is prefigured *a priori*, irrespective of the particularity of the *termini*. The judicative activity creates newer and newer objectlike formations through these syntaxes themselves that are constituted in them. As objectlike formations in general, they can also become the *termini* of syntactic forms, and these forms belong again and again to the same system of forms. *Accordingly, the difference between the syntactic matter and the syntactic form is relativized.* The theoretical-logical interest can (1) *go purely to the syntactic formation of forms*. Without concerning ourselves with this relativity, by leaving the *termini* indeterminate, one can ask: Which forms of judgments are possible for objects as such, grasped as thematic substrates no matter how they are constituted? The extent to which the objects remain indeterminate, and with them all *termini* in the forms, which is evident in the mere statement of the letters S S', p, p' etc., is the extent to which the judgment, as objectivation, itself leads to particular formations of *termini*, like to the form of the determination in contrast to that of the substrate; but this form belongs here in the very form of judgment. It forms something that abides in an indeterminate manner.

(2) On the other hand, then, one can (and one must even [43] necessarily), however, pass over to another direction of observation, namely, *posing questions concerning the syntactic matters*, particularly of such a kind that can be posed with regard to all matter in general, and which are therefore formal characteristics as well. One arrives, then, at the *general questions of the origin of the judicative objectlike formations*, at relativities of the type just mentioned, and at the absolute lying *a priori* behind them. That designates a *system of forms that is prefigured a priori for the indeterminate "objects" of possible judgments*; after the exclusion of the relativizing syntaxes that could be included in the *termini*, there remains a general typicality of the absolute syntactic matters to which all judgments in their *termini* point back as relative matters, in other words, *a typicality of the ultimate object-senses* that contain nothing more stemming from intelligibility. They are *types of the most original objects* that can be *thought* as objects of possible judgments, *as objects of possible knowledge*.

Guiding distinctions for this are the distinctions that we made previously between sensible objects and intellectual objects, or, as they are also called, *categorial objects*; further, between original substrate-objects and original determination-objects, likewise the distinctions in concert with them between the *independent objects* and *non-independent* objects, between concrete and abstract objects, etc.

We call *concepts* that grasp scientifically *the elements of determination that belong to the form of possible judgments, logical categories*. Accordingly, *distinguished* for us, and corresponding to the two directions of examination, are *syntactic categories* and *core-categories or object categories*—like, then, the one logical direction of investigation that is specifically syntactic, and the other that is object-theoretical. In transcendental logic, even the examination that is directed toward the *a priori* of the objectlike formation is naturally transcendental; it treats all of the essential distinctions and laws proper to this direction in [44] essential connection with judgment, in connection with constituting consciousness.

<§59. Whole and Part. Sets as Special Cases of Wholes>[193]

At the conclusion of the last lecture, we were right at the point of continuing the development of judicative forms a bit farther in a purely syntactic direction. We had won the first group of forms from a single center of objectivation. An affection [stemming from] an objectlike formation led to the grasping of an object *S*; the thematic interest belonging to it led to explication and was satisfied in certain partial ident<ifications>, i.e., with respect to the properties that were won. This yielded different modes of identifying interconnections here. To this we added the total identification and the syntactic forms arising through the connection of both functions. Let us expand our horizon. *Instead of one thing exercising an affection, we will take a multiplicity.*

Of course, there are many things that will exercise their affective force, partly originating from a living present, partly arising from memory. We are considering only a plurality in which the special affections are combined to form the unity of an entire affection. Such a combination is already based—in the realm of passive pre-constitution—on commonalities of pre-constituted single senses. With this, pathways are prefigured for thematic interest, as well as for courses of interest and courses of judicative syntheses that are connected in the unity of an encompassing interest. Indeed, we see a constant variation here already within passivity to the extent that the intensities of the affective forces can change; something new can push to the fore, something else can weaken and can become submerged; and likewise, when a thematic interest is activated for an object, we see a constant variation in the thematic horizon. But such a horizon will always be there; and what is alien to the theme will always separate off from what belongs together with the theme when an object breaks through to attentiveness. The former has [45]
nothing to do with our present thematic substrate, it is alien to it;

[193] Editor: See Appendix 35: Multiplicity and Judgments of Multiplicity, pp. 559ff.

the latter, which is grasped thematically, does not alienate us from the previous theme, but rather, by fulfilling its new thematic interest it also enriches our previous interest, fulfills it along with it. This is just a correlate of saying that in the process of grasping and explicating the one theme, a common element comes to light along with what is self-contained in the other theme and along with what gets explicated from the standpoint of it. Active operation and fulfillment of objectivating interests bring commonalities to light, which is to say, it consists in continuously interrelating syntheses of identity. That which is common, and in the broadest sense, that which is identical, is disclosed in judicative syntheses, and in very different forms.

Yet certainly judging as an objectivating process is an unending one, and one that is infinitely enriched because not only is an ever new objectlike formation pre-constituted within passivity (and everything that is pre-constituted can be combined precisely with such a pre-constituted objectlike formation [leading] to new reconstituting accomplishments), but because even the process of activity does not only "make manifest," but at the same time accomplishes constitution and thus fashions ever new themes in which it does justice to the previous determinative ones.

We can also make the following general statements. If thematic interest is realized, that is, fulfilled, it necessarily branches out into a manifold of thematic interests that are, however, organized in the unity of one interest. The corresponding judicative activity proceeds in judicative steps, each step is a judgment, but it is connected with the other judgments that are already carried out, forming a unity of judgment. It does not matter here whether we take as a guiding theme a single substrate, at least in fictive isolation, or whether we have a manifold of substrates. If we look at this more closely, we will recognize—precisely as an essential peculiarity of a thematic-unitary process—that no matter how many objects may thematically exercise an affection and coalesce in the unity of a theme, a satisfaction of interest is however only possible through concentrations [of interest] in which an object becomes a substrate each time and with this a subject of determination.

Every judgment, out of which the thematically unitary [46] movement of judgment is built, has the necessary bipartite structure: subject and determination. If, in the reorganization of the form, the determination (as we see in the case of relative clauses) is adjuncted to the subject, if likewise a determination itself in turn gets a similar annex through its own determination, a central caesura remains in the whole advancement of the determinative process in every judgment, namely, what separates the subject of determination from its determination.

If several objects that cohere thematically are there from the very beginning or enter into the thematic unity successively, then each one will demand becoming the thematic main subject successively, that is, a concentration of interest that pushes its companions back into the provisionally secondary position. To be sure, we have to consider the two possibilities here: Namely, the one possibility is that in a movement of judgment several thematic objects are necessarily equally valid in the thematic interest; the other possibility is that one of them or particular ones hold, so to speak, the royal interest, so that only they should become the subjects of determination, while the interest in the other objects only leads to determining the subjects or the dominant subject through them.

And this is the way it is already in the explication of a single subject. The singularity that we presupposed in our first considerations was obviously this singularity of domination from the very outset. The subject cannot remain the singular object, the subject develops; that is, precisely new objects, the subject's properties come to the fore. But these objects are interesting in this connection only as determinations; their function is to serve. But they could also have an interest coordinated in their previous substrate, or they could gain it after they have arisen. And now we would have not only the path of its explication through <which> the S would also be determined mediately, but also <the path> of a new mode of determination that we have not yet considered, not where S is determined by p as we previously considered, but where p is determined by S.

This will become immediately intelligible if, in a formal and general manner, we call the property-like determination a *part*, and say that an object—*speaking in the formally most broad sense*—is *a whole* that has its properties within it. The whole is [47] determined in that grasping that grasps each part. We had alluded to the originally unitary mode of determination with the terms "*S* is *p*"; we could just as well have said "*S* has *p*." Conversely, however, the part is also determined through the whole. The *p*, taken as substrate (and perhaps already grasped as substrate before the *S* was grasped), leads to an identification in the transition to *S*, but naturally not to a complete one. *P* is contained in *S*, the interest in *p* is also enriched with this. Every identification determines its subject. It determines it here not in a property-like manner or, as we also say, through an *inner determination*, but through an *outer* one. Already by virtue of the *constitutive community with its whole* prior to the judgment something has, so to speak, made an impression on the part due to the fact that it is precisely not for itself, but rather is constituted in an affective coinciding with *S*. This coinciding is unfolded in the thematic activity in the form of a peculiar identification. Thus, we have counterparts here. Let us alter our characterization in order to remove the priority of the place of the predicate, a priority alluded to by the letter *p*. If we have two objects, *S* and S_1, they can stand in a relationship of parts, and depending upon which one is the main substrate, we will have on the side of the determination the "has S_1," or the "contains *S*." And in both cases here, the object on the side of the determination has the *correlative form of determination*, that of the *inner* or the *outer*.

If we call the *substrate-object*, *subject*, and the *determination-object*, *object*,[194] the former will always have the same form of function as subject; that is, being such a form of function, it is not differentiated. Only other annexes, like attributions, can provide the subject with further forms without altering it as such. On the side of the determination, however, we have different modes of determination within the general form of function, determination, and then also supplementing this are possible attributive annexes. Every new form of determination *eo ipso* fashions new forms of attributes, for example, "*S* which contains *S'*." These forms of

[194] *den Bestimmungsgegenstand Objekt*

determination, however, are not the only ones, and even not the only ones looked at from the standpoint of the community of parts.

In accord with a self-evident ont<ological> consideration, two [48] objects can stand not only in the *relationship of the containing and the contained*, but can also stand in the relationship of *intersection*. Even this develops in a peculiar form of identification. In [the relationship of] intersection, S is identical with S' (according to a common part S''), which is explicated differently: "S contains S''," and "S' contains the same S''." Or in the new, plural form: "S and S' contain S''," in which the determinative object S'' only appears once, and diverging from it are two rays of identity, the one to S and the other to S'.

We can and we must understand these forms of judgment here in the most universal and syntactic generality, and this means that we leave it open whether the respective objects themselves are unitary or multiple. Surely, the concepts that appear here, concepts like *the containing* and *the contained*, *whole* and *part* are *cat<egories>*. No object is conceivable that is not part or whole, and the universality of our formation of these concepts allows us <to call "wholes"> even the loosest of sets, [including] all the multiple unities that we call, following a more restrictive concept of whole, "mere sets" and do not want to call a "whole."

We do not yet have such restrictive concepts here, but we do have the "having-within-itself" and the "being-contained." It suffices that we have in mind several objects in indeterminate generality, objects that stand in the corresponding community of parts, and then the active judicative activity yields judgments of new elementary forms we characterized, elementary forms that obviously make possible ever new forms. On the one hand, this takes place through iteration, since we conceive the objects functioning as the point of departure in turn as parts of objects that lie further back, and we can continually imagine their splitting into parts. On the other hand, the thematic synopsis requires the conjunctive connection of particular judgments, and besides this, the reconfiguration of determinations into attributions is always possible.

Let us add here at once the discussion of the differences [49] between the unity of sets and the whole in the special sense.[195] *Objects that do not stand in a community of parts* are not therefore devoid of community. Their affective and active community can be given to consciousness in an original manner; they can possess their affective and active community from entirely different means, and can accordingly constitute a judicative community. If we call such objects *separate*, *disjunctive*, it is clear that such objects are also given in separate intuitions and that in the transition from one to the other, an entirely different consciousness comes about, like when the identifying coinciding that we described here would become manifest.[196]

Now, it can be that disjunctive objects, regardless of the affective force that each one exercises, coalesce to form the unity of one affection, and that initially this unity awakens the activating interest. Then the disjunctive multiplicity as unity, a set, is grasped, and it is explicated by going through the particular members and in the conjunctive collection of them: set T is m[ember]$_1$ and m_2 ...m_n. Every single step is an identification of "the containing." If the series of the steps is finished and if every contained member is manifest, then a complete identification occurs.

The entire set enters member by member into an identification with itself, and every member is given to consciousness in a two-fold manner, in implicit and explicit form. The set as the subject of determination is resolved into the sum of it members. The set that is grasped in a single ray of attention as the subject of determination is completely identical with <the> collection that is constituted in multiple rays of attention, with the entire totality of individually grasped members, each one of which has the form of determination, "the contained." Noteworthy is that in this collective form of sets, "m_1 and m_2 ... m_n," the "and" as a summary "and" carries the sense of separation, the sense of the connection of disjunctive objects. Yet, on the other hand, we have the indirect

[195] The following presentation was not immediately successful. See the appendices that contain improvements, and see the next lecture.

Editor: See Appendix 36: Set and Whole, pp. 562ff.

[196] Does this suffice?

construction when we say that it would consist in the members not [50] standing in a community of parts. There is no talk here of negation. The set is given as objectivated accordingly in a completely identifying judgment of the structure indicated, in a plural judgment if we understand under this general rubric, every <judgment> in which many rays of identity coalesce into one member and are thereby connected—be the judgment on the side of the object or the side of the determination.

A set, explicitly objectivated, that is, completely determined, is a collective combination of disjunctive objects: Yet the objects can themselves be sets once more, only that they must be disjunctive sets. Ultimately, every intuitively pre-constituted set leads to final members, to particularities that are no longer sets. For belonging to the idea of such a set is that in the first givenness of the substrate, the set <as> an already pregiven manifold of special affections is there that we only activate. But certainly this does not rule out that "in bringing something closer," the intuition brings new affections into play, affections that were not already there, and that resolve the unites once more into multiplicities: so that the concept of the unity of sets becomes relative. It[197] hardly needs to be said that within the unity of a set different part-sets are to be delimited through affective special connections, that overlapping sets are possible here and that in this general way, sets in relation to other sets can offer all the possible relations of containment. Naturally, the concepts of set and unity of sets are also logical categories.

Sets are special cases of wholes. Now, what distinguishes a set and a whole in the restrictive sense, and in an original way? We carry out all of our considerations in such a way that we appropriate the cat<egorial> forms, forms in which objectivity is constituted on an intellectual order, to t<he> acts that give in an original manner.

In order to have a set "intuitively," in order to have the unity that unfolds its sense in a collection through total identification, the mere unity of an affection of separate elements must come first. How this unity comes about otherwise, what lends to the [51]

---

[197] Insufficient.

particularities their community—this remains fully open. If, for example, a memory reminds me of several other melodies that I had heard in an entirely different context, their similarity suffices to produce a connection. They can be of interest unitarily as a combination and can yield a set; we say here that they do not have any "real" tie; they do not form any real, general whole, no whole bound in a materially relevant manner, like what holds for tones and figures of a melody, or for houses and streets, but also for the parts of a proposition, for the propositions of a judicative framework, etc. If we compare the *self-givenness* of a set with that of a whole in the narrower sense, it is clear that the *whole* is *constituted* in the *unity of a continually harmonizing intuition* and as correlate.

<§60. Non-Independent Members of Judgment, Independent Judgments and Interconnections of Judgment>

At the end of our last lecture, I began to clarify the categorial concepts of whole and set, concepts that arise around the judgments of containment, and initially the concept of set. I should not dwell upon this any longer, and for this reason I will not pursue my treatment of it further *in extenso*; rather, I only want to say briefly that our point of departure from pure passivity and from the described multiplicity of affections that coalesce to form the unity of an affection, should not mean that like the sensible objectlike formations, the set-objectlike formations are also already actually pre-constituted in passivity and are objectivated only in categorial activity for the intellect.

A closer investigation shows that a set as such only arises at all in categorial action, put more precisely, in colligating, in the "and-synthesis" of thematically grasped, separated objects. Wherever, and no matter from what motives, the thematic interest passes over from one object to other disjunctive ones, and leads to an encompassing "and-synthesis," a new objectlike formation is thereby constituted (namely, purely by virtue of this synthetic activity); because this object has arisen from such a synthetic [52] accomplishment, it is called an intellectual object, a categorial object. Every objectlike formation that is generated from the ego synthetically, that is, with many members, has two constitutive stages. The categorial synthesis, in our case, the S_1 and S_2 and S_3, is noetically a unity of consciousness, to be sure, but not yet a unity of one object in the genuine sense, which is to say, a theme. Thematically, the S_1 is found in it, and again, S_2, S_3, but not yet the *collectivuum*. Likewise in the case of all categorial synthetic actions: The thematic element is what every self-contained step has given to consciousness as a thematic object. Synthetic consciousness encompasses several objects in a unity, but it does not have a single object having many members. However, a new object—precisely the object having many terms—is essentially pre-constituted through every synthetic, unitary consciousness, and this requires only one thematic grasping that is possible at any time in order to render this pre-constituted [objectlike formation] a thematic object, and thus to become an object[198] in the full sense.

In particular: Colligating is a synthetic accomplishment through which a *collectivuum* is essentially pre-constituted. It becomes a thematic object through a reverting grasp that takes place after the completion of the colligating process; this grasping is possible at any time and it is only with it that the set is first given as object, as an identifiable object for the ego itself. Afterward, it is an object like any other; it can not only become completely identified as the identical object of many modes of givenness, it can also become the subject of a judgment in a different way, for example, in relations of containment. It is not necessary to explain that sets colligate, in turn, with other disjunctive sets; thus, sets of a higher order can be pre-constituted and then thematically objectivated, etc. Standing in contrast to pre-constituting sensibility is pre-constituting intellect, which, by unifying thematic objects synthetically in different ways, brings about through this a pre-constituting accomplishment, a categorial pre-constituting one: an accomplishment that brings with it *a priori* the possibility of making, through a "reverting grasp," the pre-constituted [objectlike formation] into a new theme, a new object that in itself has many terms and points back to the preceding categorial action with every member and with the form. And only because this is [53]

[198] *Objekt*

the case, can one speak of synthetic categorial objectlike formations as such. Yet we have to add that we will also get to know non-synthetic categorial objectlike formations that carry out an accomplishment that is executed in several steps, not as tying them together. But otherwise, what we have said holds, namely, that a pre-constituting activity precedes an objectivating grasp.

Where the concept of the *whole* is concerned, it also has a categorial character and is very closely related to the concept of set. The restrictive concept of the whole is the concept of an object—*no matter how* it has arisen, whether sensibly or categorically—an object that can be *completely*[199] explicated in a multiplicity of *disjunctive* parts, as in the form of a set. Part means here that which is merely contained. Corresponding to every whole is the set of its parts, but it is not itself this set. We need not belabor this point here.

More important for us now is to take *the doctrine of categorial objectlike formations* somewhat further here. They are named after "predicative statement" (cat<egoria>)[200], which is the expression of judgment. What this term should actually mean is "judicative objectlike formations." It is to be understood as the objectlike formations, which, through judicative activity, that is, through objectivating activity, first arise in the shape of an accomplishment that pregives activity in contrast to the merely receptive objectlike formations stemming from pregiving sensibility. Every categorial objectlike formation is either that of the full judicative category, or it is an objectlike formation occurring in the judicative context as an element, being actively pre-constituted in it. This requires some explanation.

Once a thematic interest is put into play—be it by one or several objects exercising an affection in a thematic solidarity and being grasped—then an open infinity of thematic interconnections is unfolded from there, interconnections that are all connected thematically in the form of an open, unending unity of a thematic [54]
interconnection. From the ever new objects that break through to

[199] The "completely" would have to be clarified. The parts taken together make up the "whole," etc.

[200] Translator: *Aussage*. Husserl is referring to the Greek sense of κατηγορία, namely, as accusation or form of predication.

attentiveness, only those objects in their thematic acts, their judgments, are connected to the thematic feature that is already put into play, expanding its unity—only those objects that have something in common with the previous ones and that have something to do with them. This already shows up in the affection in the form of combinations of affection, and something is developed in the judgments that produce judicative interconnections between the individually grasped objects. The thematic objects are juxtaposed to one another and, at the same time, they are internally determined; they are explicated individually whereby the explicated objects themselves in turn have a direct or indirect thematic connection to all the previous substrates.

Thematic action is intensive and extensive, or what is also said in place of the latter, discursive (likewise we could also say, analytic and synthetic, if these terms were not so philosophically burdened). It belongs to the essence of thematic activity that it begins again and again with acts of a single ray of attention that we call graspings, with straightforward objectivatings, or rather it begins with one of them; and further that it demands such objectivations in particular "concentrations" again and again, but that straightforward objectivation cannot be thematically independent, although it can stand alone for itself. It stands alone if the thematic movement breaks off, but breaks off before it has accomplished something. Then the thematic interest is left completely unsatisfied. The first satisfaction is carried out in the form of a determination of a completed judgment, and it is carried out in ever new judgments, which for their part are thematically interconnected with those that have preceded. Every judgment is an accomplishment contained within itself, that is, thematically a self-contained satisfaction. However, the fact that ever new thematic interests are aroused in this [thematic] interconnection and are then satisfied in new judgments does not speak against this. In any case, every judgment has a thematic completion, is something thematically complete. And yet it is an element[201] of an open interconnection, and according to an ideal possibility, a

[201] *Glied*

constantly expanding thematic interconnection, and is thus not self-contained. This interconnection is built completely out of judgments, and with every new judicative step, fashions a unity of [55]
accomplishment out of particular accomplishments, a unity of satisfaction from satisfactions that have already been gained.

Accordingly, in each relative completeness, that is, with a final judgment conceived of as relatively self-contained, a thematic interconnection has divisions in a very different sense. First is the division from the standpoint of thematic independence. It leads back to judgments as elements. And where judgments in turn contain judgments, the final elements are in turn judgments. They are the final cells of the thematic system, cells which, however, <are> also functional elements of terms[202] of a higher level, relative cells for higher judgments or judicative combinations.

If, secondly, we now go back to judgments in which each thematic interconnection is divided according to the current standpoint, they themselves will have a division; they will have a subject-side and an determination-side, they will have a caesura: This new division has an entirely different character; it provides thematically non-independent elements: be there already a combination on each side, for example, through collection or attribution (which always rests on previous accomplishments that has further implications). Thus, every judicative element (every judicative element that is to be divided up through the immediate partition of a judgment) is in any case thematically non-independent. For good reasons, a house may be called an independent object, something real existing independently for itself; but in thematic activity through which such an object is first constituted for us, namely, as an independent one, the house as a judicative object, as a thematic substrate, is non-independent. Only a judgment in which the object occurs as a thematic element, for instance, as a subject-element, is independent. In concert with this, naturally, is the fact that a mere name, like this house, does not yield a complete proposition. A complete proposition is the expression of something thematically complete, the expression of a complete judgment.

---

[202] *Glieder*

If judgments are inserted into an interconnection of judgments, after they are constituted in independent self-containment, they themselves in turn certainly take on the form of the interconnection and lose their independence. Independent unities arise again and again only as judgments of a higher level, which judgments are founded in founding judgments. In this way, every [56] theoretical judicative unity is to be disclosed as a unique judgment of a higher order, which judgment is quite inextricably founded in judgments, which are themselves founded in judgments, etc. Where judgments become connections that do not fashion a judgment of a higher order, like for example, "*A* is *B*, and *C* is *D*," the connection has the same thematic non-independence as the connection of two simple substrates, *A*, *B* forming the *collectivuum A* and *B*. On the other hand, belonging to the essence of the thematic variation, which inserts a judgment into the syntaxes and makes it thereby non-independent, is the fact that the originally independent judgment can be reproduced at any time as a thematic independence in itself. Looking ahead, we can use an example whose sense is sufficiently familiar to you: If we have the connection "Because *A* is *B*, then *C* is *D*," we can gain from it in turn the independent judgments that are precisely complete unto themselves: "*A* is *B*," "*C* is *D*."

<§61. Logical Sense>

But now we want to supplement this with a fundamental distinction. We wanted to speak about categorial objectlike formations and we spoke about judgments, their elements and their interconnections. Surely, all categorial objectlike formations are contained in them. But are, or in what sense are, judgments themselves categorial objectlike formations? And the elements of judgments?

Here the *concept of categorial objectlike formations* is in need of *an essential delimitation*. If we give the name of categorial objectlike formations to all objectlike formations *simpliciter* that are constituted through judicative activity (excluding, of course, judicative lived-experiences themselves as objects of the inner sense), then naturally judgments are to be characterized as categorial objectlike formations. But if we look more closely we come up against a *radical distinction*: Not only <is> the judgment constituted in the activity of judging, or rather, not only [is it] there as a pre-constituted unity (and not only an even more encompassing noema, a variable mode of givenness of the one and [57] the same judgment), *but a still quite different objectlike formation is pre-constituted*, a categorial objectlike formation found in the *judgment*, and yet not the judgment itself.

The objectivating activity as a unitary consciousness, which we called judging in the most general sense, has a "What"; specifically, in different noematic modes, it has something identical, a unitary sense, the judgment as something judged, as we would say even more clearly, the judicative proposition. For example, judging on the basis of the perception: this (the roof) is contained in the house, is a part of the house.

Thus, on the basis of a repeated perception through which our standpoints vary, the modes of appearance can be different; or retroactively on the basis of multiple memories, the judgment can be absolutely, identically the same despite the differences of the modes of givenness; it is identical despite the noematic differences. Imagine now how the subject of the judgment executed at the outset is determined in a newer and newer way through progressive judicative activities, and through this, takes on ever new attributive contents. Let us compare, for instance, the judgments "S is a part of S_1," "S_α is a part of S_1," "$S_{\beta+\gamma}$ is a part of S_1." Here it is a matter, we say, of the same S; this is to say that in these transitions to new judgments, and in going back to the previous subjects as thematic substrates, we say and we judge "*the same*" S, we may also say, the same subject. Yet, the judgments are indeed different; in the first element, the *subject-element*, the judgments now have the S without attribution, then S_α, then $S_{\beta+\gamma}$, etc. The same interplay can also take place with respect to the object-position: the same object on the side of the determination, but with different attributions, whereby different, new syntactic matters occur which we however do not take into account when we say that it is the same object. Subject and object, even the self-determining subject and its determination itself is accordingly *ambiguous*.

The distinction is an ultimate one, at the heart of the essence of objectivating activity. Belonging to its essence is the fact that every taking notice and grasping bears on a grasped Something, a Something which, in the alteration of the progressive objectivating activity and in the synthetically identifying return to the same something, can show up precisely *as the same theme* despite the [58]
manifold changing logical shapes that have been appropriated to the theme through the accomplishments of judgment. A finished judicative structure is a system of *thematic poles, each one* entwined with shapes of significance, and all of this is connected by syntactic forms. Judging means posing *themes*, determining themes, connecting themes with themes, *giving to them ever new shapes of significance through this process* and, as it were, *guilding* themselves with the significances already made: *The themes remain the same* and yet they have varied according to the judgment; they have *taken on a varying thematic sense* of predicative significance.

The identity of the theme does not mean the identity of the *thematic sense*. For instance, thanks to the judicative activity relating to the theme, a theme can be laden with products of the accomplishment of this activity, products that signify for the sense a new, that is, a specifically *logical sense-giving*; a variation of sense, which, I say, *does not alter anything with respect to its thematic identity*. Like, for example: this, this red, this house, this red house, etc., continually the same, but with a varying *logical* sense. In a certain way, a sense is self-evidently and necessarily there wherever something is a theme, and already when something becomes a theme for a judging for the first time. Already within pure sensibility, we have distinctions of objective sense in the alteration of appearances under continual or discrete coinciding in "the same."

If this variation of sensibility takes on the mode of attentiveness, then "the same" is what remains singularly One through the variation, the thematic object; thus it already has its variable objective sense even before the determinative activity with the theme carries out logical accomplishments. But the objective sense is *not a logical sense*, that is, is not a product of intellectual activity. What arises through it under the rubric of

altering logical sense, logical significance, is something essentially new. It is something that does not at all take up into itself the alteration in the passively constituted sense. Both of them have the following in common: when an object is the theme, the sense in which the object is given as the pole of identity is not the theme. But sense as logical significance means something new and not something connected to and intermeshed with that sensible sense. If the intellect has appropriated from sensibility something [59]
constituted as the continual unity of manifolds of passive sense-giving and worked on it, then this One is a theme in the starting point, to be sure, but a logically undetermined theme. In the progression, and already by the first step, it is determined logically; it becomes the subject of judgment and perhaps later the object of judgment; that is, in the first judgment "*S* is *p*," the *S* is the One that is still logically undetermined in the subject-position—but the undetermined that gets determined in these judgments and has its determination on the side of the predicate. However, by the attributive activity adjuncting to the subject the determination carried out in the first predication, holding on to the *S* as being *p* in the following predications and subjecting them to further determinations in new activities, we no longer have an undetermined logical substrate, but rather a substrate encumbered with a logical significance.

Thanks to logical activity, every substrate takes on logical encumberments, ever new contents of logical significance in an ideally infinite number of ways. One can say that every judging concerning a thematic object alters the consciousness of this object itself and habitually imparts to it an ever new sense as well. The ever new determinations do something to it on the order of consciousness; but, on the other hand, the fact that the predicate, which has been made manifest as the sense acquired for the subject, is added to the sense attributively, and pos<sibly> joining this, even a habitual volition to want to conceive the *S* persistently as determined only in this way, as the *S* of this sense (like in the case of the *definition*)—this is something specific, it is something to be accomplished first of all in a specifically active procedure. Here, the *S* is not given to consciousness in the manner of mere sensibility with varying content; *through conceptual processes,* that is, thematic activity, it has taken on not merely expansions of its passive content of sense; but in the full light of the intellect, in the creative activity of the ego which abandons itself to the *themata* and to the work for them, *the creation of logical structures is carried out*, the creation of judgments as a whole or in particular, and thus everything that we call the alteration of logical significances as well.

But how is it that we come to speak of an alteration of logical significances in a way that is similar to speaking of an alteration of objective sense within sensibility, pre-logically of course, with [60] respect to the same object? And how will the *concept of logical significance (= logical thought)* itself be determined? We say that logical significances: "this," "this red," "this red house," "this school building," etc.—are different, and that the subject is the same. The subject is the subject of judgment or a theme functioning in one way or another in the judicative context. It is completely beyond the scope of our investigation [to consider] through which objective sense the subject is given from case to case in an extra-logical manner. The identity of the judgment is not oriented according to the objective sense.

In order to speak more clearly and to avoid entirely using the term "sense" for something extra-logical, let us now say "*presentational content*"[203] for that objective sense. Thus, differences of the presentational content are now irrelevant. Only the theme, the presentational *object*, the sensible One, the house, is consciously meant in the judgment. But when the house is the thoroughgoing theme, it is the identical object in the context of thematic actions, no longer as the unity of its sensible manifolds and of its changing presentational contents, but as the identical element precisely of these thematic actions and of the accomplishments arising through them for the theme itself. It is the identical element in the manifold of active identifications that manifest *the theme as a point of intersection of manifold judgments* and then as the point of identity of corresponding *attributions*, as the same again and again. No matter which of the attributive structures we take: "this," "this house," "this red," etc.,

[203] *Vorstellungsinhalt*

each one of these structures is a thematic judicative element. Each one (no matter how different its content may be), has, as the judicative element, its theme and each one evidently has the same theme. We take the "this," so to speak, as the What of attribution in this series.

<§62. The Distinction between State-of-affairs and Judicative Proposition>

In general we say that all judgments necessarily have in themselves a thematic division, every element has its theme. Wherever <there is> a manifold of different judicative elements and wherever <the latter> are identical according to their thematic [61] content, we say that they deal with the same theme in different logical significances, or that in them the same object is conceived in different logical significances, or is also conceived in different logical thoughts. Yet, we are not to think immediately of an individual object when speaking of the same object, and if the realm of generality, the realm of "as such"-*themata*[204] had already come into view for us, then all of this would be transferred to them, as well. The term *themata* is the most correct expression for this: the object as object, that which is judged in the judgment, does not mean the presentational object, but rather, the conceptual object, that is, what functions in the judgments as theme, no matter what shape or what level the judgments have, and therefore that which is the content of a judicative element. For a theme is nothing for itself; it is the judicative element just like, conversely, every judicative element is self-contained as an element by virtue of it possessing its unitary theme.

One could now say: A judgment is a thematic structure through and through, it is a structure made up of syntactic matters and united by syntactic forms, like we ourselves have already shown. If, in judging, I speak of a house that is roofed with shingles, the house is my theme, but the shingles are too, etc. Yet, one must not overlook what is decisive here, what necessitates precisely the distinction between logical object and logical significance, and

204 *"überhaupt"-Themata*

then further, the difference between the *state-of-affairs of the judgment* and *the judgment itself* (the judicative proposition).

Let us treat the matter at hand noetically. In every judging we have our themes, for example, an "S is partially identified with S_1"; "S has S_1"or "S is in S_1." Judging means carrying out this identification. But now an integration of acquisitions of judicative activities that were previously current (judicative activities that refer to the same themes) into the thematic consciousness of the judicative elements can be intertwined with the current judging to form a thematic unity. The annexes are no longer actual judgments, actually executed identifications, but modifications of them given to consciousness. The predicate that is added on attributively is no longer the actual theme (the theme as actually functioning) in the current judgment, but rather a modification of [62] it—a theme that points back to a previous judging that we can also restore at any time, but which now is no longer carried out. The theme that is given to consciousness in the modified manner, the annexed theme, does stand in our regard to be sure, it is indeed also something that is grasped, it also has a thematic character in a broader sense. But there is an obvious difference, a tremendous one, between a functioning theme, the theme of the determination that is actually carried out in the present judgment (with the current object-theme on the side of the determination itself), and the themes that are no longer functioning and that are only drawn into the unity of the judicative consciousness, i.e., of the consciousness of the respective element [of the judgment].

If we have attributively packed into the subject and into the predicate of a simple judgment however many objects and predicates from other judgments (newly carrying out the judgment of the previous form, under the co-execution of all of these thoughts that are packed into it) this does not at all change the fact that we carry out the same determinative identification, and only carry *it* out presently; whereas we do not carry out with the attributions any predications indicated by them. What we do carry out are precisely attributions, mere modifications in the form of supplements. What we actually judge again and again is "S has p," the one substrate contains the other, however many other contents of thought from earlier judicative accomplishments S may be conceived as having. *What we have described noetically* has its *correlate* in the judicative proposition, in the identical judgment, that is, in the object being constituted in the thematic activity. We see here *that it belongs to the essence of judicative action to constitute two types of objectlike formations in an integrated unity.* Judging as a *thematic action* is *intending*, thinking, positing something in the present as theme, drawing new themes from it through the process of explication, and positing them in a unity with the main theme, drawing in an external theme and relating one to the other, etc. Every conceptual intending has its intended object, every synthetically relating and connecting [act of] intending has its synthetic intended meaning.

The objectlike formation constituted in this way, which makes [63]
up the primordial essence of the judging as present judging, is the *state-of-affairs.*[205] It is the pure synthetic *unity of the themes* and the theme here is everything that is *currently* grasped thematically in the corresponding execution of judgment, and currently *posited* in each and every relation to precisely such an execution. If the themes are *individual things, they go into the state-of-affairs*, they are the *termini* that *"relate" themselves within it*. When we speak of relationships, relationships of part, relationships of similarity, relationships of elements, etc., here relationship is only another word for state-of-affairs. The points of relation of the relationship are single objects that have an objective unity in a certain syntactic form. States-of-affairs, relationships, are objective correlates of judgments, that is, they are originally constituted only in judgments.

Every state-of-affairs is a complete categorial objectlike formation, and all elements of the state-of-affairs or *termini* that are not simple *termini*, thus, are not thematically simple, are themselves categorial objectlike formations in their turn. Thus, when we take a plural judgment, " S_1 and S_2 are p," or even "S is p_1 and p_2 ...," a collectivity is contained in the state-of-affairs; at least we can form sets by easily reconstructing single themes, and distributive predicates relating to them would then have as a

[205] Editor: <See pp. 298f.>

thematic element a categorial objectlike formation that contains within itself single themes of a second order.

We make the following distinction: [1] *a first sense of the word "Meinung"* in which these categorial objectlike formations and above all, the complete *states-of-affairs* make up the "*intended meaning*" in a judgment, and [2] *the second sense of the word "Meinung"* as the judgment itself [understood] "*as the intending*" *of the one judging*. The intended meaning, *which the one judging intends or means*, is now the *judicative proposition* as the complete *unity of significance* that harbors all logical significances of the corresponding judging.

I do not need to explain in much detail how this is to be understood. In the transformation that takes place by drawing the conclusions of manifold judicative activities in relation to the momentary, present determination-themes into the new judgment [64]
that is being carried out, the state-of-affairs, the intended meaning in the first sense, remains untouched.

Every current theme can become the substrate not only for the current judicative accomplishments, but at the same time can become the substrate for the modified connection of previous acquisitions of accomplishment. This does not alter the state-of-affairs, which is constituted thematically, but is does alter the mode in which it is conceived. In this "mode" it is the thematic product, not only of the present judgment, but also of the implicit judgments. Or the current thematic product has an embellishment of formations that point back intentionally to the intermeshed judgments. Every such formation can be extricated, the explication of the judgment according to all of its significances leads to a connected system of primitive judgments that no longer contain anything of the intermeshings, that is nothing of the like, like attributions. This unraveling is not the resolution of complex judgments into simple judgments; for a judgment is not genuinely complex by having appropriated an attribute in a thematic element. The expanded judgment is not a judgment that contains two current judgments as parts, as intimately inherent components. Yet in a certain sense it is certainly true that the expansion is the same thing as a complex composition, for if the predication after being transformed into an attribute is also no longer currently a predication, then it is certainly found analytically in the attribution. This points toward a special form of identification and belongs to the theory of analytical deductions.

Let us now consider the limit-case of judgments that, in this direction, are analytically simple insofar as their thematic elements do not contain any co-intended meanings that exhibit sedimentations of an earlier actuality. *In the case of such judgments* it is clear that *we do not distinguish between the state-of-affairs and the judgment itself.* In the case of judgments, the state-of-affairs is the unity of significance itself. Nevertheless, we even have to speak of these judgments in two ways, taking into account the fact that they are mere limit-cases and that unending manifolds of judgments accord with every such limit-judgment in the state-of-affairs. The concept of state-of-affairs characterizes from the very beginning the identical thematic framework that precisely all judgments have in common, judgments that have identically the same current themes and that connect these themes [65] in the same categorial forms (i.e., the synthetic forms of function). The limit-proposition is the pure proposition proper to such a group, and is a correlate of the pure determinative actuality. The objective element in judging, the state-of-affairs, which itself should be a judgment and in particular a limit-case, looses its paradoxical status if we remind ourselves that we are speaking here of objects in quotation marks and thus of states-of-affairs in quotation marks, as well.

<Chapter 4: THE GRADATION OF OBJECTIVATION>

<§63. Two Dimensions of the First Level of Objectivation: Mere Intuition and the Dimension of Explicative Examination>[206] [66]

The insights that we have gained into the structure of judgments are important in many respects. They are insights into the internal

[206] Editor: See Appendix 29: Theme and Theoretical Interest, pp. 548f., and Appendix 37: <Object and Content of Interest> pp. 556ff.

construction, into the structure of judgments as ready-made formations of determinative action, and insights into the *a priori* genesis according to which every ideal possible continued construction of judgments and intertwining of judgments of higher and higher complexity is carried out systematically and according to fixed laws. Accordingly, all judgments ultimately lead back to fundamental shapes of primitive judgments and to principles of syntactic reconstruction and intertwining. What strikes us here is the variation in thematic consciousness, thus, the different way in which the *termini* in the state-of-affairs, that is, in the logical main clause, and the *termini* in the subordinate clauses are given to consciousness, and accordingly are even characterized in a categorially different way, and moreover, even the difference of thematic form of the subject of determination from the object determining it. And it strikes us that the wholly other distinction in thematic consciousness, which consists in the fact that in a certain way the entire judicative accomplishment, the entire state-of-affairs in its syntactic shape, stands in the brightest light of thematic interest, while only the *termini* are thematic objects. Required is a new attitude, which is a change in thematization so that, for example, the state-of-affairs or the judgment becomes a theme in its complex shape of significance, namely, becomes something grasped thematically and even becomes the subject-theme.

We recall here our doctrine of interest[207] that we have placed at the beginning of everything having to do with the theory of judgment; already prior to all determinative judging we have the following distinction: the object of interest—the One within the objective sense that varies in manifold ways, what gets grasped, what is thematic *simpliciter*, which we more clearly also call the thematic object—and the content of interest, that is, the varying objective sense of the What-content of the object in its persistence [67] or even in its variation, which lies necessarily in the field of interest, but is not the thematic object. If a simple, unchanged tone becomes the object of interest, if it is perceived in a simple manner, then the interested regard rests in each moment on the

[207] Editor: <See above 288f.>.

tonal phase that is sounding right now, and what has just sounded is also still held onto in a modified way; in these relations we have something different from moment to moment. But running through it is the perceiving of the one tone. It is the object of intuition, and it does this only by the interested examination running throughout all tonal phases and tonal components, throughout the intuitive content. *This difference between object and content is repeated on all levels of thematic activity* and takes on newer and newer shapes.

Our chief aim, which is to understand objectivation as a genetic gradation of thematic accomplishments, must be carried out consistently to the very end, and the exposition that we have presented up to now already makes intelligible for us a gradation of objectivation and the sense of such a gradation. The doctrine of the *genesis of logical significance with respect to the substrates of determination* is especially illuminating. It even shines back to our beginnings, with respect to which, as we now see, we would have to introduce the distinction between the activity of mere intuition and of the merely synthetic examination, and the genuine determination as a unique genetic distinction of levels. Let us survey the gradation of objectivation, and let us append the supplementation to which we alluded.

The example of the tone illustrates for us *the first* conceivably most primitive level. It is the level of the intuition that examines prior to all explication. [Examined here is] the thematic *object*, the one *enduring tone*, thematic *content*, the flowing What-content of the tone. *The second level* is that of the examination that delves into and the examination that extends beyond, two types of examination that do not have any essentially prefigured order. The former thus concerns delving into the content of a thematic object [68]
S; it is explicated. The progressive objectivation consists only in the fact that what was previously merely a thematic content now becomes a thematic object in a particular way, whereby what has become the special theme enters step by step into partial unity of identity with the *S* still being held onto, and precisely through this connection remains firmly held onto in the progression. The interest in the object running through this is fulfilled in this way in concentrated form in every partial theme.

Upon closer inspection we distinguish here: that which is the matter <of> activity, namely, the process of making the object thematic, which will be explicated afterward, then the process of making the components of its content thematic; and on the other hand that which is the matter of passivity, namely, the occurrence of partial coinciding and the occurrence of the enrichment of sense which is carried out with respect to the beginning thematic object.

We will have to say that the determinative identification, the activity of judgment is not yet carried out. It demands a new step in order to transform what passivity has prepared into an activity, into a judging in which the state-of-affairs is first constituted.

Before we make this clear, let us point to the second form of examination, the one that extends beyond, the examination that is not constricted by delving into the content, but rather expands by passing over to different objects. Commonalities are constituted here in passivity through connecting syntheses that can guide interest and can expand it into the unity of an overarching interest. By passing from object to object, passive synthetic unities are thus produced again; in this are found excitations for possible determinations, but that is still not the activity of determinative synthesis, not judging. Thus, for example, not first when an object has become a theme and after it, a whole encompassing it. The inverse partial identification is constituted passively. Likewise, where syntheses of uniformity and similarity are concerned.

<§64. The Second Level of Objectivation: Active Relating and Determining. The Task of a Universal Theory of Relation>[208] [69]

Up to this point we had considered two levels of the lower activity of objectivation under the general rubric of an examining intuition. In contrast to this, we posit as the *second main heading*, "the level *of judgment*." By this we mean now *the determining process found prior to conceptualization*, that is, prior to the function that constitutes the universal for us; we mean the activity of relating one theme to another theme, and then further all the

[208] Editor: See Appendix 37: <Object and Content of Interest> pp. 556ff., and Appendix 38: <Categorial and Non-Categorial Connections and Relations> pp. 573ff.

activities that give a higher synthetic unity to such determinative accomplishments. The *level of conceptualizing thinking is then a fundamentally new and higher level.*

The distinction of the *relating process of objectivating*, of *actively* determinative objectivating, *in contrast to mere examination*, e.g., to explicating, becomes clear if we describe clearly the actions that lead upward: If in the transition from one object *S* to a moment of its content, *m*, the coinciding that forms identity has occurred passively, then with this, as we had said, something is done to the object, it has an increase of sense in this synthesis. If this enters into the circle of the content of interest, that is, if the *S* (which is different at the end of this process than it was at the beginning, the *S* that has receded and is only being held onto, yet no longer in the focal point)—if this *S* [now] enters into the focal point of interest as the *S* that is expanded in sense, then this is only possible in such a way that *S* again becomes the primary object of the grasping in a new grasp; and now an active intention, in the active transition to *m*, is concerned with originally and actively generating what is accruing to the *S*. Only now do we have a subject-theme as *terminus a quo* and a counter *terminus* as the *ad quem* at which the activity is striving; only now [do we have] the consciousness actually constituted only in the synthetic activity, the consciousness that the *S* undergoes determination through *m* in the mode of "*S* contains *m*."

Likewise in the inverse case: "*m* is a part of *S*." If the interest in the object passes over from the part to the whole—precisely by virtue of the inner commonality and of the coinciding that forms identity, which sets in passively—then remaining in the second [70] element is the first that is still held in the interest's grasp. But what is new is the active intention toward determination, the activity that directs its aim at *m* to determine it with respect to *S*, thus, the activity with an eye to *m* and electing it as the subject-theme driving at the *S* and, in carrying out the process of identification, determines *m* as a part of *S*.

The progress of objectivation of this higher level consists in the accomplishment of determining and is seen in the active formation of new thematic forms: subject-theme and determination-theme. We no longer have thematic objects like [we had] on the first level where the thematic shape is the same everywhere; rather, we have new thematic shapes harmonizing with one another that have arisen from new activity. Each one has a syntactic formation, and they are connected to a syntactic unity, a unity of an state-of-affairs.

This is likewise the case (to hint at new judicative shapes), if we have wholes that are pre-constituted in passivity, if we have unified combinations in the broadest sense or in some more narrow sense, unities in which different particular objects are divided off from each other, now being elements that intersect, now being elements that are disjunctive. Every two elements of such a unity can enter into a determinative relation. But the active grasping of them in passing from one to the other is still not an active process of relating, although the transition can yield a passive product that points the way to active determination.

If we have an intuitive configuration before us, for example, in the most simple scenario, a pair of balls in a sensible unity of position, then lying in the possible horizon of examination is the concrete unity, the pair of balls in its position: initially unexplicated. Then grasping the single balls individually and then grasping them together. Then the form of connection as a non-independent moment. This is not a third part that is on a par with the others; rather, the grasping presupposes the explicit grasping of the balls, and in this respect it is essentially a mediate, non-independent moment.

All of this provides the foundation for different possible determinative judgments, the relationships between the whole and immediate parts, between the whole and the form of connection of [71] parts, between the combination of parts and their form of connection, finally: relationships between the elements as connected in these forms. In this relation of connection, the elements participate in the same whole and have a commonality in it. If we pass over from the one to the other, (if each one is given to consciousness with the direction of interest toward the connection), then a coinciding takes place in the elements that share something in common, and therefore an enrichment of sense that can give occasion in turn to activities of determining which form relations. Determinations arise in this way, like "*a* is in a particular position to *b*." Relations of a higher order can also arise in this way, like relationships of positions between positions, etc.

You see here that one may not confuse the connection of parts to a whole and the relation of a part in relation to another part. The relation is an state-of-affairs and originally has two elements. (A relation having several elements is a connection of several relations taking place in judgment.) But a whole is, in general, not a state-of-affairs, although every state-of-affairs, every judgment, is a whole insofar as it can be divided up.

Let us note further that relationships between whole and part, between parts of a whole, thus, in principle all types of relation that belong most generally to the idea of a possible objectlike formation as such, does not only concern sensible objects and wholes that are connected sensibly, but also categorial objects. Judicative action fashions wholes that are structured by being generated from elements. But that they are given as elements, that they are given as parts of the whole, and that the relationships of every other kind are given among them as parts of this whole—this presupposes that the constituted judgment is first made into the subject-theme by a reverting grasp, that from here its parts are brought into relief by dividing them up; these parts are being related to one another in determinative actions, actions that are entirely different from those in which the judicative whole in question would be constituted.

In accordance with the general path that we have sketched out, a systematic theory of relation would have to outline, at first in formal generality, the types of relation (types of state-of-affairs on the level of pre-conceptual judging) proper to the most general [72]
idea of the objectlike formation as such. The development of all forms of judgment and forms of judicative interconnections yielded by the intertwining of significance leads in this case to an infinity of types of wholes, which examined under the formal idea of the objectlike formation, are special forms of objects with special forms of elements and connections, with special forms of relations. They are forms of objects called categorial objects that relate back to all objects as such, that is, even to themselves. The universal investigation into the categorial realm, according to its forms, and the pure laws determined by these forms, laws of true

existence of states-of-affairs, make up a self-contained discipline, "*formal logic*."

Let us note here that when one treats as objects states-of-affairs, and the formations of the states-of-affairs themselves, the relations proper to them are not only those that can also occur with other objects, with objects of all types in general. Rather, it belongs to the essence of states-of-affairs and of categorial objectlike formations in general to ground relations of this kind as well. Thus, the relations are divided into specifically categorial relations, into relations belonging specifically to the essence of categorial formations, and into non-categorial relations. In this way, the relationship between a relation and its inversion is a categorial relation; the relationship "*a* is a part of *b*" is, as one says, included in the relationship "*b* is the whole of *a*," and *vice versa*. And this holds for every relation and its inversion, and for every instance of a judgment being contained "analytically" in another judgment, for every instance of an state-of-affairs being contained "analytically" in another state-of-affairs.

In the formal logic of the second level of objectivation, the level that relates to the *states-of-affairs*, the relations, *inhabiting* the judgments of all possible forms, the main theme is the systematic disclosure of the radically different genera of relations that are possible for objectlike formations as such, in formal generality, that is, forms of states-of-affairs—forms, since the *termini* here are conceived in an undetermined, general manner. Accordingly, it is a matter here *of a formal classification of relations*, while a *logical theory of relation as such* makes differentiations even where the *idea of the objectlike formation* is concerned, and gives an account [73] of all differences of this idea, i.e., the *fundamental types of objectlike formations* that are sketched out precisely through the essence of the objectlike formation itself. The difference between *individuum and eidos*, between *categorial and non-categorial objectlike formations*, is sketched out in this way.

We have already suggested that categorial objectlike formations have their peculiar forms of relations, and that not all objectlike formations can enter into them; they, the objectlike formations, only have sense for these forms of relations. Likewise, *individuals* have their *peculiar individual-relations*; these relations have their foundation in so-called *individuating determinations* and in the forms of connection that are rooted in them: thus, the *relations of temporal position* related to the absolute temporal loci, and likewise, the spatial position for external objects. Genera and species in turn have their specific relations. The said *types of objects are logical ones*, because every such type refers to every other one, and there is no *objectlike formation conceivable* that is not related back to all of these types. Even the *topic, "consciousness,"* belongs here, for not every object is consciousness, to be sure, but each one is an object of a possible consciousness and, constitutively, has a relation to the interconnections of consciousness. *Consciousness, further, has specific, peculiar relations*. Considered here are relations that have their foundation in *essential connections* that *intertwine* with one another the objectlike formations of these types of objectlike formation, which are prefigured "*a priori*", that is, the intentional relations between object and consciousness, ultimately between *individuum* and consciousness, the relation *between individuum and eidos* as the genus-universal, between *individuum* and the non-categorial object and the categorial object as such, etc.

Thus, the great task of a universal theory of relation consists in *establishing all fundamental shapes of relations, taking the fundamental shapes of objectlike formations as objectlike formations*. The result is a theory of the forms of relations. The establishment of the affiliated fundamental laws of relational truths like, for example: that the true subsistence of "*a* temporally [74]
after *b*" and "*b* after *c*" contains the true existence of "*a* after *c*"—all of this is yielded by the so-called *axioms* of relations that are assigned to every radical type of objectlike formation for itself and in an intertwining, for example, axioms of individuality, axioms for genus and species, etc.

But preceding all of this is a theory of the forms of those relations, a theory that has a formal universality in a still higher sense. We have precisely relations into which the objectlike formations as such can enter, be they individuals or genera, etc., so that in the disclosure of these forms of relations, the *termini* remain fully undetermined and thus indicate nothing concerning whether the objects belong to this or that radical object-type or, as one also says, to this or that category of objectlike formation.

As you see, the universal *relationship of part and whole* and *vice versa* (or better, the universal relationship of containment and of what is contained, of part to part, and much else) naturally belongs in this realm of the *most general, formal universality*, a realm contained in itself.[209]

<§65. The Third Level of Objectivation. Conceptualizing Judgment on the Basis of the Consciousness of the Universal>

<a. Relations of Comparison, Association of Similarity as its Foundation, and Direction of Interest toward the Universal>

Of interest to us now, however, is a new topic, the topic of relations of comparison, as we call them, the relations of uniformity and similarity, which serve us as the transition to the third level of objectivation, that is, to the higher judicative level, that of the logically conceptualizing judgment.

The peculiar place of these relations shows up in the following exposition: We have directed our whole undertaking up to now to [75] the sphere of positionality, even in the realm of self-givenness. We conceived of various objects as being self-given, and posited them in a relation which, in turn, was also self-given. The opposite of positionality is quasi-positionality, phantasy, be it perceptual or reproductive. If we transfer an objectlike formation (so to speak) into the sphere of phantasy, we will also have the states-of-affairs belonging to it in the mode of the "as if"; thus, it appears that we can dispense with the consideration of phantasy, since corresponding to each type of positional givenness is a quasi-type of phantasy givenness; both share the same essential content.

However, phantasy is Janus-like, so to speak. Leaving the ground of actuality, we can situate ourselves completely in phantasy, and as subjects immersed in phantasy, live in the "as if"; thus, we can have a phantasied actuality given as if it were an

[209] <See the> parallel exposition, Appendix 39: <The Task of a Theory of Relation> pp. 575f

actuality. But, secondly, we can keep our footing on the positional ground as actual subjects, as the subjects of actual, current experience peering into the world of phantasy, and then comport ourselves positionally to it in a unique way. In other words, now the phantasy formations are for us something given as existing, namely, precisely phantasies, fictions, characterized as modifications of actuality and insofar as this is the case, as non-actualities. But not non-actualities in the sense of actualities that are crossed out, like in the case of an illusion as a negated illusory appearance. From certain perspectives, then, we also characterize these objectlike formations that are constituted by phantasy, these "phantasy images" (whose essence it is "to imagine" objects together with their mode of being in a phantas<ma>tic manner), as possibilities, possibilities of objects.

I now need merely to point to the fact that relations, like those between whole and part, part and part, cannot exist in principle between actualities and such possibilities, that is, that they cannot be constituted in self-givenness, where the one *relatum* is given to consciousness as actual and the other as a fictum. If a whole is actual, then the part is too, and a fictum cannot be connected to something actual to form a whole. However, actualities and fictions can obviously be compared with one another; they can [76] stand in a relationship of similarity and uniformity to one another.[210] Closely related to this is the following distinction which likewise characterizes the essential division of both fundamental types of relations. Even if they do occur in the unity of one consciousness and in a living present and achieve an overarching unity, every intuition that gives an actuality, and every phantasy-intuition *do not, as intuitions, form a cohesive connection.* The very *same thing* holds for a perception and for *a remembering* and for several rememberings that do not make up a continual unity of remembering from the very start. *A memory*

[210] We distinguish: Relations of materially relevant ties (relations of connection) and relations of comparison. Relations of comparison can not only place actualities in relation to actualities, but also actualities in relation to possibilities. Or alternately: What they place in relation does not need <to> be constituted as connected in a continually cohesive intuition.

irrupts into the perceptual present without any connection, or [we can have] a different memory rupturing a memory.

An objectlike formation is constituted in the continual unity of an intuition as the unity of a giving consciousness. What is constituted in several intuitions, intuitions that do not form a cohesive connection, is constituted as not forming a cohesive connection in a materially relevant manner. *But there is a kind of connection that can even connect what is given without a connection. Association* can unite each and every thing that is given to consciousness together in a present of consciousness, whether connected or unconnected in materially relevant ties. A whole as whole of its parts, the connection of parts as parts of a whole, can only be given in the unity of a cohesively connecting intuition of the whole, of the connection. However, this does not effect uniformity or similarity at all. And they are unaffected precisely because uniformity or similarity have their *original source* purely in the connection that is pre-constituted by the unity of the so-called *association*. No matter how much efficacy the association may have for the formation of unitary intuitions, and no matter how constantly it is efficacious for them, like for the constitution of unitarily, cohesively connected objects, it also has [77] efficacy where objects, so to speak, are floating down into the present of consciousness without a cohesive connection; they fashion a tie between them, and specifically as an association of similarity.[211]

This connection entering into thematic view is the *fund<amentum> relat<um>* for the active constitution of a relation, of the relation of similarity and of uniformity. Now, if I say, further, that this connection is at once the source even of the consciousness of the universal, and therefore of conceptualizing judgment, of judgment in the full sense, then I mean by this that a primordial form of association is this source, which is to say, it functions as a specific *a priori* of passive pre-constitution upon which a new level of activity can build, which level of activity not

[211] Accordingly, sensible uniformity and similarity are thus also an original association and a unity of affection. Does the affection of one element pass over to the other, or conversely, does the coinciding in a unity give an increased force of affection?

only yields the universal as object, but lends to all judgments (in the sense given to them up to now) a new, a specifically logical shape.

The universality of the accomplishment of comparison and formation of concepts rests on the fact that everything, no matter how it is constituted as an object in actuality or possibility, can emerge precisely in an *a priori* fashion as the *terminus* in relations of comparison, and can be grasped conceptually by the activity of eidetic identification and by the activity of assuming it under the universal.

An excellent form of the relation of comparison is uniformity; a multiplicity of objects that are merely *repetitions* of one another precedes and is internal to the unity of the consciousness of a present, and this means an *associative form of unity of a peculiar* type, already constituted prior to the active process of relating. A coinciding that forms uniformity enters into the particular grasping of a *terminus*, with the transition to a second *terminus* that is grasped individually, and in the active relating, the one *terminus* is determined in the form of the subject as uniform to the other: $a = b$. And *vice versa*, as well.

The peculiar feature of this synthesis of two uniform objects consists in the fact that it indeed seems quite similar to a synthesis of identity, and yet is not one. It is so similar to it that, in passing [78]
from one uniform object to the next, we are virtually inclined to say that it is indeed the same thing. The uniform objects, however, are two separate objects and not one and the same. And yet, in every such duality and in every manifold of uniform elements whatsoever, there actually is a unity and ipseity in the strongest sense. It arises originally in the synthesis of the coinciding that forms uniformity, it is constituted through it originally as object. Naturally, this original constitution means pre-constitution. An ἓν ἐπὶ πολλῶν, the unity of a universal, can only first be given to consciousness for the ego *a pr<iori>*, can only first be ready for possible thematic grasping by carrying out the activity of grasping uniform objects separately in the synthetic transition [from one to the next]. A relating judgment of comparison is not required; this occurs in a different attitude. The direction of interest toward the universal, toward the unity as opposed to the

manifold, is not that of determining the one uniform object in relation to the other as being uniform to it; rather, what awakens interest is the One being actively constituted in the coinciding of individually grasped uniform objects; the One is the same, and is the same over and over, no matter what direction we may pursue in passing from one to the next. Obviously, what we carried out here is not anything like an explication of uniform objects where the One is not found as part, as something partially identical in the uniform objects; otherwise the One would only be something that is everywhere uniform, and the uniform elements would be in the relationship of intersection. The One does not repeat itself in something uniform; it only occurs one time, but is given in the many.

The original grasping of this One has a different field of interest, a different "content" that the interest must move through, different from the case of a sensible object, an object on the level of the individual. For now the interest, the ray of attentiveness must pass clear through the individual objects that are already constituted, and by traveling along the path of uniformity and carrying out the coinciding, the One that is constituted becomes thematic.

<b. The Universal as a Novel Intellectual Object>[212] [79]

It may initially seem that it does not require the open infinity with the form of the "etc."[213] Two repetitions already suffice to grasp the unity. But we note that with the pure directedness toward the unity, the existence of the objects that are elements [of the unity] is immaterial, that the potential positings of existence do not play a founding role, and that as long as we do not have the particular objects in the consciousness of the whatsoever,[214] in the consciousness of the exemplary, we do not have pure self-giving.

We achieve the highest grade of self-giving in the free passing from the given manifold into the realm of free possibilities and in

[212] Editor: See Appendix 35: Multiplicity and Judgments of Multiplicity, pp. 559ff.

[213] Yes, for pure universals.

[214] *Beliebigkeit*

going through an open manifold in consciousness, a manifold of any possibilities whatsoever as any example whatsoever in the form of the "etc." We encounter the universal as a novel objectlike formation, as an intellectual objectlike formation, an objectlike formation from the original well-spring of activity: although, of course, on the basis of sensibility, insofar as the activity of "going through," of grasping the individual, of bringing into a coinciding, is necessary so that the universal as such can be pre-constituted, and then later can become a thematic object. If we take independent, concrete objects as our point of departure, this *universal of repetition* (with open unending scope) is the *concrete eidos*, from the Latin, the *concrete species* or the concrete essence. Where the *non-independent moments* of the concrete object are concerned, the lowest non-independent species are given, the lowest abstract species: if "abstract" only serves as a counter term to "concrete." Now different species among these abstract and concrete species can themselves be compared; like all objects, they can also have their associative unity in consciousness; in it we grasp, in a still higher intellectual activity, a specific-universal as the unity of manifolds that are already species, a universal of a [80]
higher order. We arrive at species and genera.[215]

In going from uniformity to similarity, we would have to expound upon related matters. *Similarity is a fundamental shape of association*, and the transition yields <here> an analogue of partial identity, a coinciding and yet not a complete coinciding. The elements of the similarity that overlap (which, by the way, need not be separate) have a *distance*; different similarities can have different distances, in fact, they themselves can be compared. Thus, *similarity has a gradation*, and the *limes is uniformity* without any distance in the coinciding of elements merely being repeated. Nevertheless, something in common also comes to light here in the *coinciding at a distance*, or, as we could put it perhaps in a better way, it shines through originally as a universal.[216] It

[215] All of this would have to be given more thoroughly, above all, the pure *eidos*, what is purely exemplary, etc., would have to be grounded more thoroughly and be set in contrast to the empirical universal.

[216] Thus, it is apparent that by virtue of the *limes*-relation, the universal of similarity is the universal of uniformity.

achieves complete givenness in the process that we have described previously, namely, that of the transition from the lowest species of similar elements to the corresponding higher species. In this disclosive process, the universal of mere similarity is given as a higher *species*, as a *species-universal.*

If we carry out a determinative attitude, *a fundamentally new mode of determination* is given *after the constitution of the universal.* The object that is the subject[217] *a* is not determined in a manner uniform to *b*, but as *an a*. The thematic grasping of the universal requires moving through the open manifold of any singularities whatsoever lying in the circle of interest concerning content. *Positionality*, where it occurs, *is not carried out* for these singularities, *the mode of being is irrelevant* and can also be a phantasy mode of being from the very outset.[218] If we focus thematically on such singularities, they will have a *unique relationship to the universal*, precisely *that of the species to its* [81] *particularizations*. What exists as actual, as what is actually posited is then determined as the actual particularization of the *eidos*, what is possible, a possible particularization. The new judgment that arises is "*This is an a*," which is to say, it is a particular of the species *a*.

There are other essentially different forms, and not only *the forms* that arise from the previous ones, by, so to speak, *dressing them with concepts*. Up to now, we had placed determined constituted objects in relation [to one another], and had determined relations between determined objects. Each object can be grasped conceptually. If we have judged that S is a part of S_1, and in this way relating them, then "S and S_1" can now absorb conceptual determinations by attribution, and this can take place in all cases.

---

[217] *Subjektgegenstand*

[218] For pure universals! For pure *eidé*!

<*c. As Such-Judgments*[219]>

But thinking in general takes on still different and more significant forms. *Thinking, in the form of the function of the "as such,"* governs the entire judicative thinking of the higher level. The universal is constituted, we said, in relation to a field of openness, and with this we have already unleashed *the function of the "whatsoever."* Yet, in a certain way, it is actually already prepared from the very start for judging as a thematic activity: If we have objects pregiven, we can choose any "this" or "that" whatsoever as the determinative theme, we can relate it to this or that other object whatsoever. But the whatsoever, the choosing, and correlatively the indifference also enters as a thematically *formative* activity into the judging itself; it also constitutes, it fashions new forms of objectlike formations, new forms of state-of-affairs and of judgment. Not only do species emerge as objects, but correlative to the species, new thematic formations are constituted. They are As Such-objects and *As Such-relations*, i.e., *universal and particular judgments*. The concept of the *objectlike theme* certainly *also* takes on with this *an essential modification*. For it is fundamentally different <to> have this table here before us thematically and <to> determine it—even if it be determined [82]
conceptually as a table—and <to> make a judgment about a table as such. It is, I say, something completely different: instead of making a judgment about this table, even if it stand before us and stand in the illuminated field of interest as a thematic focal point, to make a judgment about any table whatsoever, regardless of which one, to make a judgment about a table as such, whereby the "whatsoever," the "as such," belongs to the thematic center. Now this table is characterized here as example, this whatsoever attaches to it, it is given to consciousness here as this table, but indifferently. It is only a "*representative*," and the theme—and *in a modified way, the subject-theme—is a table as such*. Thus, here the *judging is fundamentally modified*, it is a *positional* activity that operates on an entirely different level, on an altered and complex thematic substratum.

[219] *Überhaupt-Urteile*

If we have a multiplicity given in the manner of experiencing positionality, and by moving through it, a universal comes to the fore, we will have the determinative attitude, “This is an *A*,” “This is an *A*, etc.” But also: *anything whatsoever that has been extracted, anything whatsoever to be extracted, is an A*; in moving through it, I have the consciousness of the “*A* again and again,” each and every time an *A*, but also the consciousness of anything whatsoever to be extracted, of “*every whatsoever*.” Likewise with two universals occurring together: *An A as such is B, each one.*

But, *on the other hand* (e.g., when at first a *B* occurs now and then), the thematic interest can also be directed toward the occurrence of a *B* as such. Even here the function of the whatsoever plays a role, but in a second form. It is immaterial that *this here* is *B*, but that *one there is B*. Obviously here again is the particular, “an *A*,” and likewise the indeterminate universal, “some *A's*,” that stand in the subject-position, not an object in the genuine sense, to which object (like a determined table) one would attribute a predicate; and yet it is a theme, a theme being constituted by the higher functions of activity; the entire judgment, and all functions of judgment as relating functions take on a new shape in the consciousness of the “as such,” and by assuming the “as such” into a theme. It is evident that the function of the “as [83] such,” to which the specification is bound, is a conceptualizing function, that of universal and particular conceptual grasping.

If we sever the *tie to <a> realm of experience*, to a pregiven sphere of constituted objects, if we operate in a scope of pure concepts, in a manifold of possibility in which possible actualities remain indifferent according to their positionality, we gain *pure “as such”-judgments*, judgments that have the character of *judgments of laws*, like, “A triangle has as such three angles”; “An extended object is as such qualified [in some way]”; “A red object as such is colored.” The *particular* judgments are expressed as *judgments of possibilities*: “A triangle can have a right angle.”

All such judgments thus operate on the ground of pure possibility and say nothing about actuality. If I imagine an extended object in pure possibility, I will find it colored or qualified in some other way. But by modifying in free variation and by holding firm to the extended object, in the free transition from the one fiction to any other one whatsoever, and where it is a matter of indifference to me which other it is, I find both the conceptual determinations united, I see them in the transitional consciousness as united as such in lawful regularity. With this, the realm of universal judging is opened up, the realm of law-giving, the knowledge of law-giving for itself and for all judgments as such, and therefore for all possible objectlike formations.

<Supplementary Texts>

<Section 1.
FIRST VERSION OF MAIN TEXT PART 2 (1920/21)>[1]

<1. The Misunderstanding of Modalities of Being by Logicians and Epistemological Psychologists> [225]

Thanks to the phenomenological analyses that we have undertaken, we now understand the origin of modalizations arising in the sphere of perception. We find modal distinctions purely with respect to the objective sense of every perception, and the same perceptual object can potentially present itself one time in this modality, another time in that modality, or in variable modalities, now as existing in pure and simple certainty, now as problematically possible in conflict with other problematic possibilities, and then as not existing or as existing in fact. On the other hand, the perceptual object is inconceivable without one of the modalities we just mentioned, just as it is inconceivable without open possibilities, since the perceptual object moves into an open future. If we go from the noematic attitude to the noetic one, the perceptual lived-experience will likewise be inconceivable without what makes up the modalizing accomplishment in the perceptual lived-experience; and this is completely clear to us.

---

[1] Translator: The difference between this first version of Part 2 and the one published above under the heading of "Main Text" concerns §12 through §40. The italicized sections of this text are repetitions of the former. Like the editor of the German edition, I include them here for the sake of coherence.

The following pagination to the German text corresponds to Husserliana XI.

We are therefore in the position to be able to comprehend the bad mistakes into which philosophical logicians and epistemological psychologists have fallen by having believed it necessary to distinguish between the so-called mere presentation and judgment for every perception (and correspondingly for every other consciousness of an object), thus, in this case, to distinguish between the perceptual presentation and the perceptual judgment; here, under the rubric of judgment, one would distinguish between active acceptance and rejection as more specific differentiations. They obviously carried over the quite evident distinction between the perceptual object and its modal modes of givenness to the perceptual lived-experience and—since the sense of constitutive [226] analyses were still completely hidden, and since they were even lacking the fundamental distinction between noema and noesis—they made a real distinction in a lived-experiential consciousness out of an irreal distinction; indeed, they even divided a lived-experiential consciousness into separable elements of lived-experience. If an object appears in the flesh in the perceptual lived-experience, and is thereby characterized in this or that mode of being, this does not mean that perception consists of two elements or layers of which the one constitutes the object in its presentation in the flesh and the other, building upon it, apportions to the object "being" or "non-being," etc. According to Brentano and his school, with which Meinong was also affiliated, there is supposed to be one unique perceptual presentation that presents[2] the object in the flesh, and in addition to this a judgment that now actively accepts, now rejects, affirming or denying what is presented. In principle, however, those judgments do not have to be supplemented, and in this case we would have a mere presentation.

But according to our analyses, it is clear that there is not anything, and there cannot be anything, on the order of mere perceptual presentations, neither as particular lived-experiences nor only as sublayers in lived-experiences that are self-contained in an intimately inherent manner. A perceptual presentation would certainly be a consciousness, a consciousness that gives an object

[2] *vorstellig macht*

originaliter. But such a consciousness would not be anything other than that system of intentions of the structure described, and it would be entirely inconceivable if it were anything but that. Such a system, however, is necessarily a system in which the intentions run their course either in original concordance and are unbroken, and if this is the case then this mode is called nothing other than perceptual faith, and the object, existent. Or a rupture ensues and then we have the other possibilities that have been prefigured; the intentions themselves undergo a thorough internal recasting of their concordance, even though the systematic structure may be the same as the systematic structure that constitutes its presentation in the flesh. Belief and the modification of belief is not something added to intentions. The fact of being non-inhibited and of being inhibited by parallel and partially coinciding intentions is not something that is juxtaposed to intentions; it is not a new supervening lived-experience called belief, judgment, but precisely a recasting of concordance, a modification that makes possible the nature of consciousness as consciousness, and in fact, as we will hear, makes possible every consciousness. Thus, [227] belief as being certain, as negation, as affirmation, etc., is to the so-called perceptual presentation as the timbre is to the tone or as the tonal intensity is to the tone. One cannot separate the tone and then add the tonal intensity to it, although the analogy is naturally precarious and to be taken *cum grano salis*. One should not treat the perceptual object as such and the object as such presented in a different manner, that is, one should not treat what we termed either the objective sense or the noema, as a piece in consciousness, as generally happens in the traditional literature that is still unaffected by phenomenology. One must not ignore all the multifaceted and actually demonstrable structures of lived-experience in which sense is constituted in the process of perceiving as an intentional unity, and inseparably from this, the being-modality of sense. And one must consequently not pass off the mode of being as something that the judicative ego adds to the sense which would, as it were, already reach the ego in advance as ready-made.

But if one pursues a deeper analysis, it will be clear that corresponding to the unitary element of the object in the noema is a manifold of intentions in the course of the lived-experience; these intentions in all of their elements are a consciousness through and through, and as a consciousness they have those uninhibited concordances or inhibitions and conflict; and it will become clear that this modalizing process that recasts the hue of the whole of consciousness is what necessarily constitutes the mode of being in relation to sense. Accordingly, in the case of the modality of problematic [possibility], one will also not interpret, as does happen, indecisiveness as a cessation of the putative perceptual judgment, and will not altogether reduce this merely to positive and negative believing or even only to active acceptance and rejection. Positive belief designates: (1) the primordial mode of consciousness, consciousness that is unbroken, but that is also still unaffected by any breaks. (2) Active acceptance, confirmation is that consciousness of unbroken concordance that is restored after undergoing a break; it is a concordance after overcoming the inner bifurcation. This overcoming is carried out in the original form in the perceptual progression of concordant fulfillment as the resolution of inhibition from the side of the one party of the conflicting intentions, whereby the other simultaneously gets [228] annulled in the form of being crossed out, in the form of negation. Every affirmation also entails a negation.

The one thing that we have not taken into consideration and that will yet play its role is the participation of the activity of the ego. If the ego executes an act of affirmation (of active acceptance), it activates, it lives through those particular intentions, those of concordance, while the annulment of the opposing intentions, as suppressed, takes place precisely in the implicit form of emptiness, as a modalization in subconsciousness. On the other hand: Negation as an act is the activation of this crossing out in the transition from the activation of the opposing intentions to the living through of concordance or vice versa. There isn't the slightest reason to exclude the mode of being undecided, the mode of question, or what amounts to the same thing, the mode of the consciousness of enticement and of probability from this set in which the latter consciousness always plays a part as an intermediate stage with a corresponding accomplishment of sense, questionable, problematically possible sense. Even here we have different forms of execution on the part of the ego. None of this excludes the fact that decisions have a priority over the indecisive uncertainties for logic as a normative science. But one must first bring it to light, and then also do justice to the modes of indecisiveness and to their norms.

<2. Non-Prominence and the Prominence of Sense and Modes of Being for Consciousness>

Before we advance beyond the realm of perception in which our analyses were carried out, let us develop the results of our analyses a little bit. The distinction between the perceptual sense and its modes of being did not imply a separation for either perceptual lived experience or the perceived object. In this respect we have to consider the following. There is no rupture in the origin of perception where naive perception is concerned; it is simply a consciousness of the perceptual object. If the ego is active, that is, if it grasps [objects] in an attentive manner, it will grasp the object *simpliciter*, and here mere objective sense and mode of being are not distinguished at all for consciousness, and the objective sense and the mode of originality are distinguished just as little; and in general they are distinguished just as little as the results of our reflective analyses and our scientific conceptualizations of perception would themselves become an [229] object [of perception]. In the normal attitude of the act, of the "I perceive," the grasping bears on the unity that is constituted in a continuous coinciding as this object becoming there. In this case, [we call] the objective sense in the original mode of being that which is in a continual coincidence, that which is, so to speak, continually identified in the process of constitution; it is the correlate of the unbroken concordance of original intentions. It is only in a transition to discordance and therefore to modalizing transformations that sense and modality of being are first set apart from each other in a relation of contrast; we recognize this when we look back at the perceptual object as it was given prior to the break, and when we also look at the objects that were then given together with this and as conflicting. Seen more precisely, under the rubric of "mere" object-sense, something takes on the character of an object that previously did not have this character; it was itself not there like an object in the simplicity of perceptual consciousness. A coinciding and an identification take place that had not existed before; a consciousness of an object is established that had not previously existed: If the pure and simple external This-Here, the external perceptual object, has already been constituted, then the sense of this object and its mode of being will now be constituted, and this constitution is not itself an external perception, but rather a consciousness founded in external perception. It is likewise an originarily giving consciousness if it has arisen in the way that we are taking it here, namely, fashioned from the reflective attitude on the perceptual object as such. But even though it is a consciousness that fashions its object in its full originality, it is not a perceiving, not a grasping *originaliter* of an individual object or even of a thing. To be sure, senses are not things. We will have something to say about this consciousness later, for its objects are also the principle themes of interest for us logicians. If the disruptive discordance has been overcome, a contrast between sense and modality arises, "being" gets the new character, "it is actually so"; but then in the active perceiving that now once more devotes itself to the object and to its acquisition of knowledge, the object will once again be given as an object in a straightforward manner; that is, as we continue to perceive, a constant sameness is given in an original mode of being, but [230] without the distinction between sense and mode of being. Once again, we have the same object in a straightforward manner, just as if a break had never occurred.

One can still sense a remnant of unclarity. We have an original shape of continuous perception as a system of concordance. Perception has its perceptual object, the existing object, the unity of just this concordant ratification of the self, which proceeds in an unbroken manner.

If a break occurs, [through] doubt, [or through] crossing out, we will have a "revaluing," a cancellation of this concordant nexus that no longer proceeds in an unbroken manner; rather [we will have a nexus that] becomes modified through the break itself. What is crossed out there is nothing other than the object *simpliciter* as "being." And we see that the spared, remaining object does not consist of two components, "sense" or content and "being," but rather is precisely existing object or object *simpliciter*. A perception and a crossed out perception having the same "content" now have something in common, and that is precisely the content of the perceptual lived-experience and of the negation of perception. But this is not something general that is differentiated; it is not a part that gets a supplementary part through the "quality of belief." The noetic content of the negation of perception also "contains" the perceptual belief, but as crossed out, or rather, "being," but as crossed out. Can more still be said about this?

In the case of other modalizations, like negation for example, we do not have an object *simpliciter* from the very beginning without inner differentiations of sense and the mode of being, but rather the negated object, or in problematic consciousness, the problematic object. It seems that a bifurcation is essentially situated here in the dyad of sense and modality corresponding to the fact that in general the consciousness of such modal transformations is indeed more intricate than pure and simple perceiving.

<3. The Modalization of Immanent Objects>

Let us get first get clear about what can be used from our results vis-à-vis immanent perception. We see without further ado that the [231]
discourse of the being of an immanent object and the characterization of immanent perceiving as a certainty of being essentially leads back to the same sources as the discourse of the being of the external object and perceptual belief. Even the immanent object is constituted in internal consciousness and is constituted through the systems of intentionality, the systems of primordial impression, retentions, protentions, which pass over into each other in steady concordance. Correlative to the unbrokenness of these original intentions is again "being" situated in the consciousness of the object. We are conscious of every one of our lived-experiences as being; we are certain of them in a straightforward manner, and this certainty means the same thing as the certainty of the external object in unbroken external

perception. But now we understand the difference and the reason why we could not profitably link the doctrine of modalities to immanent perceptions. An immanent object can in principle only be given with certainty. Original constitution, which lets it arise as an *esse* in *percipi*, does not in principle admit of any modalization, of any dual apprehension in conflict. There is not an apperceptive apprehension here in the same sense as there is in external perception; apperceptive apprehension already presupposes immanent constitution as its foundation. But this is, by its very nature, a steady, passive process of intentional concordance in the succession of primordial impression and retentions.

Still, we must be more careful: On the one hand, seen more precisely, modalization is indeed possible for immanent objects. The extent to which they are already constituted in the flux of their becoming (and they are certainly objects of immanent perception only as becoming), is the extent to which there is no way of speaking of modalization. The lived-experience, which has already been constituted as present and together with it as just past, cannot be subject to doubt and therefore cannot be negated, either. How should an overlapping doubling and a reciprocal inhibition of intentions arise here? A primordial impression can only fade away in a single sense, can only sink back into retention in a single sense. It is inconceivable that it become doubled. The necessity is absolutely unequivocal. But the constancy of protention also belongs to the constitutive process; what is already constituted [232]
motivates an indication of something that is to come through its own content; it prefigures an empty horizon of the future, but one equipped with an indeterminately general sense. For example, the thought brewing in me (coming to me without my aid) has a prefiguring horizon of expectation through its style, and that issues in something like open possibilities that are being determined more closely. But also an interruption of the thought or a turning against the expectation, that is, possibilities for the consciousness of the thought's non-being. This becomes even more clear when we take a tone or a tonal formation purely as an immanent datum, as a pure sense-datum, without any transcendent apperception. It becomes questionable which tonal phrases will occur there now, how the tonal figure will be pictured—questionable insofar as an

ambiguity of protention develops. There are also, therefore, modalities here in this protentional direction. Thus, our results in relation to the latter also hold *mutatis mutandis* for immanent perception, whose indubitability, that is, whose inability to be modalized is only valid to the extent that it actually accomplishes an original constitution. That concerns the immanent lived-experience that is in the process of becoming to the extent that it moves into the process of becoming in every moment and that it has been given to consciousness as a duration that has just become by virtue of the retentional continuity. No disappointment of the anticipation can summon changes of sense in the backward direction and bedazzle what is actually constituted.

<4. "Types of Lived-Experience" are not Empirical Facts, but Formal Structures of Consciousness as Such>

Going beyond the perceptual sphere, let us now turn to other, non-original lived-experiences. We will be able to expand upon our insights here in an important way. I would like to mention a general observation at the outset. If an external, naturalistic psychology and transcendental philosophy approach the life of consciousness, they will be presented with perceptions, memories, expectations, imaginings, and then further, with judgments, feelings, desires, volitions as special names for types of experiences, and they will appear here as factual types in the realm of human and animal consciousness, similar to biological, [233] psychological occurrences as empirical facts of organic nature.

But if one has learned to see phenomenologically and has learned to grasp the sense of intentional analysis, if one has—expressed in the form of the Goethian myth—found the way to the mother of knowledge, to its realm of pure consciousness in which all being arises constitutively and from which all knowledge as knowledge of beings has to fashion its ultimate comprehensible clarification, then one will initially make the quite astounding discovery that those types of lived-experience are not a matter of arbitrary special features of an accidental life of consciousness, but rather that terms like "perception," "memory," "expectation," etc., express universal, essential structures, that is, strictly necessary structures of every conceivable stream of consciousness, thus, so to speak, formal structures of a life of consciousness as such whose profound study and exact conceptual circumscription, whose systematic graduated levels of foundation and genetic development is the first great task of a transcendental phenomenology. It is precisely nothing other than the science of the essential shapes of consciousness as such, as the science of maternal origins.

Thus, the knowledge that perception is an absolutely universal essential structure of consciousness as such has already come to the fore with our analyses of the original constitution of temporal objectlike formations the moment we had distinguished between immanent and transcendent objects: There is no lived-experience conceivable, and there is no nexus of consciousness for lived-experience conceivable, without them being subjected to the law of time-constitution; that is, it is only insofar as it is constituted in the steadily prefigured lawful nexus of primordially impressional, retentional, and protentional intentions. Lived-experience is not only given to consciousness, it is also given as being *originaliter* and as being in the process of becoming and having-just-become. External perception has a certain universality, but one that is noticeably completely different from that of immanent perception. That external perceptions arise in the stream of lived-experience of which we are conscious immanently as perceptually given is a universal fact insofar as an external world of perception is constantly there for us, namely, in the form of some kind of perceptually given external surroundings of a continually co- [234]
constituted lived-body. But the necessity that external perceptions must arise in the immanent stream and must arise from this continually integrally cohesive accomplishment is obviously not a necessity in the same sense: as if consciousness would not be conceivable at all without the like. On the contrary, our intentional analysis here leads us to understand all external perception and all organization of external perceptions; these constitute an infinitely open spatio-temporal-causal world as a developmental whole, and this leads back to a consciousness "before" this development, for which no external being-in-itself can be given at all. Moreover, we see that this development is tied to conditions that signify

facticities for every empirical ego and its individually determinate stream of lived-experience which is examined in pure possibility. Without determinate courses of hyletic data spatial objectlike formations cannot be constituted.

<5. Presentifications as Necessary Components of Perceptual Lived-Experience>

Yet we prefer to pursue the path of absolute necessities which takes its point of departure from the necessity of perception as a title of shapes of consciousness, without which no consciousness itself can be. The structure of immanent objects in immanent originality highlights for us the curious fact that we are lead back to elementary, but non-independent components, non-independent as mere phases that present the headwaters of a most pure originality under the rubric of primordial impression. The original consciousness of immanent perception and of every perception is only a most purely original consciousness with respect to a phase; next to that we have components that are not original, no longer originally giving, and here we encounter two types, both presentifications—if this term characterizes precisely a conscious-having of something that is not present in the original. On the one hand, we have the character of "still being conscious of," on the other, "not yet conscious of." The one, the retentional, becomes immediately empty, non-intuitable, but it maintains the primordially instituting knowledge, it maintains the sense in its full determination and mode of being. The other expects, anticipates the acquisition of knowledge and has leeways of [235]
indeterminacy. In this shape, presentification thus belongs to the original stock of every concrete perceptual lived-experience, and therefore to every lived-experience in general with respect to its immanent constitution; it belongs as a type of function that is non-independent and that makes possible concrete perception.

<6. Presentifications as Independent (Concrete) Lived-Experiences. Concrete Retention and its Modalization>

But we also certainly have concrete lived-experiences of presentification; we have memories and concrete expectations, and we have here under this rubric intuitive and empty consciousness. Further, we also have concrete empty retentions and not merely non-independent retentional components and continua in the nexus of a perception that is in the process of unfolding.

Let us first consider this kind of concrete presentification. It necessarily attaches to each perception; namely, we no longer speak of a perception after the last phase of the originality of perception has flowed-off, e.g., the moment when the resounding tone ceases, and when this has also become transformed in the retentional phase; we have a momentary continuum of retentions that reproduce in the Now the entire course of the preceding perception in all their phases, and this entire continuum undergoes further transformation at one and the same pace; and as it continually coincides, we become conscious of it as something that has just flowed-off and as something that is simply being pushed back further and further.

Naturally, such a retention, too, has a universal necessity for consciousness insofar as it must be regarded as a continual, immanent perceiving. Necessarily linking up to each lived-experience that has flowed-off is a concrete retention of that very lived-experience. This takes place incidentally in pure passivity in the same manner that original time-consciousness in general (to which this retention belongs), runs its course in steady passivity. The original accomplishment of retention consists solely in helping the emergent consciousness of the temporal objectlike formation to advance, even though with respect to the appearance it does permit this accomplishment to shrivel up ever so quickly and to pass over into an empty, undifferentiated distance. Just as in [236]
the change of orientation of spatial distance an outermost distant horizon always presents a "vanishing point"[3] in which all phenomenal differentiations of proximity and distance (like all

[3] *Abschluß*

other phenomenal differentiations) fade away and disappear, so too [does this occur] in the process of sinking back into the past. All differentiations of succession and the differentiations that are formed in them with respect to content—differentiations that shrivel up more and more—fade away into a temporal distance that has finally effaced all phenomenal differentiations. And yet they are intentionally contained in it in a concealed manner. Out of the distant horizon that is given to us in a non-intuitive manner (and unlike the distant horizon in the consciousness of space that is still intuitively given), this or that can exercise a special allure from something implicit, affecting the ego, steering the ego's interest in a certain direction; and now what is distant emerges once more in the form of a remembering that brings what is distant closer in the form of "re " or "again."

This can certainly occur without the [active] participation of the ego; a clear remembering can suddenly break through. As we shall soon see, this is certainly something essentially new. But we must say phenomenologically that what presents itself to consciousness in remembering within intentionality and in explicit intuitability gives itself as the same thing that is found *implicite* in retention by virtue of an identifying coinciding of sense with the corresponding accentuated components of retention. The emptiness that seems to be entirely undifferentiated only discloses its hidden sense-manifold in this way: through the transformations that occur when intentional, special moments surface in their particularity from the abiding emptiness, and then through the transition to presentifications that explicitly bring something to intuition. But all such transformations and transitions are in their very sense connected through syntheses of coinciding.

If after this structural analysis we inquire into what happens in the retentional sphere with respect to the belief of being and its modalizations, it will be clear that what holds for the modalization of concrete perceptions must also hold for the modalization of concrete retentions. The intentionality that is put into play does indeed become modified in the retentional transformation, and if perception ceases, then we will have a pure retention and ongoing change. But this change is not an inhibition of the intentions that are put into play; they continue to run their course in the concordance of coinciding; the objective sense has the mode of [237] straightforward being, but in accordance with the retentional transformation, [it has] the altered mode of being-past. A becoming-discordant is likewise impossible for the immanent objectlike formation in concrete retention, just as in concrete perception. That is, doubt and negation are essentially precluded. On both accounts we will find that a theory which would limit the so-called "evidence" of the givenness of individual being only to inner perception, and would deny evidence for concrete retention, indeed, a theory that would go so far as to allow actual evidence to hold only for the punctual Now of the primordial impression is pure non-sense. Where doubt is essentially precluded, and negation too, the evidence of being is included.

Let us now continue to consider concrete retentions that follow transcendent perceptions. What about modalization with respect to them? We see that doubt, that is, a bifurcation in sense-giving, can occur in the perceptual sphere by an immanent lived-experience; a concrete lived-experience within immanent time carrying out within itself a transcendent sense-giving, i.e., "points" beyond itself in an anticipatory manner, and is therefore dependent upon possible fulfillments in the progression to new such lived-experiences. These immanent data can undergo doubled apperceptions that are motivated from different sides, and these can be mutually inhibiting. In this case, we also speak of a radiating back of an inhibition occurring in the course of perception, a radiating back into the intentions that were previously uninhibited; put more precisely, we speak of its radiating back into the retentions and therefore into the past of consciousness that is concealed in them. Of course, this holds for the concrete retention that endures after perception ceases. Accordingly, such a retention can have all modalities insofar as it harbors a transcendent intentionality. Indeed, even if it emerges from an unbroken perception in unbroken concordance, it can subsequently pass over into the mode of doubt and therefore into all affiliated modes of being. In order to make this clear, let us point out, for example, that an external perception of a thing is not something isolated, that it does not constitute existence in an isolated manner for itself, but does so in the universal, [238] transcendental nexus of constituting being—initially in the nexus that does not only constitute this one thing in an intuitive manner, but intuitively constitutes a far reaching spatio-material surrounding. Thus, a discordance can occur in the environing-worldly apprehension of the environing-world at the place of this or that thing, and this, then, will motivate reinterpretations of meaning or doubt beyond this place and with respect to givennesses of a past that is still retained in consciousness. For example, during a break we hear several piano pieces and have the apperception that someone in the next room is playing the piano. Suddenly we wonder whether it is not really coming from a mechanical apparatus (of a baby grand piano). Doubt is immediately carried over into the retentional sphere, namely, to the pieces that were previously heard.

<7. Empty, Concrete Expectation. Its Modalization>

We can also gain insight into empty concrete expectations in a manner that is quite similar to concrete retentions, both as to the necessary universality of their function in consciousness as well as in relation to the way in which so-called anticipatory belief (as consciousness constituting existence) is to be understood and how the corresponding modalizations are to be understood in them.

We distinguish the intentions of expectation that belong immediately to the constitution of every perceptual givenness, the non-independent protentions, from the concrete expectations that present the futural concrete nexuses to consciousness in an empty manner—to be sure, always referring conjointly to other perceptual spheres. The former naturally belong to each moment of consciousness, since every lived-experience is co-constituted in inner consciousness by protentions. Even concrete expectations, as empty, have their latent intentionality that is made explicit in intuitive presentifications, parallel to remembering of something past; even here we see that the intuitive, explicit presentification, the intuitively pictured expectation, is a secondary form: The picturing already presupposes empty-consciousness.

Where the modalities of being are concerned, obviously nothing else can be said to hold for concrete expectations than what is said to hold for expectations which, as protentions, link up to the perceptions themselves. The essentially possible reproductions [239] that are intuitively presentified verify the fact that a transition to negation and to problematic possibilities can only take place through diremptions and overlappings of sense in a way that is entirely similar to what we have studied with respect to perceptions, only that [in the former case] this transition arises precisely in a reproductive form. We observed here as well the difference between [a] the implicit and, as it were, the inauthentic constitution of sense and being regarding empty expectation, and [b] the authentic and explicit sense-constitution of the corresponding expectations that are intuitive and that picture: That we attribute the same sense to the corresponding acts, that we do this takes place by virtue of the synthesis through coinciding in which emptiness is fulfilled. What is intuitable arises here as the intuitive expectation with the character of anticipatory fulfillment. This is obviously a different kind of fulfillment than the fulfillment occurring with the process of bringing empty retentions to intuition; here it is not anticipating but again presentifying. We will certainly have to treat this in a more precise manner.

<8. Concrete, Empty Presentifications of What is (Temporally) Present. Its Modalization>

We must finally point to the fact that there is still another type of empty presentification, namely, the presentification of something present, temporally speaking, but not of something present in the sense of original intuitability. We are familiar with such presentifications as components of all external perceptions; they are connected to the latter as empty horizons. But they also arise in an independent, concrete form, namely, as empty presentations of concrete things surrounding us. For example, if we glance around this room, the view of the windows and doors will awaken in us, immediately, images of the street or of the foyers, but generally in a shape that remains empty. It also belongs to the essence of such empty presentations that they can pass over into intuitive presentifications. Thus, we can at any time make intuitively present the back side of the thing of which we are emptily conscious, as well as the concrete spatio-material surroundings of which we are emptily conscious; we can do this by imagining that we are walking around the thing or passing through the door to the foyer and then that we are walking out into the street, and now that we are allowing the co-connected series of [240] appearances of all these non-visible sides of the object and of all these objects [themselves] to run their course, namely, the sides and objects in which the present actuality of the thing is exhibited. The series of appearances, which conform to every path of kinaesthetic systems that are given to consciousness as freely at our disposal, can in this case be motivated in an unambiguous or ambiguous manner, that is, the corresponding intentional nexuses can run their course concordantly and in an uninhibited way; or the series of appearances disturbing one another can overlap, thus allowing conflicts to set in; and in this way modalizations are possible. This is intelligible because all of the intentions that are at issue here are reproductions of anticipatory intentions [linked] to reproduced perceptions; these are connected to hypothetical kinaesthetic courses as demanded successions. What is disclosed by bringing something to intuition in a reproductive manner is found *implicite* in the empty presentifications of something present, and this "*implicite*" has its sense precisely in the mode of essentially possible explication.

We could still point to a shape of presentifying something present, and a quite curious shape indeed. I mean empathy as the consciousness through which an alien psychic life can be given to an ego in its life of consciousness. Empathy necessarily arises in its original form in connection with transcendent perception. It is based on the perception of the alien lived-corporeality as a physical thing-like body, by this thing being apprehended through its similarity to my own lived-body as lived-body. In a manner similar to the way in which I become co-conscious of the non-visible aspects of a thing through the empty intentions of perception, through "empathy" I become co-conscious of the alien psychic life, an alien psychic life that is inaccessible to direct perception as such, and for the most part in an empty manner. Thus, empathy means here a level of founded presentation that is connected to the perception of the lived-body-thing, a presentation which, when brought to intuition, has its own mode of bringing to intuition and its own mode of fulfillment. It is an empty making co-present, a presentification of a consciousness that is made co-present and that belongs to the lived-body, a consciousness, however, whose process of bringing to intuition certainly has to embark upon quite different paths than those peculiar to the non-visible aspects of the thing-like body. Even here it can result in modalizations by means of doubt. The lived-experiences of consciousness that are indicated through the medium of lived-corporeality and of expression that is conveyed in a lived-bodily manner emerge in an ambiguous and discordant way. We will not [241] go into this here; our only concern is to give examples of the empty presentifications of something present.

<9. Even the Presentifications of Something Present are Universal Occurrences of Consciousness>

They are nevertheless examples from the sphere of transcendence, and so it appears that we are not dealing with completely universal occurrences here. In fact, one might like to maintain that if every conceivable consciousness should contain the shapes of empty-presentifications of something present, this would also have to be manifest if we were to bracket all transcendent world constitution from the immanent realm. But an immanent present is indeed *eo ipso* given in the flesh, constituted in inner consciousness, that is, not merely in a presentified present. However, this is not a conclusive argument. Immanent perception and the presentification of something perceived are perhaps compatible in a certain way. I mean in this way: Everything that we designate by the term association is characterized phenomenologically as a connection of consciousness concerning the process of one thing recalling another, a connection that exists between the so-called associates. It is not a mere objective fact that the thought of "Vesuvius" reminds me of the thought of "Naples"; and when given to consciousness, both thoughts are not merely juxtaposed or in succession; rather, one of them refers to the other; within the consciousness of one of them, there is a pointing to the consciousness of the other. But "pointings" do not merely occur within the consciousness of transcendence, but also within immanence, and among them even those that go from one simultaneous thing to another, and from one present thing to another. For example, if color data occur together with olfactory data once or even often, these colors and these smells will not only be there together again in the new case, but will have their integral togetherness given to consciousness: Connected to immanent data are indicators[4] of something simultaneously given, and the givenness is not an obstacle to an indicating consciousness connected to another [datum]. But if the indicated data do not emerge along with the others, they will be "lacking" for [242] consciousness; the indications are then empty and at the same time, inhibited. I mean, in other words, that even this kind of presentification of something present has its universal significance and is a matter of concern for us here.

<10. Fundamental Types of "Presentations">

We have spoken of several types of empty modes of consciousness of something individual, each one of which referred us back to intuitions; for we could only genuinely speak of their intentional content and their modalization through the explication of the corresponding intuition. We will have to reflect upon these now. Let us first review the general typicality of the presentations of something individual, that is, the typicality of all the shapes of the passive consciousness of something individual, just as we have already come to see them. All activity of the ego and its special accomplishments, above all those of theoretical thought, of identifying, of differentiating, of predicating, etc., were outside of our thematic interest in our previous considerations. The following fundamental types can be distinguished:

(1) The radical distinction between intuitive or full presenting and the non-intuitive, empty presenting. Empty presenting is an implicit, inauthentic presenting; it only contains within itself sense and modes of being *implicite*, and further it contains precisely everything that can actually and genuinely be found only within

4 *Vordeutungen*

explicit consciousness. "Actually finding" is precisely the intuition-of-the-self and taking something constituted in a living manner from the vivacity of a process of intuitive constitution. What empty consciousness harbors within itself *implicite* is not actually constituted in a process of manifold intentions that carry out within themselves a unity of sense-giving accomplishment. One can therefore also designate the opposition [between intuitive and non-intuitive, full and empty] as the opposition between genuine and non-genuine consciousness or between explicit and implicit consciousness. But as we know, the mode of the non-genuine, of empty intentions, is indeed essentially fundamental for making every explicit and concrete consciousness possible. No concrete object can be constituted for consciousness without the co-functioning of empty horizons; what is required is a constant intertwining of fullness and emptiness.

(2) If we consider empty consciousness, we can distinguish, in general, between empty retentions and empty protentions. From now on, we use the latter term in general for any type of consciousness that is not only an expectation in the customary sense, but also for everything that is essentially related to it. Although we had distinguished between expectations and presentifications of something present, there is a unitary character of the term, protention. This unitary character of protention, and its radical distinction to the type, retention, will be attested to by its features which are in principle distinctive—distinctive features with respect to the corresponding processes of bringing to intuition and fulfillments, as well as modalizations. [243]

(3) Therefore, when we consider intuitive consciousness, which in its very typicality corresponds in a certain way to empty presentations to which intuitive consciousness can refer, we encounter the radical difference between perception and reproduction. Perception is a giving consciousness *originaliter* of something individual. The object is constituted and, if you will, produced in its very sense and mode of being in the primordial mode, in originality. But reproduction reproduces, reconstitutes; it carries out constitution in a peculiar modification, in the mode of "after a fashion," whereby it gives itself in itself as modification,

and refers back to original consciousness according to all its components and accomplishments.

But that is also possible in a different way: One peculiarly characteristic way is the mode of remembering that we analyzed earlier in detail and in relation to which we clarified what is peculiar to a reproduction in general. It is a reproduction in the narrow sense; its peculiar nature is to refer to something that was previously perceived in the same immanent stream. Its object is characterized as an object that has been, an object of a perception that has been in the same stream of consciousness. The more general character of reproduction as a modification of perception and something perceived in the mode of "after a fashion" however also occurs in another form, which, as we will see is essentially different: in the form of anticipatory, intuitive presentification of something futural and in the form of the intuitive presentification of something present. One usually understands by the term "rememberings" intuitive reproductions in the narrow sense, that [244]
is, more or less clear reproductions of the past, memories of the past. One could be tempted to use this term broadly to refer to all reproductions, and to speak of memories of the future and memories of the present. In any case, we lack an unequivocal, generally encompassing term, unless we use the term presentification. Or the term, phantasy. The latter, however, is dangerously ambiguous; but the former term is also the only useful term at our disposal if we should deal with all modes of consciousness of individual being, modes of consciousness, that is, that stand in contrast to perception: the intuitable as well the non-intuitable lived-experiences. All of them "presentify" insofar as they make something present to consciousness, but not in an original conscious-having.

One must become quite familiar with their essential differences, and one must initially guard against taking empty presentifications for merely obscure reproductions, as if, in the gradation of the clarity of reproduction, emptiness would only signify the lowest level of clarity and even within this level of clarity, e.g., intuitive phantasy would be nuanced. The object in question is actually constituted there if the "phantasy" is cloudy, as long as the phantasy is, in general, an intuition in a precise mirroring of the intentional process of perception. The object is structured, as it were, before the reproductive eye as an intentional unity of its manifolds. If the phantasy is intermittent, then the constitutive processes will cease, and an empty consciousness will step in in order for it then to be transformed once more into an actual reproductive constitution. Nothing happens in empty consciousness; it does not contain any constitutive structure; there is nothing to look for in it. What can be said about it can only be said by reason of its process of bringing to intuition. If we clearly recognize, essentially and universally, that every empty presentation has its process of bringing to intuition and does not at all accord with just any intuition, and if we clearly recognize what this accord means, then we will realize that an empty presentation in general is only the potentiality of what lies in the corresponding intuition as actuality. The process of bringing to intuition itself (disclosure), the transition from empty presentation in its corresponding intuition is the actualization of the potentiality of constitution which lay in the empty presentation precisely as mere [245] potency. In the latter, sense was not a given sense, not an appearing sense.

The phenomenon of transition is characterized as a synthesis through coinciding of the empty presentation and intuition, of the *potency of a constitution and of the corresponding actual* constitution; thus, with respect to sense, it is the agreement between potential and actual sense, or again, object. The presented object in its sense and being is not a doubled one, but rather a single object that, on the one hand, is now merely and emptily presented, emptily meant and, on the other, completely intuitive. Because there is in this way a possible empty presentation corresponding to each intuition, a possible intuiting corresponding to each empty presentation, and with this essential kind of synthesis, we witness not only an essential characteristic of the realm of presentation, but generally, of the realm of consciousness as such with respect to all levels, no matter how high they may be (as will be demonstrated). Here, where presentations are concerned, one must initially become completely clear about what is peculiar to this "emptiness," to this implicit sense-giving, or better yet, to this potentiality and its disclosure in actuality. This will turn out to be a point of decisive significance for logic. For (privileging only the theoretical sphere of thought) linguistic thinking in the mode of emptiness plays a constant and entirely essential role. And the logical central questions, the normative ones, those concerning verificatory grounding are related to this linguistic empty thought in a special way. But here I am getting way ahead of myself. First and foremost, what has not at all come to light (but it must now be shown) is the fact that not every adaptation to an intuition shares the same fundamental character, not all of them are giving in the strict and genuine sense.

<11. Fulfilling and Merely Disclosive Intuitions>

Let us consider the mode of remembering. Rememberings are intuitions that are classed among empty presentations belonging to the type, retention. Here, the synthesis is one of disclosive, clarificatory bringing to intuition. We can contrast it with the protentional reproductions that, on first glance, appear to be completely equivalent to the reproductions that turn backward, distinguished only by the fact that they are directed ahead to a [246] coexistent present or to a future. But here we will recognize the curious feature that empty protentions as corresponding to intuitive protentions have a dual mode of bringing to intuition, on the one hand, as a fulfilling process of bringing to intuition, and on the other, as non-fulfilling, merely clarificatory. And the expression, bringing to intuition, is appropriate only here. For example, an expectation of the future is intuitively fulfilled if what is expected transpires in perception; but even without that it can be brought to intuition. The intuition is then a mere anticipation of a futural perception. Inversely, an empty retention or, if you will, an emptily emerging memory is brought to intuition only in one way. That is, an intuitive remembering will occur if an appropriate intuition is affixed to the empty retention in general.

This distinction between fulfillment and the merely disclosive process of bringing to intuition gives us something to consider. Actually, we are already familiar with this difference from our analyses of perception with respect to their empty horizons. There is a difference between merely presentifying in intuition the non-visible side of a perceptual thing in a suitable reproduction, and bringing it to an actually giving perception by walking around it, which is to say, bringing to fulfillment the empty intentions of expectation as perception progresses. In both cases, in both the disclosive process of bringing to intuition and fulfillment, the synthesis is carried out in unbroken concordance. At least, this is the normal case. That something else is also possible, that the disclosure of the empty intention, that the explicating actualization can also bring to light hidden discordances, certainly remains to be seen. But if we stay with the normal case, we will then have in both cases—in both the process of bringing to intuition and fulfillment—precisely an unbroken consciousness of being. Then this means here as well: Where there is no break, there is a thoroughgoing consciousness of being.

But on the other hand we have the great difference that the empty presentation (as an empty intending into the future, so to speak) is "verified," ratified in the fulfillment, while in the other case, this intending has only been clarified, made intuitive. [In the latter case,] standing here before our eyes is only what was "genuinely" intended [in the intending]. The discourse of the mere disclosure of sense is the most suitable way of speaking here. At all events, intending still remains merely intending. In our way of [247] speaking, the term "intending" often serves to characterize any type of consciousness that is in need of fulfillment, regardless of whether it is an empty or intuitive consciousness. In this concept of intending there is the idea of a claim, demanding precisely fulfillment. Thus, even intuitions can have this indigence; in this case they make mere "claims." What that means is also at issue here. We grasp an aspect of it now, initially with the statement that not all intuitions can assume the function of entering into a synthesis of fulfillment, that is, of filling it in a verifying manner in their coinciding with empty presentations. Intuitions that have the character of "intendings" cannot fulfill in a verifying manner, and on the other hand, there are indeed intuitions that do not have this character. But insofar as one and the same intuition has a double faceted structure, which is a universal possibility, on the one hand, mere intending, on the other, the opposite, the statement is naturally valid, namely, that in the former instance the intuition cannot serve to fulfill if it itself requires fulfillment, while in the latter, it very well can.

This characterizes an essential distinction peculiar to intuition (and potentially to the components of intuition, too), and it is a distinction that becomes evident precisely in the dual syntheses of bringing to intuition, now as the fulfilling process of bringing to intuition, now as a mere disclosive process of bringing to intuition, and in the respectively distinctive accomplishments of these processes of bringing to intuition. In order to elucidate this let us first advance the following: Perceptions, but rememberings, too, can fulfill, can serve actual verification, but never (even if very clear) intuitively pictured expectations, a memory of the future, or even an intuitive presentification of something present, a memory of the present. What is intuitively seen is given in the first group of intuitions, in the other group, it is not given. What does this distinction mean? Naturally it concerns the essentially distinctive way in which the objective sense is constituted in the respective cases, and accordingly the way in which the objective sense is characterized noematically. It is in perception that the object, the object itself, is given, and perception constitutes the object in the mode "itself," itself in a primary and most original sense, in the [248]
sense of being presented in the flesh. In contrast to this, the intuitively pictured expectation does not give the object "itself," or as we can say in a straightforward but less clear manner, it does not bring the object to givenness. And this is just what we also alluded to when we said that "it anticipates," it anticipates a self, but it does not give it. A synthesis of fulfillment can take place between an expectation and a perception: What is expected happens. The relationship is one-sided. It is not the perception that is fulfilled in the expectation, but rather, the expectation [that is fulfilled] in the perception. Furthermore, with respect to something expected, an expectation cannot, in principle, be fulfilled in another expectation: The new cannot give anything because it does not itself have anything. What is delivered over in a fulfillment to a consciousness in need of fulfillment, as what should be given appropriately to it, is precisely the self of the object. It thus requires a consciousness that possesses this self. But

it does not only possess it in the form of an original acquisition, namely, in the form of perception, but also in the form of remembering. Even remembering is an intuition that gives. Standing before our eyes in remembering is the object itself, to be sure, in the temporal mode of being-past, and this mode is originally given here. But would it make any sense to deny the possession of the self of the object of remembering because it lacks within itself the privilege of being present in the flesh? Does not this self belong to remembering's own most phenomenological character? How then would evident identification be possible—through which the object is recognizable as identical, recognizable as one and the same subject of its predicates—if not on the basis of repetitive memories? The individual self is thus *a priori* the title for the connectedness of primordially instituting perceptions and the chains of rememberings belonging to them, connectedness through the commonality of the self that is given in all of them as accessible.

<12. Further Clarification of the Difference Between Fulfillment and Disclosure>

There are problems that arise here now. We have worked out a clear distinction in intuitions between self-giving and non-self-giving (but merely intuitively presentifying) ones. And at the same [249] time we distinguished within empty presentations (which as such have this in common with the reproductive intuitions, namely, that what they make present to consciousness is not present *originaliter*) those that are disclosed in self-giving reproductions, and those that are disclosed in non-self-giving ones. Empty anticipations are disclosed in intuitive expectations, and both of them can obtain, through suitable perceptions, an entirely different adaptation to a corresponding intuition, that of fulfilling verification, that of an adaptation to a self-giving consciousness. Where empty retentions are concerned, there is only a disclosive process of bringing to intuition, there is only the adaptation that provides a fullness, namely, the adaptation to a remembering, to a self-giving presentation.

But we must pose a question here. Is not the empty retention fulfilled insofar as it really does take on the fullness of intuition in remembering? Certainly. But according to what has become apparent to us in the meantime it has become clear that we must take hold of a more precise concept of fulfillment. Should fulfillment mean a synthetic consciousness only because one consciousness leads into a corresponding self-giving consciousness and coincides with it, then the disclosure of a retention would naturally also be a fulfillment. On the other hand, if we think of our explication of the concept of intending and the fact that "intending" characterizes a consciousness, which as it were raises a claim that is to be verified in the fulfillment; or, in other words, if we think that the fulfillment of intendings means verification—if we think of the like, then we may not say that retentions find their fulfillment, namely, their verification in the corresponding remembering. Are retentions as such really "intendings" in the sense we have indicated, in the sense of being capable and in need of verification? Certainly, they can also be intendings. But are they not so, then, merely because perceptions, which had been sedimented in them, were already not mere self-givings, but encumbered with anticipations? And is this not shown precisely by the fact that even disclosive remembering is indeed self-giving but at the same time also intending? Thus, this would mean that if a retention can be an intending, then it is also a [250] remembering that discloses the retention, as was the case with the preceding perception. Insofar as this remembering merely discloses, however, it in no way fulfills the empty retention; rather, the fulfillment would have to take place now more than ever; the remembering would have to bring to fulfillment its intending, it would have to bring to fulfillment what, in the intending, is not actually self-giving, but instead [only] points beyond, above all, to new self-giving presentations. The decisive test cases are the purely immanent, fresh retentions that are naturally free from all components of co-intending and their disclosive rememberings that we want to think of as immediately connected to them. An immanent tone that I hear right now breaks off; if I actualize the fresh retention immediately it will become disclosed, but it does not get verified, the intended meaning was lacking. And the remembering is likewise free from the intended meaning, hence its renewed repetition does not carry out any further accomplishment of verification either.[5] We also recall the fact that immediate retention can never be encumbered with doubt and therefore is incapable of modalization; and at the same time we recall that in the case of verification, and not without reason, we think of the opposite, namely, of a possible rejection; we recall that opposing the corroboration occurring through the self that fulfills the intending is also the possibility of the abrogation of the intending through a self that contests it. We keep in mind that the great theme of a fulfilling verification and thus also, so to speak, of a disappointing, abrogating crossing out stands in an essential relation to our theme of modalizations with which we began.

In any case, we have now attained a deeper insight into disclosure and genuine fulfillment. We see that only what we called protention construed in an extremely broadened manner, regardless whether it is intuitive or not, is an intending intention, that is, corresponds to the concept of a consciousness that admits of fulfillment in the sense of verification. Let us not get confused by the fact that even self-giving presentations of any kind (while precisely as self-givings, they are not intendings), can still be intendings, and we understand this by virtue of the fact that self-giving can be imbued with components of intending. Then the [251]
presentation will be precisely in need of fulfillment, or if you prefer, open to verification according to facets of its intendings; on the other hand, however, this does not apply with respect to the facets of its actual self-giving. In relation to this latter facet, as we have already recently remarked, it can itself exercise a verificatory accomplishment for the other intendings to which the self-giving can be suited.

With respect to the empty intention we must accordingly establish that while it is not a self-giving consciousness, it is still a consciousness that has the self *implicite* within itself. The disclosure discloses what we already find here in empty

[5] Thus, it was incorrect to speak in the *Ideas* of memories of the future and memories of the present. We must distinguish between memories and expectations, in the broadest sense, retentions and protentions with their intuitions.

potentiality, and here the self is already potentially in it. But where there is already precisely an "intending" in the form of *potentia* in the empty presentation, precisely this intending is disclosed and is now intending in the disclosed form: It is then a reproductive intuition, but an intending, protentionally anticipating intuition.

Having clarified the distinction between disclosure and fulfillment we also see the essential ambiguity in speaking of potentiality and actuality. In the disclosure, or as we also put it, clarification, the intentional content hidden in emptiness gets "effectively realized"; it is laid bare, clarified. An intending is effectively realized in the fulfillment, and that is an entirely different sense of effective realization; it is an entirely different accomplishment. The self that was anticipatorily meant, but that was not contained at all in the intending, neither in a concealed nor in an unconcealed form—exactly this self emerges in fulfilling intuition. The fulfilling self lies in the direction of intending, like the target lies in the direction of the arrow. But the intending must first approach what is meant; the arrow must first make its way toward the target, and that takes place in the synthesis of fulfillment. Therefore, the self-giving intuition follows the protention as a new one, while the self-giving intuition follows the retention as a familiar one, as a mere resumption of the self-giving perception from which the retention has arisen.

Intention, directed toward a retentional past, is "fulfilled" in a certain way, too, while an intention into the future, a protention, is not fulfilled. Even here we must be able to distinguish in a protention between what in it is empty consciousness, and what in [252]
it is intention. It is intention through kinaesthetic motivation.

<13. The Passive Processes of Experience>

These observations have enabled us to gain an understanding for a most universal structure of consciousness in general. All life of consciousness is constantly carried out, in a dual life-form irrespective of all intervention of egoic activity; it is constantly consciousness of something in a dual mode, now self-giving, and to be sure actually and potentially self-giving, and, on the other hand, it is anticipatory, expectant. In the first respect, it is in part

continually perceptual, and together with this it is in part held retentionally, and with regard to the latter, now disclosive retentionally of this, now of that. In another respect (where everything functions within passivity according to the essential laws of passivity), the life of consciousness develops protention together with self-givings, constitutes relative self-givings of a higher level in the connection of self-givings and intendings, as we came to understand with respect to external perceptions, and in this case lives through the passive processes of fulfillment, but also in processes of disappointment whose universal possibilities we will still have to trace out in some detail. We can characterize all these processes of the passivity of cognition as processes of passive experience, on the one hand, as processes of expanding, verifying experience, but also as processes of experience that determine more closely, and on the other hand, processes of bracketing intendings of experience that are unfitting, processes of rectifying experience. In passing through a schism, through modalization, consciousness achieves unanimity once more by means of a negating crossing out. The divergent possibilities, the privileged probabilities, get resolved through positive decisions, etc. We come to understand newer and newer portions of this and reach a deeper and deeper understanding. But we must pursue this even further in order to be able to get at least a rough overview of the main structures of pure consciousness, where consciousness is to be understood throughout as a stream that constitutes objects and that is subject to pure essential laws; and we must understand that this still takes place on the founding level of passivity. For [253]
that is the soil upon which the free activity of the ego moves and without the knowledge of which the higher accomplishments of this activity must remain for us completely unintelligible. For otherwise, what would remain unintelligible above all would be the sense and extent of the logical norm, which is nothing other than the universal legislative norm according to which all free accomplishment (like all consciousness, again unfolding along the lines of the essential distinctions between self-giving and merely intending consciousness) can be brought down the path of concordant fulfillment.

<14. Transcendental Logic (Comprehensive Reiteration)>

After this lengthy Christmas Break[6] let us go back to where we left off in our lectures. Our method was essentially Socratic-Platonic. Guided by vague, completely indeterminate and general thoughts about the sense of logic as the science of *logos*, and drawing on the meanings of this word, we took up concrete analyses and organized them in such a way that we could simultaneously get from them, along with the concrete understanding of the particularities, the general, but now quite determinate thoughts and set goals, and in such a way that progressing further these could be developed in ever new, appropriately ordered and materially relevant analyses. Logic points us from *logos* as linguistic expression to thinking, to the multifarious consciousness that is capable of expression.[7] And along with this, it points us to the intelligible sense that is intrinsic to thinking, to something meant in thinking and something that is posited in different forms of positing. By harboring sense, consciousness refers in and of itself to objectlike formations, that is, it refers to the identical element in diversely varying sense. This reference to the objectlike formation, which occurs within consciousness, however, can be a reasonable or a non-reasonable one: Logic should be the general science of reason. What is meant in consciousness, the sense and proposition, can be true or false, correct or incorrect, the meant object is actually existent or truly existent, or in truth non-existent.

What characterizes consciousness (and what characterizes the sense immanent to it), respectively, as something that bears within [254] itself truth and true being? How is this to be understood? In the beginning we engaged in concrete investigations, though in this regard still in a completely unclear manner: investigations into passive and active, slumbering and wakeful consciousness and egoic-consciousness, and then further into original time-consciousness, into immanent and transcendent perception, memory, and expectation, into the differences between

---

[6] Editor: Christmas Break, 1920.

[7] Translator: See Main Text, Part 1, especially §§1-2.

intuitability and non-intuitability, into the way in which the concordance of sense-giving, i.e., the constitution of an objectlike formation is carried out initially on the founding level of passivity, and on the other hand, into the way in which this concordance is ruptured through modalizations—all these organized, particular investigations contained and awakened ever new general insights. What became clear to us was the distinguished feature of the transcendental formulation of questions and research, and of the special attitude peculiar to them. While natural-naive knowledge and research bears on pregiven objects and regions of objects that are taken for granted in their existence: knowledge and research that bears on self-evidently existing nature, on the human world, on the self-evident givenness of the series of numbers, self-evident givenness of geometrical constructive formations and the like, we realized the possibility of a wholly different and extremely necessary knowledge and research that holds in abeyance every naive pregivenness in order to make such pregivenness problematic in the most universal generality. And we recognized this as the mode of research that every philosophical logic invariably requires, every one that makes thinking and thought, reason, reality, truth, and at the apex, scientific truth—that makes all this scientifically intelligible in a radical manner.

It became clear to us that consciousness within itself (and that means, by its very nature) carries out sense-giving and thereby the legitimate as well as the illegitimate relation to an object; and [it became clear to us] that if naive-dogmatic knowledge and science accept objects as pregiven realities, it is already a consciousness, and a highly multiform consciousness through which those objects have been constituted for the knower, and have been constituted in a distinctive way as evidently given. A study of consciousness must be possible with respect to its pure immanence through which we are able to comprehend how consciousness, within itself and according to all its fundamental kinds and fundamental forms, [255] carries out the sense-giving of objects, and how consciousness itself structures its world and its true world and true theories, theories that methodically explicate this world in its truth. It must be understood step by step, and in this pure immanence, how manifold lived-experiences of consciousness form a synthetic unity, how such a unity essentially and intelligibly maintains the identity of sense, and then further how an identical object as the substratum of varying determinations in and through manifold sense can be given to consciousness, and thus can be given to consciousness as the same, but determined differently. And further, how consciousness within itself can, on the one hand, undergo transformations of concordance and discordance and, on the other, can produce that special concordance with special essential necessities that are called here norm-giving truths, and how, correlatively, the essential features of those discordances are characterized, discordances that "in themselves" must count as falsities with respect to the norm. From there one must understand the orders of foundation concerning the accomplishments of reason, the structure of the methodological ascension from truths to theories.

The phenomenological reduction gave us the evident method for our purely and necessary immanent research, and really determined the genuine sense of this research for the first time. It availed us of pure consciousness and the pure ego of this consciousness, and the method of pure and universal research into essence had to be linked to it, an investigation into essence that does not bear on consciousness that is reduced to something that is momentarily arbitrary, as a single fact, but rather that bears on the general essential types of consciousness and the essential necessities proper to them, for example, under the noetic heading, perception as such, or transcendent perception as such, under the correlative heading, perceptual sense as such, the present in the flesh as such, etc. Thus, transcendental logic does not want to be a dogmatic science juxtaposed to other sciences, not a science in the customary sense; it does not want to be, like the dogmatic sciences, directed toward a pregiven realm of objects, naively taking them for granted as pregiven. It wants to be the ultimate science that goes back to ultimate givennesses, namely, to those givennesses that are already presupposed in all other givennesses, in all naive givennesses. That is, it wants to be the science of [256]
consciousness, the science of consciousness as pregiving in general, and specifically, the science of consciousness that pregives meant reality, and grounded upon this the ultimate

elucidating science of theoretical accomplishments and of all accomplishments under the ideas of reason. Indeed, it wants to be the universal and pure science of *logos*, the science of the essence of *logos* as *logos*, that is, of knowledge as knowledge, of the known objectlike formation as objectlike formation, of truth as truth; accordingly of science as science, too, and of all scientific types that the idea of science includes. But [it wants to be] this according to the corresponding essential correlations of scientifically cognizing consciousness, science as the theoretical system of true propositions, and the realm of science as the realm of truly existing objects and of objects determined within the theories, objects that are investigated in scientific thinking.

Thus, pure logic must yield essential insight into how consciousness as such contains sense, in which structures, in which noetic and noematic modes, how within itself it makes objects present to consciousness as its intentional accomplishment, and then how it necessarily makes [them] present to consciousness as the objectlike formation of this and that sense-content and of these and those modes of appearance. It must inquire into the essential typicality of consciousness as such, and for each one of those types, disclose the modes of accomplishment that give sense and constitute objects. And it especially aims, finally, at those essential typicalities that make intelligible the constitution of truly existing and not merely meant objectlike formations, and likewise makes intelligible for us true theory, true theoretical science and the true life of reason of every kind as a certain methodological accomplishment, with norms, whose original source is to be illuminated in every last detail. But for this we require extremely encompassing investigations that initially investigate, prior to all inquiries into truth, the general types of transcendentally pure consciousness with the types of their sense-giving, their relations to objects in noetic and noematic regard—still more primitively, investigations that distinguish consciousness with respect to the difference between passivity and activity, and initially pursue the intentional accomplishments that are carried out within passivity, accomplishments that are already presupposed as the constant foundation of all egoic activity.

Our last considerations were still carried out within the framework of passivity; these considerations dealt with the modalities of belief and following this, the considerations that dealt with disclosure and genuine fulfillment, the latter representing confirmation on the level of passivity. At issue in all of this are the eidetic descriptions of the nexuses of the consciousness of concordance and discordance, and of the modal occurrences of the "problematically possible," of the "dubitable," of the "null" which occur along with the latter with regard to their sense; then at issue are the distinguished cases of concordant or discordant annexes of consciousness to consciousness in which (like in the fulfillment or the disappointment of an expectation or of a protention in the sphere of memory), the intending is verified or annulled through a decision making process, [and this was seen] in contrast to the cases of mere disclosure in which an intending into the future is merely clarified, like, for instance, in the mere picturing of an expectation prior to its fulfillment through perception. [257]

<15.> Corroboration and Verification

Before we proceed, we must first supplement what we have said. We distinguished intuitive presentations from empty presentations, and within the intuitive ones we distinguished between self-giving presentations, as those which are alone capable of verificatory accomplishments, and those that are not self-giving, like, for example, those merely intuitively pictured expectations. They only refer, as it were, beyond themselves, to a self that is not given in them, to a presenting against which they are measured, a presenting in which the self, verifying the intending into the future, would be given. Or they are also intuitions, but merely as anticipating other intuitions. It now becomes clear that belonging to the essential character of all individual presentations, even the self-giving ones, is the fact that they are, at the very least, encumbered with components of protention, with components of intending into the future, yielding in this case possible nexuses of corroboration that we will distinguish from verification. If we consider connected presentations, regardless whether they are intuitive or empty, and specifically ones that do not undergo a rupture, that is, are in themselves concordant, then they are *eo ipso* of the mode of belief, namely, with respect to their protentional side. They anticipate in belief, be it according to certain components, like in the perceptual course of a process with respect to expectations that are directed [258] ahead, be it thoroughly, like when they thoroughly have the feature of anticipations, for example, when we expect a thunderstorm on the basis of various indices. Now, where these several intendings into the future are connected in the unity of a coinciding of sense, and where they harmonize with respect to what is intended into the future, this harmony does not provide any verification, but essentially the consciousness of corroboration. Every intending into the future is motivated; we have already alluded to this when we closely examined the structure of perceptions and memories (that is, the structure of the fundamental kinds of self-giving presentations). On the side of primordial impressions and retentions we find the original constitution that has resulted in an institution, a constitution of a self and of its original acquisition of knowledge, and it essentially motivates an anticipatory belief that is directed ahead—the further path of events is expected according to the respective constellation of knowledge in more or less determinate sense-giving. This does not only hold here, but everywhere: The course of experiential knowledge also motivates a course of intending into the future, and in this way all intending into the future is motivated belief. Now where intendings into the future of the same sense arise from several motivational sources and result in a coinciding, they corroborate each other, or the new one corroborates the current one. In a certain respect there is, by the way, also something like corroboration for the acquisition of knowledge, namely, for knowledge fashioned out of repetitive acquisitions of knowledge that have the same sense. It has a dimension of intensity, knowledge is deepened, fortified. Now, where the corroboration of belief in the sense of an intending into the future is concerned (of "belief" in the strict sense), this corroboration is not to be confused with verification, and is never to be equated with it. And this holds likewise for the opposing occurrences. An intending into the future can undergo a rupture and therefore modalization by a motivation radiating from the realm of acquired knowledge—a motivated experiential belief that, by virtue of a partial harmony, coincides with the belief already given, but on the other hand, partially conflicts with it.[8]

<16. The Question Concerning the Verifiability of Experiential Belief> [259]

But when we articulate and think through such principles, we notice first of all that we have not yet acquired concepts like truth and falsity here in the full sense, and that speaking of the norm, of a norm of correctness and incorrectness had an incomplete sense. In immanent time consciousness we have the stream of givennesses in lived-experience, givennesses that are strung together temporally with their anticipations which have the character of an anticipatory believing that is directed-ahead. A spatio-temporal world is given in the stream that is contained in immanent time-consciousness, in the stream of transcendent experiences, the intuitive and non-intuitive. And constantly referring to this spatio-temporal world are manifold lived-experiences of a transcendent believing that is in need of fulfillment. In both respects, belief is not only directed toward the present, but also toward the anticipated future and toward the memorial past; manifold memory-beliefs and expectant-beliefs emerge that can be verified or rejected.

What happens now with the verifiability or refutability of experiential belief in all these respects? What is to be meant by the axiom we tested that every such belief is either positively or negatively verifiable? To be sure, it does not just mean that the mere possibility of one or the other belongs to the essence of such a believing, and that when the one possibility is taken up as realized, the other would thereby be annulled. It is certainly evident from the principle of contradiction that positive and negative verification, fulfillment and disappointment, are mutually exclusive. But if we want to say that every belief is verifiable in the

---

[8] Editor: There is a gap in the text here.

sense that it is valid or invalid in the usual sense, as it is in the view of the traditional principle of the excluded middle, then quite a bit more is being said here.

Let us get clear about this by drawing a parallel to mathematical judgment, to a judicative believing that bears on what is mathematical. Either it is valid, it is verifiable, or it is not valid, it is verifiable in a negative manner. This certainly implies that whether we ever will or are even able to carry out a verification or not, even without thinking about whether it may ever become a decision of the positive or negative sort, it is surely decided in itself whether the judgment is verifiable or whether it is refutable, already in advance and thus for all actual and possible [260] *consciousness in the future. Only we do not know in advance how it is decided; we only first know this in the actual, intuitive verification as the current decision. It is determined in advance, as it were, how the die is cast, whether on the positive or the negative side. If we actually and positively confirm the judgment, then we will know that it was already established beforehand that only a positive confirmation could take place and that the opposite was excluded (and for every conceivable ego).*

Let us now turn to the spheres of our external experience as they had been constituted in passivity, and so far as we are able to understand them from there up to this point. We ask with respect to these spheres: Is it an actually intuitable, essential law to be drawn from the structure of the intentionality of experience that every belief, no matter how it arises in the stream of consciousness and its motivations, is decided in advance according to the possibilities of verification and refutation? How can this "in advance" be understood? Certainly, if a fulfillment occurs, then belief is decided as valid; a prehension of the self has emerged from mere anticipating, the anticipation has been ratified. But as long as the verification has not taken place, both of the open possibilities do exist. Must it be determined in itself and in advance what alone can occur there, if it is ever to be decided at all and regardless of whether a decision takes place or not? To elucidate the structure of truth or validity is to elucidate this "in itself," and perhaps there are fundamental differences there. In fact, truths of the mathematical kind and other essential truths are fundamentally different from truths like experiential truths. This in-itself is divided according to the correlations: As correctness in itself it belongs to believing, as truth in itself, truth in the strict sense of the word, it belongs to sense or rather to a proposition. The object in itself corresponds to the truth in itself. Now the in-itself belongs to the object.

Thus, we have brought the problem of the empirical in-itself clearly into relief by drawing a parallel to the mathematical in-itself where it is easiest to detect the peculiar trait of validity in itself. And we were able to understand that we are in no way in the [261] *position of fashioning in an original manner the axioms of the sort that we previously tested relating to the empirical in-itself. This concerns no less the immanent sphere despite its privilege through the evidence of the ego. In fact, if we conceive of a consciousness that has sounds, colors, and similar hyletic data given in passivity in immanent time-consciousness, and these data are being constituted in the process of becoming, then it is not clear in the least how it should be decided in itself, in advance, whether precisely this new tone, or whether a new tone at all should follow a previous tone. And even if an anticipatory belief in a new determinate tone were motivated by the preceding immanent experience, we cannot tell in advance that it should be decided in itself, whether it will actually occur, or instead will fail to appear, or will take place in an entirely arbitrary and altered manner.*

What happens with respect to transcendence, that is, the spatio-material world, at least when we conceive of it constituted purely passively in a consciousness? Naturally, proper to the constitution of a spatio-material environing-world is not only a super-abundance of prefigurings of inner horizons for every thing that is actually experienced, but also of outer horizons—which are reciprocally interwoven with one another, and ultimately all things of experience are connected in the unity of an environing-world with a unitary outer horizon—and thus a superabundance of prefigurings for the path of further possible experiences. But there are precisely prefigurings, there is motivated experiential belief, superabundantly corroborated and ratified through innumerable accordances; but in the final analysis is it not possible for the further experience with its ever new self-givings to continue as it will? Contrary to each and every expectation, contrary to all the overwhelming preconvictions and probabilities? Can it not continue such that everything becomes a confused muddle, such that the entire perceptual world-order is destroyed, such that this world as the unity of experience is no longer even maintained, such that it becomes unstable for consciousness, such that all sense-data lose their apperceptive apprehensions, which themselves only actually grasp appearances in concordant believing? But we have held that there would be the world in itself, and every experiential belief would be valid in itself, would be in itself true and false.

In the last lecture[9] we made plain the peculiar virtue of the verification of the self as opposed to mere corroborations, of the verification arising through self-giving acts of belief. This was the view where corroborations are concerned: A belief that does not arise as such, that resides in a self-giving unbroken presentation, can be connected with another one whose very sense is similarly directed; it can undergo a fortification and thus in a broader sense, verification, like the expectation of a coming event through a new *indizium*, through a new similarly directed protention. But no matter how rich this intensification may be, as long as the expectation is unfulfilled, it has, aside from all emotional interests, a deficiency; a mere ratification does not occur in fulfillment in the manner of an intensification of force; rather, the event itself is there now—for consciousness, of course. And this consciousness accomplishes a verification in the transition to fulfillment, a verification that gives itself as definitive, so to speak: And so it is; I have the existing thing itself, I do not only mean it; the being of the sense-content in question is not only a meant being, but "actual." Just as sense thus has a new and superior mode in self-giving belief, so too does being, the correlate of the unbroken character of the presentation. Further, in the transitional synthesis, belief has the character of a belief being demonstrated as correct, and its noematic correlate, the character of actual and true being as opposed to merely intended being. Correspondingly, decisions made concerning questions, doubts, or problematic possibilities [262]

[9] Editor: Beginning of a new lecture.

are also diverse according to whether they take place through self-giving unbroken presentations, or through non-self-giving ones. The self-giving decision is so to speak the one through the *terminus ad quem*. And the expression is also fitting insofar as the life of consciousness in its entire scope is permeated with continuously connected tendencies toward universal concordance, not only toward concordance in general, however, but toward one suited to self-givennesses, one resolved with respect to the self-givennesses.

In this way we had obtained preliminary concepts of truth, correctness, norm, and with respect to being, of "true being." We [263] can immediately add: We had obtained a preliminary and original concept of evidence as the consciousness of verification and decision about something self-given; but here we had already presupposed working out the radical difference between an arbitrary belief and its modalities as opposed to belief in the mode of a self-giving and concordant presentation.

I paused there in order to make clear to you that the concepts of truth, correctness, etc., that we obtained in the context of our studies up to this point are not yet those concepts that guide us in customary and scientific discourse and that we also ascribe to the traditional logical law of contradiction.

There is a moment here for which we can find no model, namely, being resolved, as it were, being prejudicial concerning what is true and what is false, in advance of [or] prior to all actual experience; what is true is true in itself, and that is to say, it is not open whether positive or negative confirmation is possible.

If it is a matter of the future, for example, then it is decided, even where I lack a decision. To our mind, every belief directed toward the future has its truth or its falsity prefigured in advance, once and for all.

However, if we remain in the framework of pure consciousness and consider the immanent and transcendent given matters that are constituted within it in passivity, then as I said, what we have shown has not yet accounted for the clarification of the idea of that in-itself. Where the immanent given matters are concerned, and especially the sense-data, every Now brings with it new ones. But in spite of all aroused anticipations, it cannot be foreseen why it must necessarily be decided which data will occur in the future. And this also holds no differently where the transcendently constituted spatio-temporal world is concerned.

Perhaps some clarification is still required here. This world is given to us originally through external perceptions. Generally speaking, they cohere with one another in continual concordance, and they are likewise intertwined concordantly with self-giving rememberings that potentially span gaps like those of sleep. To be [264] *sure, occasionally discordances do also occur. We speak of illusions, experiential belief being ruptured, passing over into doubt; but in the progression of experience, which never undergoes breaks in each and every respect, a thoroughgoing concordance is indeed restored through the changes in meaning and the crossings out just described; that is, running through our consciousness is a sustained unity of world-certainty that is produced again and again over against the disturbances. "The" one world is constantly there, only it is determined more closely and occasionally determined somewhat differently.*

The first problem arises here, however: Must it then remain just as it was up to this point, according to the testimonies of our memory? Must an external experience be continually adjoined to another external experience in this way? Can it not be that an external experience is the last one, while consciousness endures? An external experience is assuredly a complex structure of consciousness that emerges in the nexus of consciousness as naturally motivated. Still, must the motivations proceed in such a way that a perception has to be connected up with another perception? We have kinaesthetic courses with which the appearances of things are connected in accordance with associative motivation: thus, certain exhibiting sense-data (in the case of the visual appearance, that is, visual data) along with their apprehensions. By motivation we mean that certain data and their protentional horizons are demanded as co-emerging along with the emergence of other data in our lived-experience. But such associative demands can be annulled in the course of present experiencing. The series of sensation must actually arrive in a certain way in order for the apprehension of a thing to be experienced according to the kinaesthetically aroused pre-

demands, and in order for the consciousness of an existing thing to be maintained. If the sense-data were suddenly to begin appearing in a muddle, if our visual field were suddenly to be filled with a confused muddle of colors, the kinaesthetic motivations would lose their force. What was formerly linked up to the kinaesthetic courses in an expectant manner would no longer be able to occur in the otherwise firmly regulated manner in anticipatory believing, and it would therefore be an end to the [265] *play of external perceptions. Their emergence means precisely a regulated functioning and a continuous further development of cultivated motivations, and this essentially depends upon the actual course of sensations. Yet this is always conceivable as an entirely different course, and as a completely unregulated one. That it is not an unregulated course, but is such a course that makes a continual perceptual flow possible, that is simply a fact. However, if we inquire into the truth of this fact and more precisely, if we ask why what was previously so must still be or will be so, this truth will obviously not be such that it could be decided by recourse to one of those passive confirmations of which we were alone able to speak.*

Secondly, even if we do presuppose the truth of this fact, and thus assume that for us, that for the particular experiencing pure ego, external experiences will continually be adjoined in its stream of consciousness and will also issue in concordance every time, this would only be to say that for this ego the unity of a true world will be continually maintained in ratifying judicative intendings. But this is not to say that this world, beyond our present experience of it, is a world determined in advance, determined in itself, such that the decision of true or false would be univocally prefigured for every believing that is directed toward any temporal situation, or for a corresponding believing that is produced hypothetically.

This is illustrated most simply by referring to the difference between the world-view of that part of humanity that is influenced by modern natural science, and alternately, the world-view of the rest of humanity. This world is constantly and self-evidently there for all human beings, and they believe that it will also continue to endure. In their conscious lives they live into a world-future. But by far most human beings do believe that what will come depends to a large extent upon accidents that cannot be ascertained, or that gods decide the world's course according to a momentary whim. Only roughly is there an order that can be foreseen, according to which one can be practically directed, but only roughly. A conviction was indeed forged quite late in a causality that lawfully and absolutely determines each and every thing occurring in the world; and the sense of this conviction is none [266] *other than precisely this: that each and every temporal being (and in the natural attitude this means all beings in the world) is determined in itself, determined as truths in themselves. From the very beginning, nothing is open in order to have to wait and see first how the Fates of destiny will decide.*

Our question rested in this consideration, namely, whether we already attain definitiveness (in the mode of experiential ratification) should an intending into the future actually be ratified by experience. Here, then, the other, last mentioned difficulty comes into play, and in a way that is very touchy where external experience is concerned. Does not external experience evidently lead eo ipso into infinity? Each experience is still itself an open intention; it has dimensions that are unfulfilled. Must, indeed, can a synthetically progressing acquisition of knowledge come to an end?

Let us turn back to the immanent sphere.

<17.> The Problem of the In-Itself of One's Own Past. Evidence in Remembering

One's own past of consciousness with its noetic and noematic components is for us a field of possible remembering, and furthermore is a field of at least *idealiter* possible, complete memory, true and valid. Extremely large stretches [of the past] may be forgotten; they may never re-emerge involuntarily in a current memory or be at our disposal in a deliberate memory: Yet we are sure of the fact that there was actually a past of consciousness and that it can be restored unequivocally in the form of rememberings—both are obviously equivalent. Belonging to the empty horizon of the past, which brings every present of consciousness to a close, is a belief of the past that is essentially fulfillable through chains of rememberings and nothing else. Put more precisely, belonging to the essence of this empty belief of the past is the fact that as a positive belief it cannot in principle be annulled; to measure this empty belief of the past against a self-giving intuition can never lead to nothing. As the horizon of belief, it holds up [and] is always and necessarily there in an unbroken manner. A horizon of belief that is adequate to any self-giving intuition is a remembering, and yields a portion of it, that is, it yields a portion of one's own past going back indeterminately. It [267] can indeed be the case that the remembering in question does not hold up; it can be that it winds up as a memorial illusion. But this very thing is only possible because a memory is pitted against other memories, that is, because something past endures as self-giving, and this past served as the regulating measure for the memorial belief that was dismissed. It is inconceivable that there would be nothing behind my present of consciousness, that every one of my rememberings would be null, since this nullity can *a priori* only be demonstrated again through other rememberings. I might be convinced on occasion that my actual rememberings here were null; the new authoritative rememberings may also assume the character of nullity once more; but what is evident is that every remembering is either already complete, which is to say that it is purely self-giving with respect to that section of my past that revives it in a clear and intuitive manner, or that a purely self-giving remembering of this completeness is possible; and it is clear that this possibility is not a mere possibility of phantasy, and that it is not a problematic possibility either, but rather a possibility that characterizes an ideal limit to be intuitively discerned in evidence for all incomplete and deceptive rememberings, a limit that thus always prescribes an in-itself to rememberings. The intuited idea guides us, even in our futile attempts to bring a remembering to perfect clarity and concordance, even when, with respect to the repeated object, we become cognizant of the fact that remembering has taken up moments in a deceptive way, moments that did not exist and that could not have existed in this way—we still firmly maintain that the past lived-experience is to be regarded as determinate in itself and that the disclosure of a true memory is to be regarded as a goal that is to be sought in practice. This true remembering, as a purely self-giving consciousness of the past, is thus an idea that is necessary and [is something] to be discerned intuitively; experiential belief is decided in advance here; every empty protention is actually verifiable in a predetermined sense. It may be dubious to me as to how it actually was, I may now be obliged to leave it open. But in itself, it was how it was, something determined in itself.

<18. Consciousness of the Memorial Illusion> [268]

If we pursue a deeper clarification of this situation, we would initially have to view more precisely how a remembering, even though it is a self-giving consciousness, can nevertheless turn out to be a deception for consciousness, that is, how it can split into a conflict and then pass over into unbelief, into negation. The order is prefigured, the beginning can essentially only be such that a remembering in the primordial mode of belief emerges from the obscurity of memory as the retentional distant horizon. Likewise, it is clear *a priori* that doubt and negation are not possible in just any fashion, but are possible only in a certain way such that the memory that was at first unbroken comes into conflict with other memories. Memories can be more or less vivacious, more or less unclear; by being intermittent they can transmit expanses of intuitability through empty expanses of non-intuitability. At the same time (and these are essential possibilities that can be drawn purely from examples), empty intentions can arise anew in conflict with intentional moments that were first established in the initial memory, be they intuitive or already non-intuitive; but there is no doubt that they are still memorial intentions that can only pass over into clear self-giving rememberings in a fulfilling manner. In this way a consciousness of a memorial illusion can arise, one that is in itself evident and clear, and we can recognize how it must necessarily look here. Every such consciousness, every evident annulment of a remembering (which is precisely the evident consciousness: "it was only an illusion") has the typicality of disclosing a muddle of diverse rememberings, and this disclosure has the form of a certain transition of the initial memory into a multiplicity of discrete, completely clear memories that, phenomenologically [understood], are related to one another, and in this relation, are completely concordant.

We now describe this transitional phenomenon in the following way—a phenomenon that can also contain in its intermediate stages an empty phenomenon of doubt and of negation, in order then to ground the evidence of negation, of deception with bifurcation: Our point of departure is the remembering that is [269] initially unruptured. With the transition to a sufficient level of clarity, the originally unitary memorial image falls apart into several images, and finally into several clear rememberings, [that are] in themselves unbroken [and] that belong to different temporal loci. For example, the memory of Sils-Maria surfaces, and I see a young author before me, and we are engaged in a lively conversation. It concerned Gundolf's "Shakespeare."[10] But now a doubt is "stirring," and if I give into it, a second image will appear; I am now with the same young man in his apartment in a small room in the country in Fextal; he is reading to me from Gundolf's "Shakespeare," and we are speaking about it there. If I go still deeper now into the image that I first remembered, I will note that a piece of its memorial continuity remains purely self-giving and unruptured. But there was a small break in continuity, a break that was previously inconspicuous where I listened to the young man speaking and intuitively honed in on what we spoke about; the reproduction changed there, unnoticed, into the other image which as it were remained concealed visually through the first situation, falsely imputing the second conversation to the first. And seen precisely, the same person in the one situation along with his external modes of appearance was at the same time, or was actually already falsely imputed to the one in the other situation. It is evident that this duality in fact, just not noticed, already lay in the initial image giving itself in a unitary manner, an image that was then subject to a divergent duplication by virtue of

---

[10] Translator: The reference here is to Friedrich Gundolf, author of *Shakespeare und der deutsche Geist* (Berlin: Georg Bondi, 1914), and translator of Shakespeare into German.

a peculiar overlapping in which parts of the one memorial image covered over parts of the other.

The divergence into two discrete memories is not a theory, but rather a situation that is intelligible in its intentionality. The discrete rememberings arising in this process are not two arbitrary memories, but memories that are characterized in a certain way. What is evident above all is that they do not arise anew, but were already there for consciousness, and that they have assumed only an altered mode of intuition and an altered nexus of relations. There is a justification in speaking of a splintering of a combined memorial image into its combining elements. We find both of the [270] memorial situations in this combined memorial image, though certainly not both of them developed in a completely intuitive manner; rather, only a part of each one is represented by intuitive parts in the entire intuitive image; but the supplementary portions of the situations are also there *implicite*; they are only "suppressed," "eclipsed." It is entirely analogous to the overlapping of two perceptual apprehensions that stand in conflict, where they are likewise grounded in a commonality and where, when the one perceptual apprehension (e.g. of a mannequin) prevails and the other, the apprehension of a person with its own peculiar perceptual moments, is suppressed, but is however grasped in a peculiar manner, only non-intuitively, only hidden.

<19. Remembering and Association>

That which makes overlapping and fusion possible, and on the other hand, that which makes the dehiscence or the divergence of memories possible is also intelligible to us phenomenologically. If we consider the pair of discrete memories [just discussed], we will recognize that they are essentially not two arbitrary memories, but two memories that are related to one another in a certain way, and that this reciprocal relatedness, or rather, the connection of consciousness did not first accrue to them in clarity. Both of them are "associated"; the one situation recalls the other, and this is not an expression of some kind of objective psychological fact, but a purely phenomenological matter concerning essential structures. It is also contained in the fusion that forms a unitary image, only that

here it has assumed just this special shape. It is not for nothing that I say in my example that the reproduction of the conversation in Sils was the first; it recalled the conversation in Fextal, and both became conflated in a partially intermingled image for me. In separating them out, we can describe what is peculiar to association. Generally, we can say that every event of recalling something (to conduct our description noematically) is essentially a noematic nexus of synthetically united memorial givennesses under the rubric of association. This noematic nexus entails two moments: an immediate and a mediate recalling something. In every association we necessarily find a pair of immediately [271]
complementary elements: An element of the one situation immediately reminds us of a similar element of the other.[11] If we call the one situation associative or awakening, and the other the associated or awakened, then in the former instance what is primarily awakening is a prominent moment, even if not yet noticed, a moment that awakens something similar. This pair of similarity is situated in a special unification that we will discuss shortly. It fashions a bridge between both pasts or even between a perceptual present and a past. The similar moment that is immediately awakened awakens something coexisting with it, and then this has the character of being associated mediately. An *a* recalls an *a'* and thereby a coexisting *b*. Yet seen more precisely, we must say: The awakening spreads out from the complementary element of similarity to the entire realm of something that is coexistent with it, and from there further to the continuous temporal sequence. In other words, what is awakened through the similar moment is not only the individuated similar moment in the sphere of memory, but with regard to the inseparable empty horizons of the past stream of consciousness, the entire present of consciousness to which the similar moment belonged. It <spreads out> through it, but then further to the continuous, subsequent, concrete streaming course. Yet this awakening does not imply an explicit process of bringing to intuition; what is awakened can be

[11] The general theory of association, by the way, is expounded upon here. Something recalling something else that is similar? The synthesis of coinciding can also be the coinciding that forms identity—i.e., also something recalling something else that is identical.

entirely or partially obscure, and there are still special motives for the special reproduction of images. Thus, contained in every association as a primary and founding association is an association of similarity, but it is contained as a non-independent element, and then further association through contiguity, namely, at first the association according to coexistence, and then according to succession.

But there are still supplementary principles of association that come into view as phenomenologically demonstrable, namely, insofar as everything that produces a special unity (that is, produces a prominence of constituted objects) in an original consciousness of the present according to coexistence and succession, thus everything that constitutes passive multiplicities is reawakened in the reproduction that was already awakened elsewhere through similarity, and can give rise to special [272]
associative connections between present and past prominent elements. In fact, one would have to expand the concept of association phenomenologically and not only speak of association as the association involving the connection of the present consciousness and a memorially submerged consciousness, but also speak of analogous connections within a consciousness of the present. Uniformity, contrast, for example, belong here, together with everything that, in general, renders a prominent multiplicity given to consciousness as a unity within passivity.

What should be constituted as a unity in the present for itself, as present in the flesh, is a for-itself as a unity of prominence for the ego, understood as a unity of the affecting allure. Where several prominent elements are connected into the unity of a group, of a passively constituted multiplicity, there is the unity of an allure as a whole, of an affection as a whole that distinguishes it as a for-itself. In this case, either the particular affections flow into the unity of the affection as a whole such that they only function as a moment of the affection as a whole, or this or that exercises yet another special affection, its voice especially comes through the chorus. It is a special problem to investigate what creates multiplicities, what organizes multiplicities themselves into multiplicities, how uniformity and similarity everywhere play a role there in different directions, how rhythmatizations arise

whereby something uniform in relation to something else that is uniform, something relatively the most similar in relation to something else that is most similar, shows an especially awakening force, and in the awakening, an affectively binding force for consciousness. In this way, a special unity is fashioned through affective force, especially in succession for instance; it does this in such a way that a tone as affectively unitary appears for itself, and in such a way that a new tone and then another new one do not merely appear in the same way, but rather forthwith takes hold of what has just past, as bestowing an affective salience on an object uniform to it or similar to it, imparting to the pair an affection as a whole; and special fusions occur with this and summon for themselves a unity of prominence of several of them. But then, for instance, after *a b c* have flowed-off in a unitary manner, another *a* occurs once more and then *b*, and then *c*; here the uniform *a* awakens the uniform *a*, and the course, a repeated [273]
course, a rhythmatization arises as the unity of a cyclical multiplicity. But this implies that the second *a b c* does not allow the retentionally submerged element to sink down placidly into its retentional grave, but lifts it up and holds on to it (in spite of its submersion) in its particularity and ties it to life. On the other hand, the uniform element that is running-off awakens an expectation of something uniform to it, a cycle running-off, the anticipation of the continuation of a cyclical becoming. In my view, these are primordial forms of association, originary impressional associations, so to speak, and they become efficacious once more in remembering. What was connected in original (we also say, impressional) consciousness through principles of original association, those that <unite> the discrete unities into connected coexistent or successive multiplicities, is also efficacious in reproduction; this is not only to say that it evidently has a corresponding connection in the reproduction that is already intuitive—but that before the unity of the reproduced intuition is produced, it induces the awakening to become an intuition.

If we now examine the primary association of similarity more closely, we unmistakably recognize that it is its nature to produce a certain coinciding. Two pasts can be brought to a complete intuitive memory, one after the other, but never in the shape of an ensemble, never in a consciousness that lets both images run-off together in a intuitive manner. If we have a memorial situation and it remains vivacious, then another memorial situation awaked by it cannot be there completely and intuitively in a similar manner. But a consciousness of both pasts being there together is, nevertheless, carried out with the awakening, and this obviously occurs in the mode of repression under partial coinciding. What are coinciding are the similar members, namely, the associating member covers over the associated, the awakening member of the first intuitive situation [covers over] the awakened. The coinciding is intimate according to the degree of similarity, and where what is awakened is an object that is constituted as the same identically, the coinciding is also a coinciding of sense according to identity. What is connected with these terms of similarity or identity, then, exists in a tension, in a kind of struggle. One could speak here of positive and negative coinciding. If the first situation is completely vivacious, the associated situation that is radiating out from the [274] awakened term will remain repressed. Yet the force of the single partial awakening can become stronger, and then two things become possible here: Either the rupture on the part of an association of contiguity makes the entire past come to life and, radiating back, tears the term of similarity loose from the awakening term. Then the entire associative situation, the entire image of it, becomes repressed and we live entirely in a new past. Or the images are interspersed with each other, resistant elements of the first reproductive past are combined with elements of the other past to form a unity of one image; here however the interspersing must also consist in the fact that what has not [explicitly] appeared in the image is still there in the mode of a repressed consciousness. It is like what happens in the "rivalry of the visual fields," in the stereoscopic combination of pictures that do not fit together: Elements of both are connected, but they can potentially alternate such that the consciousness of one of them remains continually repressed. The unity of the image is produced through a unity of the apperceptive interpretation with a unity of the constituted sense; this is analogous to similar cases where similar components were constituted as an unbroken unity. But, in their succession, as soon as the two situations diverge from one another intuitively by the affective force of the suppressed one winning out and the development of the image progressing—after this, as soon as each of the two situations flows-off successively in an uninhibited manner in its discrete unity and concordance, the dual stratification of the combined image and the connection of its conflicting parts becomes evident. It becomes clear that only the discrete reproductions were actually self-givennesses and that they belong together in intentionality with respect to all their parts and with respect to all the combinations of their parts. It belongs essentially to every memory that it admits of being brought to progressive clarity, to the clarity of the process of the presentified present running-off again, and to the clarity of unraveling the interlacings [that have occurred] through overlappings with other memories, and to the clarity of the elimination of apperceptive superimpositions. And however much that belongs together with the accomplishment of the ego that is active and purposively occupied here, the ground of the evidence of a true past being will [275] consist in the evidence that every memorial deception can be elucidated only by means of memory, and that elements of a genuine past sense can necessarily play a part in every such deception only in the mixtures, and that the pure fulfillment of sense and pure explication of the concordant lines of memory is an idea lying in the finite[12]. Note that when we speak about an idea, we are speaking about a *limes*, about a limit that lies in the finite, and also about a limit that is to be discerned in evidence and yet not seen in the customary sense—there where we do not suspect a memory of any alloy. The sensible qualities and the entire clarity of memory already wavers, and the true qualities of the true memory are ideal limits.

We have gained quite a lot with this, notably and above all, a deep look into the origin of deceptions, of "errors" in passivity, and particularly in the field of memory as a field of the in-itself.

---

[12] *im Endlichen*

<20. Kant's Doctrine of the Synthesis of Productive Imagination>

Our problem is the clarification of the idea of the in-itself to the extent that passivity can account for it. In the full sense of the term, we [can] only speak of validity, correctness, and truth, and of evidence in which it first comes to original givenness in consciousness, in the sphere of judicative cognition, that is, in the sphere of the free rational activity of the ego. But we already have pre-levels of evidence and their correlates in the sphere of passivity as that which fashions the founding soil for all activity. Thus, fundamental investigations must begin here.

It is of historical interest to recall here Kant's brilliant insights that are expressed in his profound but obscure doctrine of the synthesis of productive imagination, above all in his transcendental deduction from the first edition of the *Critique of Pure Reason*. When Kant in his great work speaks of an analytic synthesis, he means cognition deployed there in explicit forms of concepts and judgments, and this points back, for him, to a productive synthesis.

But, in our view, that is nothing other than what we call passive constitution, nothing other than the team-work (disclosable by our [276] phenomenological method) of the constantly higher developing intentionalities of passive consciousness in which an extremely multiform process of immanent and transcendent sense-giving is carried out passively and is organized into encompassing formations of sense and formations of being, as is the immanent unity of the stream of lived-experience, and with respect to transcendence, the unity of the world with its universal forms. Since Kant was not in the position to recognize the essence of passive production as intentional constitution, and could not yet see the actual task of making systematically intelligible the essential necessities of the constitution of all objectlike formations and the path of their order of foundation, he also understandably missed the problem of evidence. Of course, this is likewise lacking in its genuine phenomenological shape for those who followed, and for the same reasons.

<21. Development of the Problem of the In-Itself for the Immanent Sphere>

Let us limit the problem of truth in itself, or rather, of evidence to the immanent sphere, and naturally for good reasons, because it must first be clarified there. Every self-giving presentation carries out a certain accomplishment in relation to a non-self-giving one [and] with which it reaches a synthetic coinciding, an accomplishment that the term fulfillment intimates: It brings the fullness of the "self" to the non-self-giving presentation *qua* mere intending. The meant being is now there as true being, as the object as actually there. But now it happens (as we have already said earlier) that a self-giving presentation in general can become dubious and can be annulled through negation; protentions are intertwined with it. Even where we distinguish between genuine and non-genuine self-giving, like with external perception, we see that disappointment is possible, also allowing for something that is actually appearing to be crossed out.

Let us directly pose the question with respect to immanence whether self-givenness here cannot be something that is entirely relative, that does not contain any definitive self at all or that does not have this definitive self supporting it as a persistent norm. Can [277] *it not be that every self-giving is to be rendered invalid through a conflict with other self-givings, and these again in conflict with others, and so on in infinitum? More explicitly, can it not be the case, when any kind of presentation is verified by a corresponding self-giving, that the latter is forthwith annulled through negation, whereby now what is presented would also be given as not actual; that then, however, the self-giving functioning as norm is also negated once again, and so the actual and the non-actual are always only something momentary, something belonging arbitrarily to the process of fulfillment? Or when we take any kind of presentation, is it decided in itself that corresponding to its meant being, to what is given in it in the mode of belief-certainty, is a definitive being of the self as true and as incapable of being crossed out for all time?*

To be sure, at first we see that immanently constituted being in its living present is not only self-giving as being, but that this being is essentially incapable of being crossed out. The moment we assert that it is not, like we can do at any time, we see that apodictically this assertion is annulled with respect to what is given. Here the indubitable, the indefeasible validity is clear. But what good is it, since its validity is only momentary? What is immanent flows off and is gone. But where we speak of a true self and of a presentation that is verified definitively, there we reach beyond the momentary consciousness through rememberings in which we repeatedly come back to the same presentation and to its same meant object; and in which, on the other hand, we can repeatedly secure for ourselves and potentially do secure for ourselves the verified self as an identical self, and one that is not capable of being crossed out. Surely, we do have the momentary lived-experiencing (e.g., of an immanent sense-datum that we see in its present becoming) in a certainty that is not capable of being crossed out. But the being that we grasp there is only meant as being in itself when we not only take it as a momentary datum in the mode of the present, but also as the identical dabile that could be given in arbitrarily iterated rememberings—that is, when we take it as a temporal datum, for instance, as the tonal datum in its temporality, a temporality that is identically one as opposed to possible orientations like those that are given through variable rememberings.[13] [278]

We see that the temporal form is the form of objects which, as objects, pretend to have their in-itself. All talk of objects thus leads back to remembering. Thus, this does not only hold for immanent objects. Even when we consider a noema, even when we consider that which is momentarily present as sense in the mode of the present and make an objective assertion about it, we grasp it as such a [noema or present object] that can be presentified to us and identified in iterated memories, namely, with this reproductively presentified character, "momentary present." We learn from this consideration that the question concerning how an objectlike formation, an objectlike formation in itself, is

[13] *I am not entirely satisfied with this. The object is surely constituted from the very beginning as something temporal, and the momentary phase is an abstraction that we must first construct. The incapability of being crossed out peculiar to the moment is thus not primary.*

constituted, how it can show itself as such originally, leads overall and from the very start initially to the problem of the constitution of an in-itself of remembering, thus, to the question concerning how remembering can be justified and to what extent it can become a source for definitiveness. We must first of all become clear about it.

Indeed, seen more precisely, we are still lacking an intermediate term for a completely systematic exposition. The living present which is structured immanently is as we said not capable of being crossed out so long as it unfolds in constitution; doubt is not possible here. This also concerns, therefore, the span of the living retention proper to it. To make ourselves explicitly clear: Every progressing retention that continues to exist in a living fading-away cannot be modalized. But once again we must add that we still cannot identify an in-itself here. I can keep a hold on a sound fading away attentively, yes even hold onto it more tightly. Here is something of the most primitive activity. But it can also be the case that I am not even attentively aware of it, that I do not actively turn toward it and toward the series of sounds, and yet it exercises a special allure. By virtue of an associative awakening, it gets the character of an intention. In both cases, and in an entirely spontaneous fashion this provocation (this affection) is possibly discharged in a remembering that not only arises in [279] *general, but arises as the fulfillment of the intention.*

Note that this remembering is something essentially different from a retention, and is not for instance a mere reanimation of it in the sense of an increase in the level of clarity. A clear retention whose essence we grasp in the retentional levels lying most closely to the primordial impression always remains a retention. Every retention is what it is and has its intentional mode only at the place of the streaming perceiving at which it stands. But remembering is a kind of re-perception, that is, it is not a perception to be sure, but a being constituted anew, a beginning anew with the primordial-Now, and a retentional fading-away, but precisely in the mode of reproduction. Thus, in remembering all retentional levels emerge "once again," reproductively modified. If such a remembering now occurs in connection with a retention harboring an allure that is just past, it necessarily occurs in a coinciding of sense and in a coinciding of being with the retention. The same sound that has just sunk back and just faded away appears once more on the scene, and I live through its being once more. This can be repeated; I either deliberately or involuntarily reproduce the tone or an entire tonal phrase once more, even an entire melody. What happens here to the incapability of being crossed out peculiar to remembering? Is it to be rejected out of hand? Are we speaking in epistemology from time immemorial about the general possibility of deception in remembering? And does this not hold as well for the immanent sphere?

<§22. Rememberings as the Source for an In-Itself of Objects>

We will obviously have to make distinctions here between close memories and distant memories, (1) between rememberings that are awakened through the retention that is still primordially living, still articulated in itself and found in constitutive flux, and (2) between rememberings that reach into the distant horizon of retention, like with those of an entire piece of music.

1) Rememberings as Arisen from the Awakening of Primordially Living Retention [280]

Where the first are concerned we will say: For what has just past and is still sinking back, which remembering intuitively grasps once again, we have what is absolutely incapable of being crossed out—and we have this even if remembering is repeated, whereby the second now creates its evidence from the first and no longer from the retention that has, in the meantime, completely faded away. In spite of all that, we certainly do not grasp the self and the identity of the self in such a repeated coinciding of the self without incompleteness and degrees of completeness. For we know indeed that remembering can essentially waver in its clarity, and can even be intermittent. The different moments of content are more or less veiled, as it were, as if by a fog of unclarity. And yet, it is not one of those concealments in the usual sense, namely, of objects by other objects. The fog of unclarity does not blacken out objects, it is not a real fog. And yet it does conceal, renders the self-giving incomplete. And yet, what is just past as past is absolutely secure, it is incapable of being crossed out, indubitable, and it is absolutely secure with respect to everything that is given from it according to quality, intensity, timbre. It is itself there passing through the fog of unclarity, in all relative unclarity, but just not in an entirely obvious manner, just not as realized ultimately. Accordingly, something is lacking with respect to this incapability of being crossed out. Essentially belonging to this situation is the necessary coinciding that forms identity, the coinciding of given matters throughout the alteration of different levels of clarity, and a certain enhancement in the direction toward an ultimate, most genuine self, the self that is completely evident in its appearance but which is only an idea to be discerned intuitively, an ideal limes.

But what is peculiar here is that it does not first require something like approaching this ideal limit in order first of all to confirm a less clear reproduction. In this connection with the living present, it has its original justification in itself, continually. And "original justification" means that it contains a self that is inviolable, although it only stands in a gradation of degrees against a limes which in its very nature would alone completely [281] *exhibit the "self." The less clear remembering is less saturated, the clearer one is more saturated, it is a "more intensive" self-givenness, but if it is an intuitive remembering at all, it will give just one self and not give any other self, or any of its moments.*[14] *However, empty remembering is not actually a remembering, but an awakening or an affective allure of a retentional sedimentation that is emerging as prominent from the immersion of memory. In a certain sense there are also gradual differences of proximity and distance here.*

One must say then that we still have another gradation, namely, with respect to reproductions that reach into the outermost

[14] *During the course of our lectures the following was added: The justification of close memory does not yield any elucidation of the possibility of the knowledge of an immanent object as being in itself. For we are still bound here to the chain of rememberings, rememberings which adhere to a living retention, which had a departure from it, and were carried by its self-giving evidence. Only when we have first justified the distant memory do we have the possibility of recognizing at any time an immanent temporal object as existing.*

horizon of immersion, even with respect to those reproductions that draw near to it. Namely: Self-givennesses arise here that are indeed actual self-givennesses and are incontestable in this connection, while it is left gradually undetermined how far the actual self-givenness reaches, and what can yet actually be ascribed to it with respect to determinate moments.

2) <Rememberings of a Submerged Past of Consciousness>

The systematic path leads then further to rememberings that do not have their retentional connection to the immediate realm of the present, but rather, that revive a distant, long submerged past of consciousness. We speak here of distant memories as opposed to close memories. Even here, for distant memories, I hold that every remembering has its original justification, and this means that we are to understand essentially that corresponding to every [282] *remembering, even to this group, is a necessary idea, the idea of a self that is incapable of being crossed out. My guiding thought here is the following: An intuitive distant memory, when it is not one that fleetingly flashes forth, but is a steady one and is synthetically iterable and identifiable, has with respect to its objectlike formation essentially only one possible way of passing over into doubt and then turning out to be null, namely, as a confused muddle of rememberings. Thus, becoming discordant, inhibition, and annulment of the belief that is initially unbroken in the self-given past, necessarily leads to the phenomena of bifurcation in which the distant memory in question splits into several distant memories. And it does so in such a way that the unitary objectlike formation of an undivided memory is disclosed as the fusion of single objects, single features and events that belong to the separated memories and are self-given there with partially different objectlike determinations. In the same way it could now happen that every one of the splintered memories lose their character of unbroken concordance, and undergo a crossing out by splitting into yet other memories that are in themselves concordant.*

But on the one hand it is still the case that the content of every memory that is characterized as false is only false with respect to the unity of the combined whole, but it remains correct with respect to its parts. What is crossed out is always the whole that has arisen through a commingling, but the elements that issued in the commingling remain self-given, only they belong to different nexuses. On the other hand, however, this process of splitting cannot continue in infinitum; it is a muddling up of discrete matters, and thus it must come to an end. Indeed, it suffices that what appears in a memory, essentially, cannot as remembered be completely empty, that its self-giving cannot be an empty title, but rather that it has its source in actual self-givings such that we are necessarily referred back to the idea of a chain of pure self-givennesses that are no longer capable of being crossed out, but are only identifiable with respect to their content and repeatable [283] *in complete identity and concordance. Naturally, even here we have degrees of clarity for every portion of genuine self-givenness, and in this respect the idea of the most complete self-givenness as limes. Thus even this type of saturation <has> differences of evidence. In both relations we are certainly referred to the active ego and its free activity in which it is guided, precisely, by the experience that memory can turn out to be deceptive, and that, in particular, clouds of unclarity can conceal the comminglings. Accordingly, the ego strives to check its memories thoroughly, to clarify them deliberately, to investigate the intentional nexuses with respect to memories' parts, to disclose the illusion by analyzing, and thus to advance to the true self.*

But still necessary for our further understanding is the elucidation of the origins of error in passivity, and in particular of error in its most original shape of commingling. This problem leads us to a radical portion of the analysis of passive consciousness, namely, as genetic analysis: to the phenomenology of association.

<23. Immediate and Mediate Awakening>

The rubric "association" characterizes for us a form and a lawful regularity of immanent genesis that constantly belongs to consciousness in general; but it does not characterize, as it does for psychologists, a form of objective, psychophysical causality; it does not characterize a regulated manner in which the emergence of reproductions, of rememberings, is causally determined in human and animal psychic life. For we are working within the framework of the phenomenological reduction in which all objective reality and objective causality is "bracketed." What is there for us is not the world taken as reality with its psychophysical beings and its causalities, but only the phenomena of them, thing-phenomena, human-phenomena, etc., in their intentionality. In this framework of pure consciousness, we find the streaming present of consciousness, we find constituted in every case a perceptual reality constituted as in the flesh. But [284] *pasts can also enter into present consciousness through remembering. Put more precisely, in the unity of a consciousness that is streaming in the present, we find concrete perceptions with their retentional components, as well as concrete retentions—all of that in the flow of retention fading away into the distant horizon of retention. But in addition to this, emergent rememberings as well. Between the noematic components of something present and something remembered we find a phenomenologically peculiar connective trait that can be expressed in the following way: Something present recalls something past. Likewise, a second remembering can occur while a remembering runs-off; the second remembering can occur along with the first one in a nexus that is characterized noematically by the fact that the first recalled event recalls the second recalled event. A perceptual consciousness, that is, a consciousness that is constituted originarily can accordingly be characterized as a consciousness that awakens, awakening a reproductive consciousness, and this consciousness can function as awakening in its turn as fetching a past of consciousness, as it were.*

Let us consider this nexus in broad strokes. If an *a* that is given to consciousness reminds us of a *b*, then the associative awakening is either immediate or mediate, and immediate and mediate associations are always intertwined with one another, even if it is only the mediate one that is able to obtrude for itself upon us. For example, in a conversation we are having, a thought is expressed that reminds us of a friend. The thought belongs to the unity of the present conversation; the conversation reminds us immediately of a previous conversation with the friend in which the same thought was expressed. The association between *a* and *b*, thought and friend, is a mediate one. What is immediately connected is *a* with *a'* and *a'* with *b*, i.e., the immediate awakening goes from something that is identical, uniform, or especially similar to something that is similar to it; that is, the bridging member, we say, connects the present consciousness with the previous one. Something uniform, and something quite especially strong, something identical in the sense, the identical thought, awakens a memory of something uniform, and the awakening goes further from here to the other previous content of consciousness. Then the [285] awakening goes still further in the steady series of memories or pasts that advance forward in the direction of the present, but also discretely toward the previous pasts or even discretely toward the future. Still, several questions arise here when we examine this more closely.

<24. Association in the Impressional Sphere. Its Significance for Remembering and Analogizing Protention>

The multiplicity is in relief for itself, it exercises an affection as a multiplicity and, at the same time, it affects a term with a special force, one that drowns out opposing terms. It is a special problem, and surely also an important one, to investigate what the general and essential conditions are for terms of a multiplicity to be capable of reaching a phenomenological union, to investigate what then allows multiplicities to connect with multiplicities and thus in general what produces unities for consciousness, unities that are not original singularities. "Original singularities" are objects that must first come into relief essentially as a whole, and then require analysis in order for parts or moments of the object to come into relief on or in them, while a multifaceted object is precisely a multiplicity that is essentially preceded by prominent singularities. The general conditions of singularity and of multiplicity lead us to certain essential relationships of uniformity and similarity, and integrally related to it, of contrast and the like. But all of this concerns, I contend, conditions of the possibility of intention and affection that are carried over in an intensified form from something that is prominently uniform to something that is uniform to it, from something that is relatively most similar to something that is most similar to it, and together with the one-sided or mutual transference or awakening for consciousness, they have a connective force that links the special affections.

"Recalling something" is an evident nexus: thus, for example, in the constitution of successions forming multiplicities as connected unities, for instance, in a connected sequence of tones. A tone initially arises for itself; that is, the tonal phases that are continually melded together through an internal continual similarity form an enduring unity for itself by virtue of their contrast in the point of departure and in the terminal point to the [286] tonal series, to the phase; they are affective as one. Then a new tone arises, and then another new one. But each new one that arises here does not arise in the same way as the first one does, as if no tone had arisen prior to it. The new one enters forthwith into the unitary relation with the one that has just past, with the one that has sunken back retentionally. By virtue of the similarity of intentional objects, the affection of each one is connected in a unity as the affection on the same ego. But there is still more. The object sinking back retentionally, and which as such loses something of its affective force, obtains through this connection an increase of force, and so forth with every new tone.

Not only do similarities and relations of similarities of qualities, intensities, in short, of moments with respect to content contribute to unification and to the efficacy that reawakens the sinking force, but also temporal and spatial relations contribute to them. Rhythmatizations and cycles are especially instructive examples. Let us take a simple repetition: a b c and again a' b' c' with the same content and the same temporal relations. The new a' reaches back to a, likewise b' to b, c' to c, and in addition to this also the connection a' b' c' to a b c. In this affective coinciding what has already sunken back in the first group undergoes a new intensification and unification; every new repetition is of benefit to the previous ones; what has retentionally grown old and fallen in the grave is, so to speak, still tied to life with the creation of a unity of a higher order, [i.e., with the creation] of a cyclical unity of succession. These are all processes of phenomenal formations of unity that seen from within are processes of affective connection, and affective connection is at the same time the awakening peculiar to affective force. This is why I am obligated to see primordial forms of association here, so to speak, originary association, association within the impressional sphere.

For in other cases as well (that is, also in the series formed out of reproductive associations at a distance), it seems to me that the most essential feature of this process consists in affective interconnections.

But first we must point to the fact that the same principles that institute a connection, principles that are originally efficacious in the impression, get a hearing again in remembering. I mean it in [287] this way: It is evident that when a remembering is developed in a completely intuitive manner, all the single prominent elements and multifarious connections are there once again reproductively, which connections were constituted in the corresponding original perception. But in the case of a remembering that is revived associatively by a present, this revival is a process there; accordingly, what is similar in the present initially provokes a reproduction of something similar in the past, and the path of the awakening of an ever broadening intuition now follows the connections between which the bridging term is trestled. If a present thought recalls a past one, there is a tendency toward awakening the series of thoughts. If the thought exists in an apperceptive unity with the person who expressed it, and the person stands in a multiple relation with other persons forming a society, then all such unities are unfolded in intuition, or there is a tendency toward this unfolding. So much for the question we are treating here.

It is a primordial law that one similar thing is originally connected to another similar thing through a "fusion," and in addition to this, there is the further essential law that every awakening of a member, every augmentation of the force of affection that proceeds from it, also augments the affective force of all members connected to it. Moments in the present that have their special prominence in consciousness and thus their current force of affection on the ego not only awaken something similar to the past, but in accordance with a further law, also awaken something that is continuously connected with them in the temporal nexus according to coexistence and succession (this is the law of association). And in addition to this, there is for us the law of analogizing protention whereby an analogue of the associatively awakened past is projected into the present, and an analogue of the becoming that is past, that is, of the past "having-become" (of the past processes)—which were in the process of becoming in the corresponding situation of consciousness—are projected into the present as those that are quasi-becoming. And in this way a consciousness of an analogue of a becoming arises in the present that is however not a memorial consciousness, but rather an analogue of a memorial consciousness; and on the other hand, just as the memory of a becoming that is past is directed ahead into a familiar future—familiar because it has already [288] become an epistemic acquisition—so too is the analogue directed ahead into the future, which is not familiar to be sure, but is analogous to one that is familiar.

<25. The Lawful Regularity of Retention>

But if one asks (and this would be our other question) how we are to understand the fact that the present retrieves the distant, submerged past through similarity, and that the past, which is certainly no longer anything, can determine the path of reproductive awakening in the past's affective nexuses (nexuses that only had significance once for the past ego) the answer would run as follows: The present turns into the past as the past that is constituted for the ego through the lawful regularity of retention; and finally, everything that is retentional turns into the undifferentiated unity of the distant retention of the one distant horizon, which extinguishes all differentiations. However, this extinguishing is to be understood in the following way: The affective force is necessarily decreased with the submersion, which is to say, it decreases the force that makes possible the special prominent elements, the unities for themselves even within the non-intuitability of retention, be they singularities or multiplicities or even multiplicities of a higher level, for instance, cyclical multiplicities. What is given there broadly in near-

retention as something extended and as a unity of continuously connected affections, and likewise, what exists there as a multiplicity of elements given together or that follow one after the other, but as largely diverse—[all of this] moves closer together; I would say that corresponding to the temporal perspective, to the phenomenal moving-closer-together of those matters that have just been, is an affective perspective; flowing is a flowing together of affections. All of this holds, we must add, as long as the opposing forces, retroactive affective animations, do not originate from the living impression that issues forth. These are present in the form of repetition, of cycles and the like. What would have otherwise already flowed-off and become unnoticeable will continue to be retained by virtue of a fortifying awakening, of an increase in the force of affection that radiates back. But not to infinity. Finally, the awakening force becomes powerless, and the undifferentiated horizon assimilates everything that is on the verge of sinking back into a steady retention. In addition to this there is the [289]
supplementary law that this distant horizon, the horizon of the distant retention that is already dead, can be reawakened from the current present in such a way that a stimulating force issuing from the present can go into the horizon in a discontinuous fashion and can effect a prominence in it. This prominence is then propagated further according to the awakening force itself that issues from the awakened element, for example, from the force of the awakened obscure memory of a lecture in a series of similar lectures. This retentional awakening, then, functions in such a way that it brings with it a tendency toward remembering that could then be realized in an actually occurring remembering.

This certainly creates new problems that pressingly await further work. It requires a phenomenology of tendencies and the demonstration of laws that govern the realization of tendencies, the fulfillment of tendentious intentions. Lacking is the association of mere simultaneity and succession as pure temporal association, which serves as the framework for all association taking place through fusion.

<26.> Expectation and Association

We have touched upon the phenomenology of association from the perspective of remembering and in my view it must be tackled from there and not by drawing on just any kind of reproductive phenomena, like the phenomena of intuitive expectation and, especially, the phenomena of free phantasy. We have already repeatedly employed and touched upon what concerns the lawful regularities of expectation. They will also be enriched and clarified through our present genetic analyses. We called protention a shadow that is projected ahead, an inverted memory. Let us now say with more clarity that as the most original protention, namely, as that of the impressional sphere, it is initially a modification of retention motivated by the near-retention, and the same holds further for distant-retentions as well. When a tendency to bring something to intuition attaches to the expectation just like it does with respect to a retention, and when this process of bringing to intuition occurs with the appropriate awakening force, then the awakened intuitive element is indeed [290] not a pure and simple remembering, but a peculiar and original modification of a remembering. When, for example, a tonal sequence, which is running-off at a sufficiently slow pace, is repeated and we become precipitate, so to speak, during its repetition; and instead of the empty anticipatory intention, an image of what is to come appears that illustrates it, then this is evidently the remembering of the corresponding element of the tonal sequence, but in a new mode, with a new function; what is given in memory gives a picture of what is to come; what is to come is not what has past, but what is expected, which has its prototype in the past. We do not have a consciousness of the past in the expectation, although the past appears here with it; it is transformed into an anticipation, and one can read that in it.

Let us examine the situation in the sphere of the living present more closely. The occurrence of something futural is expected through its similarity to what has occurred in the past, like already happens in the most primitive case of a steady protention.

<Section 2: APPENDICES>

<A. APPENDICES TO PART 2>[15]

Appendix 1: (To §§6-8) Descriptions of the Phenomenon of Conflict without Regard to Position-Taking[16] [346]

Wax figure—human-being. The human-being wins out: And the wax figure is presented in its determinate spatial orientation and not, for instance, in any random orientation. But it is a curious way of being presented. It is similar to the way in which I see an object and an object "repressed" by it, namely, visually eclipsed by it, an object that I had just seen, an object that is there for me in an "empty" manner, in its determinate orientation, in its determinate mode of appearance, and yet not appearing in "actual" colors, etc. [They are] not given, but rather, presented in an empty manner. Or like when, having familiar objects in my surroundings before me in the dark, I reach for them, go toward them, but only in exceptional circumstances [do I] have presentifying intuitions. And even when these do occur, they presentify the emptily presented object and the emptily perceived object which is identified with them (for presentification is not perception). These are similar cases. The similarity consists in the mode [of givenness] that we designate as "empty-appearance" in contrast to full-appearance. There is some basis here in saying that what is empty continually passes over into the full, and vice versa, namely, insofar as the distinctness and clarity of the full

---

[15] Translator: The following pagination to the German text corresponds to Husserliana XI.

[16] Editor: Between 1920 and 1925.

apprehension underlies graduated differences, but the full apprehension here can have a strong surplus of something apprehended in an empty manner, and in this case as something apprehended in an entirely determinate and perceptive manner. It actually stands there, just that I do not see it, or I only see a little bit of it, only a vague glimmer, like when I still have a glimmer in the dark, and what I can make out of the completely appearing thing that is barely discernible in this vagueness here is the wholly determinate and quite familiar desk.

But viewed precisely, is it really a matter of a gradual tapering off in the genuine sense? The fullness of the thing seen gradually tapers off; the quantity of the thing that is completely perceived gradually increases, so to speak, as opposed to the thing that is emptily perceived. But can one say that the full perception itself has a limit in the empty perception, that this is a limit of something less and less full? This is true in a certain respect, false in another. In such cases the appearance is always a mixture of two components of a full appearance and of an empty appearance. [347]
But the full is not itself a gradation of the empty, and likewise, the empty is not a gradation of the full. The empty is in itself something that is at the same time a limit insofar as the gradual loss of fullness, while maintaining the sense (we can suppose that at first it does not contain emptiness at all), also brings about emptiness along with the indistinctness, until the only thing remaining is sheer emptiness. This will certainly have to be rethought. The question is whether the empty appearance is only a very dark appearance, and whether it is not rather a different appearance. It is a further question whether the full appearance, insofar as it contains a pure sense, has an empty appearance in it, an empty appearance that only assumes the fullness of clarity.

But it seems to me that it is precisely conflict that shows [the following]: Two full appearances (and not only relatively, but absolutely, independently of the clarity of fullness) cannot be maintained simultaneously in a coinciding; they can only be in succession. But a full appearance can be united through overlapping with an empty appearance, and that necessarily belongs even to the consciousness of conflict. While we are conscious of the full appearance, we are also constantly conscious of the empty appearance, precisely as what it is, emptily meant. If it becomes full, then the opposing appearance must be transformed into emptiness.

To say that the full appearance would contain an empty appearance (and that this would then be the pure sense), does not seem to me to be tenable either. For the empty appearance has its empty fullness, it is the appearance of this and that side and has its "back side" which is an emptiness in another sense (one should not confuse the emptiness of something concealed with the emptiness of something apprehended); and so it appears to me untenable to say that the fullness of a full appearance would cover over an empty one, but rather, where emptiness passes over into fullness, the full replaces the empty. In place of the empty presentation, I have an "actual" giving presentation.

But a conflict can also occur in a different way in the unity of a perceptual consciousness. A perceptual appearance runs-off, and the way it runs-off can be concordant. In running-off, concordance is something different from concordance in a phase, or in a duration, as continual concordance that is not continually discordant (overlapping of the full and the empty). Admittedly, both concordances are closely related and will clarify one another in their uniqueness.

Thus, we speak of concordance running its course and of discordance running its course. Sense is constantly altered in the progression of appearances, the thing shows itself from different sides and is also qualitatively modified, e.g., through movement, etc. But this series of change is concordant; what occurred previously is not betrayed, annulled by anything occurring later; nothing occurring later clashes with the previous occurrence.

However, the series of change can also be discordant. I see the [348]
thing in a continuous fashion; it gives itself as unchanged, in rest, with such and such features. I see it from one side, I walk around it, and now I see that it is different than I had apprehended it. I remain in the apprehension, "resting, unchanged thing." In this respect, there is a stream of concordance, but resisting it is what I see now; it does not fit into it. The thing is indeed the same, unchanged, but it is different. "It is different" also means: It was different. The previous perceptual phases are no longer current,

but they have not become nil; they survive in retention, and retention holds fast to the apprehension as *a*; the retention, according to its very sense, makes a necessary contribution of unification to the constant consciousness of an enduring thing (identically, but temporally pushed back), and this retention reaches its verdict the moment I see α'. The perception of α' and of the entire appearance, particularly, the "Now-appearance" to which α' belongs, coincides with the retention of the α-appearance, but it coincides in conflict, while in the case of a concordant consciousness, the coinciding would have been a coinciding of concordance.

Thus, in passing through a thing, as perception continues to progress (and not, for instance, only insofar as it "comes back to the same aspect of the thing"), we find coincidings of Now-phases of perception (the originary ones) with retentions as originary presentifications of previous perceptual phases—and [we find] coincidings of concordance if the entire perceptual consciousness is to be concordant, but on the other hand, [we find] coincidings of discordance where this is not the case. These are obviously completely different occurrences, although they are closely related to the previous ones.

If, in perceiving, I move through the series of givennesses of the thing, then the coinciding will not at all merely take place on a particular occasion, one time; rather, the coinciding is continual, that is, as long as I have a steady consciousness of unity, as long as I see one and the same thing, I will have a continuity of phenomena coinciding with each other: Coinciding is indeed a consciousness of unity. Here, consciousness flows in such a way that a new Now, an ever new originary phenomenon is always given to consciousness as a phase, and something new is continuously transformed, again and again, into something that is not new; the primordial perception is transformed into retention, which in turn is pushed back through an expanded chain of retentions of the same [primordial perception]. Certainly, after everything has run-off, the whole is itself pushed back in relation to the new Now-actualities; from this it follows that the pushing back is a moment of the retentions themselves, retentions that could not be ordered successively if there were not anything present in them as the ground of order. With every current Now, we have an "ensemble" of retentions in their succession (in "coexistence"), and this entire sequence is in "coinciding." Not merely the ones that are immediately contiguous, but even the phases that are mediately connected interpenetrate and have an integrally cohesive unity. And vice versa. If some new [349] "discordant" thing arises, then the discordance can potentially be propagated throughout the whole continuity, or in any case through whole sections of this continuity. But how is that? Only in such a way that every "closer determination," which a moment of indeterminacy of sense has undergone, is preserved in the sequence in each phase, naturally even if only in the form of "being-co-meant." Thus, actually, each phase only coincides directly with the one next to it, but the conflict with it is propagated to the extent that, in the continuous succession, the moment of sense in question, as determined sense, is maintained in the sense. But such general descriptions probably do not suffice.

We have different cases:

A temporary concordance suddenly becomes a discordance, that is, the continual unity no longer holds out: A phase emerges where the retention of something that has just past coincides in conflict with something that is new. The consciousness of "different" arises, for example, on the very same surface, the constant white suddenly changes into black. Coinciding with conflict. The black eclipses, as it were, the white of retention.

But now we have to say: It can be that discordance arises at one place, and at the same time, it issues in a commutation of the whole apprehension; this happens in such a way that the entire series that is running-off is subject to the same commutation by virtue of which the entire series including what is new (which the explosion has called forth), is transformed into a [different] concordant series and now continues to flow on once more in a concordant manner. Then I have the current retentional series up to the Now and I have it in a conflictual coinciding with a "presumed" retentional series that has not in fact run-off. I can potentially carry out a remembering and generate through it a presentification with an altered apprehension; this is not actually a remembering, since I had grasped it differently. I had taken the mannequin with its mechanical movements for a moving woman, and now the commutation ensues. I then recall the earlier event, I rehearse the movements in memory, but I grasp them now as movements of a mannequin. Now we have here a coinciding of the entire series, coinciding in conflict, and at the same time, the altered series is transformed into a concordant series all the way up to the Now.

It is an entirely different case when the object "changes," when its color "suddenly" changes. The discordance concerns a moment, but the object remains concordant throughout, and the sudden transition does not demand (even where the moment of sense in question is concerned) that a new apprehension should occur in what has already run-off. Thus, here we see how the discordant unity at one place does not have to exercise a retroactive efficacy on the concordance of the previous series.

But certainly, what kind of efficacy is that? Why do I say: "It was not a human being, but a mannequin"—and why not: "The human being suddenly changed into a figure made out of wood, etc.?" Now, that which is new, and each and every thing, is motivated according to the way in which I alter the apprehension, [350] and when I do not, I have an unmotivated transition, and what is "lacking" is the motivated. The switch of color is a discordance, but it can be motivated, it is the result of turning on the light, etc.

Discordances can occur continually by maintaining a thoroughgoing unity; then we have a continual process of becoming other, a continual transformation.

What role now do the position-takings play? Now, when we were speaking of motivation, it was certainly a matter of position-takings. The moment we do not consider them, we only have the occurrences of coinciding in concordance and in contra-cordance, of coinciding in a continual non-change, coinciding in expanses of continual non-change, and suddenly places of discordance and coinciding in a continual deviation of pure concordance (of non-change), thus we have occurrences as they take place in continuous change. And further, [we have] expanses of the first and second type commingling with each other.

These are possible occurrences of appearances without regard to position-takings. If we do not regard position-takings and their "motivations," their "demands," we cannot distinguish free phantasy from actuality. In free phantasy I have an appearance of the thing, it begins to flow, it begins to take on shapes randomly, to change suddenly in its sense-moments; it is pure chaos.

Position-takings can accord or conflict with other position-takings. This "accord" is not the coinciding of appearances, i.e., the coinciding of their senses in "concordance" or in "conflict."

But I am not at all satisfied with this. And I think that we were already further along previously. Appearances are nothing without position-takings; appearances are merely *abstracta*. Position-takings are abstract modes of appearances; these modes as variable under the rubric of appearance, allowing something to be maintained.

Appendix 2: (To §§ 8-11) <Sense and Modality of Being in Perception and Remembering>[17]

Let us take one step further. We spoke of the fact that a perception can turn out later to be a deceptive perception. Let us now imagine such a case, namely, the case where a perception, regarded according to its content of appearance, maintains the character of perception in a certain respect, and yet takes on the essential modification that we all have probably experienced: the consciousness of illusion. Let me draw on an experience from my student years in Berlin. While I was viewing an exhibit in a museum I saw next to me among the other attendants, a young [351]
woman with a catalogue in her hand, intently viewing the same piece that I was. I became suspicious of this young woman after a while. I recognized that it was a mere figure, a mechanical mannequin designed to trick us. What happens interiorly in such a case? Now, we are at first wavering; two perceptual apprehensions are in conflict with each other: in this example, a human being made of flesh and blood, and the mechanically moving figure made of wood and wax. Although we are sure, we can still deliberately pass from the one apprehension to the other, and we can thus "see" what is before us, now as the human being, now as

[17] Editor: 1920/21.

the wooden and wax figure. But in contrast to the preceding perception of the young woman and the present way of regarding the same young woman, there now arises an essential difference—even though not only is the object the same and appears in the mode of being presented in the flesh, but also the succession of perceptual appearances, the successive aspects in which it is given, are essentially the same. Shortly before this the young woman was still standing there before us, existing with certainty, and now [she is standing there] with the character of illusion, with the character of nullity. The intentional object, the perceptual sense, is exactly the same in a certain respect, but it has a different "mode of being" for consciousness: one time as in straightforward existence, another as null, as non-existent. Viewed precisely, the earlier character is still maintained in a certain respect in the second one, the "existent" is in the "non-existent," as the expression correctly suggests: still in it, but annulled, as it were, crossed out.

We encounter the same contrast here if we give into the tendency to alternate attitudes between the young woman and the mannequin, and successively bring into view thematically the appearing objects standing in a conflictual relation to one another other. The intentional objects are different here, but not only this: The qualification of their being is also evidently different: The one object, namely, the "young woman," qualified precisely as null, the other intentional object, the mannequin, as actual, as existing in a straightforward manner. If, in addition to this, we pay attention to the intermediate phase of our wavering in doubt, that is, if we pay attention to the mode of consciousness of doubt that concerned both objects, then inhering in both of them is a third kind of qualifying character, and for both in the same way: the character of "dubitable," or what often means the same thing, "questionable." This character, too, is in itself characterized as a modification of the primordial mode, "being": Dubitable is existing dubitably—at the same time there is a modification of a crossing out in it, a modification that issues from the opposing member, though it does not make it to a decisive break-through.

Let us now also pay attention to the tendency of the ego belonging to the consciousness of dubitability or questionability as it attributes being to the one thing in the thematic directedness toward it, and then alternatively takes this one rather as existent in the thematic directedness toward the other; and let us pay attention to the fact that when we put ourselves into the one tendency, the [352] one thing entices us as being, but the other as clashing with it and null, and vice versa. It can be that these penchants of the ego have not arisen from an actual rivalry. In any case, the consciousness of two problematic possibilities does arise here, and something speaks in favor of each one. But a consciousness can potentially arise in which more speaks in favor of one possibility and in which the ego gives a preference to it without being able to pass over to a decisive belief. Then we have the new qualifications, [on the one hand,] those of one possibility among [other] possibilities, and on the other hand, those of probability, related to possibilities. Further, we have possibility and probability themselves in their essential character as a modality of "being": possibly existing, probably existing. It is easy to see that the number of mutually related members belonging to the unity of consciousness can be greater than two, and is ideally unlimited.

By showing these matters phenomenologically, we have taken an important epistemological step forward. Each one of the intentional lived-experiences that we call external perception, and thus all perceptions in general immanently, contain their objective sense. We see that this is to be understood in two ways; two concepts of sense are to be distinguished: (1) Sense can mean the full What of perception, which is to say, the intentional object with its mode of being (*positum*). (2) But "sense" can also mean the mere intentional object which is something that can be identified from and throughout the modalities of being that are potentially variable; the same tree, namely, as believed with certainty is then doubted whether it is, etc. (the material of the *positum*), the mere unqualitative objective sense. Thus, every perception has its intentional object as such, but always and necessarily qualifies this sense, and, as we say in our terminology, is conscious of it in some mode of being. The mode of "being" is characterized as a primordial mode; by contrast the other qualifications are modal modifications: being dubitable, being possible, being probable, etc. It should be emphasized that these modes of being belong to the objective sense, and not for instance to the contents of appearance, to the thematic regard, etc. For where we reflect upon these, for instance, where we make them a theme of reflective perceptions, we do not have the character of dubitability with respect to them, for example, in the case of doubt. What is dubitable is the external object that is meant with respect to its being, but not the appearance in the reflection that exists in an entirely certain manner, [i.e.,] that exists in a straightforward manner: Reflection teaches us that for all lived-experiences, the contents of appearance in general can never harbor such modalities.

We notice here that with memory, as with perception, we can in fact show that an immanent sense is something identical in manifold lived-experiences, and can show it in an evident manner. If we carry out manifold memories of the same event that were [353] given to us *originaliter* in an earlier perception, then what is given to consciousness in all memories and is potentially meant is the same in an evident manner; it comes to the fore again and again in a synthesis through coinciding as the same, as the same objective sense. It is the past event as the event meant in each of the memories. Necessarily belonging to this sense is a modality of being, and it belongs to it in precisely the same manner as it belongs to the sense of perception, either in the primordial mode of having-been, the "it was actually so," or as a modality of it. For memorial certainty can also be modalized. We can doubt whether that actually existed (or existed in some respect) in the way it is given in the presentification, whether the events in question really ran-off in such a way. We can decide negatively or we can have the remembered event in the mode of probability, etc. This concerns everything, not only the direct memorial objects, but all contents that can be extracted from the presentification-content of memory through immersion and reflection, that is, [it concerns] even the presentified ego, its egoic lived-experiences, its acts. In every direction we find the memorial sense and its inherent modes of being, just like in perception.

As I stated at the outset, however, the senses and modes of being about which we are now speaking in the context of memory are not, for instance, the presentified senses and modes of being that lie in memory by virtue of their peculiar intentional self-contained structure, namely, by the fact that it is a piece of present experience in which a concrete piece is presentified from the stream of the former lived-experience of the same subject, from the stream of a lived-experiencing that naturally becomes presentified with its senses and modes of being. This reproduced lived-experiencing of the past ego is however there for me who I am now and who presently experiences the reproduction, since I can thematically apprehend [something] from the standpoint of the Now, [that is,] can grasp the event of yesterday's perception as remembered now and presently as my past event. This present thematic intending has its current sense precisely in this past event as meant, as past. And I am actually and presently certain of it, or I am unsure of it, I suppose it, etc. That is, the modes of being actually belong to my current memorial object.

Accordingly, we distinguish here: (1) the reproduced past perception and its sense—I grasp both by transposing myself into my past "I perceive"—; (2) the current memory and its sense. Both of them can coincide to a certain degree, but they need not do so. Thus, it can be that by immersing myself in memory I find that I had such and such perceptions, and that from the standpoint of the Now I am certain that I, as the ego of yesterday, perceived in this way—but I am now of the mind that this perception was a deceptive one. In my present motivational situation, what I saw yesterday as a human being is for me today a wooden figure. It is quite usual for us "to take over" the modes of being of the reproduced perceptions that make up, not the intimately inherent, [354]
but the intentional content of our memories; the remembered perceptual belief, the remembered doubt, the remembered supposition, etc., are not only taken up as reproduction, but without further ado we believe along with it, we doubt along with it, etc. Frequently, however, we do not take over positions, but take up a new position. To be sure, we can only speak of a memory as long as a ground of certainty is there at all, a general past that is being given with certainty from the Now, that is related to my past ego with past lived-experiences, intendings, etc.; and this present general certainty has such a character that it actualizes the certainties presentified in it from the Now, that is, it takes them over. On this basis, particular deviations, non-appropriations can

ensue. – That's enough where the modalities of being in memory are concerned and with respect to how they belong both to an immanent sense peculiar to each memory and to each perception.

Appendix 3: (To §11) <Evidence of Possibilities as Such and Modal Modification *in infinitum*>[18]

If we take several anticipations clashing with one another in an ambiguous situation: A future must arrive, but several anticipatory intentions mutually inhibit one another in their certainties, giving to these certainties the character of "perhaps." Different problematic possibilities are such that they have different motives speaking in favor of them, different motives grounded in experience that allow the rays of expectation, as it were, to come to some good for these possibilities. But several possibilities can be given as incompatible with each other, as incompossible with each other, such that there are "motives grounded in experience" speaking in favor of each one, but in such a way that they have a different "weight," and finally, in comparing these weights, one of the possibilities can be characterized as the probable one, having at the same time the preponderance of weight over all the other ones.

But in a good sense all of these are, once more, occurrences of the object; namely, [they are] real possibilities as supposed possibilities, and like other occurrences of the object, probabilities can be meant but not self-given, not grasped in evidence; they can be falsely meant, they can turn up as null in the transition to self-giving evidence, or alternately, the meant objects can prove to be truly existing.

We have a special liking for speaking here of occurrences of expectation. But if we consider that anticipations are contained in every transcendent perception, we will see that this entire region belongs here. All anticipations in external perception are either [355]
actual anticipatory intentions or intentional horizons, which, as it were, are on the threshold of passing over into expectation; [they are] intentional horizons that exhibit quite enveloped potentialities

[18] Editor: 1922/23

of expectation. But that is much too large of a theme. In any event, we must designate the mode of empirical certainty here.

These are all occurrences that already take place prior to all (in our sense) philosophically oriented interests; these occurrences play their role in customary experiential and scientific life, and they leave their mark in the modal forms of judgment that have been very incompletely demonstrated by traditional logic. For us, however, they are shapes that can be exposed in the egological sphere. If we seek to turn up what can be shown apodictically in this sphere, we need not merely train ourselves on the individual facts that can be shown with apodictic certainty, on apodictically straightforward being. We must recognize that where we have a possible being in mind, or where inductive possibilities and probabilities are presented to us with respect to a futural event or to a past being in the pure lived-experiential sphere of the ego, we cannot assert anything apodictically for its straightforward reality, for the absolutely certain being and being-thus in the past or future; but on the other hand we can gain apodictic evidence for the respective possibilities themselves having an actual existence as possibilities, as suppositions, and probabilities. We can potentially have them in their self-sameness, grasping these possibilities themselves in a wholly original manner. And precisely the same holds for generalities and for particularities of generalities—all of these are objects.

For example, if I have the phenomenon of a thing that is extended, the probability that I will have these and those aspects of it given perceptually in a specific succession is of course not apodictically certain. If I make judgments here purely about the aspects and not about the real existence of the thing, I cannot make any pronouncement about the arrival of the object. On the other hand, however, I do have absolute evidence for the fact that the present change in aspects, of which I have the unannullable certainty of the present, makes its progression probable, and in the form of such and such new aspects belonging to it. The being-probable of something to come is given apodictically, and not the thing itself that is to come. An aspect itself is given as existing apodictically (or again to justify) only in perception. The probability of the coming aspect is apodictically given for its part in this motivational framework that we have described. The probability is experienced, is itself perceived in it, so to speak, as this probability itself. The probability of an *A* is an object different from *A*, the being-actual, the actual existence of the being-probable of *A* is a being-actual that is different from that of the *A* *simpliciter*. The main point is to show that all things like [356] possibilities, probabilities, generalities, state-of-affairs, etc., are "objects" and have their own ways of being given originally, and potentially, apodictically.

In addition, we should also note the following here: In relation to each object that is given as actual, i.e., certain as being, we can speak of its possibility, that it could be, or of its probability, dubitability, etc., and this can even take place in certainty and in truth. Now, if possibilities, probabilities, etc., are themselves objects and potentially truly existing actualities, then there will also be, in turn, possibilities, probabilities, etc., of them. Thus, the modal modifications yield precisely such objects *in infinitum*, if not always in truth, then at least with a meaningful portion of it. If *A* probably is, then the probability that *A* is can also be probable, and so forth, in continuous iteration. The being-actual of such an object is given in a certainty, and is given originally in an evident certainty, that is, in a demonstrative certainty that places it before us in its ipseity.

I would like to add just a couple supplementary remarks here. These modalities are meant or given in particular lived-experiences. Thus, we have lived-experiences that on the one hand themselves "are," and in the best case scenario are grasped in absolute certainty according to existence and components of being-thus; and on the other hand they are lived-experiences that harbor an intentionality, lived-experiences that within themselves intend something, and potentially intend with apodictic certainty, but in such a way that what is given in them there in apodictic certainty are not lived-experiences *simpliciter*, not concrete individual data *simpliciter;* rather, they are possibilities, probabilities of such data, or relations, state-of-affairs, generalities, or some kind of anticipatory being, like something that is arriving through original motivation, etc. Corresponding to the intertwining of intentionalities are also the intertwinings of these evidences and therefore intertwinings of these modal objects. They condition their non-independence, which for their part occasion similar evident demonstrations.

We must especially take note of the following: If we are directed purely to the change of the coming aspects as we regard a thing, then we will progressively have an evidence of what is to come; it has the character of a presumptive evidence and not of an evidence of pure and simple futural being. However, this evidence of the supposed being of what is to come is not something isolated; it stands in the context that gives to its supposition the character of a conditioned supposition. And this being-conditioned is a self-giving being-conditioned, and itself belongs as a very important type to the framework of our demonstrations.

Appendix 4: (To §§ 14 and 15) <Levels of Decision. Receptivity and Spontaneity>[19] [357]

Conversely, however, if one of these problematic possibilities that I had rejected, for instance, *B*, is altered in the actuality of experience, then my rejection of it comes to naught, my declaration of *B* as invalid is not justified, is "false," "amiss"; precisely this possibility is shown as a reality, is the one that I let be valid, the one to which I must apportion an acceptance of validity. But correlatively, because my judgment that I passed earlier, the positive decision for the previously preferred possibility *A*, is "untenable" due to its conflict with the actuality of experience, I must withdraw my accedence to *A* and transform the declaration of validity into a declaration of invalidity.

Such occurrences are disagreeable for persons passing judgments on these possibilities; they fall into doubts of legitimacy and questions of legitimacy. Concerning its most general essential structure, we gather from this that a question always remains a question—a practical intention toward an answer.

An answer always means: the transformation of the respective problematic disjunction of problematic possibilities speaking

[19] Editor: 1923

against one another into an unproblematic conjunction of actualities that accord with one another and speak in favor of one another; running parallel to this in the sphere of egoic comportment: the transformation of doubt, which is disagreeable and frustrating practically, into the comportment of the ego, which is uninhibited and satisfied, at ease, into a judicative decision for one of the possibilities; connected to this, at least *implicite*, is a negative rejecting judicative decision against the other possibilities.

In general, this is essential to question and answer. On the other hand, however, there are essentially also questions on different levels, and questions concerning legitimacy form the most significant higher level of questions. First is the question whether something behaves in such and such a way, and the question bears on the corresponding judgment or on a conjunctive connection of judgment as an answer. Then a higher question concerns the legitimacy or illegitimacy of the judgment that is already passed, or the legitimacy or illegitimacy of the earlier standing conviction. Legitimacy or, as we can also put it initially, tenability, the ability to substantiate the judicative position-taking, or untenability—this is now what is problematic. Thus, without the formal essential structure of the question as such being altered, the problematic content has altered it.

Naturally, logical judgments in the specific sense are not merely decisions in general, decisions that can always be seen in an intelligible manner as answers to questions, but are also judicative decisions that at the same time respond in a satisfying manner to the corresponding questions of legitimacy, that is, they have [358] followed from the substantiations as decisions of legitimacy.

What we learn from the latter considerations is that [a] occurrences within the sphere of the passive, merely apperceptive substratum and [b] such occurrences as the ego's modes of comportment and position-takings, go hand in hand in a curious way; and original equivocations accordingly arise that are now intelligible in one stroke.

Let us take something that is purely "aesthetic," [i.e.,] that is purely a matter of perception, only let us put into play, at most, the ego as perceptively aware, and thus the ego as opening a latent apperceptive intentionality. We could say that the genuine concept of receptivity has its seat here. The ego that behaves in a merely perceptively aware and observant manner comports itself in a merely receptive fashion. Indeed, it is a wakeful ego and lives as such in the form of "*ego cogito*"; but this form itself encompasses a mere passivity and activity. This receptivity is the founding presupposition for the possibility of the specific "spontaneity" of the ego, that is, for making possible the position-takings of the ego and what is [specifically] in question here, [namely,] judicative position-takings.

Within the sphere of receptivity, the modalities of being are modes of apperceptive intentions, modes of their uninhibited intentions, that is, here, a course proceeding in concordant fulfillment, or modes of inhibition, modes of a break, of the "negative" synthesis of intentions inhibiting one another and connected to one another in the form of conflictual "opposition"—with the corresponding break up and modalization of their intentional sense-correlates. Sheer annulment of an objective sense by another sense or a relative annulment, namely, like a problematic possibility, is "contested" by opposing possibilities, and conversely (just as it is subsequently restituted by a two-fold crossing out)—these are purely apperceptive occurrences, that is, occurrences in the perceptual noema. And [this is] likewise [the case] with respect to other intuitions, like rememberings.

We can't help using the same words that we use in the sphere of spontaneity here: We speak of negation, potentially even of affirmation, of certainty of being, of deciding a problematic disjunction, namely, by a passive restitution of the concordance of an unbroken experience. But what goes on in the egoic sphere, what is carried out on the part of the ego in free or inhibited position-takings as deciding with certainty, judging in certainty, affirming, denying, or as doubting, potentially then as questioning and answering—these are in no way the same kind of occurrences as those in the receptively constituted sphere, and the fact that they bear the same names with corresponding concepts in no way means that they are of the same mettle. The mutual suitability of decisions, their so-called concordance, and likewise uncontestedly holding to a genuine conviction for itself as holding to a position-taking, is something completely different from a concordance in passive apperception; likewise, the annulment of convictions that is carried out by the ego itself as a rejection, as a taking-a-position-against, and likewise, the rivalry among convictions—these are something completely different from the similarly named rivalry belong to apperceptive intentions: Indeed, just as the expression "intention" already means something totally different from an ego-intention taking a position (on the basis of an apperceptive intention) and as a mere apperceptive intention. To be sure, for us the main point, which we must see here with full precision, is that apperceptive intentions [either] pass into each other synthetically, concordantly, and become unified, or they diverge from one another and separate; they become separate for themselves, and this takes place with respect the apperceptive intuitions themselves. But decisions are not lived-experiences that are passively affected. Unlike apperceptions, they do not have concordance and contra-cordance as features that are, so to speak, produced in a materially relevant manner, thereby constituting objects of experience, perduring or besetting and repressing one another; rather, *I* make a decision. If I do it for a problematic possibility, then I "must" decide against the competing possibilities, rejecting them. What does this "must" and this negation mean here? The "must" does not mean that it takes place of its own accord, like everything happening of its own accord in the apperceptive sphere. Rather, for the sake of "consistency," I must carry out a rejection corresponding to the accepting, affirming decision that I carried out vis-à-vis the one side, and this "must" means: "I can do nothing else" in this motivation. [359]

And when decisions conflict with one another, like when the ego finds itself in a condition of doubt, this conflict means that I must relinquish my concession to the one position because of the motivational force of the other, and then conversely. But that is an entirely different motivational force and an entirely different motivational efficacy than the motivational force of apperceptions that stand in competition with one another and that repress and eclipse one another passively; and the latter is quite analogous to the well-known phenomenon of rivalry occurring in the visual field. I am motivated in the decision, I am motivated to make a

decision, and if the passive motivation has some analog, although only an analog, to mechanical force and causality, it in no way exists for egoic motivation. And so too it is likewise the case in the broader, most broad sphere of judgment that we have not yet taken into consideration: like when I suddenly notice that a judgment that I have made by deciding a question concerns some previous convictions [which now] inhibit them and conflict with them. Convictions as convictions, judgments as judgments do not stand together in the unity of an apperceptive framework; rather, the unity here is in the ego as ego, as the unity constituted in the egoic center of motivation; it is the unity of the ego as the unity of the spontaneous ego, affectively motivated from the apperceptive sphere that now decides this way or that, and hitting upon ever new decisions, is now also motivated to abandon judicative decisions once more. It is then further motivated to shape and to build up its system of judgment in such a way that it only contains [360]
judgments that are safeguarded from being abandoned; that is, the ego is protected from having to decide any differently. In this way it, as ego, remains concordant, concordant with "itself," consistent as ego. What I have said is what I have said, what I have decided remains decided. In this way, I am always the same, namely, the identical subject of concordant spontaneity.

One should doubtless note here that every judicative resoluteness (and thus no less every valuing and willing) is in no way merely a momentary act of the ego; rather, every act is either primordially instituting or a merely repeated act. As primordially instituting, it institutes an abiding resoluteness of the ego with the decision. The ego that has decided in this way is from now on a different ego. Something is sedimented in it as its abiding characteristic, and when the ego now repeats the judgment, it "actualizes," it effectively realizes only the decision that was in it from the previous time as its abiding resoluteness. The new explicit judgment is then given not merely as a remembering of the previous decision, but rather as the previous, but enduringly valid one, as the effective realization of the earlier resoluteness that still belongs to the ego.

In a certain respect, we had already addressed this repeatedly in our previous lectures, though now we see more clearly that the habituality belonging to the general nature of subjectivity is precisely of a different mettle in passivity and in activity, or better, in receptivity and in spontaneity. In passivity, it is the transition to retention and through it, to that seemingly dead forgetfulness that can be reawakened passively by what is generally called a reproduction, namely, remembering. This concerns each and every lived-experience.

But in egoic activity, habituality does not mean the same thing. For the ego is not a lived-experience; a lived-experience occurs first with the "*cogito*," and the ego accrues to it only insofar as it is a point of acts radiating outward, an ego-pole of lived-experiences. The polarized lived-experience having the form "*cogito*" has its possible reproductions that reproduce it in the doxic modality and normally in concordant certainty. But if I repeat my decision I made for the first time yesterday—this is not a mere remembering, but it is precisely an actualization of my judicative resoluteness. As a lived-experience, a crossing out can occur in the form of a deception of remembering. But as the resoluteness of the ego, there is an entirely different kind of crossing out. This implies that it can be annulled by the ego making a different decision according to new motives. The memory of it remains, but *I* have become different in the judgment. We could say that the ego as ego progressively develops through its original decisions and is respectively a pole of manifold current decisions that have been made, a pole of a habitual system of rays of actualizable potencies for positive and negative position-takings, and corresponding to them, it bears its entire history, to be sure, by means of remembering, which can be unfurled again.

Accordingly, we see that if we have gained clarity in this way [361]
about the levels pertaining to the life of the ego with respect to receptivity (or even the passivity prior to receptivity), and on the other hand, with respect to spontaneity, and if now we form the concept of judgment as the concept of a fundamental species of spontaneous position-takings, then naturally all descriptions and considerations of essential laws that belong under the rubric of judgment (judgment as position-taking) fall outside of the framework of a transcendental aesthetic. For this transcendental doctrine of sense wants to investigate, constitutively, precisely all

the occurrences of aesthetic, perceptual apperception, and intuitive apperception in general. If it is expanded to the transcendental doctrine of receptivity as such, then standing above it is the transcendental doctrine of spontaneity. Naturally, then, this will not only include judgment. Already with questioning, a wishing and a practical striving intervenes. There are also spontaneities of the heart and of volition, of a spontaneous valuing and of a spontaneous practical comportment of the ego, valuing and volitional resoluteness, each one with different modalities of spontaneity.

Appendix 5: (To §16) <Intuitive Presentations and Empty Presentations>[20]

We are in the habit of contrasting intuitive presentations and empty presentations. Belonging to both are horizons. Empty horizons lack an appearing content; they are potentialities of appearances. But what is the case with determinately oriented empty presentations? They have their differentiated, more or less richly formed objective sense. We can do nothing else than also ascribe to them a certain "content" as opposed to their "horizon." If we compare intuitions and empty presentations in a more precise manner we will say: The same thing that is presented in an empty manner on one side appears in and through intuitions. But intuitions (their "outer horizons" notwithstanding) are not pure, saturated intuitions. <They have> inner horizons, but also contents which, [when] determined with respect to content, are empty: empty contents. We will have to say: empty presentations and intuitions are "presentations" in the broadest sense.

Where its sense-content is concerned, every presentation is differentiated into a genuine presentational content and into a horizonal content. The latter is differentiated into outer horizon and inner horizon. The genuine presentational content has gradations with respect to fullness; it is intuitive if it has fullness, and depending upon the gradation of fullness, is more or less intuitively complete. The presentational content in the intuitive

[20] Editor: August 27, 1926

fullness makes up the appearance. The empty appearance, the empty pictorial presentation[21], designates the case of the nil of intuitability. But one wonders whether these expressions are advisable. Appearance means intuition in the sense of intuitive [362] pictorial presentation[22], that is, presentational content in fullness. The same presentational content can be emptied. Corresponding intuitive and empty presentations present the same thing with the same genuine presentational content; while they are quite different as lived-experiences, they coincide according to this content, according to the entire objective sense, which is actually contained in every presentation and is an analytically explicable sense (explication without further change in the presenting lived-experience as a special way of regarding the self-contained element in sense in an intimately inherent manner). The same analytic (or genuinely presentative) presentational sense has now more or less intuitive fullness, and another time the nil of fullness—this however does not harm the concreteness of the presentation; in some ways, this is analogous to the way in which colorlessness as the emptying of color does not lead to a zero-degree of intensity in which the phenomenon disappears.

Let us note further that our talk of presentation and intuition here naturally shifts such that now it includes the empty horizon, another time it does not. Indeed, the presentation of a thing is not concrete without the empty horizon.

A presentation can be called confused insofar as it has not availed itself to the form of an analytically explicit presentation, and insofar as its presentational content (thus, straightforwardly expressed, [is] always understood as the genuine presentational content) is not identified and fixed in the possible shape of explicative syntheses. In this shape, the identically presented object as the substratum of determinations is distinguished from the determinations, and in such a way that the form of unity arises: O[bject] is α, β, ... – and this is prior to all conceptual predicates. The production of such an analytically explicated shape yields the analytic clarity of the presentation. – One cannot say that an

[21] *Bild*
[22] *Anschauungsbild*

appearance (intuitive pictorial presentation) of a sensible object is of the same mettle as the phantom, which is the object itself becoming intuitive.

By tracing out the occurrences that belong to the specific appearances and to the empty presentation, there is here in the "transcendental aesthetic" a relatively cohesive set of investigations: initially, how continual perceiving, continual intuiting of other intuitive modalities, continual presenting, reach a synthetic unity, and by means of this how specific appearances and presentational contents in general become unified, constituting unitary appearances, unitary presentational contents. Belonging to every perceptual phase in the continuity of a perception is, for example, a momentary appearance. This can be completely uniform in every new phase; but this is only a limit case. In this case, the perception of the same perceptual object is such that the latter appears in a continually enduring and completely unaltered unity of appearance. The normal case however is that every momentary phase offers another appearance, whereby these appearances coalesce to form the unity of an appearance and not, for instance, merely through summation, but rather in a [363] synthetically constitutive manner. Etc. Belonging here then is the doctrine of near and far pictorial presentations, etc. In considerations like these, we operate with a thematic abstraction insofar as kinaesthesis is intrinsically involved everywhere, though initially it is left unthematic.

Appendix 6: (To §16) <Sense and Intuition>[23]

In the case of adequation, like everywhere, we have to distinguish between perception and other modes of intuition. In perception, the object is given in the original, and on this assumption, it is adequately given; in the other modes of intuition it is not given originally, but it is (adequately) presentified, adequately given in a pictorial presentification, etc. = it is adequately intuitive.

[23] Editor: Most likely between 1918 and 1921

The sense in which the object as such is given to consciousness is completely determined in adequate perception; it does not leave anything open for the object, it is the sense of the object itself and completely so.

But the sense is not an empty sense, albeit fully determinate; rather, it is a "full" sense through and through, a sense replete, as it were, filled with the fullness of intuition. We cannot juxtapose, e.g., two components—sense and fullness—in the intuitive object as such. Only by comparing empty and full sense can we gain the difference between them, that is, through the synthesis of intuition and empty consciousness. Perhaps we could say: The abstractly identical element, which we call sense with respect to the different acts of consciousness, is an essence[24] (sense-essence) that is differentiated in a peculiar way, and according to two basic modes: in the mode of intuitability (and in the perceptual sphere of originary intuitability) and in the mode of non-intuitability, in the mode of emptiness. There are still other modes in addition to these, modes about which we do not say, for instance, that they are specific differentiations of the same type like those, e.g., of color, etc. At all events, we must distinguish the intuitive (here originally intuitive) content of the object in the mode of adequate perception (and even in adequate perception) from the sense that lies in it, insofar as something identical belonging to the essence remains with the non-intuitive acts, whose object can be the same and, ideally speaking, can be fully determinate. The fulfilled sense, which in the case of perception is obviously the object itself (and when we allow the thesis of existence to become codeterminative), harbors what makes the object specific, that is, the essence that is understood individually, and on the other hand, the individualizing spatio-temporal situation (whereby the "spatial" should only be a term for what is potentially individualizing with respect to coexistence beyond the successive temporal situation). Both come to originary givenness in adequate perception. (Naturally, we do [364] not say that every object is adequately perceivable; we know that transcendent objects, according to their nature, exclude such a perception.)

---

[24] *Wesen*

In the case of inadequate perception there is an ineluctable expanse between sense and fulfilled sense. The object is the idea of the fulfilled sense (as fulfilled completely). This also holds, of course, for an expanse between completely determinate and indeterminate sense; and the fully determinate sense is an idea that is contained in the idea of the object itself or of the originally given, fully fulfilled sense.

We have to emphasize important distinctions, however, in the case of non-original intuitions. In these intuitions, sense and object, everything that perception offers noematically and that perception harbors as a telos as something ideal, is given to consciousness in a modified way, and this modification brings with it new noematic components.

Appendix 7: (To §20) <Belief and Intention>[25]

When we speak of belief, of a consciousness of being, we indeed have in mind within the lived-experience in question, a directedness, an intending directedness toward the object. *Doxa* is intending. Intending can mean: *I* intend, *I* am directed toward something in a believing manner. But already within passivity we find an intention, as belonging to the structure of intentional lived-experiences as such without the ego coming into play as a subject who is [actively] directed, as a cognitively striving subject potentially making a decision. It is what in truth transforms a concrete lived-experience, perhaps like a remembering or an expectation already stirring in the background, into an intentional lived-experience. If, from the ego, this same lived-experience later becomes patent, the intention takes on the act-mode of the "*ego cogito*." Intending, understood in its current sense, signifies something wholly determinate, something we obtain in a completely clear manner when we think of our analyses of experience; it is that intending that is ultimately an intending constituting a unitary objective sense, and that makes up the formal fundamental structure of a consciousness as a consciousness directed toward an object.

[25] Editor: 1923

In the broader sense, belief is nevertheless the form of the concordant consciousness of the object in general; but consciousness is directed toward an object only by the fact that an intention as a whole, unified in and through harmonizing intentional rays, is actualized, while the other consciousness of the object remains a milieu that can take on the form of a specifically directed intentionality at any time under motivating circumstances. This determines belief as a living belief, and all modalization takes place in relation to it. What is occasionally inhibited is the *intentio*, what clashes is an intention with an intention, belief with [365] belief, whereby belief, in the mode of rivaling and being contested, is however already modalized. Likewise, it is the belief-intention occurring in the process of synthesis that is ratified or annulled, shattered with respect to the unbroken, purduring counter-intention as a whole. It is a special case, then, when the synthesis is a synthesis of confirmation, when an empty belief-intention that terminates in a self-giving is ratified in the form of confirmation, and in being self-given takes on here the character of a legitimate validity of something attained as the final attainment.

Thus, we have advanced to the primordial sources of objectivating consciousness and thus to a genuine theory of judgment and of cognition.

To be sure, only now would this undertaking have to begin in all the higher levels; and, on the other hand, we would have to make the great distinction, a distinction that runs through all consciousness and that Hume had brought to a head under the ambiguous rubric of impression and idea. It is the distinction between pure phantasy and positionality. All our analyses operated within the positional sphere. In the pure life of phantasy there is no belief, but only quasi-belief, belief-imagination, just as there is no volition, no valuing there, but only a phantasizing into something of the kind.

I have presented what I wanted to offer in my lectures: a basic feature of the work elucidating and leading to an ultimate understanding; in this work alone is [disclosed] the sense and the accomplishment of that life of consciousness that is completely hidden from us *because* it is our living life. Ultimate self-knowledge that we gain here, however, is not a specialized subject matter, but the source of all true and ultimate understanding of the world. After we make intelligible the most general How of constitution of objectivity in consciousness, we must, ascending from here, clarify the constitution of all special regions of objects: the distinctive constitution of "nature," lived-corporeality, sociality, culture.

Appendix 8: (To §§ 24 and 25) <The Apodicticity of Remembering>[26]

<1. The Consequences of Assuming that Remembering is Dubious>

Indeed, a specter of skepticism appears and grows ever more threatening, namely, the specter of the dubiousness of memory. I have spoken of my stream of consciousness without hesitation, and I have hastily used not only memory as a phenomenon in the present, but also as the gateway into my transcendental past lived- [366]
experience. But if memory is no longer a source of apodictic certainty for my past *cogitationes*, then I may no longer speak of my unending stream of life, I may no longer speak of my past ego and my past intentional lived-experiences; in this regard, I must also allow the phenomenological reduction to reign. I have only the momentary present "*ego cogito*," and only while I am reflectively directed toward it; and while it is elapsing, if I make a statement about the "*ego cogito*" that would adequately suit what is experienced phenomenologically, I can never repeat the statement. Thus, I have neither the particular "*ego cogito*" in question in the actual sense as a fact, nor do I have the corresponding proposition, "*ego cogito*" as a truth in the normal sense of an iterable truth that can be verified again and again. If the "*ego cogito*" has past away (if I can even assert this much apodictically), I can recall it, to be sure; but even if I am also absolutely sure of the current memory as I am of the present lived-experience, I am not absolutely sure of the remembered lived-

[26] 1922/23

experience. Thus, I cannot be absolutely sure whether that lived-experience that I have in mind as a past lived-experience was actual. Accordingly to my guiding principle, if I cannot be absolutely certain, then I may not make any claim about it—and just as little may I make a claim about the statement concerning it, the statement that was formed when it was still present. If I repeat the statement, I have a new statement that I could only verify by recourse to the remembering that is unfortunately of no use.

For this reason I may not speak of my unending stream of life, of my life spanning an endless past and extending into an endless future; I may no longer speak of phenomenological time as an actual form of actual life, etc. Thus, I am arrested, so it seems, at the absolutely sterile "I am": I perceive—now while I perceive, I think, namely, while I think now, I feel, and only while I am feeling, etc. During all of this, I can by reflecting make observations and can make completely useless assertions, none of which have even the slightest tinge of enduring truth; they only have the barren, fleeting relevance bearing on the fleeting life of the present. Yes, actually barren, for fecundity is precisely something of abiding value and not something merely existing in a moment of growth.

<2. Two Types of Transcendental Reduction with Respect to Remembering>

I am given to myself apodictically as a transcendental ego and *ego cogito*, so it seems, only in the transcendental self-perception as the ego that is perceiving now or remembering now or feeling now, willing, and perhaps I must even look for limits here. But I am also given to myself through reproductive acts, e.g., through [367]
rememberings as the past ego and with my past perceiving, remembering, feeling, hoping, etc., and not only as the past empirical ego, as the past human being in the past time of the world.

It is necessary for us to clarify the following issues, and they will easily become clear to us after we have gained a little practice in the phenomenological way of seeing. There are two phenomenological, or better, transcendental reductions for a remembering, and the same holds for all presentifications. Let us assume that we have a remembering of a song. Then we have (1) a phenomenological reduction of the actual remembering that is evident now as lived-experience, the remembering that has the past song of this or that singer as an intentional object. The point of departure in this case is natural reflection: "I now have this remembering." (2) But curiously, there is not only a reflection on the present remembering, but also a reflection within the remembering. For belonging to the essence of remembering, as we ourselves have observed earlier, is the fact that it not only brings a past to intuitive givenness in general, but that it presentifies it as something perceived earlier by me. The song is not only the temporally previous song, but (in accordance with the very sense of remembering) the song heard by me; and that this is the case, this I myself find in a reflection, a reflection that I carry out by delving into remembering, into its intentional content.

Now, if I take as my point of departure this reflection that remains completely unphenomenological, this "I have heard the song," I can now, as phenomenologist, carry out the reduction on it, bracketing the spatio-temporal mundane actuality of song and singer. Then I will gain the transcendental phenomenon, namely, the past transcendental phenomenon of my—of the ego's—earlier acoustical perception of the song, whereby the real song itself is only the intentional object of the hearing.

In this way I can gain the entire realm of my memories as phenomenologically reduced, thus the realm of all memories of objectively mundane things and processes, and then of all memories in general, e.g., memories of mathematical proofs that I have carried out; and I gain them not only as present facts, but according to their remembered intentional content. And what finally results from this is my (i.e., the Ego's) past transcendental life with all of its past poles which, insofar as they are object poles, are bracketed, while the ego that is everywhere identical is the transcendental ego, and should not be bracketed, just as little as the past transcendental lived-experiences.

One can also put it in this way: In the case of rememberings and with all other presentifications that we will take up later, we deviate from our earlier principle of putting out of play all positings that are carried out in the lived-experience itself. I only put out of play the positing, the memorial belief in the past [368] objectivity, but not the belief implied in it, namely, the belief in my past ego and my past lived-experiencing, and my past perceiving in which my past life was given perceptually. We emphasize this because the transcendental subjective element that is past is also precisely transcendentally subjective, and because we initially want to appropriate transcendental subjectivity as a whole in one stroke as far as the unity of the ego and its lived-experiences can reach (regardless of whether they be present or past). In doing this we follow the evidence that itself lies, in part, in the phenomenological reflection on the present, in part, in the phenomenological reflection on the past (that is, in the reflection penetrating into the intentional content of memories). But we do not ask whether or not this evidence is apodictic, whether or not it is better than the evidence, e.g., of external perception (that we had to put out of play to begin philosophically).

We can treat memories of the future, expectations, in the same way as memories of the past. The expectation of a future course of a natural occurrence that I await while perceiving becomes, through the phenomenological reduction, an expecting of the futural transcendental egoic life. All in all, the transcendental or phenomenological stream of lived-experience and the concrete transcendental subjectivity that I gain in this way, accordingly, has its mobile phase of the present and, in addition, the endless stream of the past and future. Only by letting the presentifications be valid does the transcendental ego have an endless life with an endless immanent temporal form on both sides.

<3. Evidences within the Flux of Perception and the Evidence of Expression Belonging to It>

(1) An immanent perception is apodictically evident with respect to the object's enduring individual present, thus, in our example, the phenomenologically reduced tone toward which we are directed as if going along with it, the tone as the current tone and as the tone continuing to endure. This "going along with" is at the same time a "swimming toward"; the grasping bears on

something lighting up now and constantly toward something that is newly being lit up, toward the new Now, and it receives it with open arms: a constant grasping that grasps something constantly new and in the constant grasping, grasps something enduring as such. This is the enduring tone *qua* the tone that is enduring and that continues to endure. The apprehending intention is an intention that is being constantly fulfilled to the point of saturation, and in the fulfillment it is an intention that is continually repeated, namely, as continually directed toward something new, once again being fulfilled, and so on perpetually. It is precisely in this process that the being of the tone is given adequately as enduringly present, as original being in the enduring. We must say here that the incapability of being crossed out is the result of adequation in the sense of a self-giving that is [369]
fulfilled, actually complete. It is a constantly fulfilled intention, as we just said.

(2) We also have a different kind of evidence of the enduring tone, and it becomes apparent that the duration and the evidence of duration have two senses. For this reason we distinguish between

(a) something continually enduring, the enduring tone itself,

(b) the tonal expanse, the past one and the one culminating in the "enduring" present. And in this expanse, every phase is given retentionally in a different, and thereby constantly varying mode of "just past."

We also have here the incapability of being crossed out, but no longer adequation in the genuine sense. Here we do not have full, namely, fulfilled self-givenness, but precisely only still-having-a-hold-of, still-having-in-consciousness in the mode of certainty, a certainty, however, that is still incapable of being crossed out to a certain extent.

Just like the tone itself that is given to consciousness as enduring has the mode of certainty, so too do we see here, generally, the necessity of the tone that is given to consciousness as just past having the mode of certainty. The mode of certainty extends unchangeably throughout the entire continuum of intentionality. But it will also be good to consider the content, what is certain there, and the evidence of the description that is suited to the phenomenon here. The perception of the tone *simpliciter* is the prehension of the continually enduring tone, and this is given in its entire fullness precisely as constantly fulfilling. If we call it the "violin tone," or still more generally, the "tone," then the meaning of the term fits the corresponding moment that is found adequately in the given tone itself and that coincides with itself continually throughout its endurance. The extent to which we have such a constant, stable moment in the en- "durance," a moment that finds a steady coinciding in and through this; the extent to which we have a term that fits <it> exactly with its meaning, to this extent we have evidence of the statement, to be sure, only in the endurance itself. But at least the universal "tone" as such is necessary here; the unity of the perceptual givenness of something enduring is inconceivable as the unity of a continual synthesis were the entire unity not supported by the existence of a thoroughgoing coinciding, that is, by a universality of essence that all phases must have in common. In this way, we can speak of "tone," or more specifically, of "violin tone," of a sharp tone, a loud tone, etc., with absolute adequation. This original self-coinciding in endurance occurs in the intuitive realm of original presence. This realm of the living intuitable present is not a mathematical point, but already has a thoroughgoing, intuitively fulfilled extension whose apex is the absolute Now. In this realm we also grasp continuity and alteration, modification of intensity, leaps in quality, etc., in their most original forms. Even an alteration can continually persist in the endurance of the tone, and can generally be stated with fulfilled evidence.

Therefore, if we glance at the empty parts of the concrete present, then all such descriptions of them can have an evident [370]
content despite the emptiness, namely, by a kind of transference occurring in each instance where, in the current present of the genuine perception (in which the fulfilled intuition takes place with respect to all that is stated) something is given that can coincide with something else of the same sort in the retentional sphere. For example, the empty retention of the tone is a continuum of coinciding that terminates in the intuitive givenness of the tone; and in this way I can speak precisely of a tone that has just been; the general term is fitting for the entire continuum, even if I only use it now for the first time. What has just been given in

an intuitive manner is the tone; the past, as coinciding with the intuitable element according to its empty intention, is also a tone. Thus, the evidence of the description would be derived in a certain respect from the intuitive sphere of perception.

Remembering need play no role in those cases where contents, which are drawn from the enduring perception itself, give a "transferred" evidence so to speak and give an evident interpretation to retention by coinciding with the empty retention. The evidence of the description, then, rests on and presupposes the fact that the retentional objectlike formation in question is grasped for itself in clarity, even if in an empty manner; and it is interpreted by a comparative coinciding in the sense originally drawn from the original intuition. Where such an original "fitting" of the expression to something given originally in the present follows the thing given, holding on to it (for example, when it concerns a momentary scratching tone and while this scratching sinks into the past), there is no doubt that the expression sinks along with it and necessarily maintains its belief of expression. Repeating the expression would already be a matter of remembering.

All of these modes of evidence that we have described up to now only have a momentary certainty, a certainty that, as attached to the flux of perception and retention, is incapable of being crossed out; but with them we do not have the same apodictic certainty of the unending past and future of life; we do not have such an apodictic certainty of the identical ego that is subject of this unending life and is subject of certainties, certainties that the subject can verify again and again, even after the original living certainty (arising from the original perception) has passed away with it. The "again and again" is only possible because of remembering, and only from it does there stem the possibility of facts that are in themselves and that can be originally experienced in perception, but also that can be experienced again as often as we please, re-identified as the same, and accordingly can be re-described in an identical manner and re-described with identical truth as often as we please. Thus, this is to say that there is an abiding truth in contrast to the momentary truth. But the question will concern how this can be clarified and the manner of its relationship with apodicticity and adequation.

Retention is a certainty that is incapable of being crossed out, a certainty of what has just past; but the apprehending ego that [371] strives to seize retention's object as just this object is itself, the ego that strives to know it by delving into it, reaches into emptiness. The intention directed toward it has its form of fulfillment in remembering. It gives the past itself as being fulfilled.

The fact that remembering can be deceptive is the unanimous teaching of the philosophers, and who in fact would want to deny the possibility of deception here? This also holds for transcendental remembering, that remembering belonging to the transcendentally reduced sphere. Every transcendental reduction of a naive-natural remembering that turns out to be a deception yields, as one can easily see, a transcendentally reduced remembering that is shown up as deceptive.

Yet, I must also depart from the tradition here; I must deny the unqualified rejection of all apodictic evidence in the sphere of remembering, and must clarify this rejection by showing what is lacking in the [tradition's] analysis.

<4. Remembering as Reproduction and its Relation to Retention>

The fundamental character of remembering is "reproduction"; this implies two things, and its sense is two-fold. Reproduction can signify presentification. This is a general characteristic that is peculiar to remembering like to other modes of presentification; each phantasy, emerging haphazardly or freely generated, is a presentification, but for this reason it is not a remembering. An intuitive presentification is essentially given as a modification of perception. To present something in phantasy, but also to present something in a remembering, is "to perceive after a fashion," but precisely only "after a fashion." The tone that is perceived after a fashion begins and endures, and the entire constitutive shapes belonging to perception, the entire play of retentions and intentions of expectation that are directed ahead, together with the transitional sphere of the primordial impression—all of this also belongs to the stock of presentification. But all of this occurs in the mode of "after a fashion." Corresponding to the perceptual certainty as certainty in the present individual being of the tone is the memorial certainty. But the memorial Now that thus has the mode of the "after a fashion" is not believed, is not certain as Now *simpliciter*; rather, this memorial Now, like the entire content of something remembered, has the characteristic of the Now that is re-presentified, the Now that is presentified in the form of renewal, of re-perception, of perception being played out once again "after a fashion." The most original consciousness of the past is the retentional consciousness belonging to every perception like a comet's tail. Should the remembering that has such an essentially different character also be called the consciousness of the past, [372] then it must have an essential relation to retention, namely, it must be in relation to it in a synthesis of coinciding that forms identity, or it must essentially be able to take on such a synthesis.

Where is such a synthesis produced? Now initially, while a retention is still running-off, a fresh past can become prominent, even if emptily given to consciousness; a corresponding remembering can emerge or can possibly be generated deliberately as a remembering of the same thing. A tonal phrase has sunken back, the same thing is heard "after a fashion" once again, it takes place once again after a fashion from beginning to end in the mode of reproduction. "The same tonal phrase"—that is given to consciousness here, that is, the empty retention, which of course continues to do its own thing, namely, letting the past appear as further and further past, but in itself as the same, this empty retention, I say, is synthetically one with the tonal phrase which is, as it were, sounding anew; and in the coinciding, the emptiness of the retention is fulfilled with the fullness that is reestablished in the renewal. The intuitive element is given as the fulfilling or true self of what in the retention is presented in an empty manner. In the fullness of intuition, in complete remembering, the entire fullness of the latter's inner moments and articulations comes to the fore, which moments and articulations had become indeterminate and blurred in the retention.

<5. Levels of Clarity of Remembering>

Certainly, we must still take into account here a peculiar feature of remembering shared by all modes of making present. Remembering can be a presentification with many different levels of clarity; while it is running-off it can vacillate with respect to clarity. Once it has elapsed, it will become empty analogous to the way in which a perception becomes empty after it has elapsed; but then it is not an empty retention *simpliciter*, but an empty remembering that in its very emptiness has the peculiar feature of being the remembering of an empty retention. But at the same time it is the actual retention of the lived-experience of the intuitive remembering that has just elapsed.

Yet we also come to know the gradation of clarity with respect to the peculiar feature of rememberings, namely, that rememberings are "iterable" as repetitions of the same past, and in our case, as the repetitions of the same retentional past, as well. This is disclosed, so to speak, through the first remembering, but also through this, it remains held onto, and it is by means of a new remembering that it remains held on to now more than ever, and is disclosed once again. For after the course of the first remembering an empty consciousness was indeed there again. From this we see that from the perspective of the object the different rememberings do indeed coincide in the transition from one to the other, yet they must not be entirely uniform; we see that the one discloses more of the object, the other, less; the one greater richness with respect [373] to prominent and intuitive traits, the other, less. Thus, in accordance with its essence, there is a gradation of inner fullness and emptiness in remembering *qua* presentification, a gradation that has an upper *limes* that we call complete memory; the latter reproduces the perceptual objectlike formation and implicitly the perceiving itself in a complete manner, and it provides the most complete explication of something that is emptily retended, and reawakens everything that has become unclear and blurred in retention.

Concerning the question how we know whether this is not just a constructed fairy-tale, we can respond by referring to a series of repetition (possible series of repetition) of rememberings of the same thing in which we can deliberately arrive at higher levels of clarity. In the transition we see that the same thing reaches intuitive givenness more and more completely—the same thing that was intended before, though still partially given to consciousness in an empty manner—and we see that we even gain the evidence of a *limes* lying in the direction of this progression, a *limes* of saturated fulfillment, of complete fulfillment. <We gain> the knowledge that there certainly has to be a *limes* since every possible remembering, if it is fulfilled at all, is fulfilled univocally, precisely as the identity of the object.

<6. Deception and Apodicticity in Remembering>

But we also recognize the possibilities of deception within remembering, initially the possibilities of "retouching." The remembered object is the object intended originally in an empty manner; it is to find its fulfillment in the content of the renewed intuition, in the perception that is modified "after a fashion." The image that is being generated and formed intuitively coincides with the image that is intended in an empty manner. But a clear image can arise which, while being on the whole an actually fulfilled presentation of the thing intended, nevertheless has other features painted in that do not belong there, i.e., those that are not the fulfillment of the corresponding features of the intended meaning. One often notices this interiorly. The empty intention is enriched with the process of bringing to intuition; when the new features of the empty intention are awakened, it is possible to become conscious of the fact that intuitive features have intruded, features that conflict with such newly awakened components of the intention; they are not its fulfillments, but false retouchings. Indeed, it can turn out that a fusion has come about, a fusion in the unity of a memorial image of something that has stemmed from different pasts, and that was not initially noticed <as such>, since the empty retention is quite indifferent, and the anticipatory process of bringing to intuition through "association" had strayed into a different sphere of the past. When one enters into a profound consideration in this way, it is quite understandable that remembering can be deceptive.

But, to stay at first in the present sphere of the disclosure of [374]
retentions, it is equally doubtless that remembering, too, has apodictic contents. It is absolutely evident that I have just heard a tonal formation, that I have just seen a countryside, and that I do not erroneously posit the tonal formation that I have <in> remembering instead of a perception of the countryside that has elapsed, etc.; and it is absolutely evident here that I have a past object, an individual object, a temporal object, having a certain general character of countryside, and the like. Of course, insofar as we conceived remembering as being in a fulfilling coincidence with a retention, belonging to remembering is the inviolable certainty of the latter; and precisely with this certainty, we have apodictic certainty that there is a true self in the memorial image, a true self that can be approached in the *limes* and <that> can possibly be reached in the consciousness of fulfillment. But it should also be noted here that the "having just been" of retention is disclosed as a represented Now, and in every repeated remembering this Now is given as identically the same represented object by virtue of the synthesis that encompasses the repetitions; accordingly, this Now is at the same time that which is "just" past. Remembering essentially characterizes the memorial object as the perceived object after a fashion and as the re-perceived element, that is, as the enduring present after a fashion. Since it is fulfillment, the "just past" of retention is disclosed in it.

By the object that is given in the original Now in perception passing over to the "just past" of retention, and then again appearing renewed in the "now after a fashion" of remembering and possibly in new rememberings to be repeated as one will, we have, with evidence, the given object as the same object in manifold modes by virtue of synthetic identification, namely, as the same individual object, as the same temporal object with the same temporal locus and temporal duration. The perceptual or original present, the memorial present or the "re"-present are modes of givenness, modes of appearance of the same individual whose original being as an enduring that is constantly being formed (unity of a duration that is constantly being generated) is reproducible and re-cognizable (that is, "re"-experiencable) precisely as absolutely the same again and again.

<7. The Modes of the Past of Something Repeatedly Remembered>

Every new remembering as lived-experience is itself a new present in repetition; and although each one reproduces the same thing, and each one has a uniform content with respect to uniform levels of clarity, there is still an essential difference of irrevocable necessity. The element that is repeated in the repetition in a completely uniform manner—the same enduring tone—necessarily has a new mode of the past in each remembering.

By bringing the same non-Now to consciousness in a new Now, [375] in which a new present is developed in the original "enduring," every remembering gives a new mode to the memorial object that stands in relation to this present itself. But it does this insofar as each remembering harbors an intentionality as yet undeveloped, an intentionality that is modified for every new remembering. Naturally, one only sees what kind of an intentionality it is and what the changing past ultimately means as the development that fulfills.

More precisely: In every remembering as such there is an intentional tendency pointing beyond its own remembered content. Its fulfillment leads steadily into a continuum of progressive rememberings so that a continuum of presentified presents, a continually fulfilled time is represented again. This continually unfolding remembering finally terminates in the continually enduring perceptual present. For example, I just remembered a conversation I had in my office. I let it play out; remembering it, I follow the tendency in the direction of the rememberings that connect up and that are continually fulfilling. Then I recall the stroke of the clock, the awareness "it is time to go to my lecture," then walking over there, and finally, I am here and now, in this current perceptual present in which I am now actually holding my lecture.

<8. Remembering and its Horizon of Expectation>

We must expound upon the following in broad strokes: Every original self-giving, every perception harbors, as we know, a constant protentional tendency. It has a horizon of expectation that points into the future. Every current primordial present occurs as the fulfillment of protention that continually precedes it; analogously every remembering harbors a protentional tendency (which remembering is characterized in itself as a modal modification of perception, as the perception in the mode of "again" and of "after a fashion"); and in the same way every phase of the present occurring in the mode of "again" and "after a fashion," that is, the past Now, occurs as fulfillment. But the situation is not so simple here, already because remembering is at the same time a present lived-experience, that is, itself occurs as a perceptual present, and at the same time by virtue of its intentionality; it is a presentification of a past. In the final analysis, it reproduces the continual fulfillment of the intention of expectation.

But more than that. While in perception what is arriving is new, and only in the arriving is it determined with respect to content, (and is possibly determined in contrast to an all too determined expectation); while here something completely different than what had been presupposed can arrive (insofar as the unity of the object temporally breaks off, and now some entirely new objectlike formation is perceived), in remembering, on the other hand, what [376] arrives there in certainty is not at all new, but is rather already familiar. Indeed, it has already been there, and is only remembered. This implies that the remembered object is in itself the expected object as determined with respect to content, and in the unity of a certain and clear memorial sequence, it has the character of being thoroughly in accordance with expectation and of necessarily-having-to-arrive-in-this-way with respect to content. The memory is confirmed in the necessity of the sequence according to the determinate content, since when it is complete, it of its essence offers nothing new, but only what is long familiar. On the other hand, where remembering as the phenomenon of the present is concerned, a tendency that is directed ahead, a tendency of the associative expectation toward the rememberings in the sequence of the course, also belongs to remembering. Both nexuses, that of the necessary sequence of pasts and that of associations, terminate in the final perception—the former in the object's present, the latter in the present of the perceptual lived-experience constituting this object's present.

<9.> Remembering Distant Pasts

Up to now we have considered <the> necessary features of remembering being established in the sphere of fresh retention. If we now pass over to the peculiar features of rememberings as such and now consider rememberings of distant pasts, we will then obviously have to say that everything that was essential to the special function of this retention remains unchanged. Only now we are referred to the undifferentiated empty horizon. A closer inspection would show here that the undifferentiated, empty retentional horizon undergoes its first form of awakening by some prominent features unfolding an associative tendency within the life of the subject living in the present. In their fulfillment, these prominent features evoke a prominence of features that have already become indifferent in the empty horizon. If there is already a remembering, and if it is proceeding intuitively, the present that is re-presentified in it can for its part function in an awakening manner, precisely for new moments of the empty horizon, in other words, <it can> summon new moments that were forgotten. That would be a prominence in the mode of "again," but not for instance a memorial return of the previous retentions in their original retentional flux—such a retention is something abstract that can only be in the concrete flux—; rather, it is the concrete albeit empty intention that is awakened and that now bears its own affective force, for its part it now takes on fulfillment through a process of remembering that develops the fulfillment. In this way, the empty content is brought to the fullness of self-givenness. The remembering itself, then, brings with it its intentions that point ahead, thus, demanding new fulfillment, and in this way the series of remembering is reproduced up to the current present. This emergent, older [377]
intention gives itself as emerging from the empty horizon as emerging from the night of forgetfulness (and therefore precisely as something undifferentiated becoming differentiated, namely,

something undifferentiated into which differentiated retentions passed over).

Every repetition of original remembering, and the running-off of it, yields identically the same objectlike formations, i.e., the same events with identically the same particular temporal points and temporal durations, and all in all, entirely the same expanse of the past. But yet not entirely the same. For the present that is in the process of becoming has progressed, and although the repetition of the process of remembering of the same point of departure leads once more to the current present, this present is precisely a new one, and the previous present has become a past, that is, it now forms the final element of time elapsing in remembering. Accordingly, it is clear that every remembering, even where it does not get developed with respect to its protentions, does indeed intrinsically imply that it bears intentionally the temporal series up to the present, albeit in an undeveloped manner; remembering itself belongs to this as lived-experience, and it does this in the mode of elapse, as a "development that fulfills." Several rememberings of the same object necessarily bring this same thing to consciousness in different ways, characterizing it as something past with different distances of the past, as the past that is always in relation to the living present [understood] as the final goal of the mobile temporal expanse; the mobile end presses forward, and accordingly, the same past becomes a more distant past with each new remembering. Here, the entire content of the previous remembering and of the series of remembering is necessarily maintained in the content of each subsequent one; and insofar as they coincide, the temporal series is identically the same; it is the same series of individual, enduring objects, i.e., events.

<10. The Immortality of the Transcendental Ego—The Impossibility of the Transcendental Ego Being Born>

Let us take an important new step. Let us consider, in addition, the necessity of the endurance of the present, which will already give a portion of a critique of expectation. The present is necessarily the fulfilled present. Even if the presently "enduring" unitary object or event can cease, the process of the "enduring" itself cannot come to a halt. The enduring is "immortal." When the tone ceases, precisely something else is there in its stead as the enduring present. It could be the case that the world does not exist—this is a possibility as we have shown. In contrast, it is absurd [to say] that immanent being (the present being that is being constituted in the enduring) would cease: It is inconceivable that everything would come to a halt and that then there would be nothing. As soon as one conceives of the "then-not-being," one presupposes a "then-being," which conflicts with the non-being. [378] One imputes the possible cessation of every conceivable particular being to a putative cessation of the stream of life. The cessation itself as the cessation of the object presupposes a non-cessation, namely, consciousness to which the cessation is given.

Thus, the determinate expectation may turn out to be deceptive, the structure of the progressing time-consciousness and the structure of the constitution of new presents is certainly a fixed necessity. This implies that the process of living on, and the ego that lives on, are immortal—*notabene*, the pure transcendental ego, and not the empirical world-ego that can very well die. We do not at all deny the latter's death, its corporeal decomposition, and thus the fact that it cannot be found in the objective, spatio-temporal world, its non-existence. To be sure, an unending futural time is not yet posited with the immortality of the ego as immortality is now given, namely, as the incapability of crossing out the present that is being ever newly fulfilled. This is something that still has to be established. But we have not at all established unending time in the direction of the past, and we are still in the process of doing this.

But if we now consider the present by looking back rather than looking forward, we see that every present arises as the fulfillment of a past with absolute necessity. Namely, every present, every "enduring" being does not only have in itself, on the one hand, a protentional form that is incapable of being crossed out: "A new Now must arrive"; it also has, on the other hand, a retentional form that is incapable of being crossed out. And not only that every Now leaves a trail of retentions; we cannot conceive of a Now that does not already have retentions. The new tone that arises, which is beginning anew, surely does not yet have a milieu of its own retentions, but there is necessarily a preceding just-having-been, a preceding perception. Just as the cessation is conceivable only insofar as it is in process, though the cessation of the process itself is inconceivable, so too is the beginning only conceivable in process, though not conceivable as the beginning of the process. The nothing prior to the beginning already presupposes a something with which it could conflict. There can be an emptiness prior to the beginning, an undifferentiated, monotone, mute stupefaction, but even this is something past, and has the essential structure of something temporal.

Corresponding to this is the fact that every remembering necessarily has an intentional horizon belonging to the beginning, to the commencement of the remembered "enduring," a horizon, so it seems, that can be reawakened, and in this way we arrive *in infinitum* at new possible rememberings. Yet, that would be too precipitous. We have not yet come to know the essential conditions of possible reawakening, that is, of remembering. Upon closer inspection (which is beyond our scope), we see that remembering is a modification of perception as an act, that is, presupposes a wakeful ego. Even the awakening of backgrounds through association presupposes prominence, which implies the affection on the ego. Accordingly, the ego is also awakened. Association is not the least bit possible where there is no prominence, where the ego is completely asleep. But that has not [379] been stated correctly, and time-constitution cannot be grounded without further ado in the possibility of reawakening remembering *in infinitum*. Is this any different in the case of intersubjectivity?

After what has been said about the possibility of iterative remembering of the same object, we arrive at an identical unending time in the necessary mode of the unending past, at a necessary variable mode, since all past times must be given in incessantly varying, and of necessity, constantly varying modalities of the past. Time is only possible as the original present or as the past and as the oncoming future. But the original present is the enduring present, that is, a constant change of the present pressing toward the future. And accordingly, every past is an enduring past that varies with the present to which it belongs. But with the change of these modes, there is the one unending time to the extent that it is already past, and every position, every expanse of this time, is absolutely fixed and identical, namely, identifiable again and again with complete certainty as the same. Consequently, transcendental life and the transcendental ego cannot be born; only the human being in the world can be born. The ego as transcendental ego was eternal; I am now, and belonging to this Now is a horizon of the past that can be unraveled into infinity. And this means precisely, the ego was eternal.

Finally, we can easily see that the future signifies unending time. Remembering teaches us that what is expected in every past present occurs again and again and necessarily as the new present and has become the past; and we must now see the necessity that the protentional horizon attaching itself to every present has possibilities of fulfillment, but only in the form of an expected present, and consequently, of an expected past. What is futural, what will be, is something identical that is initially identifiable in repeated rememberings of the future, rememberings that have the character of anticipation of perceptions, i.e., of presents, and which can only find their fulfillment through the occurrence of these very perceptions and through the identifying [process of] remembering after the perceptions have occurred. Thus, what will be must become present and past, must become identifiable time.

Following from this is the necessity of an unending immanent time, which is to say (I would not know how one can escape this absolute evidence), the infinity of past transcendental life. But this in no way means that transcendental life is always a background of different acts and lived-experiences that can be disclosed, which is to say, the transcendental ego [has] always [passed] a wakeful life, a life in which all sorts of different things took place. Rather, a [380] mute and empty life, so to speak, a dreamless, empty sleep, is conceivable as a life that also had this necessary structure and that appeared in perception in a passive and interior manner, but without any prominence, and therefore without any apprehension [of it] by the ego, without any play of single affections and acts such that the ego did not come on the scene, so to speak, and the slumbering ego was mere potentiality for the *ego cogito*. There is always a possibility that prominences will occur through the modification of life, and this entails the possibility of awakening.

Let us forego <a> deeper discussion of empirical experience within the sphere of remembering (perhaps it could be shown that every remembering reproduces its remembered content with some apodicticity such that even a negated and false remembering has a necessary content of truth); and let us turn to the sphere of expectation. Here we content ourselves with the determination, which can be easily clarified, that every Now has its futural horizon, or as we can also put it, necessarily passes over into a new Now. The cessation of the tone means a rupture in the intentional unity being constituted, but a new fulfilled Now is necessarily there and is changed again; or rather, it is an abiding form whose intentional sense, which is primordially instituted, immediately passes over into retention, while a new primordial institution ensues in the Now-form. Expectation is never apodictic—and yet with respect to its form, it is apodictic. The ego lives on; it always and necessarily has its transcendental future before it; the expected element having this or that content need not occur, but a different content is there in its stead; something always takes place. And there is a forward directed "always" for me as the ego. But this future has temporal form and is the same as that of the past, though it is constituted in quite a different manner. What is arriving has its intuitive presentification in the form of a presentification of the future, in the form of an expectant image that anticipates a Now and therefore the entire flux and its streaming-off into retentions, that is, a past belonging to it—which however is anticipated as the arriving past. What is futural will be past after it was present, and it will coalesce with the current Now that will correspondingly be a past lying further back, joining everything that is now past. Even the latter past will have correspondingly been pushed back.

This structure of the future thus fashions the futural bent of subjectively oriented time, oriented toward the mobile zero-point of the temporal orientation, toward the Now, in relation to which I stand as a perceiving ego, as the ego of the present. Again, it is inconceivable that the transcendental ego ceases. You will easily see that we do not mean by this that the human being has lived and will live for eternity, and that birth and death, the emergence of human beings in nature and their disappearance from nature, say, through creation or destruction, is quite compatible with the transcendental infinity of life. Strictly speaking, the soul of the [381] body is not immortal, i.e., it is not necessarily conceivable as immortal, and it actually perishes since it is a part of everyday experience. But in a certain respect every human-ego harbors its transcendental ego, and this does not die and does not arise; it is an eternal being in the process of becoming.

<11. Apodicticity of Remembering on the One Hand, Expectation on the Other>

We have seen necessities in all of this, necessities that no one can alter willy-nilly. And while a remembering may be incomplete, while it may be deceptive, it still has a share in these necessities; it is remembering, and this means that it cannot simply be without something apodictically evident. My absolute life with the necessary form of immanent time in the mode of the past lies at the basis of remembering. The deception peculiar to remembering means: "The intended past can be retouched," but behind this lie the past and identical time and identical individual life-contents of time in the necessary mode of the changing past.

Where the future is concerned: Expectation is directed toward the future; it is merely anticipatory, and like all anticipation, it can be deceptive. But life is a living on, and the law of time also attributes an apodictic content to expectation.

But we could still advance further in the same direction. A lawful regularity of expectation under the rubric of associative or inductive expectation, which allows certain contents to be inserted into the empty temporal form, also belongs to transcendental empirical experience. Just think of the transcendental turn from natural perceptual belief to the certainty of the course of those transcendental phenomena in which the same experiential object, the same physical, natural object is presented. This can yield an extremely vast class of examples, just like the transcendental turn from the natural, naive certainty of empathy, i.e., of the certainty of the objective existence of animals and human beings.

Obviously, the possibility of the non-occurrence of something expected belongs to the essence of expectation; accordingly, no such inductive empirical experience can have apodictic validity. Even here there are apodictic contents that surely lead into modalities of belief, e.g., into real possibility and probability; and this holds everywhere when the certainty of expectation plays its role.

<12.> Recapitulation

At the close of our investigation, we can characterize our results in the following manner: The universal phenomenological reduction had <disclosed> to us our ego, the ego unique to each one, with its stream of lived-experience as the center of an egoic [382] totality with the streams of life belonging to it. The apodictic reduction yielded the mere ego as the realm of possible apodictic experience; and practically the entire investigation was actually devoted to circumscribing the scope of the *ego cogito* with its apodictic contents. I am. As soon as I reflect upon myself, I cannot posit myself as not-being, and not only with respect to the living, streaming present. And not only is the streaming *cogito* itself incapable of being negated here. I exist with an unending temporal field in its variable and firmly formed mode of appearance; I exist with an unending sphere of the past and with an open infinity of an approaching future.

Certainly, I must bracket a tremendous stock of my unending temporal life for the apodictic reduction, however much this infinity itself is apodictically certain. Hence, I must bracket every determinate being-thus of the future (beyond the temporal form and the form of its variable mode of givenness). The past, the realm of what is finished and done with, already offers very much more to me. On the basis of the peculiar feature of remembering and of my evident faculty to retain something, to strive for clarity, to repeat a remembering of the same object, etc., I can gain the evidence of the identity of something experienced, even with respect to its being-thus; and therefore in the realm of immanence, specifically, past immanence, I can carry out "objective" experience, so to speak, by observation, fixation, and intuitive determination, assuring myself of what possesses temporal existence and [determinate] being-thus. But we only have apodictic evidence for the rememberings of the retentional near-sphere with some degree of completeness with respect to the concrete content of something remembered, namely, security against overlapping and confusion. And even here is the *limes* of absolute clarity that allows the full individual self of the past to emerge—a limit-case that is not entirely free from doubt; and yet it is not such that it can be arbitrarily generated just anywhere. For example, should we want to repeat an unclear flowing phantasy or even an unclear flowing remembering as such as this lived-experience that it is, and now a second unclear reproduction occurs, how should we become certain that both of the flowing unclarities each have absolutely identical contents of unclarity?

In general, we will accordingly say that while immanent experience is in no small way objective and apodictic with respect to what is experienced, what is experienced with respect to the determinative content is only determined typically, and moreover, is referred to the idea of a completely determinate individual past datum that is not to be characterized merely in a typical, general manner. This is similarly the case where the distant past is concerned, but here the typical generality is such that it even leaves open the possibility of confusions, deceptions with respect to the special features in which the typical generality is given as differentiated. The method that possibly confirms this typical generality again points to the idea of a true being and gives the [383] apodictic security for the being of something true and of what can be disclosed *idealiter*. But every actual remembering will have its element of insecurity, although it will also always and necessarily have a certain general content that is incapable of being crossed out.

Appendix 9: (To §25) Both Variations <of Modes of Givenness:> (1) of Proximity and Distance within Clarity (2) of Obscurity as Veiledness, Nebulousness[27]

Nebulous givenness, the unclarity of intuitive givenness. A memory, an expectation, a possible experience: The memorial object is before me "unclearly," nebulously, but it appears as if in obscurity, as if "drowned out" by obscurity. Likewise in perception, which is external perception in obscurity, in the obscurity of twilight, in the fog or like when my view is befuddled, and yet I still make things out.

Here is a distinction peculiar to phenomenology concerning the modes of givenness between (1) the normal, clear givenness with its distinctions of proximity and distance or of approaching and moving away from; these are distinctions of a gradation that is not a gradation of relative clarity, but rather is a gradation of a greater or lesser fulfillment and filling concerning the seen content with regard to every identical seen moment of such a series—within clarity. It is a gradation that concerns the possession of the self [of the object], the appearance of the self, which is pure possession of the self within the framework of clarity (of pure clarity), but without regard to purity in the gradations of completeness. We can say that appearances are purely self-given, but that as reduced to the genuinely appearing thing, the appearing thing itself is transparent, [since it] itself appears through the appearances; and this self of the object, or the self of the moment in question, appears all the more complete the more of it (the more a greater fullness of it) that appears through the appearances. The optimal appearance, that of absolute proximity, is the absolute maximum in which the transparency ceases and the appearance no longer points through to something new, but is itself the *terminus ad quem*.

(2) In contrast, where the nebulous givennesses are concerned, the appearances are themselves nebulous, "unclear." Here we find ourselves in a different gradation that in general is not—yet in a different way is—subject to our free variation.

---

[27] Editor: Between 1920 and 1926.

The particular reduced appearance, reduced to the genuine appearing, is not a self-presentation without further ado, [it is not a] pure self-presentation of the object; it is precisely obscured, concealed by a haze. To be sure, the object does reappear through the haze or fog, but precisely because its genuine, clear self- [384] appearings appear through it, and in and through this, the object.

Appendix 10: (To §25) <Possession of the Self and Concealment in Remembering. Reproduction and Retention>[28]

When we said in our conclusion to the previous lecture that in near remembering the self of the memorial object is given with all moments in the mode of being that is incapable of being crossed out, this holds with the constant limitation that is implied by the gradation of concealment that is called relative unclarity. Fundamental here is the insight that we cannot emphasize strongly enough, namely, that this concealment is not a masking by or an overlay with moments that are alien with respect to content, or even by or with moments of the object. The fog of unclarity within the deeper levels of its penumbra is not on the order of an object, it is not an objectlike blackening. That would only make sense for optical data. When tones become unclear, the unclarity is not a tonal confusion; an obscurely reproduced loud tone is not a soft tone, nor is it, in place of *c* a mixture of another tonal quality or timbre. It is a concealment of an entirely unique sort: The self [of the object] is given to consciousness such that it is incapable of being crossed out, but it is pushed back into a peculiar dimension that is distant, and this "distance" and "proximity" have a two-fold *limes*: absolute clarity and absolute obscurity—the latter is a *limes* analogous to the zero-point of intensity; and just as this is excluded in principle, because an absolutely obscure remembering would no longer be a remembering, so too, a tone sensed or heard in absolute silence is no longer there in a manner conforming to sensation; the sensation of an absolute zero-point of intensity is no longer a sensing.

[28] Editor: 1920

One should also not confuse genuine remembering (consciousness's process of reproductively re-constituting something being renewed in becoming), and what we call empty memory. Or better, we distinguish between empty near retention and empty distant retention, and we no longer even call this distant retention a memory. If a musical motif dominates the further musical movement, and in this case continues to be held on to, if it exercises an affection on the ego again and again, with the ego also possibly coming back to it with its own rays of attentiveness without it actually reproducing it, then this can serve to make the contrast clear. In actual reproduction, the motif is steadily played out from beginning to end, or at least in sections. And this is also the case in all higher spheres of consciousness. If I conduct a proof and carry out the first premise step by step: positing the subject, the predicate, etc., then the further premises, then the mode of [385] conclusion demands that I return to these premises. Then I say: it follows from it. But not as if we had to carry them out once again in the steps through which they were constructed. The judgment is concluded; and the retrospective examination goes back to the unity that is ready-made and only retentionally given to consciousness in the conclusion, [i.e.,] the unity that is in no way intuitive, the unity that likewise bears here its self in such a way that it is incapable of being crossed out. In this way, a retention relating to a past that reaches way back (I mean, a retention that we cannot apprehend otherwise than as the continuous duration that is indissoluble from the corresponding near retention) can also achieve a special prominence; that is, its objectlike content of the past can, in one stroke, exercise an affection for itself and possibly determine the ego to turn and glance at it without inaugurating an actual remembering. A remote past suddenly dawns on me, the thought that just came to me comes into relief from the so-called unconscious in which the object given to consciousness, in the specific sense of a special prominence, is merely an island. Every present flows once more into this undifferentiated subsoil of the distant retention. The subsoil itself is without any prominence—though once and awhile something does come into relief. It comes into relief: That is, a completely non-intuitive affection is there in entirely the same way that a chord that has just faded away emerges in a non-intuitive manner, possibly drawing my attention to it—albeit a chord that I (perhaps entirely in vain) want to make intuitive again. It can also be that a remembering flashes forth momentarily as an intuition, but that it immediately, that is, after a minimal expanse of intuitive constitution, breaks off again precisely by disappearing in obscurity. But in this obscurity it does not continue, for instance, as if the constitutive process would run on unnoticed. To assume this would be a completely empty hypothesis. Rather, remembering then, takes on once more the mode of the corresponding distant retention; and the constancy of the connection (despite the fundamentally different phenomenological mode) is evidently understood as the coinciding that takes place according to the identity of sense and according to the general law that every intuition can take on a diminution of clarity to the point of having no intensity at all, and it does so in such a way that there is a sudden change in the empty retention that has an identity of sense within the zero-point itself.—But can we really assume this so easily?

Appendix 11: (To §26) <The Concept of Associative Causality>[29]

Lawful regularities of reproduction, of memory (expectation) of "free" phantasy and apperception: We ask: What is essentially necessary, and accordingly, what is intelligible in this? How far can one "explain" development here, the formation of memories, of expectations, of the playful course of phantasy and then [386] understanding the development, formation of transcending apperceptions in accordance with their motives, from "motives" and according to rules of "motivation"? And of course the concept of motive here is an entirely inauthentic one since the genuine concept of motive refers to egoic acts. It is a kind of "causality," but it operates within immanent time-consciousness, and already in its lowest levels (even already within inner consciousness), and <it is> of course nothing less than causality in the sense of

[29] Editor: Between 1920 and 1926.

"category," which requires its own expression: substantial (or natural) causality.[30]

Now, what should we call this causality of the sub-personal psychic dimension? Should we say: sub-personal causality? But that is misleading because one could also think of psycho-physical causality with this expression, while it should be a matter of a causality that runs purely in psychical "interiority." Or should we speak of associative causality (no matter how inadequate this expression is)? Let us use this term. This associative causality dominates in the framework of original time-consciousness, but also in a certain way in the framework of constituted immanent time and temporal objectlike formations of sub-personal interiority. But seen more precisely, the sense of causality and the sense of the regulations proper to it are different in each case here. And it is also evident here that regulations as well, necessary interconnections, must reign between [a] the constituted unities and [b] the streaming that constitutes them and [c] the multiple modes of temporal lived-experiencing that are originally constitutive. When we speak of "association," we usually have in mind unities of different levels that are already constituted. Sensual data recall other sensual data, but even objects of external experience recall other objects of experience (as appearing in this or that way, yes, even as emptily presented in this or that way). This is a relationship that is not meant as "natural," and is not to be understood in this way.

All objects being constituted from original passivity (without any participation by the active ego) arise "associatively" (that is, according to the rules of this sub-personal, purely immanent causality); arising "associatively" are all the objects pregiven to the personal ego (possibly already on its lowest level of personality), above all, objects alien to the ego, thus first and foremost all those objects that do not refer back to an "intervention" on the part of the ego in structuring their intentional constitution. Living egoic acts arise precisely from the ego,

[30] If one takes the standpoint of psychophysical parallelism, one could say with Schopenhauer that motivation = causality, viewed from within. But we do not recommend this point of view.

according to a personal regulation, and do not arise associatively. But is not the ego, the person itself, an apperceptively constituted unity? And does not this apperception, like all apperceptions, refer back to "association"?

Appendix 12: (To § 27) Note on the Fundamental Founding of the Doctrine of Original Time-Consciousness[31] [387]

A completely homogeneous, completely unchanged tone—this is a tone that is prominent, but it is a limit-case. How can something become prominent here in the past and the future? But we must interpret this phenomenon as a limit-case according to the other cases, and in cognition, the evidence that the past and the future are extended and filled out with uniform fullness arises synthetically. So too are complete rest, an absolutely fixed thing-body, etc., limit-concepts. And in this way, the entire doctrine of time-consciousness is a product of conceptual idealization! Even this idealization would have to be described constitutively, and the point of departure would thus be the realm of concrete and discrete phenomena—and in fact, taken as primordial phenomena. Hence, begin by describing the types of primordial phenomena.—Method of interpreting phenomena through primordial phenomena.

Appendix 13: (To § 27) <Primordial Present and Retentions>[32]

The sphere of the present as the sphere of constituted data, living streaming identification through multiplicities of appearances. – The "absolute" sphere of the present = the unities in their multiplicities. The multiplicities themselves "impressional" – as momentary multiplicities in primordial coexistence. Strictly ordered. Two-fold concept of impression. The new sensible datum, the new sensible "point"; the coexistence of retentions with the sensible point as momentary primordial coexistence. But many sensible points with <retentions> possibly belonging to them.

---

[31] Editor: Between 1920 and 1926

[32] Editor: Between 1920 and 1926

Within the primordial present this concerns the ensemble of sensible points and the derivations of different sensible points [that occurred] previously. Several retentions of one and the same sensible point cannot be there in a single primordial present.

Organization: (1) Connected unity of all of the continually constant retentions that belong cohesively to one punctual-linear temporal object. A point of color enduring, becoming altered. Without such a tail in a primordial present, [we have] only the points of departure of new temporal objects in original streaming, living transition of this structure into a new one. New points, constitutive beginning points for paths, thus, the streaming becoming of new retentional transversal series. On the other hand, vanishing – ceasing: Retentional transversal series without a head, finally, the becoming nil of the retentional transversal series. Living primordial present – the fusion of series functioning in a synthetically, identifying manner in the [sphere of] vivacity.

(2) But also, on the other hand, fusion in the same primordial [388]
present according to continuity of the non-identical (coexistence) within the local ordering, as well. The "affection" = vivacity <as> the condition of unity. – In the streaming, both orderings are one identical form that can only be occupied once. Thus, every primordial present is repressed by a new one, and the longitudinal series of retentions that belong together arise in this way.

The law of retention holds again in this streaming. I do not merely have primordial constancy in a primordial present, but also the retentions belonging to the primordial presents that have just past. But I believed to be able to show that it only requires a change in attitude in order to see this, and that this does not lead to any infinite regress. The streaming itself is a primordial phenomenon and must be presupposed.

Now, when a synthesis of identification progresses, the paths and canals of identity run together into a zero by virtue of the progressing diminution of clarity. The displacement that takes place through coinciding is the coinciding of elements that are continually similar, i.e., fusion in which the similar element is continually fused with what has preceded and yet is extended, since the repression is a modification. It is not a concealment—it was absurd to say that. It is a continual modification in a fixed system that is always ready there, and such a modification that is always what is in the form of the Now overlaps in the form of the just-past, etc. The form of the just-past again cannot be occupied a second time, etc.

We can say with respect to the primordial present that "unconsciousness" is consciousness in the primordial present; the sensible object of which we are unconscious along with all the other sensible objects of which we are unconscious are "given to consciousness" in an undifferentiated manner in a zero-consciousness. All the retentions that were previously still differentiated in the primordial present flow together and do this in such a way that the paths of identity are no longer differentiated, to say nothing of offering an internally differentiated object-sense. The only thing now remaining is a horizon-consciousness, a consciousness of an indeterminate, undifferentiated, completely obscure past as a whole. Thus, it still remains consciousness, empty consciousness, whose object is without affection, and it embraces each and every thing in an undifferentiated manner that was there in the undifferentiated form of the one endless past as well. Affection and association concern the constituted objectlike formations.

The awakening of the horizon: Within the living present we have the final sections of the retentional paths in which clarity has already been lost, and there is only enough affective force for the empty "Something," as what is still held onto in this instant with certainty.

An affective fortification from elsewhere can rouse this empty consciousness, can awaken it; it is clarified as a more or less clear [389]
remembering, or as a remembering that is quickly becoming clear. "Re-self-giving." –

How, in the original temporal flux, do all modes of appearance of everything that appears merge together into an undifferentiated zero, into an emptiness in which everything that appears, like the appearances themselves, remains without differentiation? Is that actually a zero, or a mere "point," a "datum," a "content" that is actually in itself empty of content, indivisible, undifferentiated?

Appendix 14: (To § 27) <The Accomplishment of the Association of Simultaneity>[33]

Non-integrally cohesive intuitions, initially perceptions, and integrally cohesive intuitions in *one* stream of consciousness; thus for example, the integrally cohesive unity of perceptions of a sense-field = an optical perception –among several such unitary perceptions that are not connected to one another; that is, they do not give one perception that would have <its> correlate in *one* perceptual object, in one continual (affective and possibly thematic) unity of a multiplicity that is continually flowing-off (in general, a unity of one continual multiplicity). But within one continual unity, special unities can come into relief, i.e., they can exercise special affections, and they can do this because within the overarching unity, that is, within an overarching continuity, discontinuities are possible, the discontinuities of intensity, etc. Thus, this yields unitary "wholes" with prominent "parts." The parts have an essential unity in the whole, and in other connections, a non-essential unity. Uniformity, similarity are not unities of connection; for this reason, they do not come into consideration. Wholes are homogeneous connections. "Mere" manifolds of individuals are manifolds that do not have any essential connection, they are manifolds that are heterogeneous. What is heterogeneous is not connected temporally by the individual beings that have a temporal form.

But where continuous unities that are constituted for themselves are not conveyed by continuity, like in the case of prominences of different sense-spheres, where the intuitions are not integrally cohesive, they are indeed ultimately "connected" in a certain way. The separated intuitions are unified in inner consciousness as time-consciousness, they are associated "in the current Now," i.e., every intuition necessarily belongs to a continuity, to the continuity of consciousness that originally constitutes time and temporal objectlike formations. But two such separated continuities are uni-vocally associated with one another. Put more precisely: Every momentary present with its primordially

[33] Editor: Probably 1920/21

impressional Now and with its tail is "connected" to a parallel present, to the other series of retentional modification. This [390] connection is that of association of simultaneity. It is association that produces within passivity a unity of a higher order between separated consciousnesses of unity, more precisely, it produces a connection between the separated consciousnesses of unity; this unity or connection as the consciousness of the intentional objects constitutes a unity of an object, of the simultaneity of separated objects. One can also say that primordial association is the connection of a non-essential sort, a connection that is not grounded in essential structures.

The simultaneous separated objects possibly exercise an affection in a unitary manner, or can exercise an affection in a unitary manner, because they are connected; and they are not connected because they exercise an affection communally. We also have simultaneity constituted within a unitary intuition. But intuitions that are not integrally cohesive, even perceptions constitute simultaneity, just like those that are in themselves integrally cohesive; and they must constitute it because they are synthetically connected, and in a broader sense, because they are interconnected. An original association connects everything that is not integrally cohesive and that arises in the mode of the Now, and this primordial connection of elements that are not integrally cohesive makes out of the whole momentary consciousness precisely a wholeness, a unity; as such this unity is unitarily one perception and originally constitutes one temporal series, and thus constitutes one connection for the Now of each separated object; it constitutes a Now of separated Nows such that the Now is constituted as connected, thus [it constitutes] their simultaneity; and with this, too, [it constitutes] temporal relationships in general, relationships from things [that were] constituted separately.

If we have a non-integrally cohesive perception and a remembering, then the remembering as a lived-experience of inner consciousness is "connected," associated with the perception as the lived-experience forming an originary unity of lived-experience, that is, inner consciousness is the unity of a "perception" ("inner" perception). The present and the memorial object are constituted simultaneously as such through association, but the memorial object is itself not originarily given as now and therefore is also not given to consciousness as simultaneously now. The most original connection of non-integrally cohesive intuitions is thus the connection in inner consciousness as inner perception that institutes an original nexus among the intentional objectlike formations of the intuitions in question—insofar as we take the intuitions as inner perceptions and take them with respect to this intentionality; but this is not the case if the intuitions are presentifications insofar as we take them with respect to their presentifying intentionality. Thus, just as the memorial object is not given originally, so too is the connection between the memorial object and the present object—the temporal connection—not given originally. But neither is it reproduced, it is not self-given in a reproductive manner. The present *A* and the remembered *A* is given as such simultaneously, but the present object and the remembered past object are not pure and simply given in a simultaneous manner. An intentional interconnection is produced between present and past through the association of [391] intentional lived-experiences in inner consciousness.

Association = every connecting consciousness that is founded in separated consciousnesses and that connects them to a consciousness of a higher order. Correlatively: An interconnection on the "object side" of intentionality between objects that are either constituted materially without coherence or, if they are constituted coherently, have constituted an overarching immediate nexus as an extra-material nexus of separate prominent features.

What role does the unitary movement through such continua play (continua such as fields) through an entire life?

And further: The fields are heterogeneous to one another. What fashions the homogeneity for the reciprocal connection and awakening of heterogeneous fields? The unity of time-constitution fashions a formal unity of phenomenal continuity. Fashions? Indeed, through "original time-association." But does it not already presuppose homogeneity? Is everything that is "Now," everything that is heterogeneous in the form of the Now, formally homogeneous precisely through this form? But what does the opposition of form and content mean here? It can certainly mean nothing other than this, [namely,] that consciousness is connected in the most general way to another consciousness by a commonality that is correlatively noetic and noematic; and all connection is connection through "commonality," through uniformity and similarity. This most general uniformity and similarity is not the similarity composed piecemeal; rather, it is a similarity that makes all other commonalities possible in a unique way, and on the other hand, co-determines them. With regard to everything that is otherwise given as content, similarity provides the basis for saying that content is necessarily given in a temporal form, in a different mode, in a continually varying mode of givenness, etc.

Thus, it provides a universal connection that connects the lived-experiences to one consciousness; all of them have the unity of a single stream of consciousness. All lived-experiences of immanent time are constituted in an immanent time as temporal unities and are constituted as unities of sense in the modes of givenness in the modality of time, but not in such a way that each one is constituted for itself and that each one stands under the same law of time-constitution; rather, the mode of the Now of every primordially impressional occurrence is certainly its Now in each one of the lived-experiences, but it is at the same time *a single* Now, *a single* mode that connects all of these lived-experiences. Indeed, the concrete primordial phenomenon is that of the unitary "flux" with the unitary primordially impressional occurrence arising ever anew as Now and with the modification being retentionally altered again and again in a unity, in the continua that we described which form a unitary phase (even a punctual unity) according to each cross-section. It is an original unity, or better, an original form of unity. The content is what can change; namely, when we compare the concrete unity in the living flux to the memorial fluxes, or [392] when we imagine something, this unity can be variable; but this form of unity peculiar to constituting time-consciousness is necessary, and its continuation is necessarily and continually attached to a living unity according to the law that governs the sinking into an empty retention; this empty retention can be awakened by rememberings, whereby every living present, <every> current actuality of the impressional flux is attached to an implicit series of possible true rememberings in which a unitary past is disclosed as the past stream. It belongs to [the essence of] the living present that it is freely possible for the ego to carry out a continuity of remembering that is livingly present and true. Ultimate descriptions and analyses are still lacking here. The fixed form of the living present (of the full "present") and what flows through this form: a timelessness of form in which time is constituted.

Appendix 15: (To §27) <Unitary Consciousness and its Correlate: the Identical Object>[34]

We now direct our attention to a new universal theme, to what is alluded to under the obscure Kantian rubric of "synthesis." Let us link this to our previous considerations by reflecting on how we had actually won our last series of results. Thus, we hold firmly to an object, for instance, a perceptual object and allow varying perceptions to be referred to it, but then also another, a different kind of consciousness to be referred to it, a consciousness that at the same time stands in contrast to the other one. Thus, the same object is perceived, perceived differently again and again, namely, according to its different sides, aspects, etc., it is then remembered and, for instance, remembered in changing rememberings as well, then presented pictorially, etc.

But now we must not overlook the fact that when we speak of "the same" object that is given to consciousness in different *modis cogitationis* (and speaking within the framework of the phenomenological reduction), a varying continuity or discretion of multiple consciousnesses does not merely run-off; rather, we must not overlook the fact that this multiplicity is also a unified multiplicity of consciousnesses, indeed, the manifold of consciousnesses in its unity is also a single consciousness. With this, we come across the fundamental fact of consciousness as unity and identity, whose correlate is the one and identical intentional object.

[34] Editor: 1922/23

Every single perception and every phase in the continuity of a progressing perception (regarded for itself and *in abstracto*) has its intentional object. The entire continuity of multiple and varying perceptions is however *a single* perception and has *a single* [393] intentional object. And if the perception passes over into a so-called fresh memory, or better, into a retention, and if a clear remembering then arises (a remembering relating to the previous temporal phase of the object), then the single and identical intentional object runs throughout all these modes of consciousness in their unification. – Or as we then also say in comparing the different phases of unitary consciousness: Each one of these phases in itself has its intentional object, but at the same time the object is the same in all of them, and it is the same in the entire unitary consciousness. Unity, ipseity, is given to consciousness through a higher consciousness founded in multiple consciousnesses, precisely in "synthetic" unitary consciousness. Thus, this can connect quite differently shaped consciousnesses, phenomenologically very different consciousnesses, and it connects them in such a way that what was given to consciousness in other and very different consciousnesses, is given to consciousness overall as one and the same. Naturally, not just any consciousness can be synthetically connected to just any random consciousness to form a unitary consciousness. But, on the other hand, no consciousness is isolated. Moreover, there is no consciousness that would not in <itself> already be a continuity and that would not already allow the distinction of phases and parts which themselves are already consciousnesses, and which intentionally constitute the unity of the object in the mode of synthetic unity. The phases, and likewise the different concrete intentional lived-experiences as we can also say, fuse by "coinciding" in the same lived-experience. Already here, where we consider the most simple shapes of intentional lived-experiences, like the simple perceptions of the object, presentations of the object of every kind in their functions as unity, we encounter a marvelous feature of consciousness in its progressing stream that we will call polarization. Let us consider once again: Whatever can be examined for itself as a "*cogito*," as an intentional lived-experience, has for itself its intentional object.

But as we look into a consciousness being synthetically unified with another consciousness that in turn has its intentional object, either one of the consciousnesses can, in a truly striking manner, claim the same object.

I say: in a truly striking manner. For the stream of lived-experience to be sure exists in the form of a temporal juxtaposition. What is sequential <is> separated in all of its elements; what belongs to every phase in parts and moments, belongs exclusively to it. And yet the intentional object of one lived-experience that is intertwined with another lived-experience in a unitary consciousness must be identical with the intentional object of this other lived-experience; and this identity is even[35] in many cases an absolutely evident one, like, for example, in the unity of a continual perception and retention. We see here that a curious difference comes clearly to the fore, namely, between [a] what inhabits the lived-experiences "in an intimately inherent manner," what is contained in them as an intimately inherent moment, and [b] what inhabits them in a manner that is not intimately inherent.

At first, it is necessary to pay attention to the essential form of phenomenological time belonging to the *ego cogito* and to its flux. [394] Objective time (which, like space, is the form of nature) is bracketed for us. But the intentional lived-experiences themselves have their temporality which remains for them even if we have put out of play every positing of the world with its objective time. A lived-experience, as a lived-experience in this "immanent" time (as we also say) expanding, beginning, and ending, has its individuation in its temporal locus; if it is now, it cannot be later; each one of its phases has its individual temporal locus. Thus generally: Everything that is a part of the lived-experience is firmly and individually bound by means of the temporal locus. For this reason, temporally different lived-experiences cannot have just any individual part in common. In the broadest sense, the individual non-independent determinations also belong to the parts. Let us call everything that is bound in a phenomenologically

[35] Translator: Reading *sogar* for *sage ich*. This reading is based on a correction by the Husserl Archives in Leuven of a transcription error.

temporal manner to a lived-experience an intimately inherent moment of the lived-experience. Then intentional lived-experiences will however not only have intimately inherent moments, but also moments that are not intimately inherent, thus, their intentional objects and everything that falls under the rubric of "sense."

In a certain way, the intentional object is also given in a phenomenologically temporal manner, to be sure. If I see a house, the intentional object "house" is the appearing object during the temporal expanse of the perceiving lived-experience from phase to phase. The phenomenological temporal expanse of perception also conditions a certain phenomenological temporal expanse of the perceived object, while the objective temporal duration of the object always belongs to the intentional object itself. But the phenomenological temporal locus binds only the intentional lived-experience and not the intentional object. The latter is not individuated through time. If we take several intuitions of the same object, for instance, several rememberings of the same object, they are indeed completely external to one another temporally; they do not have anything at all in common in an intimately inherent manner, but they are rememberings of the same intentional object; they do not have anything in it that is identical in an intimately inherent manner, but something that is ideally identical. We can say forthwith that the relation to the intentional objectlike formation means a certain polarization of lived-experiences that can be disclosed phenomenologically. Every lived-experience is an intentional lived-experience insofar as it intrinsically has a pole as the consciousness of something, that is, can enter into syntheses of identification with certain other lived-experiences and with an ideally infinite number of lived-experiences, whereby this ideal pole (as identical, not however as identical in an intimately inherent manner, but rather as the identical pole of the intended meaning, of the sense) achieves a unity of coinciding in the unitary consciousness, i.e., identity consciousness.

However, the coinciding in the same unitary consciousness can be a very different coinciding, and can give a very different sense to the ipseity [i.e., the identity consciousness]. The intentional object can be a concrete object, and thus the identity can be the identity of this same concrete object. We then have complete identity. But the intention can also be directed toward a property [395] of the object or toward a relative feature in the relation to a different object. In the transition from the consciousness of the concrete to the consciousness of the property or the consciousness of relation, however, the complete consciousness of the concrete object coincides in a certain way with the special consciousness of the part or of the relative feature, and in this way we have disclosed new modes of identifying syntheses. Such syntheses come to expression later linguistically in predicative propositions, propositions that certainly always already contain, as enveloped in themselves, quite intricate syntheses of identity. Every "is" points in an indeterminately general way to a synthesis of identity, and this holds for every position, every inflexion that is to be explicitly expressed by an "is."

We also distinguish here: (1) the simple synthetic unitary consciousness that lies in every simple enduring, concrete perception for itself, or in such a remembering for itself, whereby a self-coinciding takes place continually, that is, in an interiorly undifferentiated manner; (2) syntheses that are precisely genuine, where concrete, independent, and separate perceptions or rememberings, and lived-experiences of any other kind, enter into a coinciding that forms identity, and where the articulated identity consciousness, "this or that is the same" arises. We distinguish here between continual unity and discrete syntheses of identity.

Only in the latter case are we accustomed to speaking in the strict sense of the synthesis of identification or, more simply, of an identity consciousness. But every intentional object is genuinely an object *simpliciter* only as an intentional object of a concretely finished act, and such an intentional object already has its phenomenal-temporal expanse, and thus already has a continual coinciding of the self in this expanse. Only through a distinction that we make by abstraction or through an extraction of single expanses of continuity that are made prominent do we become attentive to the moment of continual coinciding in the same intentional object, a moment that is found in each consciousness.

All specific accomplishments, participations, affections, position-takings of the ego, then, relate to the "intentional objects," to these ideal poles of unity and poles of the identity of varying syntheses of consciousness; they relate to these poles that are not intimately inherent, to these poles that are already of particular intentional lived-experiences. Unfortunately, we do not have a name for what is specifically egoic. But in particular instances we can point to the fact that the ego is being affected by what is given to it, that is, affected by the object given to it appearing each time in such a manner, in these or those modes of consciousness; then we can point to the fact that [the ego] turns toward it in an attentive manner, and with different modifications of attentiveness; likewise we can point to position-takings, like those of believing, of valuing, again in differing modes; to modalities of striving like regretting, desiring, willing, driving at; to explicating, relating activities, colligating, grasping two things together in a single plurality, one referring to the other.

Appendix 16: (To §28) <On the Connection of Similarity>[36] [396]

The greater the similarity, the closer the data stand to one another, the stronger is the coinciding, the more that unanimity outweighs difference. Two data that do not differ, that are without distance and yet become united through synthesis, fuse. If we conceive of a multiplicity <of> data that form a unity by means of a thoroughgoing, constant mediation of similar elements, all of them will fuse to form an internally unbroken unity, a unity without discontinuity.

This is the case either in coexistence or in succession, as well. For example, a steady tone—[a tone that is] steady in itself, qualitatively uniform and, at the same time, steadily modified in its intensity—, if a coinciding that is constantly without distance is produced in the continual successive synthesis, but where the increase of intensity becomes salient as the tone continually streams on and on. If we take away an intermediary element, the increase becomes a distance and therefore a contrast. We lack the

[36] Editor: Between 1920 and 1926

phenomenon of increase in pure uniformity, and steady coinciding does not lead to the steady increase of an implicit difference.

Now, what makes up the unity of an object for consciousness, how does the object become prominent as one among others in coexistence and succession? How is it that there can in turn still be a singular object of a higher level—a closed group, a configuration, a whole composed of many elements—arising from several elements that are prominent in a singular fashion, that is, from several particular features? How can it be given to consciousness as the singular object? And in the other direction: What produces the unarticulated singularity that is not given to consciousness as plurality, although it does indeed imply something plural, even something continually plural, and is able to lend itself to such a plurality by a division that is ideally possible at each time? Obviously such questions are to be posed initially and originally for the immanent sphere.

It is clear that unity, homogenous and non-homogeneous unity, and plurality as unity, are governed by essential conditions, and that with such questions we will initially be led back to the syntheses of similarity and contrast, those that, within the form of time, connect-up with respect to content and separate with respect to content. The prominent feature contrasts with its temporal surroundings; the extent to which there is fusion is the extent to which there is no prominence. Something prominent can then again enter into a special unity through uniformity or great similarity, synthetically producing a connection in the unity of consciousness, be it in coexistence or in a sequence. From here on, however, the multiplicity of directions in uniformity and all relationships of a higher level would have to be studied further.

Awakening and the connection of similarity in an impressional present—the impressional present does not need to be constituted in one blow as a multiplicity of objects—a first object is privileged for instance and gets apperceived in a certain way because as the [397]
first it is the quickest one to fulfill the conditions of apperception. But meanwhile, similar features can also become prominent at the same time, similar features that invite apperceptions similar to them and that favor them in their very transformation during the impression (coming on the scene involuntarily in the central

sphere of the visual field). The latter must first be "dismantled." Return to the primitive limit-cases. A manifold of sensible data in a synthesis of similarity. A pair. An affective interest—do we have to consider this at once? Here we immediately have the problem of the relationship between prominence and intention and affection. If the pair is such that it is completely uniform, it can arouse an affective interest only in a unitary fashion, provided that we rule out apperceptive moments. Can one then say that such an affective whole necessarily motivates a passing through them and a cultivation of overlapping? In the case of mere similarity (and uniformity) there can also be moments here and there that arouse different affective interests, but that are held in balance.

The unitary connection of similarity can be given to consciousness as the connection of uniformity or as the mere connection of similarity. In the latter case, one can disclose the incongruence of single moments that are coming into relief with overlapping by passing through them. Generally speaking, the connection of similarity has two forms:

(1) the primordial form of overlapping in pure passive coexistence;

(2) the form of "fixation" carried out from the side of the subject, the "fixation" of one and then the other, whereby the overlapping with respect to content comes on the scene—the focus of the ego does not only concern the first, but <also> the second (can one say: as something given to consciousness at once both impressionally and retentionally?). The second is still held with an egoic grasp, and here congruence and incongruence come to the fore in terms of parts and moments.

But this action does not fashion agreement and non-agreement. In the primordial form, unity is already constituted as the unity of affinity (unity of fusion) and difference, prominence, separation as the *negativuum* of fusion, but is constituted as a form of unity. In addition to this we have, to be sure, the accomplishment of explication and thus of a comparison attained after explication whereby the special uniformities, special congruences and special differences "come to the fore."

We then have connections of similarity in primordial form and in the consideration of transition, and we have comparison for successions, e.g., series of repetition.

"Awakening" in the impressional sphere: Something occurs that already has my interest. There is one thing or are several similar things in coexistence within the impressional present—interest extends to the similar things.

Syntheses, like those of coexistence in an original present and those of succession in an original sequence, in streaming. The synthesis of coexistence is not one that generates; it does not later form a connection out of an entity prior to the synthesis; it does [398] not first have one entity and then another one that links it up to the first and through this generates a whole. The successive synthesis constitutes succession, but it links up impression and retention. We can say here that it essentially conceals what has just occurred (belonging to the previous Now) by the occurrence of the current entity, the previous Now is causally modified by the new Now. These are constant causal necessities peculiar to streaming. They are necessary causal syntheses. If two similar elements occur in a present, it is not the case that they first exist [separately] and that then their synthesis follows; rather, we call "similar" what occurs in such a synthesis <as> coexistent. Being in a togetherness, the founding and the founded are inseparable, they are necessarily one. But is it not likewise the case with a pair in similarity, which becomes a "pair" because the one member has newly occurred? To be sure, something similar can emerge, can endure in a present, and then something that is similar to it can emerge anew. But then the synthesis does not first occur afterwards, for the synthesis is there with the existence of both. – Like when we consider awakening? Is it really different here? A synthesis occurs as soon as an *a* has a certain force through an interest, and something uniform with respect to this interest [arises] in the empty horizon. In particular: The emergence from emptiness arises with the characteristic of "consequence" because in the present the interest has fallen to the content or because the content of this interest has arisen.

Appendix 17: (To: §28) <Sensible Connection of Similarity. Sensible Uniformity and Eidos>[37]

Similarity and fusion; similarity, prominence, allure to run through [the individuals of a group]:

A totality of hyletic data (coexistent or successive, and that arise in one consciousness) possesses a sensible unification that combines the data by means of the similarity held in common, and this is the case to the extent that the similarity is sufficient.

This initially means that wherever we find data that are already prominent in one consciousness, data that we apprehend in relations of similarity, we find them connected sensibly in a certain manner, in a manner that is called the connection of similarity. And inversely, where we have a group of contents given in the connection of similarity prior to running through individuals and prior to a cohesive reciprocal relation, there exists the ideal possibility (the essential possibility) of an apprehension of similarity that forms relations, an apprehension of similarity in which the relationships of similarity that belong to the pairs of contents are constituted for us, that is, originally bring them to givenness for us.

We speak of a gradation of similarity; this points back to a [399] gradation of "sensible similarity," that is, to the gradation of the sensible connection of similarity that is "more" or "less intimate." (Here we do not yet need to discuss how this bears on connected contents, whether or not uniform contents can be connected in different modes of intimacy). Connections of similarity, groups of similarity can in turn possess connections of similarity ("coalescing" to form groups of similarity), etc.

The connections of similarity that we had in mind here refer to connections of contents that are already prominent, separate, and these can be a connection of concrete wholes as wholes, or a connection of wholes by virtue of the parts or moments—as analysis shows. And "analysis" plays its role in the constitution of relations of similarity which are relations between objects. And here both *concreta* and *abstracta* can be objects. If we ascertain,

[37] Editor: Probably from the period of the lecture.

analyze, and explicate similarities here, we "observe" that certain connections of similarity take place for moments, and then that the wholes themselves possess connections of similarity in a different way: The wholes are connected "through" moments and their groups of similarity.

We conceived of similarity as a relationship, and the nexuses of similarity as relations or unifications of prominent data. But how do data become prominent? When are data prominent in coexistence and when in succession? Here we are reminded of Stumpf's introduction to fusion which (as he proceeds) relates exactly to this point.[38] Fusion is what prevents the consciousness of a manifold, it thus prevents a prominence; only that he actually already has the following in mind here: If, as is usually the case, the nature of contents is such that the contents form a manifold, that is, such that they would be set-off in relief in a relation of coexistence (which he alone has in mind) and that would yield a sensible nexus, then an obstacle arises now at the very core of the contents, an obstacle that works against the formation of the "manifold." As one sees, and this will be sensed even more strongly upon closer consideration, there are enormous difficulties to overcome if one wishes to analyze clearly all the relationships that are intertwined here.

What is it that initially determines separation? Non-similarity (heterogeneity), that which is "without" materially relevant "interconnection." Similarity is the very first thing that fashions "interconnection," "interconnection" in the sense that is at issue with regard to relations. (Without interconnection = that which has nothing to do with the other, that which is alien in a materially relevant manner, heterogeneous).

Thus, there is something like homogeneity, and homogeneous contents form a nexus in coexistence and succession, and belonging together with this nexus are certain relations, relations of similarity and relations of uniformity.

[38] Translator: See, for example, Carl Stumpf, *Erscheinungen und psychische Funktionen*, (Berlin: Königl. Akademie der Wissenschaften, 1907), which is also to be found in Husserl's personal library.

Instead of similarity, perhaps it is still best to say throughout: [400]
connection and relation of homogeneity, as opposed to connection and relation that is not produced by homogeneity. Homogeneity refers then to a commonality of a materially relevant essence, and secondarily to a commonality of characteristics that are intertwined in a nexus.

Thus in a certain sense homogeneity is "fusing" in coexistence and succession. But in what sense?[39] The nexus of homogeneity, of "similarity," has its counterpart in the non-nexus peculiar to heterogeneity. Color and sound are (in accord with their essence) "separate," and this "separation" means that non-homogeneity is one of the (not necessary, but sufficient) conditions of "plurality," or it is one of the conditions of that prominence that makes possible an "affecting" of each one of the contents "for itself," namely, in coexistence. Non-homogeneous coexisting contents stand in a relation such that in the case of prominence, one content does not disturb another one, or it does not disturb another one because it exercises an affection for itself, and possibly gets apprehended as a prominence for itself. (But one of them can be privileged where attentiveness is concerned, e.g., one of them is quite intensive and therefore obtrudes, the other one does not. Here each one can have an equally sharp prominence, but to be prominent does not mean to be obtrusive.) This does not mean that other "disturbances" are not there, that either of the contents fulfills the conditions of prominence differently. And on the other hand, such conditions can be sufficient in the case of homogeneity like, for example, for sensible data in the visual field or for coexisting acoustical data.

In the case of coexistence, and all the more so in the case of succession, one cannot say of homogeneity that it disturbs separation with the degree of homogeneity's increase if the individualizing determinations behave in a corresponding manner. Spatio-temporally contiguous contents become all the more prominent the greater their distance in homogeneity, the more "dissimilar" they are, and they necessarily lose this prominence

[39] Why have we not taken into consideration spatio-temporal phenomena in this respect? They do indeed belong to "homogeneity."

when the distance becomes zero (ideally speaking). [The expressions] "prominence is all the greater" and "fusion is all the smaller" can be characterized as equivalent.

Prominence as separation does not actually admit of degrees. If we nevertheless do apprehend it as admitting of degrees, this means that their essential difference fulfills the condition of the "being-for-itself," and this goes hand in hand with the condition of the possibility of grasping an individual under the given individualizing circumstances, but [it also] means that the allure for grasping an individual is all the greater (or the "ease" by which it gets grasped individually is all the greater), the greater there is a distance of similarity. But all of this is a matter of necessary essential interconnections.

If similarity approaches the zero-point, which is to say, if similarity is either nil or if the essences continually pass into one [401] another and the continuity coincides with that of the individualizing moments, the unification taking shape through coinciding in similarity attains a fusion of continuity shaped by affiliated laws, laws according to which the continuity of individualizing moments is sufficient to ground the "extension" of one and the same essence *via* this continuity and, on the other hand, in the case of varying essences, it is sufficient to ground extension in the form of an expansion of the essence-continuum *via* the individualizing element. The parts of such continua and the phases are united in an "unseparated" manner, and are only separable if other motives of prominence are produced. Again, similarity serves the following: Juxtaposition of the same individualizing form (expanse of space-time) in two ways of fulfilling essence; of them, one provides for separation, the other does not. Through this, a "figuration" arises in the extension that is internally undifferentiated, as coinciding goes down the path of forming relations.

But there are many difficulties here. Duration, for example, also comes into consideration where similarity and coinciding are concerned, as does spatial extension, but not absolute temporal locus. Does not absolute location, however, play its role here with such uniformities and similarities, i.e., comparisons, at least with space, even if not with time? Perhaps not. We come to comparative overlapping, and that is not an identification of locations and times themselves.

All the talk of contents that have these or those features, that have these and those internal predicates, this or that concrete essence, generic moment, specific moment, etc., points us back to comparison and to the nexus of similarity. Yet we still have to say here that we need a more general term in relation to individualizing determinations. We can say: communities of essence and differences of essence. One speaks of similarity with respect to objects, and then the question arises as to what we have posited as object, what we have set into relief. If we take individual objects in space and time, even immanent ones, we find that they are the same objects through the change of the spatial situation. Thus, this should also be taken into account. First for us is the nexus that is determined by homogeneity, but we describe the relationships as well as the laws that are found here by means of essences and relationships that we gain through comparison and through predication. But how could it be otherwise?

If we take a group of separated data from a sphere of similarity, there is the possibility that these data are completely uniform. That is, they are not only similar, and they are not only "very similar," but in relation to each other they are "repetitions." We also say that "the same thing" <is> there many times as "the same thing" again and again. It may be that this uniformity of repetition is an idea, an ideal limit, but the relation to this limit lies at the basis of all the talk of such repetition.

Phenomenologically, uniformity (in this sense of mere [402] repetition) is distinguished from mere similarity; mere similarity is gradually removed from this uniformity through the way in which coinciding presents itself in coherence and in the mental "overlapping" that takes place in the "process of comparison." The possibility of comparative "coinciding" extends just as far as similarity, just as far as this peculiar form of connection unique to "sensible" similarity. One could say that the extent to which a manifold of separated contents is pregiven—let's say, a duality (which in the final analysis perhaps presupposes the very broadest commonality)—is the extent to which we have the possibility of coherence and comparison; in this instance, a certain synthesis is carried out that brings the so-called elements of the comparison into a mental overlapping, into a line of vision in which one of the rays of vision traverses both contents. But only where there <is> similarity, and the extent to which we have similarity, is the extent to which we have a particular occurrence that we characterize as "coinciding." In a similarity-consciousness (as the consciousness of the relationship of a similarity), the similar elements that were previously merely juxtaposed and that formed a sensible manifold in a sensibly connected manner (a manifold that is especially characterized as a manifold of similar elements) come together into a new situation; and through this they gain a commonality; the overlapping data "coincide." In the case of uniformity, the uniform elements coincide completely, they become "congruent," they form an undifferentiated unity that allows the essential components of separated uniform elements to coalesce into a single essential component that is only there in two "editions"; that is, from which the essential components only one coinciding essential component has emerged through the process of comparison (of the bringing-to-coinciding). In the other case, the essential components of both of the similar elements also enter into a relationship of coinciding, but they continue to remain as two in the phase of coinciding; each one not only belongs to one of the separated elements, and is thereby separated, but each one is also in itself separate, and as the essential component, it is removed from the other one. In this distance, however, the two do have a certain unity, they have the unity of a commonality, and in this phase of coinciding, the very commonality of the objects related to one another becomes apparent. This duality with its unity of commonality can approach the unity of complete commonality more and more, precisely the uniformity and essential coinciding without distance, and it can approach <it> so closely that we speak precisely of an approximate uniformity, of a similarity that is practically complete, only with minute deviations, etc. But the phenomenal difference still remains despite the continual transitions.

Let us take as pregiven a sensible manifold of uniform data, one of them as the "repetition" of the other. In carrying out an ideally possible comparison, there arises a congruent coinciding of the same data according to their "common essence." These essences come into relief here, and by forming an undifferentiated unity [403] without distance, two aspects possibly come to the fore, and from the perspective of grasping.

On the one hand, the One can be achieved by an ideating abstraction as the absolutely identical *eidos* that is grasped as the object for itself, but as an object that is not the individual moment here and there, and not merely as the identical object of the essential moment coming into relief, the essential moment of this or that *individuum*; rather, it is grasped as a "universal" essence that "is individualized," and only "coincidentally" in this or that *individuum* such that it could be individualized "just as well" in any "arbitrary" *individuum*; in other words, just as <it> has *hic et nunc* these actual individualizations, so too does it have possible individualizations in arbitrary possible individuals being presented in phantasy (repetitions in phantasy of the given actualities and in relation to other ones exactly in the sense of such a repetition), and persists in an absolute identity in relation to all these individualizations, and is itself not disturbed by temporal coming into being or passing away, by the formation and transformation of "corresponding" individuals in phantasy, neither arising nor disappearing. One may call this Platonic *eidos*, this pure essence, a "hypostasized" possibility, but this is the absolutely "unique" in which this "hypotasization" in itself has its absolute justification; it is the absolutely "unique" that, as it were, liberates the essence (coinciding in repetition) from its individual subsoil in free phantasy-modification of given individual-actualities or from the beginning in the phantasy-givenness of pure possibilities (to which no present actuality measures up); the identical can be brought to pure identity and can be posited as an object for itself—an object that implies an open horizon of all possible particularities in which the object is "repeated," individualized, particularized, but not as a fact, but rather as pure ideal possibility that allows new such possibilities (an "unending scope") to be freely construed *in infinitum*.

Correlatively, standing thus in opposition here to the *eidos* itself which implies the horizon, is the particular "as such," the purely possible particular or a plurality, an open-"unending" scope of particulars in which the *eidos* can be individualized. The inductive grasping of the *eidos* is carried out on the basis of a "free exemplification," of a horizon of freely presented particularities that are conceived in the consciousness of "whatsoever"; these particularities are connected by sensible uniformity and are transferred by being transposed through the comparative process into the coinciding of the essence. The constitution of the *eidos* as object is inexorably connected to the constitution of the "as such"-objects of this *eidos*, a constitution included in the same consciousness; depending upon the focal orientation and the grasping, the *eidos* is posited as object and as identical with itself, or the arbitrary particulars are posited as possessing the *eidos*. But here the first [level of] objectivation[40] is sketched out as primary. For although the objects must be given in pure possibility, that is, [404] not as actualities, but as free possibilities, and although in this case they are objects of the essence in question, they are not however already constituted as individualizations of the essence, as "participating" in what is ideally identical. Rather, only first in coinciding (overlapping) is each one there as the individualization of the same ideality; and only first in being liberated from the tie to determinate individuals and their determinate actualities or quasi-actualities in the consciousness of the whatsoever does there arise the correlation of *eidos* and the scope of the *eidos*. If I proceed, for instance, from actualities (of perception or memory), and if through this its actuality becomes irrelevant for me, if I vary them in any which way in phantasy, if I proceed to new individuals that would only repeat what has been given up to now, if what is constituted here is the consciousness of the open "*et cetera*"—if we characterize this as a consciousness of the possibility to progress in an unlimited way in such a repetition (and this is a completely unique consciousness); if we progressively carry out the synthetic coinciding in which the identical element comes to the fore, then what is first is the grasping of the identical element as the pure *eidos* that is referred to this horizon of the *et cetera*, and what is second is the consciousness of any kind of arbitrary particular "as" the

[40] Translator: cf. Main Text, Part 3, §65 c.

particular of the identical element as its individualization—ideally possible individualization—, and again the consciousness of the particular as an element of this open infinity, of this "scope," of this totality of arbitrary particulars, "all" of which have the same pure essence.

Moreover, something else that is special is the grasping of a particular as a particular actuality, whereby we carry out the thesis of the consciousness of actuality, or the thesis of an imagined particular, whose actuality we carry out in phantasy in the "after a fashion" placing ourselves on its ground, putting into play and holding firmly to the particular as a possible actuality. Then, the actual object or the object that is put into play as a possible actuality is there as an actuality or <as> the onset of an actuality, in which the *eidos* is actually individualized (or would be individualized); and in each case we have here a consciousness of relation, a consciousness where that posited particular is determined "predicatively" (as the foundation of a possible predication—not only from the perspective of verbal articulation, but from the point of view of meaningful expressions, of apprehending significance), as being *a*, where *a* designates the essence (the "concept" in a certain sense).

The individualization of the *eidos*, of the eidetic essence—as a rule characterized simply as essence—offers in addition a two-fold concept of individualization in the individuals: The individuals themselves are called particular individuals, the individualizations of essence, but more properly speaking, the essential moments in them are called individualizations of essence. The essence is individualized, it is rightly said, in each *individuum*; in each *individuum* there is a moment of individualization, and in this respect, the *individuum* itself is called the particular of the essence; this moment is not the *individuum* that rather has different moments. To be sure, this requires its own exposition.

Yet the coinciding that overlaps still leads to something else. [405] Still before the consciousness of the universal and the consciousness of the *methexis* of the particular with the universal <takes place>, a reciprocal prominence arises in one stroke with respect to particular data that are included in the process of overlapping. The common element comes into relief with respect to each particular and distinguishes itself from that which is different; with respect to each one, moments come to the fore, moments that do not coincide; and this holds not only for our case of the coinciding of the repeated element (that of uniformity), but for every similarity. Even here, what we call "coinciding" concerns only the genuine foundation of similarity, the essence entering into the relation of unity here and there, which is still connected to a different element.

Every comparison is *explicite* or *implicite* also a differentiation. If individual uniformity is a complete and strict uniformity, to which belongs the identity of the whole individual essence, the uniform elements are differentiated according to their individual difference. It is evident that the uniform element is differentiated, and that the different element is differentiated by something,[41] and this Something—the differences of the uniform elements—comes to the fore in the respective focal orientation. In this connection, the necessity of the differences of the uniform elements is naturally given in the pure consciousness of universality, and in the knowledge that the sublation of an individual difference or of each one, at least of the individual difference, also sublates the separation and therefore the uniformity. Separated objects are "differentiated from one another," they necessarily have different determinations, something in or with respect to them that distinguishes them. The relationship of the difference stands in correlative relation to the relationship of congruent coinciding, and is regarded in and for itself as a positive relationship; this however stands in the relationship of exclusion (of incompatibility) to the relationship of coinciding.

---

[41] different through something: the differentiation of uniform qualities through their intensities; the differentiation of uniform individuals (*concreta*) through their spatio-temporal situation; the differentiation of uniform concrete parts through their connection. In the case of partial uniformity, a whole can be uniform to a part (piece) of the other one, or a uniform part can be connected here and there to different supplementary parts.

Appendix 18: (To § 28) Association and Synthesis[42]

Synthesis in its different shapes as universal unification of the life of an ego = association in the broadest sense. The apprehension as this tree, as this white piece of paper, etc. A fragmentary observation of the tree, an onset of actual experience. Synthesis in the same. Coinciding of the intentions in the direction of the optimum. Coinciding of the style of apperception. In distancing [oneself] from the object, it becomes obscure. As [406] opposed to approaching, which in one respect always leads to the optimum. But despite this, coinciding in the same. The modes of appearance of the same thing running-off, continual modification of modes of appearance insofar as there is a progression in similarity; but commonality here in the intentional relation under continual coinciding in it.

The consciousness of similarity without active relating, without the constitution of a relationship of similarity. Correlatively, similarity as a kind of nexus, connection (what I called sensible similarity), but a subjective nexus, constituted in the subject as the connective form of subjectively constituted objects.

Similarity as a special connection "becoming prominent," as a special nexus exercising an affection for itself. Perhaps one can say: Where similar apperceptions occur in a field of consciousness, here we also have a connection of similarity before us "sensibly," but therefore not yet "coming to the fore," being separated-off, coming into relief. What presents separation, what brings about prominence? These are questions that must also be posed for particular objects as constituted sensibly in the unity of one consciousness. The unity of the field of consciousness is always produced through sensible interconnections, in a sensible connection of similarity and sensible contrast. Without this there could be no "world." We could say that it is resonance as sensible similarity and sensible contrast (that for its part presupposes a similarity) that grounds everything that is once constituted. It is a universal law of consciousness that a resonance proceeds from every special consciousness or from every special object, and

[42] June 4, 1925

similarity is the unity of the resonating element. In addition to this, [we have the] special law of individuals in prominence. Resonance is a way of coinciding in distance, in separation. Belonging essentially to resonance is the possibility of transition and of the production of overlapping coinciding, while the elements that are coinciding here (elements that are brought to the place of congruence) are posited separately and meant in special acts. Constitution of a relationship of similarity, possibility of analysis and comparison, of intuitively extracting something identically in common as the universal, etc. Sensible unity as continual similarity in coexistence and succession. The momentary "fields" of coexistence and succession. In contrast to this: awakening at a distance.

If I am actively oriented toward an object, grasping it and examining it, all uniform elements are in resonance in my field of consciousness, indeed they are in a special resonance that grounds the special affection on the ego and makes the synthesis of similarity prominent, even if this resonance does not bring it to my special attention (not yet to a special affection). Contributing to this is the active turning of the regard toward something similar through which overlapping occurs; this takes place while both the objects remain grasped and meant in the separation. The pair emerges as a sensibly connected unity, as plural, which can become the singular. Proceeding from the A that is noticed first [407]
there is an awakening; in itself it has the relational character of awakening and of the substrate of being similar to B. Analysis and explication in the examination of A; and this then determines, in turn, the division within the uniform and the similar insofar as similar special moments must be awakened in the division through resonance. If the special interest is attached to a, then with respect to A' α is also attached to it; α and α' enter into a special similarity, they are connected sensibly in the form of a special unity of similarity. Etc.

Comparison, the way to the universal, is determined by the similarity that is awakened. If interest is attached to the *individuum* during its concrete explication, then the concrete similarity will be awakened and determinative. Concrete uniformity is there when "similarity" is there, for wholes as for explicated parts.

Let us note here that similarity is a unity through resonance, thus, through "harmony," "agreement." Harmony, however, does not exclude degrees. "Complete" harmony as resonance, coinciding at a distance, is repetition of uniform elements, i.e., the harmony of uniform elements. Nor does harmony exclude disharmony; but disturbed harmony is itself only possible on the basis of harmony. Disharmony is the correlate of conflict in a coinciding that overlaps. This yields an entirely different incompleteness of harmony: One of them is merely a lower gradation and has its contrast in the graduated *limes* of iterative uniformity. The other is the disturbance of harmony, regardless of the gradation of harmony, in any type of conflict concerning parts or moments.

Seen more precisely, one must say here that only something that is already there can resonate. Resonance is "association according to similarity." The expression, association, is fitting insofar as a passive-sensible unity is produced. Reproductive association according to similarity, reproductive association in the broadest sense, also belongs here. Even the reproduced element was already there, only not in the "effective realization" of the intuitive presentification. Something that is not yet presentified already enters into connection with something that functions as awakening. Awakening at the same time means a tendency toward reactivation. An object in the background becomes reactivated by being grasped and explicated. Even this is a "reproduction," reproduction in the form of making present again. (If I look at an object for a second time, I re-cognize it and also have the perception as a making-present-again as one with the non-intuitive, emptily awakened memory—but here the process of making-present-again does not emerge from resonance, namely, an awakening uniform element is not already there, and the process of making present is not an awakened resonance.)

If an empty distant retention is awakened by resonance, for instance, as proceeding from the perceptual present, and in this way becomes an intuitive reproduction, then the original making present will be renewed in the new mode of the “as-if,” it is reactivated in this mode of re-membering.

Thus, we have different forms of unity that belong together under the rubric of association according to similarity: [408]

(1) the association of similarity as completely passive in continual retention and in continual coexistence (fields), likewise as completely passive;

(2) the association of similarity in coexistence, the connection of similarity produced by resonance as discrete in a field of the present. Here the awakening element is either already noticed or it awakens by virtue of an especially strong affective pull that it already exercises on the ego;

(3) the association of similarity by virtue of the discrete awakening proceeding from a prominent perceptual present or from a present remembering, resonating into the realm of closer and furthest retention (into the realm of what is passively disappearing and “the forgotten” in the natural process of sedimentation);

(4) even such occurrences like perception under steady re-cognition (re-perception) and (intuitive) remembering, a remembering that *eo ipso* is a steady re-cognition—referring to the distant retentions that are continually awakened and co-connected in passive synthesis.

The association of similarity, however, is indissoluble from the “association of contiguity.” Association is awakening. Awakening is not only awakening through resonance. The resonating element awakens its entire nexus of its coexistence and succession according to the measure of its immediacy and mediacy. The awakened element necessarily awakens its surroundings and what is or was awakened in them gets privileged.

Yet, this description of association is insufficient. In the broadest sense, association is nothing other than synthesis most broadly understood, the unity of the whole of the ego’s consciousness (as the first unity of consciousness as a whole, which is original, in contrast to social consciousness). Whatever an ego may experience, every special lived-experiencing is inserted into the entire unity of lived-experiencing, and all lived-experiencing is consciousness, all “lived-experiencing” is either consciousness or an object of consciousness. But consciousness is itself no less given to consciousness. Nothing can filter into my life, occur to me, dawn on me that does not befit the unity of the nexus. This nexus is the nexus of consciousness in which a single nexus of the constituted objectlike formation of consciousness is given to consciousness.

Naturally, I know about this nexus from the reflection on the unity of my life, and through the method of induction, I recognize its *a priori* of the nexus.

(1) The first nexus—the realm of primordial association—is the nexus in original time-consciousness. Constituted here is the primordial level of the concrete present of consciousness, i.e., the primordial level of the primordial objectlike formation as a closed universe, which is essentially necessary for the ego. Namely, [409] streaming consciousness itself is constituted as a synthetically unitary consciousness for itself in continually streaming association, in continually immediate intertwining, in fixed lawfully regulated passivity; and what is given to consciousness in it is constituted as the ever new concrete present; what conforms to perception in the form of the Now along with its retentional, vivaciously streaming past, constantly passing over into a dormant horizon of the distant past, and of an open future that, however, is only the horizon of expectation.

Nothing can be introduced into such a concrete present as element or moment that is not fitting ("concrete," insofar as we do not take as a present the merely punctual Now, the moment of genuine perception free from retention by means of abstraction or idealization). The respective entirety of the present, which itself is a streaming from present to present, is a whole.

But here one will ask: Is it not possible for there to be something completely without interconnection, absolutely separate—just like in the case of completely separate world-regions in a constituted world of things, so too in the pure immanent "world" of hyletic data, the sense fields (the visual, the acoustic, etc.) which are indeed without interconnection, and which indeed do not have anything to do with one another in terms of content and are not connected to one another? Let us leave the objective world, for considering its essential possibility would lead us too far astray. But with respect to the immanent sphere, the answer is easy. Within each field we have an inner connectedness of the field; such an inner connectedness of the field can only have the optical as the optical, etc. But beyond this we have a universal unity of form, which as such makes connections, but also only makes connections in the life of a single ego. It is the immanent temporal form, initially as that of connection in a concrete present for this ego with its relationships of coexistence and succession.

(2) Belonging to an ego's complete life of the present outside of the sublayer of the original perceptual present with perceptual hyletic data and fields is a superior layer of founded synthetic accomplishments, passive and active, founded insofar as they already presuppose the sublayer and its "objective" accomplishments, i.e., its accomplishments that are objectlike in general. The relation of foundation can be an immediate or a mediate one; that is, we can again find distinguishable layers here, and all the more so since every new constitutive accomplishment in turn is subject to the law of original time constitution, and thus, the life of consciousness here, like everywhere, is subordinate to the grand principle of iteration. If we stay at the superior level that stems from mere passivity or receptivity (as a lowest form of activity)—and not like colligating, comparing, distinguishing, relating, and even explicative, relating process of making connections out of special connecting activities with affiliated peculiar synthetic products—then we will encounter association in the usual sense (in which reproduction plays the essential role), [410] but we will also encounter an association with a broader sense. What is in question here quite generally is a mediate passive synthesis that is adjuncted to the original present as a secondary one. Here is the place of the association of similarity as a connection of similar elements, a connection of a new, higher level that is produced by resonance; these are similar elements that are already connected in a field of the present from immediate, continual synthesis of similarity. Here we find the synthesis of coexistence in a lawfully regulated continual fusion of similarity with respect to what is originally occurring in a Now, and inseparable from it, we find the synthesis of succession in an original modification in which the phases of modification

successively fuse together as continual similarities. In the higher level association we have the resonance-awakening as something new, and we have the awakening being connected to the resonance-awakening by the law of contiguity, according to which the resonating element co-awakens its nexus.

Contrast is still another possible phenomenon that appears here, namely, contrast as dissonance in relation to the element awakened by resonance, and in relation to the element awakened on the part of itself and in the coinciding with the awakened element through association (retroactive awakening).

A passive synthesis of association as fusion and coinciding is also carried out with the activity of mere reception and with the individual passing through special affective moments of the object that belong to this activity of mere reception: the object itself and its specially noticed moment. An active synthesis, which can later become a synthesis of predication, first arises if the thesis of the object and the thesis of the moment (that is otherwise merely passively unified with the object in a coinciding that forms unity) are joined in the "synthesis of judgment": This is α—and if, progressing in this way in the judicative supplement and in the higher synthetic unification of the state-of-affairs in question, the series of predicates are tied to the subject in the identification of the subjective element. In passive synthesis, they are merely passive in an order of succession, in a coinciding with the object that is being continually held onto.

In the receptivity that examines as it runs through [special moments], every special moment that gets grasped individually immediately awakens every uniform moment and brings it to a special resonance. But where the non-uniform element is there instead, this non-uniform element will be given to consciousness as belonging here "differently," as disharmonious, as contrasting. If *A* awakens *B*, which is "similar to it" in the manner of resonance, then through this, *A* and *B* will enter into the unity of similarity, and that is a unity of consonance, that is, they "fuse," they enter into a kind of coinciding at a distance while remaining separate. A partial dissonance is made possible here through the consonance of the corresponding moments that produces the ground for the discrepancies according to a more universal

moment. The "otherwise than expected," however, comes about through the fact that [on the one hand] the uniform element points to the contiguous uniform elements in resonance or in the coinciding that is produced, while [on the other] something else is [411]
there instead, occupying its place in the given nexus, suppressing it.

Appendix 19: (To §§ 28, 29, 31-36) <On the Phenomenology of Association>[43]

Awakening and the unification—within the living sphere of presence. Awakening into the sphere of the distant past.

Awakening as reanimation—that is, being introduced into the "living" sphere of the present. Being roused from slumber in the distant sphere.

In the sphere of the present: Distinction between affective background and foreground. In the affective foreground: Distinction between the thematic sphere—what the ego has grasped and has held onto (possibly still holds onto)—and the unthematic sphere. The "foreground" is determined in such a way that the affective ray that has reached the ego excites [the ego]; it already knocks on its door to awaken the ego, but [does so] still before the ego follows with (or has to follow with) a "Yes."

"Vivacity" in "degrees of consciousness"—"consciousness" in a special sense; non-vivacity (not lesser vivacity)—what is "unconscious."[44] Hence, no gradation? But yet there is a *limes*.

Synthesis at a distance (in contrast to continual fusion as melding one to the other). Synthesis at a distance as melding one to the other, distant elements being connected to one another (non-contiguous elements). Thus a distinction: immediately contiguous connection of separated elements as direct separation within a continual fusion (contrast)—and mediate connection of separated elements, but a connection of elements that are non-contiguous: connection at a distance. Contrast at a distance: Something connected at a distance is continually connected, connected in the

---

[43] February 12-14, 1926

[44] *das "Unbewußte"*

unity of a continuity, and insofar as this is the case, mediately. The continual course of connection founds possibilities of contiguous and non-contiguous special data. What is separated as non-contiguous is the synthesis of connection (coinciding, fusion) under the prominence of moments that are not being melded together in this unity, but are being separated, being contrasted. Precisely thereby the *concreta* in contrast—contrast at a distance—"with respect to," the colors red and green for instance—in uniformity, similarity with respect to the shape. (What kinds of phenomena necessarily arise when, on the one hand, these contrasts at a distance decrease, and on the other hand, the contrasting in proximity, which found separation under contiguity, level out?)

Questions: (1) How is it with the "vivacity" of the specific "consciousness" of presence with respect to the degrees of this vivacity? In particular (a) the degrees of the entire sphere of presence taken as a whole (general freshness); (b) the degrees of the particular sense realms and the particularities in them, or the special connections in them?

Can the degree of vivacity be altered in one stroke for the entire [412] presence? Distinctions of freshness. The degree of vivacity can be altered unitarily for a region of unity, a region of sense. And for the particular features. What determines the alteration, and since there is obviously something like the propagation of the augmentation of vivacity, of the "awakening" to higher levels of vivacity, what are the motives and laws of this propagation?

(2) The scope of presence is the scope of the sphere of vivacity. Can the scope be altered? It is necessarily altered insofar as the sphere of presence is in flux and continually makes deposits in the sphere of the "unconscious" and takes up something new—new as something impressional, that is, not as taking it up from the sphere of forgetfulness. But together with this process that has its laws, it is still possible for there to ensue the appropriation from the sphere of forgetfulness—according to new laws.

Expansion of the scope by "awakening" that which lies dormant, by "awaking" the unconscious.

The scope and the extent of contrast is the scope and the extent of fusion as mixed fusion, continuing, interlaced.

Continuity has three forms: (1) continuity in the usual sense of a gradual transition, of a gradual distancing, of that distancing that no longer allows contrast in the special sense to arise by steady diminution of contrast (itself a phenomenon that belongs here), that is, ultimately its zero-form; (2) continuity as arising from the coinciding of uniform elements, as fusion of uniform elements, fusion of elements that are not merely assimilated in diversity; (3) Mixture of very similar elements—but not in the continual transition of phases—a mixture that allows a new similar element to ensue. Uniformity as the limit-case of mixture, as the zero of mixture. Naturally, the continuity of what is discrete and distant cannot ensue in the sphere of presence. What is discrete and distant is separated in the framework of the thoroughgoing continuity of gradualness, and is "connected," bound in the special manner of the unification of separated elements. Binding connection is the counterpart of continuity, and presupposes continuity. In contrast, rivalry of visual fields and fusion of visual fields in the continuity of mixture.

Contrast is the affective unification of opposites, of elements that are being separated within a binding connection on the basis of a continuity, or on the basis of a synthesis of similarity of elements that are not integrally cohesive as the contrast of elements without interconnection. Rivalry, conflict, is the dissension of opposing things. Mixture is the unification of "different things," opposing things that stand "close" to one another, but not by being bound together.

Where the constitutive lived-experiences are concerned, especially the appearances-of, these lived-experiences are contrasting ones, "integrally cohesive" or not integrally cohesive, integrally cohesive in the concordance of a constitution—the unity of the constitution of a living present, and especially the unity of [413]
the impressional sphere of being is the unity of a nexus. Conflict is constituted by intuitions that are not integrally cohesive. Fusion as mixture is a special form of the nexus of the "penetration" of intuitions forming a single unity of intuition. (In this case, a special repression takes place, a repression of elements, which were previously in conflict, into the "unconscious," but not into the integrally cohesive sphere of the distant past; by contrast, in

the living conflict, repression takes place as a suppression, as a suppression into non-intuitiveness, but not into non-vivacity—on the contrary, the vivacity gets augmented in the conflict, as analogous to other contrasts).

Awakening as the augmentation of vivacity, that is, of affectivity, radiating out from a place: Temporal awakening as propagation, that is, presupposing that the vivacity [or] affectivity has undergone augmentation at this place.

But must we not say that what takes place here temporally is in action in a non-temporal manner in connection to a present that is being augmented. The unities have interconnection and have the unity of vivacity—this kind of interconnective affectivity is determined by the functional interconnections of the awakening. Every special nexus is the nexus [formed] from special relationships of awakening, which relationships, however, are determined by the content[45], by the inner conditions of the unity of continuity and contrast.

But that is transferred into the continual "structure" of the impressional sphere of the present. In succession, in structuring the processes, this structuring is such a continual becoming, continual fusing and coming into relief. But what is presupposed here is the "timeless" structuration, the structuration that is not becoming in every momentary present. Only because the present "is" in continuity and discretion can it engender effects in its becoming, in its transformation.

Still it is necessary to consider in what sense this must be taken, indeed, to what point this is correct at all. Being born blind and the moment of gaining vision—is an ordered present already constituted in this moment? Can one not inversely say: Only first in the becoming continuity do unities concress and separate-off from other unities and now also constitute a coexistence of something that is enduring. Only later can an articulated impressional "world" be immediately seen with a glance, with mere awakening, opening one's eyes. The "chaos" of "impressions" becomes organized—the impressions are still not objects, elements [won by] the reduction, genetic primordial

---

[45] *Inhaltlichkeit*

elements; dismantling intentionality and the genesis of intentionality lead back to them.[46] Unarticulated affinity, unarticulated "objects." The syntheses produce the objects according to principles that can only be gained through "analysis." The "ready-made" structure of the impressional world, with its [414] unities and with its affectivities conditioning it, is always given beforehand and with evidence. How can one sketch from this an ideal genesis out of chaos? The fusion continually becoming fulfilled in the successive path in this direction, the conditions of fusion and of the formation of contrasts whereby the special unifications, and then further, higher unifications, become possible. The synthesis of identification with respect to retention and protention dominate here.

But can a chaos of such successive momentary paths persist (every path belonging to a "sensible point" in the momentary present)? Is it at all <a> possible lawfulness of coexistence—as necessary creation of coexistent objects?

The parallel unities of succession can only be "concrete" unities, and they must be concrete in the concrescence of enduring coexistences of "concrete objects of the present" that are maintained in the continuity of succession as enduring concrete objects of the present. There must be able to be a more or less enduring present as the concrete present with objects. The conditions of concretion and of contrast must therefore also be fulfilled in the "enduring present." And now the system of parallel successive momentary paths are fused.

Conceptual distinction: chaos of "sensible points" in succession (no constitution of concrete objects of succession)—chaos of sensible points in momentary coexistence, and in each coexistence.

The sense-fields are chaotically connected to one another. We have up to this point a chaos only in a qualitative sense. But commonalities exist with respect to the shapes of time. But here the sense-spheres do not have to worry about each other, and yet

[46] Translator: That is, both static analyses [i.e., "dismantling intentionality"] and genetic analyses [i.e., "genesis of intentionality"] lead back to the primordial elements. See the supplementary essays included under Section D.

insofar as this is the case, there is chaos. But every sense-field is a harmonious, cosmic unity. One can also begin with this chaos and then ask: If such order (inner unification) already predominates in the primordially impressional present, then temporal synthesis belongs to each impressional present; however, if every such new present suddenly changes over into something new devoid of system, if the presents that are repressing one another do not go together, can then the law of temporal synthesis, of identifying synthesis predominate at all? Every new momentary impression pushes aside the one that was just there. It has "nothing in common" with the one that preceded it, and this one has "nothing in common" with its predecessor. (We can imagine different cases and types here that are less extreme and "extremely extreme.") No conditions of fusion and of concrete unification are fulfilled, and therefore everything that is repressed sinks without support into the "unconscious."

The original vivacity of impressional momentary-sensibility can only be maintained as "retentional" if it is supported by the new impressional momentary-sensibility. Steady successions must be able to proceed out of every sensible point of this impressional [415] togetherness of the present. Now, can all of these points of the present be as alien to one another as the sense-fields are? Is each one a species unto itself? Then there would be no connection in coexistence, no concretions of coexistence, no enduring being.

Such analyses, constructions of all ideal, abstractive possibilities that belong here must be carried out in order to understand the structure of affective formation of impressional "worlds" of the present.

A mistake in the lectures and also in the above: The sense-fields are not chaotically connected to one another, or this does not have to be the case. If each one constitutes objects for itself, it is indeed constituted as enduring in the shape of time. Since the fillings of uniform, i.e., parallel duration are not homogeneous, they cannot contrast with one another, they cannot conflict with one another, and this is precisely due to the fact that the duration is completely fused with the parallel duration—every simultaneous element is the simultaneous element of one fused simultaneity, a simultaneity that only diverges according to different, heterogeneous fillings.

Yet, this does not suffice to make intelligible the identity "of parallel" durations within duration. They each derive from a single primordially impressional present.

Thus, we also have affective nexuses of heterogeneous elements through the homogenous shape of time.

Thus, [we have] the appearances of rhythm that are begun by the simple repetition of unarticulated contents; [we have] commonalities of the form of time, which commonalities can occur in different spheres of sense as uniform and can ground the affective nexus. A rhythm of light signals can "recall" a rhythm of tone signals—the uniformity can be the bridge leading over to affective salience. Precisely for this reason, the usual association of sense-regions can also spread from sense-region to sense-region. Naturally, the awakening is the strongest when material similarity is connected to formal similarity, like in the examples of series of light.

But this also holds analogously, or could hold analogously, for spatial forms, spatial shapes in these forms. One cannot say here that the simultaneous feature of the uniform form is fused in an undifferentiated field.

There is still a question concerning how affections relate to one another—apart from propagation, or in addition to it, i.e., as awakening salience or rousing from the unconscious. Affections can play to each other's advantage here, but they can also disturb one another. An affection, like that of extreme contrast ("unbearable pain") can suppress all other affections, or most of them (not without first having to undertake a more precise consideration of the interconnections!)—this can mean to reduce to an affective zero—but is there not also a suppression of the affection in which the affection is repressed or covered over, but is still present, and is that not constantly in question here?

In particular: Affection of the *modus excit<andi>* of the ego, [416]
being irritated, conflict of affections. The one winning out does not annihilate the other ones, but suppresses them. (In the sphere of feelings or drives: feelings, strivings, valuations, that come to naught due to certain motivations, just like the absence of value becomes evident through clarification, and the affection of value comes to naught through an appropriation coming from the inside.

On the other hand, feelings, valuations that are overcome, suppressed from the outside, suppressed in conflict, while the conflict does not lead to any settlement, to any actual "peace.")

Perseverance. Affections can be there, i.e., progressing from the "unconscious," but suppressed. Intensive attentiveness—suppression of affections of interest, but of another interest.

In the mobile present something new that is advantageous to the thing suppressed, and awakens it.

Appendix 20: (to §30) Time as the Form of Individuality and Subjective Transformation[47]

Individual being as temporal being: It has as its original mode of being, the mode of the present, and it has as its non-original modes of being, the "no longer" (the modes of its past) and of the "not yet" (the modes of its future). (The present is correlative to the life-present of the subjects, as subjects of possible experience of the respective *individuum.*) The individual concrete being in the concrete present is an enduring being. It is in a present duration and is as enduring in the necessary transformation from the Now that flows as a primordial source (primordial present) into the past and into the further and further past. It has before it the future that has not yet arrived, the future that approaches the present as the more distant future, while every becoming past of the Now clears a place for what is coming next, which itself has become the primordial present of the Now.

The mode of the concrete present is the current being in duration as enduring together with the sinking into the having been. The duration of the individual is the ready-made duration in the Now in which the duration has been completed and is no longer. That is, the *individuum* is no longer current actuality. A completed duration has a beginning, an original Now of the individual content, a Now that does not have any past behind it, namely, the Now of this content belonging to the *individuum.* The completed duration has its end that concludes it. But the *individuum* in its current duration is in constant transformation of

[47] Editor: Probably between 1922 and 1926

duration, and if the duration is completed, the completed duration is transformed into the mode of a having-been that goes further [417] and further back. The past is finished time, the finished duration which as such always gets transformed in the modes of the past, but remains identical as the same duration. Finished time is the time of durations that are to be identified by the constant transformation of modes of the past. The future also has its identity in the transformation of the modes of the future. Futural being, futural enduring being (assuming here that it will in fact be effectively realized in a current present and from here will become an ever more distant past—but that belongs precisely to its being future)—futural being, I say, is identical being in and through the transformation of the modes of futurity, and in and through the transformation that goes into the distance, and that goes into the distance that is further and further back, into the distance of the modes of the past that link up in the transformation.[48] What remains identical is the form of the individuality of this respective duration, [and this] in abstraction from the presupposed fillings of content that coalesce in this form of individuality to form a unity; such presupposed fillings of content fulfill the conditions of unity. This is the form of time "being in itself," time *simpliciter*, as it is regarded when we speak of objective time. But time is what it is only as the form of identity of temporal objects and <is> only what it is in the "eternal" transformation of finished time, of the current present and the future that is still to come—modes that are themselves only in relation to one another and in the process of transformation.

Appendix: 21: (To §§33 and 34) Sensible, Multi-Radiating Affection. Sensible Group—Genuine Collective Objectlike Formation[49]

In the immanent sphere: Several disparate data, a tone, a color, etc., simultaneously exercise an affection, each one is prominent

[48] Further modes to be considered: current-being in the mode of eternity, in an "eternal duration." To what extent is this conceivable? Eternal past. And future?, etc.

[49] Editor: Probably from the period of the lecture.

for itself. The affected ego then follows, individually, the color, then the tone. A succession in grasping the individual thing emerges from the coexistence of affection (several affective special rays directed toward the ego, meeting in the ego and becoming one within the ego through this process); through this process, however, what is already grasped either gets abandoned or is possibly maintained, depending upon the circumstances; so that a collective succession results.

Such a collective succession is still not a genuine set (genuine set—everything is *one* object). Lacking is a unity of theoretical interest and, let me initially say, a unity of interest in an object—a unity of interest in being and in being-thus, etc.; this grants a unity to the special interests being played out in the graspings of the particular things. In the transition from tone to color, etc., what has just been grasped individually is still held onto because the special interest, because the special intention directed toward what is [418] there, what is there to be seen, is not yet completely fulfilled, and the interest bearing on the next element becomes more pressing and more overpowering. The new case, however, is such that a synthetic intention encompasses particular intentions, and that *one* intention is there, as it were, an intention that is indeed fulfilled in each special grasp, but is fulfilled in each one simultaneously; by being fulfilled in the one special grasp, it is not yet satisfied with it, and <this> is because it still requires the other special grasp that is above and beyond this one.

If we imagine a situation that is still prior to all apperceptions, then the most primitive situation will consist in the different affections being homogeneous. For otherwise one would conceive only of instinct fashioning an integral togetherness. Then that would also be seen as a kind of anamnesis and as a kind of obscure background apperception. Thus, if we rule this out, we will have the case of sensible groups. Already prior to all turning toward, the particular elements have a community exercising an affection in their particularity; they stand in a relationship of "resonance"; one promotes the other, that is, the affective allure on the ego by the one promotes the affective allure of the other, and vice versa, but in such a way that these affective allures do not remain separated, but rather go together to form the unity of one multi-radiating affective allure in which the augmented affective allures are unified, and in this unification, each one bears the character of reciprocally furthering and resonating with one another ("recalling" each other).

Does the ego follow this? It can and it initially wants to follow the whole, the synthetic unity of affection; it grasps the sensible group as not separate; but by means of special graspings in the chain of succession, the intention inhabiting the affection, the intention directed to the whole presses on in the fulfillment toward the fulfillment of the affective intentions included in the fulfillment. Coinciding here in the transition through synthetic coincidence is the common element; it becomes prominent; what is different is separated-off. In this way, not only does a succession of special graspings arise, graspings that hold onto what was previously the object of a special grasping, but new unifying interests and intentions are now motivated: What is grasped, element by element, is not interesting for itself; rather, they have very much to do with one another; they stand in materially relevant relations, they participate in the same identical essence, they are distinguished by differences coming into relief, and what is newly constituted there in the transition through coinciding, namely, the common element, exercises an affection, and in this way exercises a tendency toward a new, higher level grasping, etc.

Let us then pass over to apperceptions: Just as such processes were initiated and carried out, so too are experiential apperceptions necessarily formed. That is, the sight of a group that is still not explicated (indeed already the background perception of the same group, the mere group-affection) awakens an apperceptive horizon, an intention toward the formation of a collection as the substratum for plural explications, predications, etc.

But belonging to the essence of the unitary, materially relevant interest that runs through the collection as unity (or with respect to [419]
the unity of the intending, the unity of the collective intention) is the fact that we must distinguish between the collection itself as the meant plural, and the succession peculiar to [the process] of running through [the elements], that is, peculiar to the temporal

sequence of the givenness of the colligated elements. Collection is an intentional unity that can only be constituted *a priori* in successive sequences, so to speak, as aspects of collection.

We must distinguish here [a] the intention toward the explicit grasp of the individual and the synopsis, that is, toward the formation of the explicit collection, of the explicit ensemble, and [b] the intention toward the fulfillment of the further apperceptive horizon, <toward> comparison, differentiation, reciprocal relation, and determination. In the first respect, that of genuine collective intention, the intention is fulfilled if each thing grasped has been grasped as an individual in a non-explicit unity, in any kind of sequence of a grasp that passes through [the elements]; whereas a second passing through all elements in a different sequence produces the consciousness of the same collection. In the former, each special grasping coincides with the group's guiding grasp of the whole "in partial identification," and the chain of special graspings, i.e., partial identifications, yields a total identification, namely, in the form of a total fulfillment of the multi-radiating unitary intention. But repetition serves clear and "clean" fulfillment, the completeness of the evidence of the identity of the group-whole that is given as progressing in a unitary manner, as well as the completeness of the evidence of the totality of the particulars.

We should not overlook the following: If we have a lasting sensible object, its affection does not, for instance, cease with the turning toward. The allure is still lively, and it continually draws me toward it, even if I am there with it and I am already "occupied" just with it. Likewise, the group is constantly there in the unity of attraction, as the multi-radiating unity of the allure, as the multi-radiating unity of the tendency toward collective special knowledge, as an explicit, manifold knowing that encompasses the particular knowings, and then knowledge. Affection does not end in grasping, either in its single or multi-radiating forms.

The group is grasped as group, it is apperceived as set. A unitary intention toward an object, directed toward the ensemble, toward the ensemble of these objects that are implicitly included in a unitary manner in the group-intention—and toward all of them. But is a unique constitution of the thought of totality required here? Prior to the apperceptive intention that already bears on the particular objects, I do not yet have a group; I only have it when I first have a unity of the apperceptive intention that is fulfilled in the particular graspings and positings of objects, not in the particular ones however, but rather in the unity that is collectively synoptic, the unity of synthetic intending.

Appendix 22: (To § 35) <The Empty Horizon and the Knowledge of It>[50] [420]

How do I know that the empty horizon of retentionality still signifies the consciousness of the past, and just what actually is "empty horizon"?

We ask how the distinctions that we initially make, and that we have reasons for making, relate to one another, i.e., the modifications with respect to retention about which we speak:

(1) the modification of retention as retention, as the transformation of the mode of the primordial present into the primordial pasts. Can we speak here of a *limes* of nil?

(2) the fading away of intuitiveness with the *limes* of nil: emptiness.

(3) the change of clarity into unclarity, of prominence into non-prominence.

It is now questionable to what extent one can speak of steadiness here. If I hear the sound of car wheels rolling, or the sound of a uniformly paced pounding <or a> warbling, I have clear distinctions in the beginning; I can pay attention to each particular "piece" that is coming into relief in its cyclical process. But soon I only have an appearance characterized as a unitary process of uniformly articulated events, an appearance in which I can no longer extract particular things; it is only a typical rolling, warbling, or pounding coming into relief as an entire series, but which is "unclear" in its particularities. Even if there are cycles in it, I have precisely a chain of cycles, not however the determinate differences within the particular cycles, but only something typical that I can notice. But even that very quickly comes to a halt. It is a

[50] Editor: Probably 1922.

unity generated from a continual self-coinciding. Every new rolling is a rolling and continues the previous one in uniformity, and this uniformity in succession is itself a phenomenon of intentional coinciding. But differentiation is lacking; only the phenomenon of the whole is precisely separated-off. In this case, the point of departure, the beginning, or the particular thing that is distinctive, that which interrupts the coinciding, i.e., the separation that is imposed (through the "other than expected") can continue to adhere as a distinctive point of prominence in the unity of the whole. The beginning, however, can become completely indeterminate, the uniform progression of the rolling does not have any clear, prominent beginning; it has a "mobile" beginning of intuitiveness, it advances in the prominent unity of the whole under the disappearance of prominence into emptiness; an indeterminate "horizon of beginning," no point, endlessness.

This is original forgetfulness, the retentional element that has become "unconscious," the just-past that has become unconscious. Another kind of unconscious is that which is already unclear from the beginning even though it is intuitive, the perceptual sphere—and already the primordially impressional sphere—which is lacking the force of its own affection. To be sure, one must [421] distinguish here between that which possesses its prominences, its original distinctions, and that which has lost its distinctions in the process of change, and possibly takes them on again, but without exercising any "efficacious" affection on the ego. But then does it still possess prominence? Is prominence not equivalent to affection, only that the ego is not there for each prominence through the slightest momentary turning toward or a momentary "yielding a little bit" that is not yet a grasping? Something that is given as unconscious here would be something that is not grasped and that toward which the ego does not let itself to be drawn even one step of the way. Something forgotten however is something that no longer has any prominence.

But the situation has its difficulties. What does the re-emergence from the background mean? If the melody continues to go on and something from the submerged beginning emerges out of the empty horizon, then the entire nexus that has become empty emerges in a certain way with the Now, i.e., with the concrete present that is still basking in light of day; the anterior element of the past (which is one with the just-past that is found in affective consciousness) continues on in it. If I sing the first verse of a song to its end, the beginning of the song is prominent for a while, but finally it "disappears." But by the end reminding me of the new beginning, and by the latter coming on the scene and awakening my earlier beginning and its connection that has faded away in the verse to the end of it, the verse is there for me now as a unity without it being remembered, and even more so as I continue to sing the new verse.

What kind of awakening is that? Likewise if a sequence of tones arises and is repeated after its beginning has already faded away and has sunken into emptiness, and then gets awakened by a repetition and then a new repetition, then obviously none of these repetitions are rememberings. Certainly, the affective force diminishes in the series of repetition, and *qua* series itself, it is lost in an emptiness.

This would have to hold similarly if a musical phrase or a very large nexus enters into a musical relation with a nexus that is being played anew, and by means of this nexus becomes salient as a unity. Indeed, one can say nothing else than that the emptiness is also precisely a modality of intentionality, that it is the form in which the retentional modifications "get blurred," in which their special affective force "is forsaken," but which, according to certain essential laws, can be won back. The awakening is certainly a new mode, but we will surely have to say that this awakening does not create a new lived-experience, like, for instance, when a new sense-datum arises; rather, we will have to say that it brings the progressively changing retentional transformation to the form of affectivity in the retentional mode that exists precisely in the change. By something being awakened by a repetition, by something that was submerged being hauled up out of "forgetfulness," the mode of the past is also affected, which mode belongs to what has been hauled up through the awakening. The "forgotten" in the original sense of having become empty is [422] not a mysterious nothing for which one would only affix a real possibility that a new phenomenon can be causally effected under certain circumstances; [it is not] a blind, external lawful regularity behind which nothing is to be found. Rather, there is internally a submerged life only in the form of phenomena that are "not prominent." Thus, one would have to say that the unconscious object has in principle the same style everywhere, even in the sphere of the present. And one will place prominence and unconsciousness in the following relation: Speaking from the side of the object, prominence for example is something that is presently available[51] (being constituted in intentionality) having qualitative differences, etc. But that is not itself prominence; rather, prominence is special affection, and special affection presupposes precisely certain conditions such that one must always speak of one prominence in opposition to another (prominence of something).

Since every transcendental, subjective life is consciousness, intentional life, and since everything that we place as a *cogito* into immanent time itself corresponds to the primordial law of consciousness, to the primordial law of time-constitution, then the "immortality" of every retentional flux means the same thing as the immortality of every particular consciousness, that is, each particular consciousness is situated in the eternal temporal transformation as temporally modified, and each transformation maintains its individuality, its place in time, its individual content, its sense. The unending realm of forgetfulness is the realm of "unconscious" life that can be awakened again and again.

But certainly in addition to this there is the phenomenon of the overlapping of several pasts that are awakened together, and moreover, the phenomenon of reproduction and the phenomena of association and of associative fusion.

(1) Every perception has an empty horizon of retention that is itself an undifferentiated empty retention. How do I know that this empty horizon persistently swallows up the retentional component, that every one ultimately sinks into it? Only what remains in relief can be grasped once more. How do I know this? And if I turn toward another one, and turn toward it exclusively in such a way that I let loose of the one I had grasped, then it will sink into the empty horizon. How do I know this? I have the

---

[51] *ein Vorhandensein*

phenomenon of "letting loose" and turning toward something new, and I have the phenomenon of turning back toward it again and then, further, the phenomenon of grasping that there is still "something" there; likewise, when I live in the perception of the present object and remain with it and a turning back does not ensue. If it ensues afterward, then I will still grasp something that is prominent and I grasp a background as something or other. But I know that it must have the same structure; I have something that has been; I know that there must be a temporal series. I have an empty, indeterminate, non-intuitive past. But surely, do I not know this only through the reproduction? In any case, it is something retentional—not just any appendage to retention that is constantly there, but rather, retention. [423]

We should also note that if I have seized something twice, and after a prior turning toward, I turn back again, I remember these prehensions and com-prehensions, while I no longer find anything different in the retentional field, namely, in the field of the retentional tone. Even the fact that there was a different multiplicity, a plurality, can be grasped by the regard turning back, and nevertheless, this plurality is not something that can be grasped in particular; but this is also likewise the case for the most general and indeterminate aspect of a retentional series as the necessary form, or better, the most general and indeterminate aspect of a progressing past in the process of being modally modified. But with all of this, I have at most an indeterminate infinity or something ineffable that I discover later, through remembering, as the infinity of the past (in the genuine sense).

Thus, it remains that the retentional field consists almost entirely of empty presentations that have a flowing continual nexus, that harbor indeterminacies; and it remains that this field passes over into an empty presentation that is in itself completely indeterminate, a presentation of an "endless" past whereby this endlessness of the past is not to be conceived on the order of a clean line, and not as finished by a horizon-point as an apparent end.

(2) Belonging to the essence of the empty retention is its capacity to be fulfilled, and this fulfillment is remembering. Not every remembering is the fulfillment of the retentional empty

horizon of the present. For this empty horizon is "the forgotten" that carries on the differentiated retentional path of the past; and what becomes disclosed there continues to be what has "just" been or what has been "just a bit ago," and its memorial elapse along with its memorial protentions leads into the just-having-been of the concrete present such that the remembered or the disclosed temporal expanse gives itself immediately as one with the living present.

(3) An emergent remembering, let us say a distant remembering, in accord with its temporal content, initially has the peculiar feature of positing a time in a re-presentifying manner (a temporal objectlike formation) as a *processus* with a constant protentional horizon; the *processus* is given as being situated in the direction of the retention of the present and in the direction of the latter's horizon, and it is <a> reawakening of the horizon; this is similar to what we have become acquainted with above. But we see with this that the horizon of emptiness possesses a proximity to the present, a just-a-bit-ago, and a previous past; and remembering fulfills an especially awakened empty place through a *progressus*; like in the region of proximity, and at the same time it exercises an awakening [force] in the *progressus* (association directed ahead), and what is awakened is fulfilled in further rememberings. Finally, the near horizon, and the realm of the retention that is still living, and the concrete present in general, will be reached, and with this the distant horizon gets disclosed in a path up to the present as the process of remembering.

(4) With respect to its quasi re-constitution of the present, every [424]
remembering has a past; as every perception itself arises from the past—another perception has preceded it. Every perception has a background of the fresh past.

Every past (the retention attached to remembering) can be disclosed, etc. Finally, every past is disclosed as the trait of an unending time that terminates in the mobile present.

(5) The horizon of the past is disclosed by remembering, the horizon of the future by perception, and every perception itself is the fulfillment of an empty horizon of the perception that has just preceded it.

Remembering can occur, the past can be disclosed step by step, and can be entirely disclosed through the insight that a remembering is possible again and again, a remembering that uncovers the retentional horizon of something already disclosed.

Does the future have to be disclosed? Does it always yield a present? The question is whether the present can not be reduced to a completely empty horizon (the absolute black of intentionality). Is such a thing possible without a "new present"? The limit-cases are unique and not unimportant problems.

Appendix 23: (To §35) <On Questioning the Potentiality of the Empty Horizon>[52]

The question for me is how one should interpret the potentiality of the empty horizon from which something emerges, and whether one should at all speak of a single horizon of forgetfulness. If something emerges, it evidently takes place only through the occurrence of an empty intention and fulfillment. And the empty intention already has an intentional relation to the Now and, for instance, to everything else still occurring in this manner. The intentional tendency toward fulfillment finally terminates in the unraveling fulfillment that awakens an ever new tendency toward fulfillment up to the current Now. Indeed, it belongs to the essence of every such intention that it can be fulfilled in such a way, possibly deliberately fulfilled.

On the other hand, belonging to the present is the fact that I can inquire into its past, that I can delve into it despite its limitation. But [I can] only [do this] in such a way that I awaken the past from the present; I cannot continually move down the path backwards. And if life were a "monotone" existence, for instance, a tone that is continuously elapsing in a homogeneous, undifferentiated manner, then I could not go back. This is really quite significant. If the "beginning" of life, the beginning period, were an endless monotony, then it would be a period of impenetrable forgetfulness. And if various periods of such monotony were conveyed by content-laden multiplicities, but by [425]

[52] Editor: 1922?

ones that did not fulfill any conditions of reciprocal association, then only one non-monotone life could be surveyed at a time, and not a synthetic unity of life above all periods. That would only be possible if the good "God" were to create for us such a new life after death with such a multiplicity of contents that associations could extend over all periods of life. But can a "continual infinity", e.g., an unending life without sleep be comprehended *in infinitum*, thus, can a springing back be possible, a springing back that however would then only proceed "*in infinitum*" and would not lead to any end, and never to a remembering that would remember the entire life? For remembering can only run forward, and where there is no beginning, there is no complete re-living running forward.

Thus, it must be said that if one should speak at all of an association that is continually turned backward, it cannot find any privileged motive either in the uniform continuum, or in a uniform elapse (for instance, "of the same" tone, again and again, in uniform distance); thus, it cannot motivate an efficacy of tendency in a passive reproduction, and therefore it cannot motivate any privileged affection, any attentiveness or any will to renew.

Only where non-uniform givings emerge can awakening take place, and accordingly remembering must have the form of springing back.

Appendix 24: (To §37) <Effect and Cause of Awakening>[53]

But what can "awakening" mean here, and what can it accomplish? Breaking through the fog—a metaphor? Is it not a swinging back to a stage in the process, a stage that is still distinct in which the concealment has not really progressed that far? Finally, even the re-establishment of perception? The process cannot actually be established a second time.

If an interest is turned toward something sinking down and the latter remains prominent, the process will in fact not be halted, and yet a "prominence" is maintained. Can one speak of prominence in the same sense with respect to the prominence of the datum that

[53] Editor: Between 1920 and 1926

has sunken down, like one speaks of the prominence of a figure on a white background? The latter is available or not, and it can only be produced by the production of differences of color, etc. Does the awakening make the retentional fog weaker? [Does it make] the fusion with the other retentions weaker? Or is not a synthesis produced between the present and a continually concealed past, a [426]
past [constituted] in and through all continual concealment, but which is not a Nothing, but rather [is a] concealed past. The modification of concealment, the present retentional distant horizon (namely, the horizon that has not halted its modification) does not come into relief in the sense that an earlier stage of the process would reappear, a stage of process that indeed does not lie in the distant horizon in an intimately inherent manner, and that would also only be able to be produced through remembering and by letting the "re"-perceived object fade away "again" in the mode of remembering.

But this horizon is precisely a horizon that harbors *implicite* a concealment and a continuum of concealment, a concealment to the nth level, let us say, in which something is given to consciousness as the present to the nth level. If now the distant horizon is also a nebulous fusion <in> which everything is uniformly non-differentiated as "forgotten," [if we have] an eclipse of a present by another present (this in turn being eclipsed by a third, etc.), then precisely all presents are implied within the entire fusion (which is now a lived-experience), and each present is in a relationship of eclipse. These hidden presents can now be awakened. An awakening punches through all coverings as a single synthetic ray that bears on the respective object of a present. Can this mean that that empty retention in which the object is given in the present of awakening, in which it is "now" "still" given to consciousness is brought into relief (although not given to consciousness for itself in a consciousness that is brought into relief in the present)?

We have to consider the following here: When a melody is over and fades away, I turn my attention back to it in "a single ray," into the unity of retention that bears the melody within it as a successive temporal whole. As elapsing in the impressional stage, the melody is in the process of becoming. It is only finished with the concluding phase, and then it is a melody that has become. The focal ray directed toward what is becoming is different from the focal ray that is directed toward the whole that has become. It is not a "focal ray" directed toward the last tone or toward the melody that is becoming in the concluding tone; it is directed toward the whole and can pass over to explication. This intention is fulfilled in a successive re-perception in which the melody elapses over and over "again."

Belonging to the unity of an object as such is a unity of constitution running its course in a successive manner. This unity is also at issue in the distant horizon, and the awakening is a "focal ray," an intentional ray toward the One that is there as the whole empty retention of the earlier melody, the whole empty retention of the whole melody, i.e., in this empty presentation, and is something that is awakened through this ray. Thus, my previous conception was correct, namely, that the awakening goes all the way into the distant horizon and awakens retentions there (albeit the respective retentions of concrete things, processes, etc.) in the form of the associative "rays": The awakening object is a present similar object, which, for instance, is perceived or in a fresh retention, etc.; and the synthesis is then a synthesis between [427] perception and empty distant retention.

But *must* a coinciding be produced between each and every thing that is capable of a coinciding, since the "distance" is there precisely as horizon in the present at all times? Where awakening is concerned, however, it is a matter of a special mode of synthesis as one that is causally becoming, and in this case it is a matter of a new accomplishment, namely, pointing from an *a* to *b*, that is, together with *b* being awakened by *a*. The *a* in the "present," the *a* that is first salient, must have a peculiar feature so that it functions as awakening, and does so in various degrees of intensity; and so that it awakens precisely this *b*, this must also, in turn, be somehow grounded in this *a* itself. To be sure, experiments are useful here. They supply us with examples, and therefore possibilities; through intentional analysis we can then see what can come into consideration here and to what extent essential laws are in play here, [and we can see] what in the final analysis must be the case. They are motivations.

If we assume that everything that is susceptible of coinciding within passivity *eo ipso* is in the synthesis of coinciding, i.e., without genuine causality, then the associative synthesis is not a mere synthesis of coinciding, but rather, something new that the synthesis of coinciding only presupposes: as the association of similarity. The present object that is to function as awakening, one could say, has a special interest, and not everything that is similar, but rather, something that is similar in the relation and that would correspond to a similar interest is at issue. The interest in an *a* depends upon its type in accordance with the complex of features (α, β, γ). What is uniform or similar or even identical to *a* with respect to $(\alpha', \beta', \gamma')$ gets privileged. Certainly, it is not merely inner features that come into consideration; rather, the interest belongs, for instance, to *a* in its situation, in its nexus (figurative synthesis). A nexus originally yields quality-like characters prior to explication and prior to the formation of relational predication. This yields external features for the *a*, which features also come into consideration. And, perhaps, even in a very significant manner. And so it is also the case within an impressional present. If an *a* arouses a special interest—a "pebble" turns out to be a piece of a fossil bone—other similar things immediately come to the fore, [and they do so] in special syntheses that favor a similar apprehension.

A person [looks] suspicious in a certain a situation—he immediately reminds one of similar situations. And then the association through contiguity arises (which we have not yet touched upon); the interest and the awakening passes over into that situation. But what do we understand by the term "interest"? Must it be an active interest, or a passive modification of an active interest?

Appendix 25: (To § 40) <Kinaesthesеs and Potential Expectations>[54] [428]

As we saw previously, the path of the intention of expectation, where external perception is concerned, is only a path emerging

[54] Editor: From the period of the lectures.

from the multi-dimensional continuum of the whole horizon – the whole horizon that characterizes the entire co-present. This path is the path that is actualized and especially motivated by the current course of kinaestheses, the subjective eye movements, etc.; precisely through this motivation, it gets the character of an actual expectation.[55] What is remaining in the [whole] horizon is a system of potential expectations. It is the potentiality that corresponds to the subject having at its disposal systematically constituted kinaestheses as a whole. Other paths of this horizon can, so to speak, be awakened associatively from their slumber without losing their non-intuitability, and they become awakened with the empty presentations of the kinaestheses belonging to them. Every such awakening fashions a special empty presentation that is not an expectation, but is essentially related to such an expectation precisely as a potential expectation. The kinaestheses belonging to them bear the consciousness of an ability, namely, of an ability to orchestrate them, and thereby consequently <the consciousness of> the ability to put into play the courses of appearance motivated by them (thus, the "co-present," unseen sides of the object). Thus, it is a matter of modified shapes of protentions that essentially belong together with them.

Likewise for every external co-present, those co-presents of surrounding familiar things. If our regard runs along the room and lights upon the door, then the initially empty presentation of the foyer outside gets immediately awakened. It is not an expectation; we do not go out into it. By going out into it, we would naturally have the series of appearances as expected, and the path would become visible in this series of appearances (in the corresponding subjective modes). But the awakened presentation does indeed have the character of a modified expectation, of a potential expectation. The empty presentations that occur here, among them also the expectations (protentions)—all of them have a common, fundamental feature. A current present object always refers ahead, always points on to something else, and then further, something

[55] On the transformation of the original "hyletic sphere of the present" through kinaestheses. The constitution of "transcendent" objects is carried out with the formation of systematic protentions, that is, expectation also becomes constitutive.

given to consciousness that is possibly already empty refers further on to something else that is given to consciousness in an empty manner, and in the movement of progression, refers further on to something that is especially in the process of being given to consciousness. Every empty presentation that we encounter here stands in a peculiar connection such that one thing points to another, and this again points to another (now continuously, now discretely). <That> which is being referred to itself has a character [429] of consciousness, the character of the "something being pointed to," the character of the "something being intended"; just like the element doing the pointing has the counter-character of pointer, the character of the point from which the intending radiates out. The element from which the intention proceeds and which stands within such nexuses is at the same time the point of departure and at the same time the end point of the intention. But it can also be the case that in the beginning there is a free element, an element that is given, e.g., perceptually, toward which no indication is produced, but from which an indication radiates out.

We obviously need a term for that consciousness in which something has the distinctive characteristic of being intended in this sense. Unfortunately, language fails us, namely, a language that cannot be interested in phenomenological distinctions. Even the language of phenomenology in its beginnings did not do justice to this peculiar feature of consciousness, whose special place and function was not immediately recognized. In what follows, I will speak of "associative intention"; what also gets expressed here is that it concerns a fundamental feature within the passive sphere, and is not a kind of intending on the part of the active ego.

Appendix 26: (To §45) <Repetition and Essential Identity of Rememberings>[56]

Two clear rememberings essentially belong together in such a way that they are to be conveyed in the continual unity of a single clear remembering. A clear remembering can be repeated in an

[56] Editor: 1920/21

arbitrary manner (like every memory), but it can only have this conveying function in a complete, essentially identical, and clear remembering in one and the same nexus and at the same place. The repetition of a clear remembering is the repetition of an idea insofar as each one, irrespective of its contingency as lived-experience, would be completely uniform, and [would be] completely uniform insofar as it would give the same self in it, the same past of consciousness and the same intentional sense that was constituted in this repetition. The idea of a complete remembering of past consciousness and of its intentional objectlike formations—.

We call two rememberings memories of the same concrete essence or essentially identical memories, if the one is a mere repetition of the other, if they are distinguishable only as repetitions, or if they can only be distinguished as this or that.

A remembering, taken concretely, is a becoming in which a temporal objectlike formation becomes intuitive once again, and in this way becomes self-given as past. No remembering can contain [430]
two completely essentially uniform rememberings as elements; each remembering of the same essence gives a past and each memorial element gives a different past. Every memory has its horizons that belong together to its essence. And the horizons of memories, which are parts of a memory, are necessarily different. Every concrete part of a memory can also occur as a memory for itself. Every remembering is to be multiplied through repetition, but not through repetition in an expanded memory. Every adequate remembering as essence is an element of an encompassing clear remembering, and in this way, every adequate remembering is ultimately an element of an all-encompassing remembering that is eternally operative, is a operative idea insofar as this idea is a new idea for every current Now: namely, all adequate rememberings embraced in a continual unity, a unity that embraces all pasts in truth up to this Now. Of course, just as there is only one repetition for every adequate remembering, so too is this the case for the all-encompassing remembering; and the following law is also valid, [namely, that] two adequate memories, which connect the same starting and end points in the temporal objectlike formation, are only one single one. Two adequate

memories cannot be connected by two adequate memories, which are different from mere repetitions.

Appendix 27: (To §45) <Two Fundamental Concepts of Evidence: Self-Giving as Such and Pure Self-Giving>[57]

An encompassing theory of rememberings as self-givings for the stream of consciousness, as immanent objectivity, that is, for the multiplicity of the objectlike formations of lived-experience being constituted in the form of immanent time, and included in this as self-givings for immanent time itself—an encompassing theory, I say, would have to articulate in a systematically ordered manner the essential laws of rememberings, and among them, the essential laws of their *limes* forms that have to be intuitively exacted, that is, the essential laws of complete self-givings. These would obviously be the correlates of the pure laws of time themselves.

By means of the progress that we have made, we can draw positive consequences having the most universal import, and initially for the formation of the concept of evidence. It is now divided into two fundamental concepts. The *first* is that of self-giving as such. Properly understood, it extends as far as the concept of consciousness in general extends, namely, consciousness as the consciousness of something. Each consciousness is consciousness of its Something in the mode of belief; however, it can be a consciousness of something that is bifurcated, somehow modalized, and refers back to a different [431]
consciousness that is not modalized. Every sense-giving in the mode of belief is either self-giving or not self-giving, and what is not self-giving has to be verified or annulled. Both of them lead back to self-givings. Self-giving consciousness is evident in the broadest sense, it is insighting consciousness as opposed to a blind consciousness, or (with reservations) merely anticipatory consciousness, as in the case of intuitive expectations. Evidence in this sense admits of degrees. It is already a correctly giving consciousness insofar as only something that is not self-giving can

[57] Editor: 1920/21

be directed, can be normatively regulated according to it, can be fulfilled in it. This evidence is the foundation of evidence in the higher sense. Only a consciousness that already harbors a self [of an object] can be normatively regulated in a higher sense, that is, can be approximated in a pure self [of an object].

The *second* concept of evidence, the narrow one, is then the concept of evidence as pure self-giving—belonging to the essential law that self-givings can be more or less clear and thereby more or less complete. The correlate of this evidence is the true self, true being. Rigorous evidence is an idea, the idea of a complete self-giving lived-experience, which an unclarity possibly approaches. Its correlate is the idea of a true being that is given in approximating consciousness in which the idea as *limes* is seen in intuition, and it is given as absolutely incapable of being crossed out.

All of this holds not only for individual objects or even only for those of immanent time—objects to which we have devoted our studies. When we said, for example (as we spoke of the essential laws of expectation), that the futural occurrence is motivated with evidence if the motivating factor gets clearly remembered, then the groundedness of the anticipation of the future to be expected is precisely self-given for these and those reasons, and in the case of complete clarity, it is precisely evident in the higher sense, a higher sense that determines all talk of Reason. If, in the case of evident motivation, two mutually rivaling anticipations are motivated, then a conflict between two futural possibilities and probabilities are given with evidence. Even non-being (likewise, probability, and the like) can become objective, or rather, it has its objectivity; it has its idea of pure evidence, of pure verification or annulment, and belonging to verification, its indestructible identity. We customarily speak in such a way that we understand evidence in a pure sense, and we quite often claim to possess this rigorous evidence. Though, one would probably have to confess that we come very close to the *limes* in many cases, and that this "coming very close" is also something given. Indeed, it seems to me that the true self possesses a *limes*-character in all cases, even for axioms, a *limes*-character that we can reach by "touching" it. But this already goes too far here.

You already see how very far we have progressed by our investigation into the constitution of the first in-itself, that of the immanent sphere, and how we have already clarified within this [432]
sphere the fundamental character of all the inquires into the constitution of an objectivity as an in-itself. Just as in that instance the idea of the in-itself leads us back to the distinction between self-giving and non-self-given lived-experiences, and to the essential possibility of the verification of self-giving by approximating pure evidence, and how pure evidence is an idea whose correlate is the idea of an in-itself that is absolutely incapable of being crossed out and is always able to be identified with itself, so too must this chain of nexuses obviously be valid in every instance. To seek out the system of self-giving lived-experiences in which each lived-experience would come to pure self-giving therefore means to construe concretely the idea of pure disclosure, [and to do so] to the extent that it can—and must—be conceived prior to all logical activity of the ego so that we can have the founding support for all possible knowledge of the world (as actively determinative knowledge). The thought that "what is" must be able to be brought to disclosive givenness, and that true being and possible disclosive givenness, namely, whole and complete givenness, are correlates, is a thought that has been achieved early on in our transcendental considerations. But this thought only first gains the full height of clarity with the knowledge that consciousness is such a stream of sense-giving accomplishments according to essential laws, that all its self-givings stand under ideas *qua* norms, which are not brought in from the outside, but are born from within, as it were. Thus, essentially, and not as an incomprehensible fact, but as an intelligible law that is given in insight, all sense-givings are subject to possibilities of confirmation and annulment; belonging to them are firm possibilities of ultimate clarification, of pure evidence. But the latter characterizes in each instance a *limes* that can be obtained intuitively, a *limes* which, brought out by the active ego and toward which the active ego strives, designates an absolutely unshakable pole in which all paths of consciousness converging in it necessarily preserve concordance and therefore necessarily preserve the incapability of belief being crossed out.

Practically speaking, all striving toward completeness on the part of the ego tends toward concordance with itself, and that is to say that it itself seeks its eternal poles and seeks to direct its paths of consciousness in such a way that they are absolutely directed toward a goal and, which is to say, such that they are absolutely concordant. In its own way, every consciousness is belief. To remain concordant in itself means: What is given to consciousness as an object in the manner of the objective sense is maintained in belief in an unruptured fashion. But it could be ruptured in the further march of consciousness. What cannot be ruptured is an absolutely self-giving belief, and all [modes of] belief, all [modes of] consciousness, refer in themselves to a possible absolute self-giving according to laws of truth; they can be normatively regulated according to it, and in this absolute self-giving they have their norm as the norm of ultimate fulfillment that cannot be augmented any further.

Yet, transcendental logic must not remain content with such formal, general conceptions; where the objectivities of the [433] transcendent world are concerned, and foremost where physical nature is concerned, it must especially bring to light the types of consciousness, their essential interconnections and essential laws that make necessary and intelligible the objectivity of nature in the immanence of pure subjectivity. Not only must it work out the general types of intentional lived-experiences that are intentionally related to a spatial world, and not only must it study in general the self-givings according to their noetic and noematic structures. It must also precisely construe the ideal self-givings, as ideas of course, and correlatively, in their noemas, the ideal selves that are being constituted.

Appendix 28: (To §47> <The Problems of the Definite Determinability of the World>[58]

How is the idea of a nature related to subjectivity, which is inseparably intertwined with nature? Nature is a concordant unity in and through the variation of subjective multiplicities. Belonging

[58] Editor: 1923

to the constitution of nature, and a nature as such, is a formally and eidetically determined structural system that brings to light the discipline of the constitution of a nature. Belonging to a possible singular nature, to a nature conceived as determined individually, is not only the generality of the form, which is indeed abstract, but a determinate content. But is this content one that is completely determined, and to what extent is it completely determined? Thus, for example, the content of sense-data, but also everything that individuates the respective experiential phenomena and the entire life of experiencing subjects—the entire life belonging to experiential knowledge—in the entire nexus of their communicative life.

If we take given, factual nature, it will belong to the factual universe of egos. To what extent can things and the phenomena of the world of the particular egos change, and is there an excess that remains in their individual content, an excess that is not determined on the part of the constitution of nature? Here, above all, we come up against the problem of secondary qualities and the corresponding general problems of possibility relating to a universe of egos in general, and a possible nature in general. To what extent does the essential system of forms of every constitution of nature have an indeterminate openness, an openness to which the constitution through secondary qualities belongs, and which—under the form of the constitution of a common intuitive nature for a normal communicative subjectivity—belongs to it according to its form, while, however, a content remains indeterminate over and above the formal aspect, indeterminate from nature itself, presupposed as continually constituted.

But now further questions arise. The definiteness of nature, its being-in-itself, implies an intersubjective being-thus of nature that is identifiable for "everyone" in relation to everyone—a being- [434]
thus of all that is, and according to all its things and properties; and accordingly, how one must show the necessary possibility of an "objective" science of nature, built up from "exact" truths in themselves as methodologically recognizable for everyone in a person's subjectivity and from the standpoint of this subjectivity's phenomena. In a word, nature is "calculable," "mathematizable"

for everyone. There is a *mathesis* of nature, a "mathematical" ontology of a nature as such, and in fact, a "mathematical physics" that can be produced as a general method by means of this mathematical ontology.

But now we must distinguish between nature (*physis*) and universe. In current life, I am and I find myself as the experiencing subject of my environing world; belonging to its infinity is also an open, unending multiplicity of other ego-subjects. Just as I "have" my lived-body, just as I am the functioning subject of this lived-body (and by means of the latter, I am the functioning subject for the cognizing and formative and communicative relation to my environing-world and its subjects), so too are these subjects given to me as having a lived-body, and through their lived-corporeality, as functioning subjects; and we are reciprocally and especially given to one another in communicative understanding, and we are given as functioning, referring to the same world to which we all belong. For us as functioning ego-subjects, functioning as individual ego-subjects and in social community, the world is the universe and the total unity of objects; and "object" means here whatever has individual, concrete existence for a functioning individual subjectivity or for a communal subjectivity. Seen more precisely, the sense of this objective existence is that it can be experienced and can be known experientially in such a way that it can be identified as the same by everyone, as the substrate of the same determinations. And as such, it should be able to be attested to at any time and by and for everyone.

In this respect, however, nature is distinguished as [1] the first and founding objectivity that is experiencable, perceivable in a genuine sense for all ego-subjects, albeit in endless, open infinities of experience, and [2] subjectivity itself, the ego itself that is only experiencable in perception as I am for myself, in the original, through myself (in self-experience), but experiencable for another ego-subject only in the expression of lived-corporeality and in the psycho-physical relation of unity that is instituted with this expression of lived-corporeality.

The *animal*,[59] the human being, is experiencable as psycho-physical unity only in my lived-body/psychic inner experience; only I can experience my functioning in my lived-body, and thereby my lived-body as lived-body, and in this way, myself in this functional unity with this lived-body. I cannot perceive an alien lived-body as lived-body, and I cannot experience an alien ego as functioning in this lived-body—not genuinely. I experience alien subjectivity and human beings in the world in the mode of "empathy," through the appresentation that is carried out in the expression, through the appresentation that can never become a presentation[60] for me.

The objective world is the psycho-physical world, and it is the [435] cultural world that has received its cultural predicates from functioning human subjectivity bestowing them, predicates that possess their manner of experience and disclosure, but of such a manner of experience and disclosure that it presupposes the objective experience of alien subjects and their lived-body/spiritual functioning.

What kind of significance does this have for the question of the being-in-itself of the world, the definite determinability of the world? The universe exists with all things, animals, human beings, with all cultures and histories. Nature is objective, in itself, in the special sense; we mean that a mathematical knowledge is said to be possible for it; it is said to be calculable, every question for it is said to have its answer that is decided in advance and calculable for everyone in the same mathematical method. Nature, we mean, is said to be able to be anticipated theoretically, theoretically construable from the standpoint of given finite experience, from the standpoint of finite groups of experiential facts; it is said to be a definite mathematical system, and a natural science is said to be conceivable, a natural science which, à la Euclid, is said to comprehend the totality of natural truths and their substrates by deducing them from a finite number of axioms (ontological principles) and from finite facts.

---

[59] Translator: in English

[60] *Präsentation*

But can such an ideal even be constructed for the universe, the realm of possible natural experience (including the objectivities of non-genuine experience through the expression)? Is the *universum* that comes to givenness in expression determinable in a definite manner and in accord with the mathematical principles of definiteness? Does this have any sense at all?

The appropriate determination of sense and the limitation of the appropriate sense of a definite nature already gives rise to difficulties; the essential relation of nature to subjectivity that is essentially there with nature as experiencing and thinking, functioning subjectivity, functioning in actuality and in real and free possibility, already gives rise to difficulties. Insofar as I do indeed carry out perceptual kinaesthesès, insofar as I move my eyes and hands, and through this intervene in nature itself, this functioning already gives rise to difficulties. And here we have the following general difficulty, one that has far reaching implications: What sense does the in-itself of nature have in relation to the actual and possible intervention of subjects, apart from which nature cannot at all be conceived? What kind of an in-itself is this that is included (and, as idea, is construable) in the idea of the concordance of possible experience—experiencing subjects?

This is why the sense of the definiteness of physical nature as a nature that exists in psycho-physical interconnections and that refers to subjects, especially free subjects, must already be elucidated and first of all be worked out according to its essence and possibility.

The world as such with its animal lived-bodies, with animal and human souls, with all "psychic phenomena" of these souls all acts [436] and all social unifications and cultural formations, so one tries to assume, are supposed to be definite. How is a certain "definiteness" of the world to be determined in a beneficial, rational sense? What are the essential conditions of its possibility, how is it to be construed as the formal idea, precisely according to essential content and necessary essential constituents?

Can a complete world not first be "definite" under the idea of a universe of egos that is directed toward the telos of absolute Reason, or under the idea that the universe of egos, for which and in which the world is objectively constituted, exists in a "pre-established harmony," but one that is essentially necessary; that it exists in a necessary development toward personal and interpersonal Reason in such a way that the idea of absolutely rational universe of egos is entelechy?

Must we not say: We should in no way accept in advance, even only as a possibility, that a complete world be definite in exactly the same sense that physical nature is definite. We see upon closer inspection perhaps that the demand for a mathematically definite world, for a world that is calculable in each and every respect, is in principle absurd.

One could say: The becoming of the subjects may harbor a developmental tendency, and it may do so necessarily. But subjective development can and must take place only approximately. Psychic life can, indeed, must be mathematically indefinite and yet <be> directed toward a definite being. If the absolute of the phenomenal world only has a definite structure in itself, an idea, a system of ideas of Reason, which system makes natural science possible, then the question arises concerning what makes culture and cultural science possible, what makes a rational, ultimately, a social-ethical life possible. Definiteness of nature may suffice, while streaming life can and must remain irrational, indefinite. Certainly, if nature, sociality, culture, the objective world are such that they make continual, progressive science possible, then certainly even subjectivity must be rational insofar as it must be able to be reconstrued from the standpoint of this objectivity, and insofar as it must be continually and concordantly intuitive from the standpoint of subjectivity [itself]. But this [character of] reconstruability would only be a structure of rationality, and would not yet signify the rationality initially demanded above: as if it were patently a possibility, namely, the presupposed rationality of a subjectivity that is definite according to all particular moments, the presupposed rationality of a calculability of the entire pulse of life, and thus a "mathematical" rationality.

Thus, herein lie all problems of a possibility of the knowledge of nature, of a possibility of an entirely differently grounded somatological and psychological knowledge, of a possibility of historical, social-scientific and cultural-scientific knowledge; the possibility of the corresponding sciences. The ultimate sense of a mathematics of nature, ultimate clarity of an impossibility of a mathematics of culture, despite the fact that even culture has its *a priori* and sociality has its *a priori*. The problem of the exact laws [437] of nature with the idea of the exact determination of individual psychic existence, and the problem of the inexact scientific determination of facts of history. "Historical laws," concepts proper to the human sciences as opposed to concepts proper to nature.

<B.
APPENDICES TO PART 3>[61]

Appendix 29: (To §52) <Theme and Theoretical Interest> [87]

So that you do not get confused, I must not fail to point out the ambiguity in our talk of "theme." For instance, we do not usually call the theme of a whole treatise the nexus of articulated views or intentions[62], but rather what is expressly given, for instance, in the title, e.g., "The Optative in Greek," or "The Sense of Color in Ants," etc. In this sense, the theme means the object that is to be determined scientifically, which is the task of the author. And accordingly, the theme of a whole science is its scientific region. It is thus the unitary thematic object as object for all themes to be propounded in another sense. Unfortunately however even the terms, view or intention, and all terms that come to mind, are ambiguous in the same way, and so we compromise by using a juxtaposition of both terms, view or intention and theme, each one delimiting the sense of the other here.

Let us go further. Not all acts have a thematic character; the subject as an intending, interested subject is not occupied in all of the acts—this was exemplified for us by linguistic acts. But if this character is lacking in them, they can all correspondingly be transformed; in this case, they are altered without their intentionality undergoing any other change. Whether or not they are thematic, they are judgments with exactly the same content, wishes, decisions having this or that content, only that these contents are not themes of interest. It does happen that complex thematic egoic acts are founded in one another and that thematic

---

[61] Translator: The following pagination to the German text corresponds to Husserliana XXXI.

[62] *Meinungen*

distinctions internally cohere through this relation of foundation. For example, I am happy that the sky has cleared up and that the sunshine pours over the countryside, and perhaps I even say so explicitly. Then implicit in this pleasure is the intention of expression. But this pleasure is founded for me in the conviction of the given fact that the sky has cleared up; this judicative theme is foundational, but even though it is also a theme, it is still a subordinate theme; the main theme lies in the pleasing quality of this fact. As the subject of acts, I primarily actuate an affect-interest[63] and only secondarily a theoretical interest, an interest of ascertaining a fact: namely, that only insofar as it contains the substrate about which I am happy. In this way we distinguish [88] between dominate and auxiliary themes[64], and through this, we draw distinctions in the thematic function.

Appendix 30: (To §54) <Determining as Cognitive Act>

Act and intention. Every act is a *striving* that proceeds from the ego, more or less freely at work or inhibited. By the striving being completely or less completely at work, advancing or advancing less, breaking off, it always and necessarily effects something. Every cognitive act, every judicative act is directed toward an "object," and the extent to which it is already known, it is an object determined individually or according to its app<erceptive> type in this or that manner, implied in a situation and apprehended as this or that. Beginning in this way, the *cognitive act* consists in *its process of striving, in a progressive determination of the object*; whereby this object gets determined in a newer and newer manner, and as such becomes available to us again through habit.

Yet we naturally have to distinguish between the following: (1) the mere interpretation of the object that is already determined in

---

[63] *Gemütsinteresse*

[64] The expression "dominant and auxiliary" is not fitting here. But different concepts of "dominant and auxiliary"! More precisely, simply: founding and founded themes. But even that is ambiguous. Cf. the dominant and the auxiliary in signs and in the meaning intended with it in the appendix. Here we have a means-end-whole, the sign is the means that harbors an ought to. Interest on the way is auxiliary, is interest passing on its way "toward" the goal.

its determinations. What is implicitly known is made explicit once more, thus, brought to a re-actualized knowledge (reactivation as re-execution of previous act-divisions). (2) Or by being joined to what is already known it is something new, new knowledge, new "determination" (feature), a determining process that fashions. But we can still make distinctions within this determining process.

Already in the intuitive (experiential) sphere: (a) (α) The object for instance has been perceived individually by me just now or earlier, and as perception progresses, it has been determined as this or that—then I remember and I go back again to its determinations that are already known somehow—without perceiving something new. Thus, I can go back through empty, unclear, or relatively clear rememberings, that is, the determinative process can accordingly be quite different in its mere reactivation. (β) Or I can perceive it again (as is possible with objects of external experience), I can be convinced again [89] about how it is and has remained unchanged; in this case, I have a new original acquisition of knowledge and at the same time I have rememberings of the previous acquisition and an apprehension of its identity in an unaltered duration—thus apprehending that it remained unchanged in the meantime.

(b) An object of knowledge, however, can also be new, but be ap<perceptively> apprehended in this or that way, and it does have to be apprehended somehow. The sense of the apprehension implies determinations that have not been experienced with respect to this object, but nevertheless have a similar character insofar as they point back to previous, analogous experiences with respect to other objects.

The explication here presents something that does not stem from the experience of this object as determination, but rather, something that is anticipated, and thus, the explicit determination that is gained by the activity of explication is also anticipation. The act here is thus the act of the explication of apprehension as apprehension of something new, but not of a new kind of thing.

Every "knowing," every act in the sphere of belief, is determining. But the determinative process can be a perceiving. An apprehension already lies at its basis, but the intention goes through the (more or less determined) anticipation of this or that level of apprehension toward "effective realization." Thus, effective realization is here the fulfillment of the anticipation in the original itself.

The unclear remembering can be reactivated in a remembering. But it can also be the case that the intention goes further through this unclear reactivated determination to the reactivated object itself in its clear intuitiveness, likewise with respect to an explication of an object that is somehow apprehended. The intention toward the object can mean here the "anticipation" of explicated objects and fulfillment in these explicated objects themselves, but through this fulfillment, the intention can go further to the original self; the explicated object that is produced is merely an anticipation and not itself an original.

Appendix 31: (To §55) Syntactic Formation

What is syntactic formation? (1) Syntactic formation that the judgment accomplishes anew, the separate accomplishment of the judgments in question; (2) the syntactic formations that are characterized as the sedimentations of previous accomplishments.

Every original, newly accomplished syntactic formation, like [90] "is," "and," subject-form, predicate-form is common to the judgment and to the state-of-affairs. The form as such does not have any manifolds of sense that harbor the object-theme through the process of identification. It is precisely form. Thus, on the whole, the form is only identifiable as it is. Where the *termini* are concerned, however, they are laden with sedimented sense, whereby categorial forms enter into <the> mode of sedimentation and secondary *termini* [enter into] a modification of the form of judgment.

In the attitude of actual judging, and specifically in the attitude of originally accomplishing judging, an attitude focusing on original foundations, I have primary, direct themes in their direct sense, which is to say, the sense through which the themes are constituted prior to every determinative accomplishment of judgment. "The regard rests" on these themes; the ego has them precisely as themes; sense is thematic sense and in a certain fashion, "its regard also rests" on the synthesis that is the connective accomplishment of judicative action. But it is not an objective sense that becomes thematic, but the new constitution of a sense in the bright daylight of active consciousness.[65]

This newly constituted sense is newly constituted, but not for itself; rather, the newly constituted sense is something being formed or something that has been formed from the themes. And like everything that is originally constituted, this must first be made into a judicative theme through a reflection. In thematic judicative givenness, the thematic substrata are already no longer direct themata, and accordingly are no longer given in the same ways as in the judging itself. Nevertheless: In the transition from reflection to the direct attitude, and *vice versa*, an identification of themes of the direct attitude and of the thematic positions of judgment is carried out in reflective givenness; and we say, the same theme is the judicative theme, contained in the reflectively grasped judgment, only given to consciousness in a modified way. In judgment, the theme is a theme that is formed in this way and that; in judicative consciousness, the judgment is given to consciousness and the themes are given to consciousness, and each of them in different ways. In brief, what we ourselves have already said from such a change in attitude.[66]

But now I make a distinction between judgment and categorial objectlike formation. The categorial objectlike formation that we call the state-of-affairs (and everything included in this) is the judgment made into an object, the judgment that is newly [91]
accomplished in the judging, and the original accomplishment that is carried out, the accomplishment with respect to the genuine objects-about-which, is necessarily distinguished in every judgment that does not consist of this accomplishment alone.[67]

---

[65] Is it just like appearances that synthetically run their course in bright light, but where the passive synthesis is not the theme, but instead the unity?

[66] The "reflective" collection, i.e., the collection grasped as set, thematically harbors even the single members. In this way, the judgment harbors the subjects themselves, the predicate themselves, i.e., the objects (the objects-about-which) intended in the judging.

[67] The "presentations," the sensible sense-givings prior to judging. Thematic grasping of the one. In contrast, reflection on the "appearance," on the objective sense that is a different one again and again, but in identifying synthesis "contains the same." Even this judging of identity arises in reflection: (a) the identity $S \equiv S$, a categorial object; (b) the S_{σ_I}

In every other judgment, the objects-about-which occur with sense-givings from other categorial accomplishments that have been sedimented on the object-about-which, i.e., on its pre-categorially constituted sense. Reflection discovers the themes of the main-judgment, of the one that is actually newly accomplished, with the encumberments of sense. But if the "What" that is judged is the thematic object that is made into a unity, then in the series of judgments that are compared, judgments that are encumbered with different sense, the "What" that is judged is the same: that which I intend. But I can take every theme as the theme of its sense, in particular, of the sense that was previously attributed to it, [namely,] the sense acquired by thematic accomplishments. Then I have the thematic What, namely, the judicatively given What.

Appendix 32: (To §56) <The Ideality of the Objects of Sense and the Ideality of the Species> [92]

After my lecture, indeed, during and already in preparing it, I had reservations concerning my presentation, and quickly note the

---

$= S_{\sigma_2}$, S in the noematic mode in the objective-sensual sense σ, identical to itself in another sense, and in the full sense: But that is an attributive identification!

The sense $\sigma_1(S)$ is different from the sense $\sigma_2(S)$, and in the reflection on the judgment, I have $\sigma_1(S)$ unified with $\sigma_2(S)$ by an "identical" in quotation marks; each of the two senses as thematic sense bear a thematic pole, and both senses are united by an "identical" that proceeds from thematic pole to thematic pole: $\sigma_1(S_x)$ $\sigma_2(S_x)$. But if I bring this into relief through reflection, I do not identify, I do not make S a theme, but rather, $\sigma(S_1)$ and $\sigma(S_2)$, but not even S which is given to me in the mode of σ_1 and likewise S that is given to me in the mode of σ_2, for that is itself a relation, namely, the relation between S and $N(S)$. It is a particular relation, S and mode of appearance of S (with respect to things), and this appearance is the appearance of S of the same S of that other appearance. The object in this mode of appearance, in this How, is already a relative characteristic. I cannot have the objective sense given to me in reflection without also having its objective pole given to me; I cannot reflect on the judgment-through-identity without having the synthesis of the poles given to me. But I can grasp merely the unity-through-identity of the poles in reflection, i.e., the state-of-affairs-through-identity; but I can also have the "phenomena" σ_1 and σ_2 as themes, themes that are naturally not identical and identified, and the connected unity that they have as "identical" through the consciousness that identifies their poles that come to light along with the phenomena: σ_1 presents the same thing as σ_2, they are appearances of the same thing. This already implies that what σ_1 presents is already a relation. Can one express it without relation?

following: In what sense does the judgment have an extension[68]? The judgment has no extension? Is this not due to the thematic attitude by virtue of which the judgment is identified in the thematic interest from the very start and is not abstracted from its particulars?

The species as the universal is gained by abstraction with regard to the pregiven individual possibilities or actualities. And why must that be the case? Because bringing a common element into relief presupposes bringing the particulars into relief, and <because> the species is the common element.

If I judge that "*S* is *p*," the judging is not at all brought into relief, and likewise, the judged "What" <is> not brought into relief in its theme and in its constitution, in the sense of something exercising an affection on the cognitive ego.

S and *p* and "*S* is *p*" are identified by coinciding in the repetition. Here the ideal-identical is the *prius*. But does that mean that the ideal is not particularized? It only means that objectlike formations of sense are not constituted by "abstraction," that their ideation precedes the individual intuition of sense-giving consciousness and of its noema, that these individual intuitions can only be gained through reflection.

All objectlike formations whatsoever are constituted through sense-giving, but sense-giving and all modes of givenness are given through reflection. But the species are givennesses of a second order; to be sure, is not the attitude focusing on an individual, thematic grasping, <on> individuals *simpliciter*, but rather [the attitude focusing on] the corresponding possibilities. But I also do not need any attitude that focuses on the possibilities of judging in order to grasp the judgment.

Appendix 33: (To §57) <Main and Subordinate Determination and the Division into Main and Subordinate Clause>

In the previous forms, the thematic interest in *S* was fulfilled in the determinations *p*, *q*, *r*, so to speak, in the first natural course [of determinations]. The material content of the *S*, its character as

68 *Umfang*

property, emerges in the series and is grasped together in this series.

Now, it can be that the interest in the different determinations is a uniform one or even a non-uniform one. Let us take the first [93] case, and for the sake of simplicity, only two determinations, *p* and *q*. If I repeat the judicative determining of *S*, for instance, for the purpose of communicating or remembering something for myself, I will not have any occasion to alter the form of the succession that is necessary for an initial original acquisition of knowledge. This is otherwise if the thematic significance, the weight of cognitive interest is different in relation to the particular determination. Then I will have to carry out a regrouping; I will have to immerse myself especially in the distinctive mode of determination, while the *S* and knowing it judicatively does indeed remain my dominant theme that I want to promote precisely by immersing myself in the distinctive mode of determination.

If, for instance, the *q*-being emerges as especially interesting, I will now start to judge, so to speak, going directly from the *S* to the *q*; this direction of determination becomes the main direction, the proposition, "*S* is *q*" becomes the main clause, as something that is important above all to fix for myself or for the communication. On the other hand, that *S* is *p* becomes the subordinate clause. In the unity of a synthesis of judgment, the main clause is distinguished from the subordinate clause in the form, "*S*, which is *p*, is *q*." Or "*S* (the same *S* is *p*) is *q*." An identifying connection is concluded here, but the two propositions, "*S* is *p*" and "The same *S* is *q*" are not equivalent; rather, the one determination is annexed as a subordinate determination to the *S* in parentheses, as it were, and the main determination bears on the *q*. This form can also arise in such a way that *S* already occurs with the sedimentation acquired by the previous knowledge, yielding out of itself the new determination *q* that already holds the main interest as new, and [it can also arise in such a way] that at the same time an explicating renewal of the sedimentation constitutes the subordinate clause.

That which is settled as such is the subordinate element as opposed to the main element, as opposed to what brings new fulfillment. Ideally, we can obviously transfer each one of the previous forms into such attributive forms; corresponding to each determination is an attribution, just like every attribution originally points back to a determination. Thus, "*S*, which is *p*, is *q*." "*S*, which is *p q*, is *r*," etc. The same syntactic matters now have a new form; *p*, *q* are not grasped together successively in mere conjunction; rather, the holds on them have separated.

We will also be able to characterize the differences between the *S* that is still indeterminate and the *S* that already presents itself with a prominent sedimentation from an earlier judging—linguistically, for example, the adjectival annexes to substantive subjects that are explicated in relative clauses—as special differences of the subject-form. With this we would understand <the> difference between "the big house" and "the house, which is [94] big." Where the differentiation is in dispute, the adjectival determination functions as subordinate clause, so it seems.

Let us now go back again to the original form-groups, "*S* is *p, q, r*."

Appendix 34: (to §58) <Absolute Substrates and Substrates as Determinations that have Become Independent>

We can now add an important distinction to the function of becoming independent that we just mentioned. If we already have some kind of substrate, its explicated determination can become independent, can itself become a substrate in turn, and, if we like, even a main substrate that has become independent, a main substrate for further determinations that can be freed from the interest in the previous substrate. It can operate in a similar manner, in turn, for these determinations, etc. Obviously, on the other hand, even if the substrate that functions as the point of departure has already arisen from nominalization, we always arrive at an absolute substrate; this means here [we always arrive] at such a substrate that has not arisen in this manner. Viewed more precisely, however, it is not a matter here of arbitrary distinctions; namely, there are objects that can become substrata *a priori* and in original givenness only by having occurred previously in the judgment as determinations, and there are objects for <which> this is not the case. A house is given originally, but not given as a nominalized determination. It is given independently. Its color is given in the judgment through the determination. But if it becomes independent, it is a substrate from then on, but it can only become a substrate through the process of becoming independent, and any other way is inconceivable.

In a certain way, the house can also emerge through a determinative accomplishment. For example, we see the street as a whole, and only first in the explication does the house emerge for itself: The image of the street obtruded in a unitary fashion and came to the fore before the house came to the fore for itself. The street is now determined in its houses that are grasped one by one. But however much the single graspings have the form of determinations here, the form of determination is nevertheless not something essential insofar as precisely this house could also be given independently from the very beginning; and, on the other hand, even if it were given previously as determination, even if it were to shed the form of determination completely, the house could extinguish, so to speak, every memory of it [i.e., the form of determination]. By contrast, as the color becomes independent, this origin cannot be extinguished; the interest in the substrate, house, has only become secondary, but it cannot entirely disappear from consciousness because the color is originally contained in the [95] concrete object and owes its individuality for consciousness only to the thematic apprehension through determination.

We learn here that there are original objects that are independent and those that are non-independent. The non-independent ones can only become thematic substrata (of complete and determinative identifications) by other objects upon which they arise as determinations, first being given in this way. Their first form of givenness is as a determination, and they bring it *a priori* to the form of substrate only through the process of becoming independent, which is a separate activity.

Thus, we differentiate the determinations themselves into independent and non-independent determinations. Belonging to the independent ones are all parts in the strict sense, parts whose substrata are called wholes, but further, all members of sets, as well. Belonging to the non-independent determinations are all moments whose substrata are called concrete objects. The form of determination is not essential to parts and members of sets, the form of the substrate [is not essential] to the moments. The latter have taken on the nominal form only through the special, founded activity of becoming independent. The distinction between the independent and non-independent is a radical distinction in their syntactic matters[69]; accordingly, this also conditions essential distinctions with respect to judgments. We should note that it is not a matter here of a distinction between arbitrary syntactic matters, but rather, a distinction of the universal type, one that belongs *a priori* to the idea of a possible judgment as such and of a possible objectlike formation as such. From every objectlike formation as such, from every determination of a Something and active identification of a Something, no matter how it has arisen, we arrive, depending upon the direction of the thematic activity, at distinctions of the independent and the non-independent, at determinations through the independent or non-independent. It is a formal distinction, that is, a distinction that belongs to the idea of judgment and correlatively to the idea of the objectlike formation, i.e., of the comprehension that corresponds to them. One also calls such distinctions categorial.

We arrive further at the side of the independent, at distinctions between the immanent and the transcendent, and then at the side of the non-independent, to the predicates of independent objects as absolutely real substrata: the predicates of the temporal position, of temporal duration, of the varying temporal fullness, of the identical properties running through the temporal duration, etc. The possible partition of the real into real parts as pieces yields the necessary partition of extension, of temporal extension, and in coexistence, of spatial extension, and <leads> to relationships of non-independent properties and determinations in general, determinations that are interconnected with them. But we do not [96]
yet have relationships. We have not yet treated any judgments of relation. The original constitution of relations that is carried out in relating activity will soon become our theme. We only need say explicitly that judgments of identity, judgments of determination—even judgments of determination in which parts

[69] *Stoffen*

function as determinative—are not judgments of relations, but can be changed into them in the change of the mode of judgment. Thinking is judging, but relating thinking is a special kind of judging.

Appendix 35: (to §§ 59 and 65b) Multiplicity and Judgments of Multiplicity

Original unity of thematic interest yields a commonality of separate objects (in the unity of an affection), and accordingly, original unity of thematic interest yields only a commonality of the particular features that extend beyond the particular objects (as uniformity or similarity). An original multiplicity is not merely a collection of separate objects, but already within passivity it essentially comprises a bond with an inner affinity. Corresponding to this is the new judicative formation that brings to light this affinity for the cognitive ego and produces higher levels of judgment. Each object can return as the identical one in consciousness, and can enter into connections of this kinship with other objects that are already constituted in consciousness, actively or passively, in an originally impressional manner or reproductively, i.e., no longer merely exercising an affection for itself, but in a multiple manner, in community with its kin. In this way, each judgment that is valid for the object for itself can connect up with judgments that are valid for the objects that are akin, or it can enter into a community in the unity of the multiplicity of single judgments with other single judgments—a multiplicity instituted through kinship—and new modes of judgment as such can arise, those that extend beyond the particulars, specifically, those new modes of judgment that are related to the non-independent "properties" of the particular features.

In passing over from one particular feature to another, a coinciding that forms similarity arises according to the commonality, a differentiation [arises] according to what is different. Each one coincides with its partner in the transition by virtue of it being the substrate of identity, and the substrate of moments of similarity and moments of uniformity. In the moment

of coinciding, one element melds with one similar to it according to the law of similarity, while the consciousness of a duality (a duality that is being unified in the fusion) continues, and this fusion has its degrees that are called distances of similarity, or differences in a certain sense. In the case of uniformity, the fusion [97]
in consciousness is one without distance, without difference, a complete fusion.

What influence do such interconnections have (interconnections that can already play their role in passivity) within the framework of judicative activity? If we assume that the respective multiplicity has already passed over into the collection of its single members, then this transition necessarily means a process of going through them, and through this, commonalities come into relief. Depending upon the "magnitude" of the similarity, the reciprocal awakening force will be greater; and with respect to a pair that is especially connected through this process, uniform or relatively very similar shapes come into relief; where a different pair is concerned, the colors, etc. This can also be regarded as the original source of the prominence of non-independent particulars in general.

Thus, judgments of properties "S_x is p, q, r," and "S_1 is p', q', r'" arise in this way. And in this case corresponding properties will occur here and there in uniformity or in differing degrees of similarity. But that would still not yield anything new, and several fundamentally new things occur, namely, through the fact that in the unity of a thematic consciousness, the connected unity of affection becomes effective in a thematic way.

Let us initially assume that the thematically determinative interest is to be concentrated on S_1 specifically, and is to act as if the universal interest in what is connected to it is lost. The affection, which is the drive to enkindle a progressive and synthetically encompassing interest and thoroughgoing active unification, is constantly at work.

Thus, in the limitation to S, the prominent property-moment p is grasped as "S has p." Likewise for S', where a completely uniform p can occur. But now a synthesis of identification necessarily arises, S and S' are the same, are p; in spite of S having had its moment p, and S' in turn its moment p. Just as the substrata are separate, so too are their particular features. But in the thematic transition they coincide and [this complex] becomes actively identified. But it is not the property-moments that are given to consciousness as identical, and not the S and the S', although we say that S and S' are the same. Of course, we are not speaking here of a complete identification. The partial identification to which we owe the property-moment as determination is also not at issue.

However, if we reflect more deeply on the synthetic consciousness that underlies our analysis, we will notice that a unity emerges with the coinciding in the transition from one uniform moment to another, a unity in the separated and connected duality, and that it emerges again and again as completely and identically the same if we pass over to a new element S'', then again to S''', in which we have a uniform moment p, and a uniform moment p again. We see that our more precise symbolic formulation would have to read: "S_1 is p', S_2 p'', S_3 is p'''," and now [98]
in contrast to the latter, "S_1 and S_2 is p," where p is the identically one that emerges in p and in p'', possibly continuing as "S_1 and S_2 and S_3 ... are p." For even "S_1 is p, and S_2 is also p," etc.

Within the thematically active attitude we have proceeded more precisely from "S_1 is p" to "S_2 is p," and if, now, the unity that is constituted originally only in this coinciding, the unity of the species, p, comes to the fore, then S_2 is now determined as p, turning back, S_1 is determined as identically the same, as being of the same kind, and so with every new S as absolutely the same, as p.

Instead of the fleeting and varying moment, the S is determined by something identical that is ideal and absolute, something identical that runs throughout all particularized objects and their moments that are multiplied in the mode of repetition or in the mode of becoming similar, and it runs throughout them as an ideal unity, as a unity that, as we have shown, is not at all interested in the actuality of the moments, and that does not come into being or pass away with them, that is particularized in them, and yet not as a part in them. But initially we are attentive to the fact that different judgments diverge and separate from one another here. The judgment "S is p'," where p' designates the moment, is something entirely different from the judgment "S is p," where p

designates the *eidos*, or again, from the judgment, "*p'* is *p*." Or, as we can also put it, these statements are misleading because as formulae they say the same thing. One time we have identifications between substrate and moment, another, we have the identification in which a *p'* (which, coinciding with another *p''*, allows the intellectual[70] unitary feature to emerge), is determined as the species *p*, and then *S* is determined as *S'*, etc.

Appendix 36: (To §59) Set and Whole

Our expositions dealing with the origin of the categorial object-type, set, are still insufficient. We conceived of the set as being given in a pre-constituted manner and to this extent it is given in a pre-constituted manner, namely, as having been connected from several disjunctive pre-constituted objects, and thereby as separate affections, into the unity of a single affection heading toward the ego simultaneously or successively in consciousness; and it can thus solicit a turning toward as an entire turning toward and as an entire grasping of the set. But seen more precisely, we must not say that a set is originally constituted as set. For our reference to a whole composed of disjunctive elements, a whole that is not merely a set, already shows a deficiency. Even such a whole can be given as pre-constituted, and can likewise be precisely given as pre-constituted.

Further insufficiencies can also be seen with respect to our first [99] description and points to the need of supplementing it. If the simple grasping of set *T* takes place, which necessarily takes it as a whole, or if you like, in a single ray of attention, and through this, the set *T* becomes the substrate (even a substrate, thus, every object in the explicit sense, is necessarily something given thematically in a single taking hold), the interest in this substrate lives out its life in the chain of explications. But even if these are prefigured, and we can speak of a complete satisfaction of the drive toward grasping that was first awakened, namely, when one has gone through all members, it is indeed not yet clear why we must stop here and why belonging to the explicit consciousness of

[70] *geistig*

the set T as set is only the formation of the collection of m[ember]$_1$, ... m_n, and not the further canvassing of the single members of the set according to their inner particular features and their relations.

Finally, we spoke of an identification of the set T that is not explicated as it has become substrate with the obtained collection. But how does this identification come about? For two members, we have the figure:

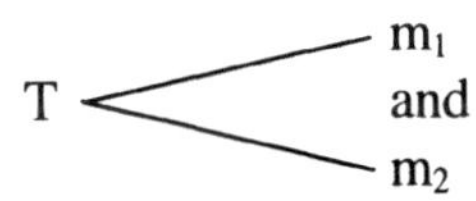

but grasping them together, which takes place on the side of the object, does fashion a unity, but not an object that can become the object-member of an identity.

If S_1 is identical to S_2, then the object S_2 is given to consciousness exactly like S_1, as given in a ray of grasping. And we can simply reverse it, S_2 is identical to S_1, subject and object are exchanged and have altered their syntactic form, but not the givenness in a single ray of attention. In fact, we note that the collective connection that has originally arisen in the plural process of explicating member m_1 and m_2 first requires a change of view in order to become the substrate, thus, to become the genuine object, to become the identifiable object. But this means that as long as we have a mere collective grasping, we have with it even more so only a pre-constituted object, multiplicity, and only in grasping that reaches back to the active formation do we have the multiplicity as unity in an objectlike manner. What is curious here is that an activity synthetically pre-constitutes an objectlike formation, which as object however, can become a theme only first in a grasping that reaches back after the objectlike formation is complete. This is the case for all objects generated through thematic activity, thus, for all judgments and connections of judgments, and all elements of judgments occurring in them, elements of judgments that are not themselves judgments, and for [100]
all objects (among them, sets) that are being constituted in judgment's [noematic] sense first through manifold judicative

action. For as special products of thematic activity, they are now easy to show. Passivity can only create preconditions here, and it is not necessary that the many objects are already pre-constituted, already finished in advance, as disjuncts and that they exercise their joint affective force. The objects can also come into thematic view successively, and while we are already occupied in a different way with the previous objects in judgment. In the succession, they fulfill the described conditions of collection, the unity of affection is formed successively. It paves the way for the transitions of interest, and if the emergent objects are disjunctive, collection can begin. But it can also arise from the very outset in activity, like when an S is explicated step by step in disjunctive particular features, and these particular features form a collective connection from the very outset. A change of view is possible here at all times, a change of view that objectivates the *collectivuum* as object.

A set is nothing other than an objectlike formation that is originally pre-constituted by a collective activity, linking the disjunctively constituted objects to one another; actively grasping it consists in a simple reaching back or a seizing of what has just been constituted. As pure formation of activity, the set designates a distinctive form in the sphere of judgment into which thematic objects of every conceivable type can enter as members, and with which they can then enter into determinative judgments of every syntactic shape. The thematic objectivating process has its syntaxes, and one of these syntaxes is the "and"; and one of the syntaxes of relation, which obviously belong in an entirely different direction, is the "disjunct."

These are the fundamental elements of the syntactic special form of collection, the set. Every objectivated *collectivuum* is identifiable like an object, and can be identified in newer and newer processes of explication, and the explication is again and again colligation. Two sets can also be given by the fact that the members are constituted in different modes of givenness. Then, the identification of a set with another set is a mutually univocal ordering and collection of single identifications. We do not have to expound upon the fact that collective unities are in turn colligated, i.e., that sets can in turn be built out of sets, and that all relationships of containment can exist between sets.

Now to the concept of whole.[71] If we form the most general [101]
concept of whole, so that it encompasses every object insofar as it contains parts, then every set, but also, every object as such, is a whole. A more strict concept of whole is characterized by every object that can be explicated [and done so] completely[72] in a disjunctive multiplicity, that is, in a set of immediate parts (things that are contained). Every set is then also a whole, since once formed, it can be dis-membered again into its unities, which means nothing <other> than that it is generated anew. But a whole is not merely a set, but an object that contains a set. A genuine whole is not self-given through collection, but self-given in a different way; but here the whole is fashioned in such a way that the object allows a set, the set of its parts, to be formed from the object itself through division and collection.[73]

While the concepts of set and of whole can themselves be ranked among the concepts that have the distinction of having their origin in the syntactic sphere of judgment or in the most universal sphere of objectivation, which does not go into the ultimate cores and their categories, it is already quite different with the concept of the whole in the narrowest sense of the real whole, with the concept of real unity in general. A real unity is something constituted in the continually integrated unity of a sensible intuition. A real whole cannot have its parts divided among non-integrated sensible intuitions in self-givenness, something that is quite possible for a set-whole.

---

[71] I will speak about connection later <pp. 342 f.>

[72] What does that mean, "completely"?

[73] Still unsatisfactory. Indeed, why is the concept of connection avoided and it is avoidable?! The set is united through collection, but a whole has parts, and they are connected. The connection can be a categorial one or a real one. Every explication of an object in an interconnection of disjunctive connected parts, which are equivalent to the whole object (disjunctive parts <are> no longer possible outside of connection)—every explication of an object is graspable as a set, whose members have "connection."

Appendix 37: (To §§63 and 64) <Object and Content of Interest> [102]

Determining S, concentrating on S and the interest in it as theme in the process of examining it in *explication* or <concentrating> on other affiliated themes <in> a broader *examination* and in an examination fulfilling multiple determinations. The question is what takes place here. In the transition from *S* to one of its parts, we surely have a coinciding in the "overlapping"; *S* becomes restricted in S_1. But if, holding on to *S*, I pass over to its parts, that is, experiencing this coinciding, this unification, through this process, have I already posited *S* as subject, and [have I] determined it as having the part in the following manner: "*S* has S_1"? And likewise, vice versa with respect to the delimitation of the examination beyond the *S*: Do I already have there "*S* is the part of S_2"? Must we not say that in this transition, which is an overflowing of the interest in a containing or a contained, *S* takes on a content, which alters nothing with respect to its identity, that is, nothing with respect to its *x* in the "objective sense." It is one thing to have attention directed toward the object given in the objective sense, directed toward the one in the varying, intuitive, presentational content, the attention as grasping, and it is another to have the attention directed toward the presentational content itself. In the grasping, in the thematically examining directedness toward the object, we find the following: We have here a giving oneself over to this content and to its change in the form of an interest in it, but in the thematic consciousness of the One. As I have explained, "the thematic regard" directed toward *x* is the content, but in the light of interest.

Must I not go further and say that if the transition from *S* to S_1 has taken place, an "enrichment of sense" takes place for the *S* as a result of the coinciding, a content of sense that has arisen from the coinciding, and now after the transition, a new interest, the genuine thematic interest (the interest in *S* as determined in such a manner), turns toward the *S* as object; but being fulfilled in the determination, i.e., the content that has newly arisen and that is being actively and newly realized. Then we would have two levels: (1) the transition from *S* into the parts S_1, S_2 ... that are coming to the fore in coinciding, the S_1 and S_2 get grasped for themselves. The interest that followed the objective senses of pre-constitution, the what-content of the object given in them, flows-off into the parts, but the S and each one of the parts that are already grasped, remain held onto.

(2) But then this is something new, namely, the ego is directed back in its interest to the S and—for instance initially taking S_1 in a special hold, directing a new focal ray to it—becomes aware of the enrichment of sense and is satisfied by generating the enrichment of sense again actively in an original manner in the new transition to S_1, and thus for every S_k. Determination is always [103] bipartite.

Thus, the most simple case is the case in which the explication (as the examination of the object) does not go any further at all to ever new moments. Let us assume that the examination would immediately halt and lead only to S_1. And through the transition under coinciding, the S has necessarily undergone something; through this, it has taken on a new "content" and this "arouses our interest." Thus, I go back to S, identify it with itself, which is only to say that in going back, it is there "again" as it, as S, and having turned toward it in the manner of the thematic object (of the center of the object), I "actuate" my interest in it in the direction toward the content accruing to it, and this presupposes that I do indeed pass over again to S_1. For originally it only comes to the fore in the systematic transition, in the coinciding. Previously it had come to the fore in the S as a result of the transition, but it was not yet in interest. Afterward it gets sensed (noticed), but the S is only "still held onto." But the transition has been carried out with its coinciding, but as transition, it was something that resulted passively. Now, however, turning again to S, it is grapsed in a primary fashion anew. I have its enrichment of sense as a mere protention, in connection with the retention of the transition that has just elapsed. As active ego, directed toward the S in its accruement and focused on the interest in the accruement itself, I now carry out the transition and the partial coinciding as free activity, and in this way I bring to fulfillment the determinative intention, the intention toward the S in its sense that is being augmented from the transition and the coinciding: I have S as the theme of a determination, and I determine it actively. One must not be led astray by the fact that after the determination is carried out and has been actively carried out often enough (e.g., as it is here) as the determination of an object as having a part, then in the new case of a transition of partial coinciding, which coinciding has not taken place actively at all (maybe the *S* had not even once been held onto; interest in it can be awakened subsequently). The "possibility" of a determination to be carried out can come to light as "associatively" grounded, and the determinative statement can also immediately arise without the determination being realized after the fact, without it being actually realized. The increasing, subsequent interest in *S* in its determination is empty, anticipatory, on the side of the determination; and it is therefore likewise empty and anticipatory with respect to the determinative action, which alone the determination can constitute originally, like the content of determination, the state-of-affairs, the judgment.

Further, we should not overlook what the *rudiments of the doctrine of interest* were, and accordingly, the rudiments of all active praxis that follows interests: the fact that we constantly [104]
have the distinction here between the thematic object as the object of interest, the presentational content from passivity or the content that is accomplished by activity, which as the accomplished content "fashions" for the object a new content in it [namely], the content of interest, but even with this, it is not its object. Indeed, here, where activity is in play, we would actually have still more to say.

Already when a sensible examining interest and not a determinative interest is being actively lived out, that is, already when we have the first level of active objectivation, i.e., the level of pure sensible examination, we will not only have here the one presentational object (the One, continually One of the varying presentation) and its content that is being modified. We modify the presentational object itself on our part, even if it should be varied or transformed with respect to content; we do not carry out alterations in the object on our part, but rather, transformations of content. We delve deeply and actively into it, we adapt ourselves to it, penetrate into it (the explicating process also belongs here, prior to the determining process). If it is an external object, we move our eyes, etc., we modify the "appearances." Consequently, we do not have only the respective presentational content, but also its subjective variations, possibly parts, divisions, the kinaesthetic transformations that belong to them immediately and that serve as the means to transformation, etc. All of this lies for the ego in the circle of attentiveness, all of this has exercised an affection and the ego is directed toward it. But it is not the theme for the ego, the theme is the One in and through a multiple content; content is this multiplicity of the One, continually and discretely changed, but standing in its regard as content, as the What of the One; and this is thus a "third thing": the fact that this content is constituted in the active process of objectivating as seized, as appropriated, as gone through and unitarily connected in the unity of interest by the ego, and that the content as content, the transformation of content as generated is precisely a product as the unity of several contents; and as such we [have] with every generating "doing" the crest of an "act"[74] that has something multiple, a multiplicity that is naturally not a product. An act is not action. Belonging to action is the active directedness of the ego and the noetic constitution of the stages of the act (eye movement, and the like) that are constituted in their own way in an objectlike manner.

The "in consequence of" that connects the kinaesthetic feature with the appearance, all of this is something constituted: the examining act is in view, it runs-off as such, put into play and directed by the ego. It yields the course of the content of interest that is in view, but it is in the vicinity in a special sense, though not at the center of the thematic regard. But the theme is the object, it is the object that is "grasped" in a thoroughgoing manner, showing itself in its content; the interest bears on it as showing itself in such a way and is lived out practically in the [105] deed of examination, in the deed of calling forth ever new contents.[75]

As can be clearly seen, we must then, however, distinguish the active determination of something examined from the active examination. In a first sense, all examining is a determining. In

[74] *"Handlung"*

[75] Active examination = intuition – distinct from determination

active examining, the object is determined for us as object, that is, as theme in its multiple content.

An interest of a higher level in one and the same object arises on the basis of active examining, springing forth from it as the interest in the determinations arising in the special sense from *S*. The examining is *experiencia vaga*. The object is intuited in its content, and the intuiting interest is actively satisfied. But by it being explicated into special objects, being clarified in them, the object undergoes something in the partial self-coinciding, it undergoes a clarifying determination, and it does this initially in a passivity that is based upon the subsoil of examining activity. An interest in the determination leads to the activity of determining in which the object that was examined only previously becomes the subject-theme of an explicating-determinative judging, a judging in which the state-of-affairs as the judicative content is constituted, a judging in which the determination originally accrues to the subject-theme in effective activity as belonging to the subject-theme, as being originally apportioned to it in the process of generation. The judgment and the subject as subject of determination is identified in the iterated, free generation of coinciding and in the iterated thematic identification of the object as the subject of the determination "attributed" to the object, apportioned to the object through the process of generation.

As the interest expands, the unitary examination can lead from one object to another, and in these transitions interest varies its object; but it is borne by a unity of connected interest because there is already a previously constituted "materially relevant unity" within passivity, or it gradually comes to the fore as constituted. A united satisfaction of interest, then, runs through the unitary examination, and it does so in such a way that the materially relevant unity here need not at all be thematic, even though it can be thematic. In the latter case, we have an object that is explicated, and—abandoning the theme of the object as a whole—the interest passes over to the explicated elements, and they get examined individually and are possibly held onto in the transition. But now these transitions also lead to determinations that are different from those that we initially considered; [they are] related to the transition of object and part.

Already the transition from part to whole allows a determination to arise in the part, and the part can become a subject, and in actuating action, the determination and the determinative judgment are constituted in a higher activity. [106]

Further, in the transition from an object to one that is separate from it, whereby the first belongs, along with the second, to one and the same sensible unity—a unity that is both salient and noticed for itself (in the higher level it belongs all the more to a connected unity). If now both terms also bear their passive augmentation of determination from the part of the whole, this augmentation will not have to lead to an active determination; on the other hand, however, in the transition from one element to another, a new augmentation of determination arises in the first one, and if the latter becomes activated, while the whole is in the field of interest, the judgment of relation will emerge in a manner analogous to the following: *S* is similar to S_1, etc.

We have as many ways [or types] of constituted wholes, connections as we do ways or types of relationships, namely, external relationships.

A whole is given, e.g., and initially so, if it is a sensible whole as a unitary object in which other objects, parts come to the fore: a unity of affection with special affections included in it. We assume here disjunctive parts, special affections are disjunctive. If a part is examined, grasped, then the whole does not have to be grasped, and if the whole is grasped, then the parts do not have to be grasped individually for themselves. It will be objectivated and given in full clarity as a whole if it is first grasped and examined in a thematic prehension, and then if it gets grasped according to its parts and examined step by step, and is held onto here as a unity, in this way becoming identified with itself. As the examination of the whole and the examination of the individual follow upon one another, and according to the way they are grasped, they coincide [insofar as] they are held onto [in a unity]. This is the case for every object that becomes subject to an explicative [process of] examination.

Now, how does the form of the whole and the form of connection of parts come to the fore? Two parts can form a special whole within the whole, can come into relief as a special whole, that is, as a part that is itself a whole of parts. Let us take from the very start a connection of two objects, a whole of two objects. It is called a whole insofar as it only has these two immediate parts, [insofar as it] is "resolved" only in these parts. The whole is from the very start only equipped with these special affections that go together in the unity of a single affection, and these special affections yield the immediate parts and their connections of sets, the sum of parts. The form of connection is a non-independent moment that comes to givenness on its own according to the elucidation of the whole with respect to its parts, that is, in the divided whole. And it can probably do so in the following way: Every element is grasped for itself (merely "examined"), and both [107] the parts come to the fore together and are grasped together in the whole in a "divided attentiveness." Now a connection comes to the fore not as a third part which the whole would still have in the same sense as these two parts, but as a mediate determination of the whole or initially a mediate moment that is not an immediate moment of one or the other part, but their ensemble. It can only come to the fore after the ensemble is there.

There can be several connections, and in the corresponding transformations, for example, they can turn out to be "founding" in moments of parts. The several connections are fused into one single connection, just like in the case of founding parts, the several moments [are fused] into the unity of one member.

There are different possibilities of determinative judgments here: "The whole has such and such a form," [and] "the ensemble of parts has this form as connection." Furthermore, [we have] the external relations [that are] in transition from one part to another. To be sure, every part is something for itself and is something grasped for itself, and insofar as this is the case, it is thematic. But each one is a part, and it participates in the whole, which, even if it is not the theme of determination, still lies in the scope of attentiveness and grasping, and the form of unity is salient in the givenness that is clarified. S_1 and S_2 share something in common when they participate in the same object, and if we pass over from S_1 to S_2, (and if each one is given to consciousness precisely as part, if each one has been grasped with the sense that has accrued to it from the direction of interest toward the grasped whole)—

then a new augmentation of sense is there in S_1, an augmentation of sense that arises from the transition and from the coinciding in the common element. If an activity is put into play, an activity that makes the S_1 thematic and relates it to the S_2 with respect to the form of the whole, and generates the determination in an original manner, then the judgment of the external relation is constituted in an original manner, then the S_1 is constituted in an original manner as in relation to S_2 (the relation of similarity, of uniformity, of situation, etc.)

This can be repeated on all levels, whether we have at the lowest level concrete sensible wholes or wholes, connections made up of “abstract,” non-independent moments, like a community, a connection of directions, of distances, etc. We have taken the term “whole” so broadly that it encompasses every kind of connection that passively connects objects or that is present through judicative activity. If the latter is the case, then I can have internal articulations within the original generating process and the product. But the whole as successively generated, thus, as having become, is only a whole after the generation, and it requires the explication into the elements first in order <to> have them as parts [108] of the whole and, in the transition, in order to be able to make the community of participation efficacious.

Appendix 38: (to §64) <Categorial and Non-Categorial Connections and Relations>

Relationship of containing and of what is contained. Let us assume that θ is an immediate relationship of parts. If “S_1 θ S_2” and “S_2 θ S_3,” then “S_1 θ S_3” can be valid: through an immediate step of explication, I go from S_3 and arrive at S_1. But it can also be that the relationship between S_1 and S_3 is essentially mediate. Like when an intuitive thing immediately has a “piece,” a piece of the piece is immediately a part of the thing. But the shape of the piece is mediately a part of the whole. The shape of the whole is immediately a part of the whole. The relationship of the mediate part (the shape of the piece or of its color to the concrete thing) to the whole is characterized differently as the relationship of the immediate part. This relationship, however, is not merely the relationship instituted by judgments.

Every judgment is itself an interconnection, and every chain of judgments is in the broadest sense a "whole" in which relationships exist (are instituted) between the parts, be they judgments, be they elements of a judgment, be they thematic objects. In this way it is a law that if "$a\ \phi\ b\ \phi'\ c$," "$a\ \phi''\ c$" is also valid. This says nothing more than that two state-of-affairs that are connected to one another produce a categorial interconnection between the first and the last thematic object, and thereby a categorial relationship as well.

But then we must distinguish: (1) Categorial connections and other connections, wholes that are categorial, and wholes that are not categorial; that is to say, there are objects that are not pre-constituted through thematic action, but separate-off from one another into a plurality of immediate parts only first through explication, parts that are "connected" in the whole, which is to say, they exist in relation to one another on the basis of the preceding unity of the whole and due to the fact that they are that which are contained in it, that is, they are its parts. They can exist in relationships beside this one, for example, in relationships of size, uniformity, degree, etc., mediately and immediately.

(2) Categorial relations and non-categorial relations. Every relation is a categorial one. It is an state-of-affairs, namely, a simple one; by this I mean one that is not a concatenation of several state-of-affairs, $S_1\ S_2$. A state-of-affairs is categorial by the *termini* being categorial or by the foundation of the state-of-affairs being a categorial one. Every state-of-affairs has a foundation, i.e., an objectlike formation that produces the community between the *termini* of the state-of-affairs. This community founds an [109]
objectlike formation insofar as it can be explicated and everything that emerges through explication is a part, that is, everything that emerges through explication has a community of partial identity with the whole and founds both the correlative relationships with the whole being determined and with the parts being determined. Further, parts and parts of a whole have their foundation in the whole. Put more precisely: Two explicated objects of a whole as such have a relation to one another; they are essentially relations

of intersection or relations of connection that can be constituted through determinative activity.

Appendix 39: (To §64) <The Task of a Theory of Relation>

A systematic development of the forms of possible objectivation on the second level of objectivation would have to progress from the simple to the complex and to the derived.

(1) Accordingly, the first task would be to disclose all the primitive shapes of state-of-affairs, or what amounts to the same thing, the primitive forms of relations whose *termini* naturally remain conceived in undetermined universality.

(2) Then, we would have to pass over to the syntactic shapes of connection, to the formal possibilities, in order to link up different state-of-affairs to one another; but, on the other hand, also in order to provide the state-of-affairs that are simple in themselves with an abundance of significance, i.e., in order to fashion lawfully regulated integrated formations of main clauses and subordinate clauses that can then, in turn, be analytically explicated at any time into mere complexes of primitive simple-judgments[76].

The syntactic unity of the *terminus* itself with its attributive significance, a unity arising in the place of every *terminus* in a judgment, makes up the concept of concept. If in traditional logic one speaks of the concept of subject, concept of object, the concept of predicate, one often has nothing else in mind than the intentional object which is the subject-theme, namely, as the substrate of the attributions added to it in the judgment in question. In this sense, an expression like "the minister overburdened with work" is a concept, and every judgment in this sense has two concepts that correspond to both of its thematic poles; and then within the attributions there are further concepts. Obviously, this concept of concept has nothing to do with concept in the sense of the generic universality.

What has been gained with respect to the products of determination by the many judgments in relation to one and the same theme of determination can be sedimented on this subject

---

[76] *Null-Urteilen*

through attribution, and can be systematically linked together and conceptually grasped together in it. In this way, the consequences of the entire judicative work up to now for the substrate in [110]
question are all there together and can be clearly seen, and are ready for all further judicative work to be done.

<Section 3. RELATED ESSAYS>

<A. Perception and its Process of Self-Giving>[77] [291]

<1. Immanent and Transcendent Perception>

First of all, it makes a most fundamental difference whether we consider objectlike formations that can only be experienced by the experiencing subject because the subject has spontaneously generated them in its thematic egoic acts—as, for example, numbers are only there for us originally as objects in acts of counting, or theories in acts of theorizing. [Or] by contrast, [whether] we have an entirely different original mode of givenness, an entirely different mode of experience, if objects are pregiven to the experiencing subject passively in experience, and are only experiencable through pregivenness in such a way that the subject merely exercises acts of receptivity, acts of grasping and then acts of explicating something that is already there, something that already appears. Since this latter realm of objects must precede all activity in general, and since, for example, cognitive objects, objects of theory, are only possible as intellectual[78] formations by having other objects pregiven through receptivity, then naturally the primary phenomena for all phenomenological investigations, and particularly for investigations that are distinguished noematically, are the phenomena of passively given objects. Normally, experience and perception only mean the experience and perception of such

[77] Editor: 1923? and 1920/21
Translator: The following pagination to the German text corresponds to Husserliana XI.

[78] *geistige*

"sensible" objects, as it is said—a limitation to which we would not want to be confined, and for good reasons.

In the sphere of passively pregiven and receptively experienced objects, we are concerned exclusively with individual objects—all universals are structures of spontaneous activity. Here, the most radical and thus the most general distinction of all is the one [292] between immanent and transcendent objects, a distinction that we have repeatedly utilized, but have not yet clarified thematically.

Thus, on the one hand, we have immanent perception, that of our own lived-experiences; here we take care to see that there is phenomenological purity in our method. In this case, "the perceived" is itself a lived-experience, and in this respect, no bracketing is required. For example, we live through[79] a joyful experience. We have here an originarily giving consciousness of this joy; not only does it exist, but it is an immanent perceptual object, and cannot but be anything other than that. We are constantly conscious of it, even though we do not have to be directed toward it attentively, and it is therefore pregiven for possible reflective cognition. Berkeley's formula, *esse* = *percipi*, holds for every immanent object, a formula that he promulgated for external objects, to be sure, though it does not hold for them, as we will soon see. Being and constituted-being for consciousness coincide where immanent objects are concerned. Both are inseparably one. I say "constituted-being." For when we appropriately limit ourselves to the sphere of objects of internal consciousness, we learn from our analyses of time that all lived-experiences in their internal flow are encompassed by internal time in which they have [both] a position and a duration for consciousness. Every lived-experience, as an internal temporal object, is initially and originally constituted on account of internal consciousness through which the temporal object is given to consciousness thanks to a flux of primordial impressions, retentions, and protentions as a thoroughgoing unity. Originally and perceptually, being and being-perceived do not coincide, but being, and in iterative rememberings, being *qua* perceivable, or being and constituted-being (which is adequate and fashioned

[79] *erleben*

from actual perception and its primordial constitution) do coincide; and that is to say, correlatively, that we can no longer distinguish here between [a] constituted sense and sense that can be construed identically again and again in a reproductive manner, and [b] the object itself *simpliciter*. The appearance of sense in the flesh is the object itself in its actuality.

Inner consciousness as inner perception carries out a purely immanent sense-giving through which pure consciousness remains with itself; in contrast to this, we find the transcendent process of [293]
sense-giving peculiar to external perception, and especially peculiar to the perception of the thing-bodily being, we find material being, which is foundational for all further modes of transcendence. The lived-experience, "perception of any kind of bodily thing[80]," is an immanent object that is given to consciousness in internal time like any other lived-experience. But what is constituted in the lived-experiences through a peculiar kind of sense-giving is precisely a bodily thing as given to consciousness in the flesh. This object that has been given to consciousness does not give itself as an immanent object, and it is nothing less than contained in an intimately inherent manner in the *percipi*. To be sure, one says with good reason that despite the fact that they are perceived, perceptual, bodily things do not have to exist: It could turn out later that the perception was a deceptive one. And conversely: Things can be in actuality without being perceived. Of course, such statements go beyond the purely phenomenological sphere. But already when we as phenomenologists pay heed to what is perceived as such within perception, we find that the objective sense of external perception transcends itself in a peculiar way, and that it transcends what it actually brings to the appearance in the flesh. By its very nature, every such perception anticipates, so to speak, an accomplishment that it is in principle not in a position to accomplish in adequate self-giving; it anticipates a *plus ultra* (self-giving is at the same time always anticipation)—while immanent perception actually makes present its objective sense in a complete givenness in the flesh.

---

[80] *körperliches Ding*

External perception concerns the entire so-called external world: houses, trees, tables, and so forth. Even when we speak of animals and human beings, among which we find ourselves as human beings, this still takes place in "external" perceptions. All these human beings and animals are themselves objects of the external world. Initially their lived-bodies are there; they are there for us perceptually only because their physical lived-bodies[81] are there. But they are for us by perceiving them as animals and human beings, not merely as thing-bodies, but as living beings, which for their part also perceive, remember, and have every kind of lived-experience[82]. Yet these lived-experiences are not given in the same way that our own are given; they are not perceived internally, but are only given to us through a transcendent sense- [294] giving a sense-giving peculiar to external perceptions that we call precisely perceptions of those human beings and animals external to us. Obviously, this external perception is a perception of a higher order: Another level of transcendent sense-giving is carried out in a first transcendent sense-giving; the latter is carried out in our perception of physical lived-bodies, and it is through this perception that they are given to us as "external," as transcending our consciousness; through this transcendent sense-giving these distinctive bodily things, which we call lived-bodies, are constituted for us in consciousness as lived-bodies, that is, as lived-bodies for a psychic life, a life of consciousness that is not our own. Naturally, there are problems here, namely, how this marvelous process of sense-giving takes place, and initially the sense-giving of the founding level of external physical perception, how we can understand that our stream of consciousness in its immanence can be conscious of objects as given, seen, touched, etc., *originaliter* in so-called external perceptions, objects that lie "outside" of one's own life-stream. What does this sense-giving look like, a sense-giving through which this "outside" gets its sense at all? External perception and exteriority in general is initially just a word. What does the structure of one's own lived-experience look like, the lived-experience that we call "external

[81] *körperliche Leiber*
[82] *Bewußtseinserlebnisse*

perception"? How do we make its intentional accomplishment clear? How does it happen, in which lived-experience (structured in such and such a manner) does it happen that "the outside" gets its available sense? Which general and necessary forms does the objective sense of external perception have, and which essential typicality do the modes of appearance have that are constitutive for these forms? Obviously, after this first question—namely, how nature in the narrow sense of mere bodily, thing-like nature is constituted through the sense-giving of consciousness—there is a second question that necessarily follows: How are psycho-physical and spiritual objectivities of different levels constituted, and then subsequent to this, the entire given world? For psycho-physical nature already presupposes that physical lived-corporeality is constituted for the perceiver as perceived.

<2. Temporal and Spatial Perspectives> [295]

We also call external things thing-bodies. This points to the form of spatiality, which in a curious way encompasses all bodily things as objects of possible perception and should encompass all thing-bodies, not only actual ones, but also possible ones, reaching into infinity. As individual objects, thing-bodies initially have the form of time, they endure, and all their durations belong to the one unending time. But where the fullness of time is concerned, they have spatial extension in every point of time, and are therefore integrated into space as a multidimensional continuum, as an absolutely firm system of locality that stretches through unending time, as a system of locality in its identity, invariably encompassing all possible bodily things through all their durations. All animated thing-bodies, human beings and animals, get their variable position mediately in this fixed system of locality, and thus at the same time in the *one* world, in this unity that is encompassed by the forms, space and time. Every object of the world, and the world itself, is obviously only there perceptually for an ego and its consciousness through the process of sense-giving that is carried out precisely in this consciousness through special accomplishments of consciousness. And this sense-giving is the problem.

As always, let us take up the phenomenological attitude. We consider pure consciousness according to essential necessities. Questions concerning the reality of the world remain out of play; we take the world only as the immanent sense of the correlative consciousness [of it]. Further, we work within the limitations of the transcendental aesthetic, we exclude all judicative knowing, and we exclude altogether the entire sphere of determinative and predicative thought that is grounded upon intuition. Thus, we restrict ourselves exclusively to intuition and more precisely to perception, that is, even to the phenomenon of the world only insofar as it is a phenomenon of perception. We restrict ourselves even further: We consider particular perceptions of thing-bodies.

We also take the noematic perspective. Let us have a look there. We said that objects of possible perception have the necessary sense-form[83] of time, objects of possible external perception also have the sense-shape[84] of spatiality, and this as the thoroughgoing [296] form of their temporal content. We find both forms in a curious analogy. For both, the object is given in a "shaped" extension: in a temporal shape (duration) and in a spatial shape. For both, the shape is necessarily given in the multifarious modes of appearance, in modes of temporal, and on the other hand, of spatial appearances and perspectives. For both, the adumbration of shape is followed by a secondary adumbration, as it were, that of temporal or spatial fullness.

Let us take any kind of bodily object as it is given in perception, like this table at rest. We direct our gaze now, not toward the alteration in temporal givenness and the sense-moments of duration, but toward the content. We observe it, and in this process our eyes move involuntarily, and with every position of our eyes we see the same object, but in a different mode of appearance. We also turn our head or change the position of our whole body, we move closer to it, or step back from it again, and the mode of appearance continues to change, even though we are conscious of the same object in this unflagging alteration of the modes of appearance. Looking at the thing, we are directed toward the

[83] *Sinnesform*

[84] *Sinnesgestalt*

modes of appearance in a way that is not at all thematic, and we do not think about regarding its transformation as the transformation of the object. This becomes especially clear in our example, namely, in the example of an object at rest; its particular determinations of sense come to the fore in the alteration of perceptions and in the alteration of its phenomenal contents, but they are only transformed in them with respect to their appearance, though they are not themselves changed. The same surface and edge of this thing presents itself in different perspectives, like it presents itself successively in perceiving. It is evident that it is not even conceivable as perceived without such perspectives. A perceptual thing is only conceivable as perceived in such a way that its thing-bodily shape, which belongs to the objective sense of the thing, initially presents itself in some way as appearing perspectivally in this or that manner, and in such a way that the identical and even unaltered shape (and according to each moment for itself determining the shape, according to each line, surface, combination of surfaces) is a unity of infinite possible modes of appearance. The same holds for color. It presents itself in a [297] parallel manner with the perspectival adumbrations of the thing-bodily shape, likewise in ever new modes of appearance. Let us take as an example a certain surface here, for instance, a visible rectangular table surface, as well as its color. With each new appearance we distinguish the color itself and the mode of appearance of the color. Of course, when we focus on the thing itself, looking at it, we do not at all distinguish the color itself and the mode of appearance of the color. But, we find this distinction the moment we reflect. This holds likewise for all other determinations of the object given in intuition that are spread over the spatial shape and that qualify it perceptually as spatial fullness.

It is evident here that the perspective of colors—and all similar adumbrations of the qualitative determinations of the external thing—does not characterize for instance a manifold of modes of appearance that is on a par with the perspective of shape; rather, it characterizes a manifold of modes of appearances that is dependent on the shape, and is necessarily subsequent to it. This is just like what happens with respect to temporal adumbration, with the modes of appearance of the temporal perspective, something that we could have already noted earlier: The temporal shape of the object, that is, its duration, appears with the alteration of pasts, is oriented anew to the living Now in newer and newer perspectives, and is pushed back further and further into the distance. But even here, following the change in the appearances of shape is a change in the mode of appearance of what fills the shape of time, of what qualifies the form: As a consequence of the fact that the appearance of the shape alters, everything that endures there also alters its mode of appearance with respect to content.

It will serve us well to pursue this incentive of drawing a parallel between temporal and spatial form still further. It will help us to make progress. Even the relationships of temporal orientation have their analogues in such relationships of spatial orientation. Within each perception a temporal present is constituted, and within the latter, an absolute Now. All past and future, even [the temporal givenness] of memorial objects is oriented to this flowing Now. Corresponding to the absolute Now as the zero-point of temporal orientation is the absolute Here as the zero-point of all spatial orientation. Every external perception brings with it its current spatial present and within it, the absolute zero-point of the Here. The latter is located in the very lived-body [298]
of the perceiver, and this zero-point "appears," and yet does so in an inauthentic way. The zero[-point] is itself nothing visible, etc., but a *limes*.[85] Thus, miraculously a perceptual object that we call one's own lived-body is distinctive in such a way that with each perception of an object, whatever it may be, the lived-body is always there and always co-constituted. And this object is entirely unique by virtue of the fact that it always "bears within it" the zero-point, the absolute Here, in relation to which every other object is a There. Just as universal, unending time is constantly and inexorably referred to the absolute Now, so too is the entire unending space inexorably referred to the absolute Here and to the coordinates of orientation attached to it: We have to distinguish

[85] The zero-region of the right-left, the zero-region of the above-below, the zero-region of front-back. The head is above, the legs below, one hand is right, the other left, the chest is on the higher side, etc. The zero-point is the ideal reference point of right-left, etc.

Translator: See also *Ideas II*, §40 ff.

the continuum of pure distancings in depth from the absolute Here, the continuum of right-left, the continuum of above-below—and these three continua in their mixture yield the entire three-dimensional continuum of orientations, or rather, the shape of orientation in which space as a whole always presents itself. The distinguishing feature of one's own lived-body as the bearer of the absolute Here gives every other thing, and the entire thing-world appearing respectively to it, the character of a world surrounding one's own lived-body or the indeterminate orientation character of "outside," of external world, understood here as outside of the lived-body and its zero-thing-bodily nature. However things may move and however my lived-body may move, nothing changes with respect to this most general mode of appearance: The lived-body remains the center, and the other things remain outside. While the latter can in principle take on every and all orientation with the exception of a zero-position, one's own lived-body, however, can only take on very limited variations in orientation, precisely because of its tie to the zero[-point]. For it, only changes in appearance are possible, changes that occur through an oriented turning, changes that emerge, for example, when the perceiver, in perceiving, turns his head and bends; here, the rest of his lived-body takes on a variety of altered aspects and accordingly changes in orientation in relation to the zero-position. Thus, one's own [299] lived-body is characterized according to the objective sense in a fundamentally different way than other things, and is constituted phenomenologically in a fundamentally different way than other things. The question in all of this is how "it gets on as it does."[86]

Many new problems radiate out from here: like the problems of change and above all the problem of movement whose possibility belongs to the fundamental nature of a bodily thing. A bodily thing can be altered in different respects, especially in relation to its thing-bodily shape; for example, it can become deformed. Mere movement and mere rest refer to the unaltered shape. Only through movement is the shape dislodged from the space that it occupies by merely changing place to newer and newer places. We

[86] Translator: Husserl has similar reflections on the lived-body from the 1930s. But now the lived-body is itself rooted in the Earth. See "Translator's Introduction," fn. 48.

also notice the following here: The same thing can appear with an unmodified shape in its different orientation and thus in its different perspectival mode of appearance; this mode of appearance can change and the object can nonetheless be at rest, and it can remain unchanged and the object can nonetheless move. The lived-body plays its role here again, namely, depending upon whether it itself moves or rests. The movement of an external thing does not necessarily influence the modes of appearance of other things; the movement of one's own lived-body revolutionizes the mode of appearance of each and every thing. The lived-body can, so to speak, run along especially with a movement, so that every change of appearance is neutralized. Moreover, lived-bodily movement plays its role with respect to all perceiving: The hand moves in touching, the eye moves in seeing, and the rest of the body assists in manifold ways here.

All of this makes up a region of phenomenological problems; in solving these problems, we must make intelligible what belongs essentially to the noematic stock of an external perceptual givenness transcending the empirical, human fact of lived-corporeality that is fashioned in just this way, with organs formed and organized in just this way; that is, we must make intelligible what typical function a co-constitutive lived-corporeality exercises for every external perception, and how it makes indispensable contributions. We must especially clarify what necessary constitutive function the so-called apparent-movements[87] have, why they must always be there, and how it happens that the phenomenon of the movement of each and every alien thing is [300] inextricably intertwined with the entirely different phenomenon of the movement of one's own lived-body.

I would still like to say a few words about the way in which the thing-bodily shape, as the thing's own spatial extension which moves along with it, functions as the formal fundamental determination of the thing. In fact, all other properties of the thing are related to it; in the broadest sense, they are that which fill this form. The thing-bodily extension or shape is that in which the thing "is extended"—as I already said a moment ago. The thing-

[87] *Scheinbewegungen*

body is divisible, and every division partitions the thing with all its properties in such a way that the particular qualities of the parts contain, as divided, the qualities as a whole belonging to the thing. It is not without reason that Descartes defined the bodily thing as *res extensa* and thereby esteemed *extensio* as the essential attribute of thing-bodily nature above all the other attributes as dependent attributes. Thus, for example, the color of the thing is nothing for itself, but it is extended over the thing-like extension; likewise, roughness, etc. Here we would have to study the stratification of the extending properties according to which the appearing thing has a layer "visual thing," with a visually appearing spatial thing-body, and a layer "tactile thing" with a thing-body appearing in a tactile manner, which is nevertheless given to consciousness as the same and is given to consciousness in numeric identity as the spatial shape that appears now visually and, at the same time, now tactually. Otherwise we would have two distinct thing-bodies, and not one thing-body. We must therefore study how other layers are related to these fundamental layers, the layers of warm and cold, the tonal determinations, the tone radiating out from the thing, even the radiating warmth that is not a warmth spread throughout the thing-body, etc., likewise radiating light.

We would then have to distinguish between the primary perceptual thing that has its intuitive content purely with respect to the thing-bodily form and with respect to its immediate fullness, and the specifically physical and chemical properties, the causal properties, all the properties in relation to which one speaks of forces. Naturally, these properties must be taken in the very manner we find them within the framework of perception: When, for instance, the hammer is given in intuition as forging the iron, and the "what follows," the result, the effect of the force is "seen," and with respect to the hammer, its weighty force. It is obvious [301]
that every thing-perception is only the perception of something material by virtue of such properties being co-intended in perception, that is, they belong to the sense. A thing that would have no mass, no dynamic force, that would not be elastic, etc., would be a mere phantom. Thus, in phenomenology we call the mere visual thing, the mere tactile thing, what appears to us as merely filled visual or tactile space, a thing-phantom; it is a mere

sublayer within the full sense of the material thing; the latter is a causal thing through and through, possessing causal properties. Thus, such differentiations must be made within sense itself, and with respect to it, the noematically constituting modes of appearance must be studied for every distinction.

<3. Time and Space as *principia individuationis*>

Now that we have sketched an entire horizon of highly ramified phenomenological problems pertaining to sense and appearance, let us augment our observations in another direction. Initially, the feature of both fundamental forms for the entirety of the structures of external thing-bodily nature, temporality and spatiality, must be understood in their function as *principium individuationis*. The bodily thing, like every individual object, is a temporal thing; it has its duration with a property-filled content that extends throughout this duration. The duration here, and every temporal point of the duration, is absolutely unique. It is called an individualizing moment because it confers its uniqueness, as it were, on the entire object. Namely, the temporal determination is unique in the following manner: Two durations can be equally long and be completely uniform with respect to their content. But they cannot be identical. Thus, not even two filled durations [can] actually [be] identical. Completely uniform objectlike formations can appear in succession at different positions in time, each one of which has its uniqueness: Completely uniform—what distinguishes them, at any rate, is the uniqueness of the temporal locus, that is, the uniqueness with which their temporal points have temporal fullness. What determines each object with respect to its properties is repeatable at any time. This also concerns its temporal shape. Different objects can appear one after the other in the same duration and with the same properties, filling the [302] duration; only the temporal points themselves and the durations themselves are absolutely singular. It is precisely this singularity that is not repeatable.

Insofar as there can be objects of perception that are simultaneous and thereby completely the same with respect to all properties, their temporal locus (which only occurs once) and the system of their durations are indistinguishable: When something is simultaneous it has the same temporal loci with their same singularity, and it does not have different ones, as is the case when something follows something else successively. Thus, time does not individualize that which is simultaneous. Taken by itself, it only individualizes the entire fullness of content that belongs to each one of the respective temporal loci and durations.

Now, the individual duration encompasses each individual object as the form of all its determination. Thus, this also holds for the bodily thing. No matter what the thing is, it is a thing as the unity of its duration, a duration that it fulfills in a successive manner. We already know, by the way, that the spatial extension, what we call the thing-bodily shape with all its properties being extended spatially in it, is thus the fullness of duration; this is to say that every temporal point of this duration is already a spatial thing with respect to content, but precisely the momentary phase of it. However, every spatial point, and therefore space as a whole, is also absolutely singular with respect to its spatial fullness. The thing-bodily shape does indeed remain in movement with the object, but it has its individuality in every temporal point with respect to the absolute singularity of the fullness of the piece of space in which it is momentary. Many completely uniform things can reoccur in space, and to be sure, simultaneously; their thing-bodily shape can also be the same, like all other properties. But what can never be the same (this is ruled out by the intuitive sense of the bodily thing-like nature and spatiality) is the spatial position, that is, the system of places that makes the two thing-bodily shapes into this singular one. Sameness of the spatial position necessarily means identity. But the same place, the same singular piece of space, cannot have multiple spatial fullness, cannot have multiple fulfillment with real content. Naturally, this is tied to the principle of impenetrability. If two things, let us say for the sake of simplicity two uniform things, could penetrate each other with their movement, thus coinciding in a temporal point, they would have to be mere phantoms. The same spatial points, the same spatial surfaces, etc., would have to be doubly occupied:
visually and in some other sensuous-material manner. Which is [303]

absurd. Every spatial point, every piece of space can only be real one time, and cannot be doubly occupied.

Now, how does the absolute singularity of space as the form of possible real thing-like nature relate to that of time? Obviously, time is the higher form; what is spatial (which is constituted in the sense of external intuition as form) is constituted as the fullness of time, constituted as belonging to the enduring content. The singularity of every spatial point is thus a singularity within every possible phase of the duration, the singularity of something simultaneous over against something else that is simultaneous. In every temporal point that we might happen to fix upon, we thus have the multi-dimensional system of absolute spatial positions, and so space becomes the individualizing form of coexistence and already presupposes time.

Everything that is constituted in the transcendent process of the sense-giving of external perception with respect to the thing is in space (and already with respect to the structure of the phantom), and has within it its absolute uniqueness according to each one of its positional points. Even if it were an immaterial phantom, outside of causality, it would have to appear as existing in space; it has within it its individuation through the spatial position by virtue of its absolute singularity. This rules out all repetition, dual occupation of the same thing-bodily shape with dual things, since precisely every *individuum* can only exist through the individuality of the site. Space is the abiding form of coexistence, which is to say, it is the individuating form of all transcendent temporal objects with respect to every transcendent experiencable sequence of coexistences that stretches through time; and space has this character because it is constituted as the constant form of temporal content and as the identical temporal content of all possible temporal contents.

The preceding analyses of perception cannot yet give us insight into how the accomplishment of transcendence actually comes about in the immanence of lived-experiencing. We must work out such an understanding at least for the sublayer of the perceptual thing, for the phantom.

<B. Consciousness and Sense—Sense and Noema>[88] [304]

<1.> Perception and Memory

Let us now consider clear memories instead of perceptions, like when, for instance, we presentify to ourselves here and now, in a clear intuitive memory, the theater with these or those events that we experienced when we were there. Everything we said is repeated here in a certain, though modified way. We look at the theater, we observe it in memory in a thematic attitude, but taken in a purely phenomenological sense this intuition is still something quite different from a perceiving.

What makes it different? What makes memory and its entire intentionality so closely related to perception, and yet what distinguishes it so radically from it? What is distinctive to and different for each of them comes clearly to the fore in and through the contrast. We are conscious of an object in perception as being there, so to speak, as in the present and in the flesh, as given *originaliter*; we only have it in mind in memory as a presentification of something that is not itself present. Perception is that consciousness which, so to speak, seizes a present with both hands by its shock of hair; it is a consciousness that makes present *originaliter*. In contrast, there are different modes of presentification. In and of itself, a presentification refers back to a making-present, though it is not a making-present. It allows something presentified—in our example, the memorial object—to appear "as if" it were present once again. This "as if and again" is a feature of consciousness that occurs in the remembered object itself, a feature that characterizes the mode of givenness of the object as a modification of the original mode of givenness, presentified precisely as merely memorial.

---

[88] Editor: 1920 ff.

<2.> Presentification and Pictorial Imaging[89] [305]

One is tempted to say that we have a pictorial image[90] of an object internally, that memory is an internal pictorial imagining. But that is absurd. And no less absurd is the naive interpretation, so tempting to primitive thought (as it was all too often at work in Ancient as well as in Modern philosophy) that explained perception itself as having an inner image[91] of what is out there in reality, in the original. But the process of perceiving is in and of itself original consciousness, whether the perceived really exists or not; the perceived object is given to consciousness as there in the flesh, that is, precisely as there in the original. In the concordant, synthetic progression of perception, the perception is thus confirmed as what it is, as an original prehension of the self of its object. Independently of the question concerning justification, whether it is legitimate or deceptive, memory is likewise in and of itself a presentifying consciousness; the thing remembered is characterized as the presentification of an original that is not present. Even it is not a pictorial imagining. Opposed to a straightforward memory, as opposed to straightforward perception, a pictorial imagining characterizes, rather, a new kind of consciousness. For an image, like a bust, is given to consciousness as a thing that serves to presentify something that is non-present. If we see it we have a perception, a perception of the thing, "bust." But in a peculiar way, a presentification is carried out in this perception, one that presentifies to us a different object, the head of a human being that the bust-thing resembles. Memory, however, does not harbor the perception of a first object that [presentifies] to consciousness a second object resembling the first. It is also clear, nonetheless, that since the pictorial resemblance, which is contained in each apprehension of an image, is itself a presentification, it would be absurd to reduce all presentifications to pictorial ones, and in this way to make it supposedly intelligible. At the same time it is clear that memory is

---

[89] *Abbildung*
[90] *Abbild*
[91] *Bild*

a completely primitive form of presentification, namely, that it is a pure presentification, while every ordinary apprehension of an image, every consciousness through which we intuit things as paintings, statues, in short, as pictorial images, is composed of perceptions and presentifications (closely akin to straightforward memory).

<3.> Self-Forgetful Remembering [306]

If we stay with memories and settle into a nexus of memories in a living manner, we will find that there are different ways of carrying them out. In the first instance, the peculiar form of memory, which is completely self-forgetful, is possible at any time. That is, we can allow the current present and ourselves as the current subjects of the present to become completely submerged [in memory], and we can live intuitively, purely in the world of memory. As in perception, a normal focal orientation and an orientation of action is initially prefigured when memory is carried out as an egoic act—we have those kinds of orientations in which we are directed toward objects, apprehending them, objects that are called "remembered" in the natural sense: in our example, directed toward the theater and the events that took place in front of it. All these objects and events have the character of a memorial, presentified past. But we can also reflect in this submersion; then the apprehending regard turns toward the changing past view of the objects (for example of the theater), toward the modes of appearance that were given with it, toward the identical element in the modes of appearance, toward the ontic modes of certainty, of doubt, etc., but even toward the past ego itself that was doing the viewing. What kind of discoveries are these that are found through reflection?

We do not actually view, we do not actually perceive, we do not really have modes of appearance as actual ones, at least not like they are contained in perception. In fact, even these phenomena given in reflection have the character of presentifications; they are presentifications of perceptions; and just as the memorial objects of the normal viewpoint are given to consciousness memorially as things and events of the past, so too [are] the perceptions, the appearances of them [given to consciousness] as memorial presentifications, as past perceptions and past appearances. If we are consistently in the attitude of self-forgetfulness here, that is, removed from the actuality of the present, then each and every thing that reflection lays hold of is given to consciousness in the mode of presentification—even the ego that encountered these submerged views, further, the remembered ego, the ego of yesterday that viewed the theater—but [given to consciousness as] [307] presentified in memory. Accordingly, it is also clear that whatever the analysis has offered to us with respect to perceptions, especially appearances, objective sense, and modes of being, we find once again [with respect to memory]; but that which has the character of being originary, of an original making-present in perception, has the character of presentification here. Thus, in contrast, we see that perception, according to all its straightforward and reflective phenomenological components, is a place of originality through and through. On the other hand, memory is through and through a place of presentification. To be sure, the latter only holds so long as we dwell in the attitude of the complete submersion in memory and cash in on its holdings in a normal and reflective direction. Then we will find a past (presentified) ego, past perceptions, past senses, modes of being, etc.

Now the one carrying out the memories could object: But I live now and have my lived-experience of memory in the Now; these are actual lived-experiences, original, and not merely presentifications. That is of course absolutely correct. But the presupposition that takes the form of carrying out this submersion of the ego in the world of memory says quite a lot about the legitimate sense of this objection. "I live now" must not mean that as a central ego of the present I am "awake" and in action. The situation will become clear when we let the ego "come to," as it were, when we let ourselves as the ones remembering awaken from our dreaming self-forgetfulness: Our ego, conscious of its present, now casts its thematic gaze into memory, that is, from the Here and Now to what it has in mind in a presentified manner in the current actual lived-experience of memory, and this is memory in the second sense: An entire piece of past egoic life is presentified with what it encountered at that time as the experienced external world, as the theater, etc. From the standpoint of the Now, the wakeful ego looks directly at this thing or, staying in the framework of presentification itself, carries out a reflection on the past acts, appearances, etc. The wakeful ego's gaze slides from the past and from the presentification to the current present as the realm of the originary and now also finds, reflecting in the Now, its present lived-experience and acts of the "I remember" in which those pasts are centered from the [308] standpoint of the wakeful egoic center. Then it also finds, looking back to the previous stages, the past actual lived-experiences of submersion as egoless background lived-experiences. The contrast is now clear: If we, as a wakeful present ego, actually carry out the "I remember," then the apprehending ray toward what is presentified will proceed from this wakeful center. But when we were submerged, we were precisely not coming on the scene as the actual ego actively occupied now; we were dreaming, and making its appearance was only the dreamt memorial ego of the past. All egoic acts were given to consciousness as quasi-acts, they all had the character of presentifications of the previous acts belonging to the previous ego, in relation to the past objects of the past perception.

Thus we have clearly and firmly this most important distinction, the distinction between acts that are given to consciousness as acts actually carried out now, acts peculiar to the ego that is actually occupied now, and the quasi-acts carried out by the ego lost in memory, presentified acts of the presentified ego. In this way, there is at the same time a new determination of the concept of the wakeful ego, a determination that was not available earlier, since then we were only concerned with current acts of the present and were only in a position to contrast these with dreamless sleep and with egoless backgrounds. Now we have become acquainted with being lost in sleep in the special form of being lost in memory, which is not egoless like the background phenomena, but whose ego, however much it is actually my ego (namely, is given to consciousness as truly my ego of yesterday), is still only a presentified ego, and not an ego currently active now. And precisely for this reason it is now said that the ego is not awake.

The wakeful ego is the ego that carries out actual acts now and actuates in them a life of acts that is constantly arising in an originary manner in relation to a perceptual present. I live entirely submerged in memory, dreaming, which is to say, my present stream of lived-experience is egoless in a peculiar way, without actual egoic acts that proceed from the Now, the perceptual present, going toward [another] Now, and by passing through a current presentification in consciousness, going toward what is past; no matter how full of life the submerged life in memory is, however much this life takes place in present acts, this present is still not a thematic present; the one who is lost in dream does not [309] have a thematic present at all; he is not awake for it and for himself. Only upon awakening does the sun of the central ego, as the radiating center of actual acts, first dawn in the stream of consciousness, and only now are the memorial lived-experiences transformed into those that are centered in the current, present ego, in actual acts of the "I remember."

<4. The Complexity of the Ego>

But we learn still more. We encounter here a wondrous dual stratification in the intentionality of all memory, and it has become patent through the emergence of the wakeful ego as the subject of a current present, and at the same time, as the subject of present acts of memory. Living in the present, the ego is thus related to a past. But the past is itself a past present. I, the wakeful ego, am the subject within a realm of the present given to consciousness. But that means phenomenologically that I perceive all kinds of things and still much more is ready to be perceived by me; I am the ego of an actual stream of lived-experiences that are original lived-experiences in every Now, and of which I am conscious in perceptions, even if they are latent. But presentifying memories, among other lived-experiences, arise in this stream of lived-experiences, and however much they are themselves my present original lived-experiences precisely a non-present present, with a non-present ego and stream of lived-experiences is presentified through these memories, even with non-present external perceptions and with a non-present external world perceived in perceptions, etc. I not only am, and I not only live, but a second, entire egoic life is given to consciousness, is mirrored in my life, as it were, namely, is presentified in my present memories. And that's not all, not [merely] a second life, but an infinity of such lives, insofar as the past is a continuum, and belonging to every point of the memorial past is another presentified present with the presentified ego and egoic life. The respective ego, however, is continually identical throughout all these reproductions, identically my ego, and I am conscious of it in the current memory [310] with its past actuality in a secure certainty.

<5. Memories of the Future and Memories of the Present>

We were afforded several insights into the wonders of memorial consciousness. Deeper forays would continually elucidate such wonders. To be sure, we must restrict ourselves to what is really necessary, specifically, to what is at stake for the sense-giving accomplishment of memory. For on our present path, which concerns the relationships of consciousness and sense and the clarification of the idea of sense, in force is the entire investigation into the intentional nature of memory as well as the previous investigation of the intentional nature of perception, of the demonstration of the contents of sense.

Before we pass over to this, let us expand our idea of memory. We previously understood by memory intuitive remembering, a word that not only characterized presentifying modifications of perceptions as such and in complete generality, but only those that re-presentify the pasts, "as if" they were once again presents of perception running their course. But there are also presentifications that go toward the future and even toward the present: thus, when we expectantly look forward to a futural event, and even already allow it to take place in intuition as if it would run its course. Finally, where the presentifications of the present are concerned, we will only need to think once more of our example of the theater or the streets and the passages in the city, we only need to view them in intuition, but as here and now where we do not perceive them, and on the other hand, not as the objects in the mode of the past, as we saw them yesterday for instance, but rather as existing now. What we have expounded upon (and are still expounding upon) regarding remembering as remembering-back obviously also holds to quite a large extent for these new presentifications, for these memories of the future and memories of the present. This can be recognized from the very beginning. On the other hand, we would certainly learn from a closer analysis that remembering is presupposed for an understanding of these new presentifications because they have an intentional structure that is necessarily founded in rememberings. Thus, they do not [311] have the originality and primitiveness like rememberings do. But we should not go into these questions, and we must let our rough considerations suffice.

<6.> The Clarification of the Immanent Sense-Structure of Memories

We had ascertained that everything found in perception, whose nature it is to be originality-consciousness, is found again in every memory, but [here it is found] in the mode of non-originality, of presentification: thus, presentified things and events, as those of presentified perceptions, and belonging to them, presentified senses, namely, the corresponding objective senses, and on these senses, the presentified modes of being. Concerning the latter, we are conscious of the objects concerned in memory as those objects that are before us either as actualities of which we are certain or as dubious, as probable actualities, etc. But memory, understood as the current, actual, intentional lived-experience, still carries out its own intentional accomplishment and, that is to say, it does not only presentify and does not only harbor presentified senses and modes of being of itself as the presentified ego, but contains its own current senses and modes of being. Memory is not merely a reproduction of a previous consciousness, of a previous intending and a previous intended meaning, but related to this in a second layer, [remembering is] also a present intending or meaning of the current ego.

Insofar as memory precisely reproduces and perception does not, the talk of originality and non-originality depicts the mode of givenness, the mode of consciousness of the object from the perspective of the lived-experiencing. But if we are directed toward the object itself without any reflection, then new values, new temporal values emerge that do not say anything of consciousness and its mode, but do speak for the fact that unique features emerge in the object itself corresponding to those modes of consciousness. Namely, in perception as directed purely toward the perceived object, we find the object as existing now, as a present object. In memory, turning purely toward the remembered object, we find it as past. And now we note that forms of sense [312] arise here, indeed for both of them. For purely with regard to the object that is perceived as such, with regard to what is given to us as the identical sense of perception, we have the "now"; with regard to the remembered object as such, [we have] the "past." If we have specific act-intendings, at one time the object is meant in the temporal mode of the "now," the present, another time [it is meant] in the temporal mode of the past. Both are independent of whether it may be a matter of legitimate, verifying perceptions or memories. For both of them we have senses, and for both we have objective senses with temporal modes that belong to them.

<7.> Time as the Form of all Senses of the Object

Now what is it that is given to consciousness in any modality of being whatsoever, or most simply, that is given to consciousness within memory as existing with certainty? The object or the event as past. By contrast, in perception the object or the event [is given] as present, as now. Time, or rather, the modes of time, come into view here for us, and we must become clear what that should mean with regard to sense or for sense. We realize that our analysis of sense pertaining to perception was incomplete and that what we had gained for it under the rubric of "sense" was not its full sense. Lacking in the sense itself was a universal dimension, a sense-form. If time is objectively the universal form of all individual, real objects, then it is accordingly a universal form of all possible individual object-senses for consciousness. Furthermore, it is a universal form in which all objects in general and object-senses in general are embedded in a certain way and necessarily. If senses are the great theme of the logic of senses, and if, as we can demonstrate, they are senses that are expressed in expressions and propositions, then the discussion of the form of time has universal, fundamental significance for us. In the Middle Ages and in the Modern era, one was helpless in the face of the interpretation of temporal modes within the doctrine of judgment; one was unable to discern whether the temporal modes belong to the so-called copula "is" or to the object, and what that should [313] mean—all this goes together with the complete lack of profound investigations into time-consciousness and its structures of sense.

Let us then go back once again to perception and reap the benefits of our analyses of memory. In these analyses we got to know presentifications in their primary and most impressive shapes of all. That will have made our mental eye receptive to understanding other and still more original shapes of presentification, shapes that are connected to perception itself. The extent to which we had to contrast perception as non-presentification with presentifying memory is the extent to which we must now also insist that, by its very nature, perception can only be a concrete process of making present by also necessarily being presentification.

<8.> Now and Originariness

In order to shed light on this paradox, let us consider what is contained in the form of this Now, in the form of the present, in which what is perceived is necessarily given. This moment corresponds to the mode of givenness of originality, to the mode of being presented in the flesh in which the perceptual object is given to consciousness. But does this mean that the thing is Now, that the event, the melody runs-off now? To be sure, it endures! This Now is obviously an entire expanse of time that is productive in an originary and wondrous way in a steady flux: An initial tone sounds and continues; it comes to an end, then a second one begins etc. But we easily recognize that within this broad Now, within this broad present, we can again distinguish presents and pasts, and each time a distinctive Now is noticeable, only to slip between our fingers just as quickly. A Now arising originally is filled with a tonal content. But the Now has already become a non-

Now, and the new Now has a new content, here a uniform one, here a qualitatively altered one, and so on, without stop. The Now changes into the just-Now. But the just-Now has not disappeared from consciousness and from the thematic hold. Not only is the Now and its content affected by the flowing change, but the just-Now is itself once more transformed into a just-Now from the just- [314] Now, and so on. Only in this way can the tone and the melody as a concrete object in perception can be given to consciousness, not just because this continual process runs its course, but also because as the process of consciousness, a unity of sense-giving is carried out in a constant coinciding with itself, a constant coinciding, naturally, according to the immanent sense.

<9 > Time-Consciousness

In our following reflections, we can make intelligible the necessity of a process, a process like the one that we have to describe now; and we can make intelligible the necessity of its intentional structures, and through this we can make ourselves more receptive to the very seeing of them. In the natural attitude bearing on the object, we say: An individual object is inconceivable without a duration; it is the identical element that endures throughout the continuum of the phases of the duration.

If we focus on the successive content of this duration, on what fills up the duration anew from temporal point to temporal point, we will have (for example, with the enduring tone) the momentary phases of the sounding tone, of the tonal process—we will have momentary phases that are newer and newer, now changing with respect to content, now [remaining] the same with respect to content. The tone itself is accordingly called altered or unaltered. Now, what in principle will enable such an identical tone to be given to consciousness, and thus the temporal object in general, as the identical temporal object of a process extending throughout the duration, and initially, such that it should be able to be given as original, given perceptually? How must such a perception be constituted with necessity? Every perception, like every lived-experiencing, is likewise and necessarily there for consciousness and its subject as something that can be experienced in a possible reflection, and it is evident that it is itself something temporal; it begins and runs-off, it has its time, its duration, its process. Now, how must this process of consciousness be so fashioned—a process in which the objective process is given, the process of a melody or of some other object that we become cognizant of through perception?

If one poses this question as my teacher, Franz Brentano, has already done, it is tempting to say at once, of course, that what is primary is the fact that in every momentary phase of the [315] perceptual process, a corresponding momentary phase of the perceptual tonal process is perceived, that is, is given in the mode of consciousness: "in the flesh." But that does not suffice. For if in the course of perceptual lived-experiencing, the respective momentary lived-experiencing has gone over into a new one, thus, past by, then the new one brings to consciousness precisely a new tonal content: Every momentary phase of perceiving [brings to consciousness] its content and no other. By no means could the consciousness of a tonal process, of a melody arise. But we have this consciousness, and during the perceiving, [we have it] in every moment; we are not only conscious of the momentary resounding tone or even of its momentary phase, but [we are conscious of] the protracted tone itself and the melody—to be sure, as a constant becoming, constant flowing and elapse. But fundamentally belonging to this is consciousness' possession of what has elapsed. That is, a perception is inconceivable without a continuum of so-called "fresh memories" being constantly and closely intertwined with the actually and genuinely perceived phenomenon of the momentary Now. The continuum of the tonal data, the tonal data that have been given to consciousness in the past perceptual phases, is further preserved in consciousness for a while, with a continuity, in the form of these memories, and naturally in the mode of "what has just been."

All of this is a good introduction, but it is still not a phenomenological elucidation of the structures of time-consciousness that would make possible an actual comprehension of the original sense-giving of time-consciousness. We now turn to a memorial content of perception; and it is evident from this (and entirely apart from the considerations of the kind we just employed) that the memorial content is at no time and by no means separable from perception.

Let us reflect once again in a more precise manner. Belonging to the nature of perception is the process of presenting something to consciousness as there in the flesh, as Now. But a Now cannot be presented to consciousness without something that has just been. Thus, memory fundamentally belongs together in a unity with a perceptual process. This memory, which is connected to the Now-consciousness in a wholly immediate manner, we call retention. But we must now note that the same modification that the Now sustains in [becoming] the just-Now, the same modification that the momentary phase of an actually original [316]
making-present sustains in [becoming] presentification, is also sustained by this presentification itself. The just-Now is transformed into a new just-Now, a just-Now from the just-Now, and so on, *in infinitum*. Let us mentally fix any moment of the streaming perceptual process (whose streaming cannot actually be inhibited), that is, let us take a cross-section of the process, right in the middle of its course: for instance, when a melody sounds, the moment a certain tone begins anew. How does the affiliated perceptual consciousness look, the perceptual consciousness belonging to this moment? Here we do not only have the onset of the tone given to consciousness as now and in the strict sense as resounding originally; but in a whole expanse, we also have the past tone with all its phases in a fresh memory, and potentially even the past tones further back. But all of that is not given to consciousness as having-just-been in the same way; rather, corresponding to each phase of the process given to consciousness retentionally and having run its course is a new mode of "having-just-been," a new mode of the past. In other words, the consciousness of the retentional presentification has a new structure of sense for each one of these phases. In the direction of the thing perceived, our cross-section of the perception of the tone offers a phase of momentary present and a continuum of just-pasts being differentiated; in the correlative direction of the lived-experiencing there is distinguished a moment of pure making-present, limiting a continuum of presentifications that are undergoing transformation. When we put the accentuated cross-

section of the living perceptual process back into the flux, so to speak, and observe how it continually changes, we experience which kind of presentifications these are and how they determine the unitary structure along the unity of the process. The newly resounding tonal phase loses its originality, the original making-present becomes a presentification, the original Now becomes a having-just-been. But the presentification is transformed into a presentification of a presentification, the just-Now into a just-Now of the just-Now; the same holds for this one, and this takes place continually. In this way, we have now obtained a longitudinal section related to the certain tonal phase. Accordingly, when we go along the perceptual process we find, corresponding to it, a continuum that begins with the original onset of the tone, and lets [317]
this pure, original consciousness overflow into the continual network of presentifications of presentifications, *in infinitum.*

But what we said about the phase of the onset of the tone also holds obviously for the entire cross-section of the perceptual process that we previously had in mind and that belonged to the moment when the tone had just begun. This momentary cross-section was an entire continuum, and the continuum of all the living presentifications of the previous tones in this moment. This entire momentary continuum, I say, succumbs to this law of sinking back into memory; it is constantly transformed. And while it is being transformed, it constitutes a steady consciousness of the momentary consciousness that has just past on, and does this in the continual mediacy of all moments that have past on. Thus, each lived-experience of perception is exhibited (in a marvelous intricacy, and yet in an understandable necessity) as an unflagging streaming process of becoming in which a successive continuum of continua of retentional presentifications is generated in a living manner, while a momentary-Now arises *originaliter* livingly in every full present, and is superseded in the streaming of a new one, but its comet's tail of retentions follow upon it. Precisely through this, a continual unity of consciousness becomes possible in the progression of the flux, a unity that has its correlate in the unity of the temporal objectlike formation with the temporal duration belonging to this unity itself. A unity of consciousness as the consciousness of one and the same datum, and therefore a continual unity of the objective sense, runs thoroughly through every continual series of successive presentifications that brought into play any datum at all (for example, the newly resounding tone) arising in the original Now. However, the mode of consciousness may alter in the retentional modification; however much the lived-experience as such may change, it is the one and individually the same tone and the same tonal phase that remains given to consciousness in the change, and only the temporal mode of givenness necessarily becomes different. The identically same tone is constituted for consciousness precisely because on the whole an alteration of the lived-experience has not taken place; [318] what has taken place, rather, is an alteration in the steady coinciding that forms identity peculiar to the intentionality that is instituted with the onset of the tone and that is then modified. In the steady, successive process of coinciding, one intention coincides with another intention according to its very sense. The modification in the form of a steady presentification maintains an identity of sense: As a steady modification that abides in consciousness retentionally, the modification maintains the identity of sense in consciousness in a steady and unitary manner. That holds for every new momentary phase of the object in its original emerging and fading away, and thus holds for the concretely full temporal object with respect to its constitution as a concrete object for consciousness, as concretely enduring.

<10. The Structure of Perception and Consciousness in General>

If, in this way, we immerse ourselves in the structure of perceptual consciousness with its intentionality that is systematically connected and regulated according to continua, and continua of continua, we will understand how it happens that all the presentifications, which stem from ever new points of originality (for example, from the ever newly resounding tones and tonal phases), cannot get confounded with one another, neither the reverberating residues of our tone with one another, nor with the other tones or noises and their retentional residues running their course along with them. Indeed, we are standing in an indefatigable and an exceedingly multiform alteration in the lived-experience; but by virtue of a synthetic coinciding of sense, an objectlike formation firmly regulated in itself, one also firmly regulated in contrast to other objectlike formations, is necessarily constituted in this alteration: with a firm duration, whose temporal loci are fixed, and with firm coexistences and successions of concrete objects becoming distinctive. Even if the one tone of the sequence is completely uniform with another one with respect to content, it is the tone of this duration, like that tone of that duration. The layers of the modification of retentionality are distinguished; they cannot be conflated. And this also holds for the particular momentary phases of one and the same enduring tone. For what has flowed from a newly resounding tone with respect to retentional presentifications is distinguished without confusion from what has flowed from every other one: The continuum of these derivations has its unity in the coinciding of sense into [319] which something that is alien to the sense cannot intrude.

That designates a primordial lawfulness of perception and therefore of all lived-experiencing in general, insofar as all lived-experiencing is constantly constituted by perception as being-conscious[92].

(1) The study of the general structure of perceptions, and of rememberings standing in contrast to them, provides us for the first time with the insight into how lived-experiences, so to speak, not only can be, but can be conscious of something, how they carry out within themselves a sense-giving, how through a continual coinciding of sense a unity of the objective sense can be formed and be maintained through the alteration of lived-experiences. It is completely evident that individual objects are inconceivable unless they are temporally formed objects. But such objects cannot be given to consciousness *originaliter*, for instance; they cannot be perceived through a conscious-having that is devoid of structure. Perceptual consciousness is not an empty box into which a perceptual object shows up unannounced and ready-made; rather, the perceptual object is immanently constituted in it by an exceedingly subtle sense-giving structure of perception. It is constituted in the steady alteration of the manifolds of lived-

---

[92] *bewußt-seiendes*

experience instituting sense by virtue of their continual coinciding of sense. It is built up as steadily becoming with its temporal shape and in variable modalities of time. Here we must understand the absolute necessity that an object cannot be "there" perceptually for the ego any differently at all, that it can only appear to it as constituted in such a process of becoming.[93]

(2) The lawfulness with which we have become acquainted in [320] the structure of perceptions easily reveals its universal significance as a primordial lawfulness of the life of consciousness in general. For the latter is not only a lived-experiencing continually streaming along; at the same time, as it streams along it is also immediately the consciousness of this streaming. This consciousness is self-perceiving, although it is a thematically executed awareness on the part of the ego only in exceptional circumstances. Belonging to the latter is a reflection that is possible at any time. This perceiving that presents all lived-experiencing to consciousness is the so-called inner consciousness or inner perceiving. – Among lived-experiences are also, then, especially the so-called external perceptions, which are themselves given to consciousness internally, but for their part are modes of consciousness of "external" objects, namely, perceptions of them, of trees, of houses, etc.

[93] It will become completely evident to you as we proceed that we can gain an actual philosophical logic only through such an elucidation of knowledge and its accomplishments, an elucidation that illuminates the innermost essence of the matters. According to our method, we let the idea of logic and its necessary problematic naturally develop and grow in us, and specifically by building it up by means of systematically organized founding elements. Watching the growth of logical fruits, we will show in the following reflections what logical fruit itself is. Drawing upon particular accomplishments, we should be able to show what, on the whole, can be accomplished and what the task of a proper science must be able to accomplish, a science of unprecedented significance, a science of *logos* in the most universal and at the same time deepest sense. Perception and its parallel modes of consciousness of intuition are, however, the first fundamental shapes of consciousness that are at issue for the structure of consciousness that is specifically logical; they are the first foundations in the logical structure that must be situated and understood. Thus, we are not digressing for instance; rather, we are already logicians here without knowing it. In our method, however, we must already be the logician in order to be able to know it.

<11. Noematic and Noetic Directions of Description>

It was necessary for us to have embarked upon two directions of observations in a concerted manner; and these two directions had become interwoven with one another in our minds for the purpose of clarifying the constitutive accomplishments: namely, [1] the direction of lived-experiencing with its structures, and [2] the opposite direction of its sense and what, in this very direction, is seen with respect to sense and conjoined along with it. Like every consciousness, perceiving presents something to consciousness in this consciousness, [and it does so] as always, irrespective of all questions concerning legitimacy or truth. Thus, we can take the attitude toward this something and describe in a purely phenomenological manner what is presented to consciousness in this consciousness; we describe what is perceived purely as something perceived in this perception and in the very manner in which it is presented to consciousness there. This description of perception, but also of every other [mode of] consciousness, is called noematic description. In this attitude we come across sense with its temporal form and modes of being, but as we will soon hear, we also come across something else. The opposite direction is the direction of the perceiving lived-experience with all the structures in which sense, and everything that presents itself in it (in the mode of those described continua of making-present and presentification), is constituted as a thoroughgoing unity. This direction of description is the noetic one. We must embark upon [321]
both, each in their turn, so that we will understand how in the lived-experiencing, in the noesis, which is necessarily variable and changing, a unitary noema with an identical sense and the other noematic structures can be accomplished.

<12. Identical Sense and Noematic Modes>

We make an extremely important observation here that holds for all senses, namely, that the sense does not belong, for example, as an intimately inherent component to the corresponding lived-experiences in which the sense is said to be enclosed. For if we take a chain of lived-experiences of an absolutely identical sense, we will have separate lived-experiences, and not lived-experiences that have in common an intimately inherent piece that is individually identical. For that would rule out the separation. We can be conscious of absolutely the same thing today and tomorrow, but that does not mean that we have the same individual piece in consciousness like in a box. The identical sense is thus an ideally identical moment of all lived-experiences of consciousness that accord in sense.

If we distinguish consciousness and its objects, and take the object merely as a meant sense within the phenomenological attitude, consciousness itself will sketch out for us a course of possible identifications that proceed from the consciousness in question to a newer and newer consciousness, as the consciousness of the same meant object. Thus, there are identifications (like in the example of a perceived tone) in which the tone itself, as an object with its objectlike determinations, is what is to be apprehended thematically and identified as the same object of manifold rememberings. Precisely this determines the concept of objective sense. Likewise, within a living perception, the new tonal phase appearing in every new Now is in the process of sinking back steadily into the retentional modification for the consciousness of the same tonal phase, just that it undergoes the mode of the just-having-been and suffers progressive obscurity. As opposed to the identical content that is prefigured by the respective consciousness itself under the title of its sense-giving to an object, we have differences within the various lived-experiences that bear the same sense, differences that emerge in the sense as the character of givenness: Thus not only sense and [322] mode of appearance, but also other modes of givenness that are not original or reproductive presentations. In this way, where a remembering is not completely obscure, it can be more or less lively or obscure. In the case of complete obscurity, the object is still meant, still given to consciousness but, so to speak, in a completely empty manner. All the same, it is not empty of sense, it is still given to consciousness with the same sense, as this object determined in this way or that—and likewise within the continuity of the retentional modification of the tonal phase that becomes obscure each time. It is preserved with respect to its sense, otherwise it would not arrive at the duration being constituted identically with the identical tone as a whole; but that which is identical, what we call sense, has modes of obscurity. We call the latter noematic modes of sense and not modes that belong within the sense itself. Thus, all differences have their place there [in the noematic modes of sense], differences that distinguish the perceptual sense from the corresponding reproductive sense.

There are still further noematic differences issuing from a domain with which we are familiar; where lived-experiences are concerned, these differences correspond to background-lived-experiences and act-lived-experiences and are quite manifold, since the background is in no way entirely dead. The objects that are constituted in the background come into relief, as we say, more or less intensely; they exercise a more or less intense allure on the slumbering or wakeful ego. If the ego becomes attentive, it will be able to occupy itself with the allures primarily and thematically, or only secondarily, or even only non-thematically, etc. All of that gives different noematic features to "objects" as identical senses, which as such do not belong to the series of possible identifications in which the meant object itself is determined. This object is what is identical over against all such noematic differences.

We likewise take into account the differences of temporal orientation within the mere noema, differences that have a distinctive place within the noema because they have a special feature, namely, the feature of modes of appearance in the strict sense.

<13. Primordial Impression, Retention, Protention>

We should introduce a further terminological definition right away. Perception is giving in an originary manner with respect to its immanent object, that is, its sense. But as we saw, this can only [323] be the case insofar as it is an integration of pure acts of making-present and presentifications streaming along, which as phases of the stream are non-independent. We call the momentary, pure making-present of every perception, in which there is a new making-present in every moment, a primordial impression. Its accomplishment is the primordial institution of a new temporal point in the mode of the Now, filled with objectlike formations. The continua of presentifications of something that has occurred "just" now which belongs to every moment of perceiving, we take as retentions; they fuse into a unity of one retention, which however has a new mode in every phase of the continuum.

Under closer scrutiny—and this would be a necessary supplementation—we notice that a new sort of presentification still belongs to perception, what we call protention. Protentions are anticipations continually undergoing change and, from the very beginning, are constantly aroused by the course of retentions. A futural horizon is continually awakened, even if it is obscure and relatively indeterminate; a future is constantly constituted, namely, as an ever newly altered future just about "to arrive." The tone resounding and continuing to sound, sounds for consciousness into a future; it reaches out to perception, so to speak, with open arms. No matter how empty and indeterminate this anticipatory continuity may be, it cannot be completely indeterminate; the style, so to speak, of "what is to come" is prefigured through what has just past. The sense-content of what is to come is also necessarily altered along with the change of what has actually occurred and has past on—whereby even this change lingers and is preserved in retention. If every fresh past is a continuum of the adumbration of presents continually fading away, then every future about to arrive is a second-order adumbration, a shadow cast by that first continuum of adumbration. Indeed, it is a primordial law that every retentional course—in pure passivity, without co-participation by the active ego—immediately and steadily motivates and thus generates intentions of expectancy that are determined in the sense of a similarity of style. The intentions of expectancy can either be fulfilled or disappointed. The progression of the perception of the same thing is characterized by the fact that not only do expectations follow upon expectations, [324]
but by the fact that a new primordial impression occurs again and again, a primordial impression that necessarily fulfills the last series of expectation that was aroused—namely, insofar and so long as the same thing remains perceived. Disappointment can effect only particular moments.

Through the systematic structure of lived-experience that is characterized by the general terms, primordial impression, retention, and protention, the perceptual object's sense-structure and the entire shape of its noematic givenness is prefigured for every possible perceptual object in general, and this is the case according to general and necessary features.

<14.> Retention and Remembering

Having studied in a detailed manner the noetic structure of perception and the noetic structure of remembering as the intuitive memory of something past, let us now compare them. Here the essential difference suddenly comes to mind, namely, that even presentifications, i.e., those so-called fresh memories, the complex of retentions we described, belong to the essential structure of perception. But [it is] not only [the case] that these are non-independent and thus, as they emerge, can never be made independent. In their very nature and with respect to their intentional accomplishment, they are fundamentally different from rememberings. Notice that, according to our analyses, making up the fundamental nature of remembering is a full, concrete perception, presentified once again in its concretion. Its immanent sense, its perceptual object, is thereby presentified as standing there, so to speak, once more before our eyes—and that is the memorial object. Perceiving is a synthetic accomplishment of sense that is carried out in a continual streaming. Accordingly, remembering is likewise, naturally, a synthetic accomplishment that is carried out in a continual streaming, and it has, as it were, the same contents, but precisely in the mode of the "as if." Perception is the consciousness of originality. But even though it is the consciousness of originality in its primordial impression, and in a superior way, it is not only this consciousness of originality. Certainly, we are conscious of only the new tonal phase in the pure character of its presentation in the flesh, the new tonal phase flashing forth in the pure punctual Now; that is, only for this new tonal phase is there a pure making-present. Yet, [325]
according to its other phases, the retentional ones, perception is also an original consciousness in a certain way. For if retention

only presentifies the tonal phases that are fading away, this presentification in its necessary function is endowed with making concrete perception possible in an original accomplishment. What is constituted there originally is precisely the first, most primitive shape of the past as just-having-been and as necessarily belonging to the phenomenon of the living process. But it is through the constant coinciding of the original pasts being modified, and likewise the coinciding of the anticipatory futural phenomena, that the identity of the tonal process that is being structured temporally, and thus every perceptual object with its identical temporal points, comes about for consciousness. It comes about in the continual process of abiding and enduring. Or, as we can also put it, the constitutive accomplishment of the primordial impression and of continuity of retentions and also of protentions that steadily modify the primordial impression is a unique, indivisible accomplishment; through this alone the immanent, temporally extended object, that is, a concrete individual object, can be given to consciousness. Thus, taken completely and as a whole, perception is actually an original consciousness. In contrast, what we call a memory in the normal sense is completely the opposite of an original consciousness. Everything that a perceiving accomplishes *originaliter*, namely, its entire circuitry of noetic and noematic structures, memory brings in itself to a presentification; and its unique, new mode is the "re" or "again." It re-presents or presents again a temporal objectlike formation to consciousness in such a way that it allows it to run-off one more time, so to speak, as if it were perceived, but everything: the current Now, just-Now, what is to come, is merely "re"-presentified; its perceiving is not actual, but rather reproduced perceiving. Thus, the retentions and protentions that are contained in it are not actual retentions and protentions, but reproductions of them. Reproduction is not, as Hume thought, and the sensualistic psychology since Hume, something like a poor imitation of a perception or a mere weak echo of it; rather, it is precisely a new fundamental mode of consciousness wherein the ego is conscious of the fact that precisely a re-presentification of a previous perception with its [326] perceptual object is taking place: Only from such a consciousness can the sense of the terms "re-presentification" and "memory" become intelligible. Obviously, we could also say something similar about expectation and the shape of its bringing the future to intuition.

To be sure, what we just said demands some supplementation, and practically a rectification. Retention and protention are the primitive, first forms of instituting the past and the future. But retention and protention as the primordial forms of presentification are empty forms. Only the intuitive remembering, which awakens the past, creates the past as the intuitively fulfilled shape of the present in the mode of past and, as it were, experienced once again; and likewise "fore-seeing" the future, the intuitive effective realization of protention, creates the shape of the future that is intuitively given to consciousness as the present in the mode of arriving and, as it were, in pre-enjoying, being experienced in an anticipatory manner.

Genetically, empty shapes precede all types of intuition, all perceptual constitution of objectlike formations, in all modes of appearance. Nothing can come to intuition that was not previously emptily presented and that comes to fulfillment in intuition.

<15.> Remembering and Objectivation. "Object."

These new modes of consciousness alone make a spiritual life possible, a life of knowing, valuing, willing, and doing. If there were no remembering (in the event a life of consciousness were possible without it), only the particular perceptually constituted objectlike formation would be there for the ego in its present temporal becoming. But in the full sense, there would actually be no object at all for the ego; lacking would be the consciousness of something graspable in multiple possible graspings, the consciousness of a being to which one can return again and again, and that one can recognize as the same, and further, that one can adopt as a possession freely at one's disposal. Thus, completely lacking is the presentation of a Something that is in itself, as opposed to the possible views that make something present to consciousness: in a word, precisely an object. To achieve this, as [327]
Kant has already seen in his transcendental deduction (to be sure, restricting it to spatial objectivity), one requires reproduction and

recognition. What perception presents to consciousness in an original manner as an existing unity in the process of becoming, through the flux of its primordial impressions and retentions and protentions, must be able to be remembered in repeated rememberings and be able to be recognized as the same, as the same thing that I had perceived earlier. But for consciousness itself, repeated remembering is grasped as a newer and newer remembering of the same thing, grasped through syntheses of identity. By becoming attentive in this way to certain identifying acts, cognitive acts, we take a look ahead to the acts that make up the field of what is specifically *logos*. Now we only have to recognize that the "One" that is constituted in perception itself and alone in perception, as it is constituted in pure passivity prior to all remembering and all active cognition, is not yet an "object." "Object" is the correlate of cognition, which cognition lies originally in the process of synthetic identification, which presupposes remembering.

This holds for every kind of object, even for the noetic and noematic structures of intuitive lived-experiences that have become for us scientific objects in our phenomenological investigations. Evidently, we have gained all of our determinations, for example, concerning sense and the structures of sense, on the basis of reproductions: We compared exemplarily manifold perceptions of the same sense-content, but we could only do that in chains of remembering related to lived-experiences that flowed-off by repeatedly running through them and identifying the common element.[94]

[94] Chains of re-perception and chains of remembering and their function for knowledge. Chains of perception of something uniform for the knowledge of something general. Modes of knowledge in the framework of the unity of perception articulated successively and in uniformity, i.e., modes of knowledge in the framework of the unity of a continual, even if articulated perception, like when looking around in a room, I turn around and turn back again to the previous objects; here a unity of perception takes place, but which, in turning back around is the perception of the same objects; for them this a re-perception, but a re-perception of the same objects in an altered original temporal locus.

<16.> The Temporal Extension of the Object as the Extension of Sense [328]

If we look back from our studies of memory to the previous studies of perception, we will not only recognize that we could gain all knowledge of what lies in perception only on the basis of rememberings and analyses on the basis of remembered perceptions. (Even our knowledge of memories required us not only to carry out rememberings, but also to carry out repeated lived-experiences of memory, and to regard even them in remembering, bringing them to analytical cognition and comparison.) But we also discern that belonging inseparably to all perception itself is a region of presentification, namely, of a special form of memory, the so-called primary memory or retention. Notice that every current perceptual-Now steadily and incessantly sinks back into a just-Now in consciousness, a just-Now that quickly fades away and is lost in obscurity. What we call the perceptual present is thus an ever new living Now with an immediate extension of the past, with the extension of the "just-now." Memory, which we had examined more closely, was remembering, and if it has the character of a presentification of perception, then it presentifies *eo ipso* the perceptual-Now with the originary having-just-been belonging to the Now, which in remembering is the previous, long past having-just-been.

Due to their difficulty, we cannot go into deeper analyses of all these intentional complexities. But what we see clearly is this: That departing from a perception it is not sufficient to distinguish its perceptual sense and its modes of being, but rather that belonging no less inseparably to the objective sense of perception, and then belonging no less to the objective sense of the objects of remembering, is a variable temporal mode as well. Perceptual objects are individual objects, and their remembering is a presentification of individual objects; all such objects are necessarily given in a temporality, and this temporality is constituted as a determination of the identical objective sense. [329]
Each individual object is initially given to consciousness in the modes of the ever new Now and ever newly changing past. However, on the other hand, every past, that is, every past Now, is

identifiable as the same in any number of repeated rememberings, as the same temporal sense, and this temporal sense is already contained in original perception in its Now that originally institutes time.

We can also expound upon this here: To say that different perceptions, which follow upon one another and continually pass over into each other, harmonize according to their objective sense is precisely to say that the momentary Now-phase of perception and the retentional memories that are continually annexed [to the Now-phase of perception] harmonize in the objective sense, and it does so completely in the case of an unaltered object. Later corresponding rememberings have the same objective sense. In contrast, the temporal modes are different as well as the temporal points themselves that belong to the momentary phases of these acts. Belonging to the continuity of a perception with respect to the object is the unity of duration during which the object endures. Every point of duration is exhibited in the sense through a constantly variable modality of the Now or the modality of the having-been. But if we focus on a temporal point, or rather, on the object in this temporal point, it will remain identically the same throughout all alterations of the being-pushed-back-into-the-past and throughout all alterations of remembering with their temporal modalities. Accordingly, the object in its entire duration, and this duration itself, is also something identical in any number of repeated rememberings. Thus we find the object-sense, which presents itself in an ideally infinite number of appearances, as identically the same in a temporal extension, in the extension of duration. The latter does not lie in its appearances, which we characterized as aspects of the object; rather, it has its system of temporal adumbrations according to each one of its points; from the point of view of form, these adumbrations are uniform for every point, since every temporal point arising in the manner of an original source is constituted as Now and runs through its modes of the past and further and further pasts. The temporal duration is the duration of this object. This is to say that the temporal duration characterizes a coinciding that is inseparable from the continual coinciding of appearances, a coinciding that constitutes the objective sense, and thus characterizes a coinciding that is carried out together with them, a coinciding of temporal adumbrations with respect to the same temporal point. Object-sense and the temporal sense intertwined with it form an integral unity of sense, and the mode of being, the "being certain" or the "being doubtful," etc., bears on both components. [330]

We have therefore elucidated the logical elementary ideas, the primitive structures of senses according to their origin; we have become acquainted with them and, in their generality, with the structures of lived-experience that constitute them originally and for consciousness.

<17. Reproductive Sense and Modes of the Past>

The latter observation has put the sense-giving functions of perception and of the corresponding rememberings and memories of the future in relation to one another, and it showed that they can work together. It will now be prudent, and even quite necessary, to work out in more detail the concepts of objective sense and noema that were first introduced, along with getting to know still more closely the constitutive accomplishments of these kinds of acts in their important features.

Corresponding to every perception is an ideally infinite manifold of possible memories, all of which presentify the same perception, exactly as it is in itself, and thereby the perceived as such, the sense. Now, all these infinitely many possible lived-experiences have the same objective sense, wholly the same, and yet they are different in the so-called "noematic" regard. Let us reflect. Two rememberings that refer back to the same perception can obviously present to consciousness the remembered objectlike formation, e.g., the melody that was heard earlier only in a modally altered past. If I hear a melody now and if it comes to me tomorrow in memory, I will have the melody in mind in the mode of "yesterday," and in a memory occurring the day after tomorrow [I will have the melody] in the mode of "the day before yesterday." Obviously this holds generally for all similar cases. That is to say, just as the retentions already functioning within every perception are perpetually transformed, so too are the concrete retentions of concrete perceptions that remain in the

obscure background similarly transformed. All of the re- [331]
presentifications being carried out successively, through which something at rest in the obscure subsoil of consciousness receives a reproductive actualization, must deviate from one another according to a strict law; they must be altered in relation to one another. While they do harbor the same reproductive sense, they do so in an ever newly altered mode. In other words, the temporal object itself, the melody, the tone, is indeed individually the same in such a chain of rememberings—like the temporal object itself, its duration is also individually unique, and within the duration, each tonally fulfilled temporal point. But the mode of the past and therefore the mode of orientation to the current present continuously arising anew is incessantly transformed: a little while ago, yesterday, the day before yesterday, etc.

If we say that the object is pushed into the past, it will initially appear that we would want to say that it would alter its time and that the object would only keep the relative site of its fulfilled temporal points continually uniform in the unity of its process. But that would be false. Whenever we come back to the tonal process in rememberings, it is individually one and the same, and this implies that every temporal point of the process preserves its identity, but in and through the change of its orientation [to the present]. That is to be seen in evident, identifying modes of knowledge. Only in this way is it possible for a unitary universal time to be constituted for us, a time in which anything that can be given to us through a presentification as existing has its firm position. Time itself is the fixed system of positions in which every individual duration is strictly situated with its system of points. Because every temporal point and every temporal duration cannot in principle occur more than once in uniform orientations, there is no displacement of an object in time, there is no analogue to movement.

Thus, we must distinguish between the identical temporal point or the temporal duration belonging to the objective sense, and its mode of orientation, whereby the latter we mean the incessantly changing mode of givenness of this duration. But with this we distinguish in evidence between the objective sense and the noema as such. We called the noematic viewpoint that viewpoint with

regard to sense, and we called noematic everything that we find in and concerning sense in this viewpoint. But concerning sense, we find in an evident manner the mode of orientation, the "past," [332]
"further past," etc., that is modified in every new intuition. As being in the process of change, the mode of orientation does not belong to the objective sense, to what is meant identically. Belonging to the objective sense, to that which remains absolutely identical, is the duration itself and every temporal point itself, but not the contingent mode of the past.

We can also say that we must distinguish between the duration itself, which appears there respectively as the temporal form of the perceived as such, and the "mode of appearance" of this duration, whereby the expression, "mode of appearance," can certainly have still other meanings; belonging here are, for example, the distinctions of clarity and those of temporal perspective.

What holds for the temporal form of the perceptual object (and then further for the temporal form of the memorially re-presentified object) holds for the concrete object itself, thus including the materially relevant contents that endure in the duration, the temporally distributed contents filling the duration. The concrete object that is determined in such and such a way with respect to content cannot itself essentially be given in any other way than in the changing temporal perspective and orientation. The object itself, as the appearing object, does not change with the alteration of appearance, an alteration of appearance that the object must undergo as being constituted temporally. From the normal perspective, the object is constantly presented to consciousness as the same object in perception or memory. Phenomenologically speaking, the identical objective sense, with its temporal determinations and its qualities, is only given in continually different orientations and temporal perspectives in a changing noematic How.

<18.> The Noematic Attitude

By focusing on the object purely as the object of a consciousness (or put differently, in focusing on the objective sense), we called a noematic examination an examination that simultaneously investigates the modes of givenness in which this "object" presents itself; namely, the How of the "noematic" mode of givenness presents itself in the object itself as the object of consciousness. Accordingly, the mode of orientation, and everything else that is to be regarded as a temporal perspective is [333] seen noematically. The noematic attitude describes in an entirely general manner the objects in the How of modes of appearance. It investigates all the modes of appearance without which the object could not appear at all, could not be perceived, and thus could not at all be given in intuition. Thus, the noematic attitude studies all types of objects purely as the objects of intuitions that give them; in other words, it studies their objective sense in its structure. But together with this, it also studies the changing noematic mode of appearance of the entire object as sense and studies every one of its intuitive components. Here we had examined objects as such, purely as temporal, and we had done so in an entirely general manner; in this way the noematic modes of appearance, which temporality brings about, came exclusively to the fore. Every temporal determination that lies in the respective objective sense—temporal point, temporal expanse, temporal shape—is a unity for a corresponding, unending manifold of modes of appearance. Every individual object as a temporally formed object is *res temporalis*, and in this first sense, *res extensa*, and thus it is essentially meant that it can only be experienced in a determinate regulated system of manifolds of appearance. This lawful regularity and thus the entire system of appearance that belongs to it is a common one for all conceivable individual objects, precisely as temporal objects, no matter the genus or the region to which these objects may belong.

The modes of appearance that are determined as necessary by the temporal form of sense, however, are not the only ones that the sense as the sense of individual objectlike formations can prescribe to the possibility of the experiencing intuition. In other words, the content of this form, that which is extended in the temporal form, that which fulfills it in different temporal shapes, the content of this form, also has its modes of appearance. It does not merely have modifications of appearance that accrue to it by being adumbrated according to temporal perspectives with its temporal shape.

However, we cannot directly pose the question concerning the noematic constitution of objects irrespective of temporal extension. We know that all conceivable objects can be experienced, and along with this ability to be experienced, have a temporal constitution. But as soon as we abstract from this most [334]
general feature and inquire into the temporal contents and their modes of appearance, the investigation necessarily splits. For here is the place where the highest, but also the most empty Something in general, as the generality of something conceivable in general, as the object as such, is divided in the genuinely highest genera. But depending upon the genus in which we find ourselves, the temporal contents are heterogeneous and have as such fundamentally different modes of experience and modes of noematic constitution according to the objective sense and noematic modes of the objective sense—ignoring, naturally, the entirely general feature of time-constitution.

<19.> The Object-Pole. Whether the Objective Sense is Ideally Identical[95]

I have a chain of reproductions, of rememberings in which I am conscious of one and the same past immanent phenomenon, e.g., a datum of sensation. Two such rememberings are separated, they are themselves immanent temporal objects, themselves objects of possible, repeated reproductions and each one identifiable, existing in evidence as the same. Thus, we have different reproductions as different beings, belonging to different temporal loci. But the reproduced object is the same. Can we say that the reproduced object is found in each one in an intimately inherent manner, that is, as a momentary element of lived-experience? But then the one reproduced object that marks one single temporal locus or temporal expanse would be existing simultaneously at different temporal loci, then the time of the reproduced object and

[95] Here I stand in contradiction to the *Ideas* and deny that noematic unities, objective senses, are transcendent to the lived-experience.

the time of the reproduction would be the same time. If we take memories of something temporally different, then the temporal orderings of something different would be identical with the temporal order of their memories, which is absurd. Here the memorial object is conceived as a true being of its temporal locus, as we also speak of a true order with respect to the order of memories. Yet, if we take the meant past, that is, the meant temporal objectlike formation with its meant temporal locus, how do we conceive of the truly existing temporal loci where the memory itself is concerned? Can the temporally separated memories contain their meant objects as intimately inherent [335] components? There is indeed no difficulty here. Every memory has its meant object, and the meant object as such has the same temporal locus as does the memory in the true nexus of precisely these memories, understood as the lived-experiences occurring in the true stream of lived-experience. Thus, this shows once again that there is no basis for dissociating the "noema" from the lived-experience, and for contesting its character as an intimately inherent moment.

<Section 4. ON STATIC AND GENETIC PHENOMENOLOGICAL METHOD>

<A.> Static and Genetic Phenomenological Method[96] [336]

We must make the following distinction under the rubric of the laws of genesis:

(1) Laws of genesis in the sense of one demonstrating laws for the sequences of particular events in the stream of lived-experience. They are either laws of immediate, necessary succession for concrete events or for abstract phases, moments of such events like, e.g., the necessary connection of retentions to lived-experiences that have elapsed, or the necessary connection of retentional phases to the respective impressional phase. Or they are also laws of a mediated sequence, for instance, the laws of association, laws for the emergence of reproductions for a present lived-experience within the present and the like for the emergence of intentions of expectation—in the widest sense of empty intentions, fulfilled or unfulfilled processes of pointing-toward or pointing-back.

(2) Lawful regularities that regulate the formation of apperceptions. Apperceptions are intentional lived-experiences that are conscious of something as perceived, [but this something as perceived] is not self-given in these lived-experiences (not completely); and they are called apperceptions to the extent that they have this trait, even if in this case they also consciously intend what in truth is self-given in them. Apperceptions transcend their immanent content, and belonging essentially to this transcending is the fact that within the same stream of consciousness

[96] Editor: From 1921.
Translator: The following pagination to the German text corresponds to Husserliana XI.

whose segments are being continually connected, a fulfilling lived-experience is possible that, in the synthesis of fulfillment, supplies its self-given matter[97] as the same, and in that other lived-experience supplies what is not-self-given and the same [self-given matter]. Insofar as this is the case, there is a law here regulating the future, but a law merely for future possibilities, concerning a possible continuation of the stream of consciousness, [337]
one that is ideally possible.

Defined in this general way, apperception is a concept that encompasses every self-giving, thus every intuitive consciousness.[98] Originary apperception is perception, and every

---

[97] What is meant here is not intimately inherent, adequate givenness, but being perceived in the genuine sense.

[98] Consider how the concept of apperception is to be circumscribed. Apperception: a consciousness that is conscious of something individual that is not self-given in it (self-given does not mean being contained in perception in an intimately inherent manner); and it is called apperception to the extent that it has this trait, even if it has something in addition that is self-given in it. Namely, a consciousness can be apperceptively conscious of something, and that same something can also still be self-given in the same consciousness that extends even further than this apperceiving. For example, if in this way we call a consciousness of a sign an apperception, then the signified [*das Bezeichnete*] can also be self-given along with the consciousness of a sign in the unity of one consciousness. Or in the unity of a perception of a hexagon there appears a hexagonal plane and at the same time another; but one of them appears with reference to the other one, and the other one is itself appearing. This holds in general with respect to the components of self-givenness peculiar to external appearing phenomena.

Every motivation is apperception. The emergence of a lived-experience *A* motivates the lived-experience of a *B* in the unity of a consciousness; the consciousness of *A* is equipped with an intention that points beyond, "indicating" a coexistence. But here we must add that every unfulfilled intention, every unfulfilled horizon contains motivations, systems of motivations. It is a potentiality of motivation. When fulfillment takes place, a current motivation is there. One can also say that apperception is itself a motivation, that it motivates whatever may occur as fulfilling, that it motivates beyond itself into an emptiness. But that will depend upon more precise definitions of apperception and motivation. Moreover, one will certainly not be able to say that a sign [*Zeichen*] motivates if it is not an indication [*Anzeichen*], a word-sign, for example. But we must also ask whether one will want to speak of apperception in that case. Admittedly, we have formulated our concept in an extraordinarily broad manner. Deeper investigations are needed here. If one speaks of apperception, perception will not necessarily express a positing consciousness, for the co-perceived is then not necessarily co-posited, to say nothing of perceived in the broader sense of "perception" [*perzipiert im Sinne von "wahrgenommen"*].

Fundamental for the theory of consciousness is the universal exploration of the relations of consciousness intending beyond itself (beyond its self)—what we call here apperception—to association.

modification of apperception in imagination contains an apperception precisely in the shape of this modification. If we consider here that every present consciousness (every expanse of presence belonging to the stream of lived-experience) not only is, but is "perceived," that is, is present now to consciousness in an impressional manner, then we also mean that an "apperception" lies in every present consciousness. In fact, we cannot even conceive of a consciousness that would not go beyond the strict present in its essential flux from presence to new presences; consciousness is inconceivable without retentional and protentional horizons, without a co-consciousness (although a necessarily non-intuitive one) of the past of consciousness and an anticipation of an approaching consciousness (no matter how [338] indeterminate it may be). Thus if something "arises out of something" at all in the stream of consciousness, then apperceptions necessarily arise from apperceptions. We do not need to consider here whether there are primordial apperceptions that could be placed at the "beginning" of the stream of consciousness. In any case, there are apperceptive horizons, kinds of such horizons, kinds of apperceptive intentions (I also say appresenting intentions) that must arise at each place in the stream according to the universal lawful regularities of conscious life—like the examples given above show. But this also holds likewise for those that can arise—even if they must not arise—at every place in the stream, namely, insofar as they are bound to conditions that are possible at each place. To the latter belong the intentions that customarily come into question under the rubric of association. At each place in the stream it is possible for constellations that are similar (I use an empty term [constellations] whose scientific content is still to be specified) to be produced again with earlier ones, to recall the earlier similar ones, to point back to them, perhaps to bring them to intuitive presence, and then as fulfillments, to show them synthetically unified with the present ones, etc. Yet even these apperceptions, and likewise these apperceptive connections—which exhibit the unities of a combined phenomenon, whose combinations presuppose apperceptions and encompass them—these apperceptions can only take place when other, especially suited apperceptions have preceded them.

(Could we not also define apperception in the following way: a consciousness that is not only conscious of something within itself in general, but at the same time intends this something as a motivation for a consciousness of something else; thus, a consciousness that is not merely conscious of something, and then still something else that it does not include, but rather, a consciousness that points to this other one as one that belongs to it, as what is motivated through it. In any case, we will have to expand and give sharper contours to our previous definition.)

In addition, types of intricate apperceptions can occur, which, once they are there, are repeated in a further stream of consciousness according to primordial laws under universally producible conditions; indeed, they run through this stream of consciousness steadily, like all natural apperceptions, all objective apperceptions of reality, apperceptions which in accordance with their essence themselves have a history, a genesis according to [339] primordial laws. Thus, it is a necessary task to establish the universal and primitive laws under which stands the formation of an apperception arising from a primordial apperception, and to derive systematically the possible formations, that is, to clarify every given structure according to its origin.

This "history" of consciousness (the history of all possible apperceptions) does not concern bringing to light a factual genesis for factual apperceptions or factual types in a factual stream of consciousness, or even in all factual human beings; thus it is not at all similar to the development of plant or animal species. Rather, every shape of apperception is an essential shape and has its genesis in accordance with essential laws; accordingly, included in such an idea of apperception is that it must undergo a "genetic analysis." And what is given is not the necessary becoming of the particular, single apperception (when it is understood as a fact); rather, the mode of genesis is only given with the genesis of essence; in this mode of genesis any kind of apperception of this type must have arisen originally (in one stroke or piecemeal) in an individual stream of consciousness. And after it had arisen (as primordially instituting, so to speak), individual apperceptions of the same type were able to arise in an entirely different manner, namely as genetic after-effects of the earlier ones already formed—in accordance with intelligible laws of a primitive form. The theory of consciousness is directly a theory of apperceptions; the stream of consciousness is a stream of a constant genesis; it is not a mere series,[99] but a development,[100] a process of becoming according to laws of necessary succession in which concrete apperceptions of different typicalities (among them, all the apperceptions that give rise to the universal apperception of a world) grow out of primordial apperceptions or out of apperceptive intentions of a primitive kind.

Every apperception exhibits the structure of noesis and noema. Every apperception carries out in its own way a sense-giving and a positing of objects in doxic modalities. We have to undertake a unique form of analysis in order to elucidate the intentionality of an apperception, in order to describe, according to their noetic and noematic structures, the possible types of fulfillment and the systems of possible omni-faceted, complete fulfillment, or the [340] systems of a fulfillment that is continually in the process of becoming complete. With these descriptions, namely the constitutive ones, we are in no way inquiring into an explanatory genesis. In our descriptions of all the modal modifications in retentions, rememberings, expectations, etc., we likewise do not inquire into genesis when we pass from the original impressions (perceptions) as a generally typical generic character that concerns all apperceptions, over to a constitutive character, and from there pursue a principle of systematic ordering of apperceptions, a principle of ordering that intersects the division of apperceptions according to the highest genera of objects (actual and possible existing regions of objects). A universal doctrine of consciousness is thus a universal doctrine of apperceptions, correlative to a universal doctrine of the highest categories of possible objects and their categorial modifications—a universal constitutive phenomenology. The latter is preceded by a universal phenomenology of the most general structures and modalities that

[99] *Nacheinander*
[100] *Auseinander*

encompass all categories of apperceptions. To this one must add a universal theory of genesis.[101]

In a certain way, we can therefore distinguish "explanatory" phenomenology as a phenomenology of regulated genesis, and "descriptive" phenomenology as a phenomenology of possible, essential shapes (no matter how they have come to pass) in pure consciousness and their teleological ordering in the realm of possible reason under the headings, "object" and "sense." In my lectures, I did not say "descriptive," but rather "static" phenomenology. The latter offers an understanding of intentional accomplishment, especially of the accomplishment of reason and its *negata*. It shows us the graduated levels of intentional objects that emerge as objective senses in founded apperceptions of a higher level and in functions of sense-giving, and it shows us how they function in them, etc. But in these investigations we are concerned in the first place with apperceptive forms, with modes of consciousness that are conceived so generally (that is, left so indeterminate) that they must belong to the make-up of every [341] monad (e.g., perception, memory, etc.). Other ones have a different universality and necessity. If we take as our point of departure the "natural concept of the world" and the human ego as the subject of knowledge, then what we have gained through an eidetic analysis is the idea of a monad that is precisely in relation to a "world" of this corresponding concept, and in this way we have a pure range of monads in whose stream of consciousness the corresponding types of apperceptions (spatial-causal thing, animal being, human being) "necessarily" emerge, although perhaps they do not necessarily belong to the idea of a monad as such; in any case, this is not immediately certain *a priori* from the start.

Further, in monads that correspond to human beings within the natural attitude, we find factually peculiar occurrences of reason in particular shapes. We <want to investigate> the intentional typicality that is made available to us through the phenomenological-eidetic analysis of the ideas "human being" and

---

[101] Phenomenology:
1) Universal phenomenology of the general structures of consciousness
2) Constitutive Phenomenology
3) Phenomenology of Genesis

"world"; we want to investigate it systematically according to all possible nexuses of reason (that is, we want to investigate its nexuses and ultimately the entire world of these monads most basically in the possible nexuses of "concordant," ratifying experience of the respective objectlike formations), and we want to gain its essential shape. Likewise, in the free realm of possibility we investigate the essential structures of the formal lawful regularity of a reason in general as formal-logical reason, etc. Aside from the fact that we form the corresponding thoughts and realize truths in ourselves, we recognize through them how possible rational subjects would think; through this we construe in an indeterminate generality subjects of pure reason and their shapes of rational activities in which they live toward and attain true being and truths, as well as true values and goods. But even with all this, we do not gain knowledge concerning how a monad, as it were, looks in its completeness, and which possibilities are prefigured for such complete monadic individualities, and through which lawful regularity of individuation [this takes place].

Let us note that we remain here within the sphere of reason, within the realm of the active ego, and that we cannot describe a shape of active apperception, that is, any integrally cohesive unity of active configuration (which as a unity of consciousness is [342] intentional and accordingly is an apperceptive configuration) without also constantly speaking of genesis. Every inferring is an active apperceiving, and as an active process of configuring, it is a judging, because another judging has preceded it—one judgment is passed on other judgments that have been passed. The conclusion follows from the premises, it is generated from them; the lived-experience genetically issues from the grounding lived-experiences, even if other genetic interconnections play a founding role there. Thus, every activity is motivated, and we have pure genesis in the sphere of acts as a pure act-genesis in such a form that I, who execute acts, am determined by the fact that I have executed other acts. Further, we have acts that are motivated through affections and that stand in a genetic relation to spheres that fall outside of the sphere of activity. We have, finally, genesis in the sphere of pure passivity, even though formations that have their origin in an earlier activity may play their part in them; but now they themselves emerge passively.

Accordingly, in the doctrine of genesis, in "explanatory" phenomenology, we have:

(1) Genesis of passivity, that is, a general lawful regularity of genetic becoming in passivity that is always there and, without a doubt, has origins that lie further back, just as apperception itself does. Special types that belong to the general idea of passive genesis.

(2) The participation of the ego and relationships between activity and passivity.

(3) Interrelations, formations of pure activity; genesis as an active accomplishment of ideal objects and as an accomplishment of real generation. Secondary sensibility: general laws of the consciousness of habituality. Everything habitual belongs to passivity. Even the activity that has become habitual.

(4) Once we have gained all the kinds of genesis and their laws, we will then ask to what extent one can assert something about the individuality of a monad, about the unity of its "development," about the regulative system that essentially unites all the particular geneses in the form of one monad, and about which types of individual monads are *a priori* possible and construable.

(5) And connected to all of the preceding we ask: In what sense can the genesis of a monad be implicated in the genesis of another, and in what sense can a unity of genesis, according to laws [of genesis], combine a multiplicity of monads [?] On the one hand, passive genesis, which in the case of the constitution of an anthropological world (or rather, an animal world) refers to the constituted physiological processes and to their conditions in the unity of the physical world with the lived-body of another; on the other hand, active genesis refers to the form of the motivation of my thinking, valuing, willing through that of others. Thus, considering the individuality of the monad leads to the question of the individuality of a multiplicity of coexisting monads, monads genetically connected to one another. With respect to "our" world [343]

it leads to the question of making understandable monadologically the natural psychophysical world and the communal world.[102]

(6) Again, all this relates to the question concerning the genetic explanation of a monad within which a unitary nature and a world in general is constituted genetically, and how a unitary nature and a world in general remain constituted from this point onward throughout its entire life, or through an exceptional span of life, and further how a world with animals and humans is constituted according to a constant process of attestation.

What precedes this [genetic analysis] is the static elucidation of world-apperception and of the sense-giving that is carried out in it. But, it seems, it is only possible to undertake an absolute consideration of the world, a "metaphysics," and to understand the possibility of a world first through a genetic consideration of individuation.

(7) My passivity stands in connection with the passivity of all others: One and the same thing-world is constituted for us, one and the same time [is constituted] as objective time such that through this, my Now and the Now of every other—and thus his life-present (with all immanences) and my life-present—are objectively "simultaneous." Accordingly, my objectively experienced and ratified locations and the locations of every other share the same locality; they are the same locations, and these are indices for ordering my and others' phenomenal systems, not as separated orders, but coordinated orders in "the same time." That is, my life and the life of another do not merely exist, each for themselves; rather, one is "directed" toward the other. Not only have sensations occurred in me in this or that order such that, in [344]
accordance with the laws of genesis, a nature had to be constituted for me, and not only has this nature endured, but a typically stable lived-body is conveyed in this process. Realized is also the

---

[102] Translator: The expression "our" world designates a first person plural world constituted through various historical and intersubjective processes of appropriation and disappropriation; as such it becomes for Husserl in the 1930s a term for the generative phenomenon of "homeworld" [*Heimwelt*]. However, it is interesting to note that the expression, "*Heimwelt*," first appears (to the best of my knowledge) in a manuscript dated ca. 1920, that is, about the time of the first version of the *Analyses* and Husserl's explicit distinction between static and genetic methods. In fact, it occurs in a context where Husserl speaks of static and genetic relations of foundation. Cf. Ms. A V 10/1, 127-29.

possibility that there are things similar to my lived-body in the nature that is given to me. Furthermore, not only has empathy ensued, but this empathy has been ratified by the fact that the interior life of the other ego has expressed itself in a regular manner, and from then on, newly determined and ratified my appresentations again and again.

Primordial laws of genesis are the laws of original time-consciousness, the primordial laws of reproduction and then of association and associative expectation. In relation to this there is genesis on the basis of active motivation.

If we compare static and genetic nexuses, then we will have to ask whether one can achieve a systematic phenomenology of static nexuses (like that of noesis and noema), that is, whether the genetic dimension can be completely suspended here. On the whole, the question concerns how the investigations are to be ordered. It is clear that one will initially proceed from particular fundamental types, some of which, as I already said above, will occur necessarily, others which will be presented as possibilities. The question concerns the leading clues of the system. As leading clues, we have types of objects, that is, leading clues from the standpoint of ontology. And with this [we have] constitutive teleologies. Here ideal possibilities of concordant modes of givenness are elaborated, ideal possibilities of monadic streams in which the unity of an accomplishment is constituted, and other possibilities outside of these are to be considered as opposing forms.

Another leading clue is the unity of a monad as a unity of a genesis, and then the investigation of the typicality of possible monads, namely, of possible types of the unity of an individual monad, of an individual ego, and of what it had to find [in its environing-world], and how it had to encounter itself, or how it bears within itself a rule of individual character traits that are then recognizable (perhaps through others).

Beginning with the natural attitude, one can also take the "natural concept of the world" as a leading clue. One raises the natural world to the eidetic level, analyzes it according to its strata, extracts types of constituting objects and describes constituting consciousness, and finally the constitution of this type, world—all without paying any attention to genesis. [345]

Perhaps I can be more clear by writing:

Necessary successions in the open sphere of lived-experience: What is arriving is then not only arriving, but necessarily "follows upon" according to the evident law of necessary succession. Naturally, one can call that a law of genesis.

All "horizons" or all "apperceptions" naturally arise in this way. But in a "static" regard, we have "finished" apperceptions. Here apperceptions emerge and are awakened as finished, and have a "history" reaching way back. A constitutive phenomenology can regard the nexuses of apperceptions in which the same object is constituted eidetically, in which it shows itself in its constituted ipseity in the way it is expected and can be expected. Another "constitutive" phenomenology, the phenomenology of genesis, follows the history, the necessary history of this objectivation and thereby the history of the object itself as the object of a possible knowledge. The primordial history of objects leads back to hyletic objects and to the immanent ones in general, that is, to the genesis of them in original time-consciousness. Contained within the universal genesis of a monad are the histories of the constitution of objects that are there for this monad, and within the universal eidetic phenomenology of genesis this very process is [explicated as] accomplished for all conceivable objects in relation to all conceivable monads. And conversely, one gains graduated levels of monads corresponding to the levels of objects.

I must now go through the *Ideas* once more to become clearer about what still distinguishes the doctrine of the structures of consciousness from the constitutive considerations if I also regard everything immanent "in a constitutive manner."

<B.> The Phenomenology of Monadic Individuality and the Phenomenology of the General Possibilities and Compossibilities of Lived-Experiences. Static and Genetic Phenomenology[103] [34]

(1) Phenomenology of possible "phenomena" and of phenomenal interconnections and their constitutive accomplishments that can occur in monads, in general.

(2) Phenomenology of monadic individuality, the investigations of laws that are included among the laws of lived-experience, and that establish what the individual unity and discreteness of a monad requires, [that establish] what belongs necessarily to an individual monad as its proper nature, [that establish] which universal form it necessarily has, which species of elements or moments this form necessarily contains, and what in this form guarantees to them precisely unity and discreteness. If the monad necessarily has the form of the unity of becoming, of a unity of unflagging *genesis*, then its concrete structure is only made up of "elements" that are themselves unities of becoming, and like the entire monad, these unities of becoming have an abstract structure with respect to their phases. Every phase has its own necessities and not merely compossibilities; in this way, every lived-experience that is being "delimited" for itself demands its "background," a horizon; every moment in a phase makes its demands with respect to becoming: thus, for the continued genesis of every streaming that constitutes the demand of temporality, etc. We must certainly not proceed with naturalistic concepts here. The monad is a living unity that bears within itself an ego as the pole of being effective and being affected,[104] and a unity of wakeful and concealed life, a unity of abilities, of "dispositions"; and what is concealed, "unconscious," is a peculiar modality for the discreteness of the monad, a modality whose necessary sense must be fashioned originally in ways peculiar to it.

But the title for (1) above is not sufficiently clear. We [35]
investigate the phenomena in the transcendental bracketing of

[103] Editor: June 1921.
Translator: The following pagination to the German text corresponds to Husserliana XIV.

[104] And as the pole of personal characters

"transcendent" reality. Belonging here in quotation marks is the thing-world with respect to the necessities and possibilities that it bears as the intuitive thing-world, and [belonging here is] experienced nature as such. I describe the mode of givenness of orientation according to time and space (of perspective), the modes of givenness according to sides, the appearing sides and the mode of appearance of the sides, the sense-data as adumbrations-of, the apprehensions, the nexuses of perceptual appearances as such that constitute unity and self-sameness, etc. I describe the relation to the ego, the grasping, relating, explicating, "comprehending" (thinking under universals, conceptually), predicating; I describe the meant states-of-affairs, propositions, syllogisms, the modes of attentiveness of the ego, of affection, of [attentive] turning toward, cogitating activities of the ego. I judge the premises and motivated by it, as a consequence of it, I draw the conclusion, and so forth. These are all occurrences in "immanent time," in the time of "lived-experiences." And in considering the monad, we have precisely its inherent nexus of immanent time and its lived-experiences and the unities constituted in it. And this entire nexus itself (a further step!) has its constitution in the original flux of time in the corresponding primordial lived-experiences.

All of this sketches a certain path of phenomenological considerations—after one carries out the phenomenological reduction, which forms the point of departure. I must proceed step by step; at first I still do not even see that a stream of lived-experience is constituted internally; I have not yet fixed this stream scientifically at all, to say nothing of monadic individuality [or] the ego of abilities constituted in it, etc.

Do I not have to develop this consideration until I show that there is a unity of genesis in immanent time, and constituted within the unity of genesis, a unity of the monad being constituted for itself temporally? Do I not need to show that this unity of the monad must however be brought back to the analysis of the primordially living monad whose absolute being consists in a multifarious streaming, and that constituted within this streaming is the immanent phenomenon of filled immanent time, of the phenomenal immanent monad?

The investigation of the *individuation of the monad*, then, bears on both: on the individuation of the immanently constituted monad, and by going back to the lawful regularity of the primordially constituting streaming, on the individuation of the absolute monad. Here, the inquiry bears on the necessary form of this unity of filled immanent time, on what gives a necessary unity to all content in the succession and simultaneity of every phase, and gives to all individual components, moments within this unity, a singularly unique nexus that cannot be rend asunder. The monad is a "simple," indivisible being; that is, what it is as continually becoming in time, and everything that belongs to it, is at some [36] location of this continual becoming, and has its being as temporal fullness in this immanent filled time and is nothing for itself, since this fullness is continual and is related to one and the same identical ego-pole. Everything that is related to one identical ego-pole belongs to a continual stream of becoming of a unique filled time, a time that is one unique time with one unique ego. Where it is a question of two monads we then have in mind two streams of becoming having a uniform temporal form, but not two streams of becoming having the same temporal form with two egos. The immanent time of one ego can never go unfilled, can never have gaps, can never crumble into several separate streams or be separated by pauses. Everything is connected to everything else in the monad.

But under the rubric "monad" we have had in mind the unity of its living becoming, of its history. But it also has its living present and it has become in this present, and directly continues in this becoming. It belongs to the nature of this present that, on the one hand, it is a primordially impressional present as the newly surging, actual moment of life having the shape, "impression"; on the other hand, as the heir to the past, so to speak, together with the impression, this present has its obscure backgrounds that can be illuminated; in every Now, the present carries its history as the horizon into which it can peer, which it can run through once more and, as it were, can live through once more in the shape of isolated or interrelated rememberings. It belongs to the nature of monadic being that every phase of its becoming has this structure with all the accompanying marvels. We have a filled unity of immanent time through the sequence of primordial impressions, but that is not everything that was or is in the process of becoming. In all phases, we also have the sedimented history of these respective phases, in each one the monad had its concealed "knowing," its habitual structure. And now this, now that was remembered in the present; the past became alive once more, and became related to the present: The monad not only is what it is now, it is also as having been, and it can gain knowledge of its past in the present, can endeavor to dwell upon its past, can have acts that connect present and past, etc.[105]

Let this suffice. In this direction, we can thus regard the unity of the monad in itself and what the essential demands of this unity entail, although there is nonetheless something contingent in the stream; all sense-data, even if they occur through empirical motivations in expectation, are contingent; for something different can still occur.[106] But however much there is contingency here, [37] and however much the idea of a color-sensation does indeed indicate that it belongs to some sensating ego (but in its ideal generality, leaves open an indeterminate infinity of possible egos as sensating), it is indeed different for an individual color-sensation. It is not the case that its individuality would be a trait, a moment, which comes to it via the general traits; rather, the sense-datum is what it is only as a sense-datum being constituted in this monad in its regulatively formed temporal context, and has its

---

[105] Does all of this not concern the mere passivity of the monadic stream and, for the ego, have a general potentiality, the general "ability," to be able to have within it a field of affection and action? But in a special sense, the ego also has its individuality, i.e., a principle of regulating acts from the side of the ego, whereby new lived-experiences are integrated into the stream. Is this individual ego, in its unity of individual egoic abilities, not the counterpart of the unity of the thing, whose individuality is also not circumscribed by general laws of constitution?

[106] The sense-datum, contingent. The fact of regulation of sense-data, and in the direction of forming the apperception of a thing, in the direction of constituting a nature and world is, as fact, contingent. How is this with the individual ego that is determined by what is contingent, but in its individuality is certainly not contingent in the same sense? Am I not a "necessary fact," and is my contingency only determined by what cannot be grasped with respect to the material codetermining my psychic (monadic) development? The necessity consists in not being able to be crossed out, and in the intelligible unity under these presuppositions, but a unity which under other presuppositions would still be the same individuality and never a different one.

being as the unity of a streaming life, as something that is intentionally unitary within it and as something that is identifiable over and over in this unity peculiar to the ego of the monad; and as the form of its individuality, it has the unique temporal location, the index of originally constituting life. Whatever is constituted in a monad does not belong to the monad like something that could be for itself and could then enter into this monadic nexus as a member and, in the final analysis, could just as well occur in a different monadic nexus. Everything immanent is indeed individual, but non-independently individual, and only the monad itself is independent. Through its phases, through its immanently objectivated order of lived-experiences, the stream yields the individuality that makes the monad distinct in the monadic nexus, that is, in the nexus of what is constituted with respect to immanent time. But all of these special individualities are just as non-independent as the individualities of each phase in relation to that of an independent *concretum*; everything concrete in the monad is non-independent, and we see that one cannot identify the concept of "the independent" with that of "the concrete" like I did in the *Logical Investigations*.[107]

Now, I can however regard the structures of the stream of lived-experience noetically-noematically in their general typicality; I can [describe] their possible modifications, their interconnections of essence, etc., without pursuing the inquiry into the lawful regularity of the individuality of a monad. The phenomenological-eidetic reduction places me on the footing of a possible monad in general, but precisely not of a monad thought individually and identically, and under the charge of circumscribing the individual identity according to its possibilities and necessities. But I can also set this new task and, of course, do so by using the doctrine of the essence of acts, of structures being constituted, etc. One can even say that I can also describe individuated geneses and the laws of genesis without systematically tackling the problem of the universal genesis of a monad and the nature of its individuality. [38]

---

[107] Translator: See the Third Logical Investigation, and specifically §17; and see *Ideas I*, §15.

I can doubtlessly designate phenomenological investigations as static, investigations that attend to the correlations between constituting consciousness and the constituted objectlike formation, and exclude genetic problems altogether. I have to distinguish from the latter phenomenological investigations that consider the typicality of different self-exhibiting shapes of lived-experiencing and of genesis according to their essential possibilities, compossibilities, etc., but without the problems of individual[ity] in this connection. Finally, we have the phenomenology of monadic individuality, and included in it, the phenomenology of a genesis integral to it, a genesis in which the unity of the monad arises, in which the monad is by becoming.

A systematic phenomenology, as I have conceived it, attends to the levels of possible modes of constitution, at the lowest level, the continual, necessary constitution of the immanent temporal stream and the constitution of monadic being as an immanent temporal unity; then the genetically higher levels, the levels of transcendence, phantoms, etc., the constitution of a nature, the constitution of animals in nature, everything "aesthetic." Then the accomplishments of thought that could be set to all levels, and to its different shapes according to these levels (activity of the ego). Accordingly, these are genetic considerations, and as the description of already constituted structures and their modes of constitution, are placed into the framework of genetic investigations. One can also describe these correlations for themselves in their typicality and necessity of the integral relatedness of such correlates. It is through genesis that we will be able to understand their [i.e., the monads'] process of becoming from out of the constitutive founding levels.

As we proceed systematically, the foundation will also be laid for a systematic doctrine of the levels of monads, depending upon whether or not they carry out higher developments, that is, advance to higher modes or constitution. And every higher monad is developed from a lower monad; it was lower in a previous developmental level. But then that still requires its own consideration of the individuation of a monad, just that it is questionable whether it would have to be an encompassing theory. At all events, we must keep this problem in mind.

Which problems motivate the entire investigation? I must distinguish the questions:

(1) What belongs to the possibility of a monad, to its unique nature with respect to ideal possibilities and necessities?

(2) What belongs to a monad that is to be able to constitute a nature?

(3) What [belongs] to a monad that is to have other monads given, is to be able to experience and recognize a plurality of [39] monads as coexisting, and what belongs to these monads themselves if they are to stand in *commercium*?

(4) Among the essential possibilities of a monad as monad are those of conceptual knowledge. What kinds and forms of conceptual knowledge "are there," which are concordant possibilities with respect to possible concepts, judgments, and interconnections of judgments that are to be constituted, and which [are the concordant possibilities] with respect to truth? Here we consider, in all generality, possible knowing as such, possible meaning, possible true being as knowable for the knowing monad, and we continue to remain in the context of the possible monad as such. Thus, we do not gain knowledge of the monad here in the way that we gain general truths for all numbers as such, as valid for every single number. But just as we know that it <belongs> to the nature of a pure number as such to be integrated into a series of numbers, and just as there is a system of special laws for prime numbers, sums, products, etc., which do not have to be laws that concern every given number, or just as we inquire geometrically into possible spatial figures and find laws for the types and species of figures that do not express the essential features of every figure, so too are the essential laws that we find for possible monads not expressions, or not necessarily expressions of features that every monad must necessarily possess. Not every monad must be a logically thinking one, not every one practicing moral acts, and yet the essential laws of logical consciousness and of moral consciousness do certainly belong to the general realm of the science of possible monads as such.

(5) Another question concerns the systematic possibilities of monadic consciousness, concerning these or those possible

fundamental shapes (genera), concerning the essential laws that regulate the possible occurrences, acts, states, formation of systematic interconnections of the constitution of objectlike unities, of contents of thought, etc.; and it is still another question that concerns the laws to which is subject the individual identity of a monad; then further, the laws to which is subject a compossible plurality of monads, monads that are to be able to motivate each other reciprocally, that are to be able to be determinative in relation to one another spiritually, etc. Naturally, both lawful regularities go together. But not every essential possibility is compossible for the ego and the lived-experience of the ego within the individual unity of a monad. Every imcompossibility in the essential [possibility] also excludes something in the individual unity of a monad. But there are also laws that positively prescribe what belongs to the necessary formal structure of a monad, and moreover that prescribe what must become if a certain individual content is already there. Thus, the primordial law of genesis is the law of original time-constitution, the laws of association and reproduction, the laws through which the monad is constituted for itself as a unity, etc.

Are not the specific laws of genesis the laws of individuality, or [40]
only a branch of these laws, namely, related to the becoming of the monad, while the other branch would concern the laws of coexistence? But is that not a poor approach?

Every law of compossibility in coexistence also prescribes a law for possible genesis. Laws of compossibility concerning temporal coexistence already presuppose the constitution of time, and also have along with them laws of compossibility in succession; these are general laws of compossibility in simultaneity and succession. But *in addition to this* we have laws that do not merely concern compossibilities, but necessities of succession. The former implies that if an *a* exists, then a *b* cannot exist (hence, coexistent); the latter implies that if *a* exists, then a *b* must exist, in temporal simultaneity or in succession. But what is temporal is constituted, and we encounter primordial nexuses of the stream in which, once again, both kinds of laws play their role, only in an altered sense.

These are fundamental questions concerning the distinction, but also the ordering of necessary phenomenological investigations. Where they are concerned, I will always speak of *static* and *genetic* phenomenology. What was actually the leading perspective here? My point of departure can be external perception; I take this type of lived-experience, I have the relation to the meant object, hold firmly to this, contrast it with meant features and with what is found with regard to sensations, adumbrations within and relating to perception itself; I can pursue the possibility of further perceptions, perceptions that are continually unified with the initial one and that are all the perception of the same thing; [I can] describe the changing sensations, the forms of apprehension, the forms of the synthetic nexuses, etc.; I follow the correlation: unity of appearing object and manifold of appearances being united harmoniously, noetically, etc. Here I construe essential possibilities for such lived-experiences and the nexuses of lived-experience, and therefore also for a monad in which they may occur. A monad is possible as bearing such possibilities within it. I do not inquire here after the genesis of the monad, after the way in which such phenomena arise. I pursue the idea of a concordant nexus of experience related to an object of nature perduring identically, but also at the same time, as another possibility, [I pursue] the branches of discordance [occurring] at any point, and naturally with this I alter the monad and its inherent genesis. Or (like the physical things before) I have given purposeful objects, spiritual formations, books, etc., and ask how they are given. I proceed entirely from objectlike formations, even ideal ones like conceptual thoughts, mathematical principles, and ask how the consciousness of them can look, how a manifold consciousness of them is possible, and how they are "constituted" as self-given in consciousness.

All of these questions here are constitutive ones, and [41] constitution concerns the essential correlations between the object of knowledge and knowing, the consideration of the noetic interconnections in which are constituted ontic interconnections, even those between objects and concepts, truths, etc. By ideas "being for me," I have certainly always understood "objects," even if I am directed toward something immanent; and I regard modes of consciousness, or more clearly, noetic-noematic correlative modes that function constitutively there, or modes of activity, grasping, observing, comparing, etc., which once again are also constitutive for higher objectlike formations. We persistently attend to possible modes of consciousness in relation to objectlike formations that we had in mind and thought under the idea of true being; they still remain before us as intentional in the phenomenological bracketing of their existence, and guide the composition of the interconnections.

But attending to constitution is not attending to genesis, which is precisely the genesis of constitution and operates as genesis in a monad. *Is not static phenomenology precisely the phenomenology of leading clues*, the phenomenology of the constitution of leading types of objects in their being, and the phenomenology of the constitution of their non-being, of mere illusions, of nullities, of contra-concordance, etc.? I have here the integral relatedness of essences as those of correlation, but genesis is not conditioned by that; we are not making the conditioning into something conditioned here. By the phenomenology of genesis attending to original becoming in the temporal stream, which itself is an originally constituting becoming, and by attending to the so-called "motivations" that function genetically, a phenomenology of genesis shows how consciousness arises out of consciousness, how constitutive accomplishments are also continually carried out here in the process of becoming; thus it shows the relation of conditionality obtaining between the motivating and the motivated or the necessary transition from impression into retention, in which is constituted the consciousness precisely of this becoming, and correlatively of the alteration of the Now into a Now that is just past.

However, I do describe statically not only the constitutive possibilities in relation to an object as a leading clue, I also describe the typicality of the nexuses in consciousness of any kind of developmental level: thus, in the *Ideas*, the structures of pure consciousness as structures of possibly appearing phenomena in the unity of an immanent phenomenal nexus.[108]

But if we are to hold fast to the individuality of a monad, then all possibilities must be selected; there are demands within existence for individual unity, and individual unity can only be demanded according to laws. It is also a law that what occurs within the form of unity precisely fits into the unity according to specific laws, and that through the law of unity, what fits is [42] something demanded by the framework (cf. *Logical Investigations <2. Vol> III*).

Is it therefore not the case that, on the one hand, we have the laws of possibility, of compossibility in the monads as such, and distinguish from them the laws that belong to the unity of a monad as an individual unity? But individual unity is subject to the laws of genesis. Thus, the phenomenology of absolute individuality, of the monad as individual unity, must clarify precisely the development of individual phases arising from one another, each one of which has its law of individuality. And general laws of the individuality of these phases?

[108] Question whether from the very beginning one must view the structures of pure consciousness as constitutive occurrences.

<C. The Intersubjective Validity of Phenomenological Truth>[109] [305]

I as the "phenomenological" ego, which is to say, I who have carried out the universal phenomenological reduction, can still say afterward, and in absolute Cartesian evidence, *ego cogito, ego sum*, and I can only say it after the phenomenological reduction. I, the *human*-ego, am then in brackets and am in the field of judgment only as correlate, just like things appear only as "correlates," that is, as meant objects (*noemata*) of particular acts or nexuses of acts. What I initially have in my field of judgment in such cases, and in an unmodified manner, is the "I perceive myself, the empirical human-being," "I perceive these things," I have this apperception, and in this [particular] doxic positing, and therein the "perception of this and that."

From here I can attend to doxic motivations that this belief indicates to me with this phenomenological content; indicated are the intentional self-contained matters with motivated doxic theses, with doxic presumptions, which include the possibilities of being crossed out in *doxa*. *Therefore, presumptively prefigured for me here is a structure of the pure consciousness of my pure ego, while the form of the pure stream of consciousness and of its pure ego remains non-presumptively prefigured in an absolutely (apodictically) certain manner*. Required, then, is carrying out this distinction in a careful and theoretical manner. In any case, what [306] must be demonstrated in pure phenomenology first of all is this general and constant necessity of the form of the pure ego and its pure stream of consciousness *that can never be annulled.* But afterward, the possible forms of the "empirical contingencies," of the special apperceptions that genesis has brought about; here it must again be shown *which special genetic forms must occur in an a priori necessary genesis*, that is, which system of forms of genesis must occur and in what order in the unity of a progressively developing ego and egoic stream, and which essential laws govern genesis there. In contrast to the *abiding*

---

[109] Also on the doctrine of genesis and on the relationships of the static to the genetic, but in a certain sense, static: the abiding structures of the monad.

Editor: Probably from the beginning of 1923.

static form of the monad and the necessary forms of genesis, which particularize the static form on every level, we then have what is actually and empirically contingent; for its part, it is the material fact as opposed to the genetic *form* of facticity, which is *a priori* something in principle, necessary.

The Egological Validity of all Truth and "Objective" Truth as Intersubjective

After the phenomenological reduction, and continuing now as the phenomenological ego, the ego *can* be solipsistic in a certain sense, namely, insofar as there is said to be no essential necessity that the ego encounter other human-beings and animals.[110]

Here we must note that all truths adopted by the phenomenological ego, <in other words>,[111] that can appear in *its* possible field of truth, cannot accordingly have the sense of intersubjective validity. Essential truths have logical general validity, the generality belonging to universality of the "as such." But the fact that belonging to this general validity is also the validity for every possible knower points to a different direction of investigation. This latter general validity, the universally subjective one, does not belong, like the first one, to the meaning-content and ontological content of the respective truths. Here one [307] runs up against the more general question concerning universally subjective or intersubjective validity of truths as *truths*, as well as the difference between this validity for essential truths and factual truths, and furthermore, within essential truths, between formally *a priori* and materially *a priori* truths.

All truths are referred back to possible doxic consciousness and to subjects of this consciousness, and thus referred back to phenomenological essential truths. The intersubjective (universally subjective) validity of phenomenological essential truths is indubitable and is in principle prefigured in the purely phenomenological sphere. If we stand within eidetic-transcendental phenomenology, then belonging in the dominant realm of

[110] I should have proceeded: Supposing it were so, then the following holds.

[111] Editor: inserted later by Husserl.

transcendental phenomenological *possibilities* of the respective ego, who is investigating phenomenologically, is also the *possibility* of phenomena of alien ego-subjects as animal subjects in open, unending multiplicity. If I carry out the phenomenological reduction on these pure possibilities, I will gain with the method of 1910/11[112] an open unending multiplicity of *possible* pure egos (phenomenological, even if not themselves necessarily phenomenologically investigating egos), which stand to me in a possible relation of empathy; or rather, I do not gain *an* open, unending multiplicity, but rather, an *a priori* range of "unendingly many" groups of such open ego-multiplicities; that is, when I leave out of account the factually, phenomenologically investigating ego, and form the idea of an ego as pure ego as such through the eidetic variation of myself, the fact, I will find the eidetic nexus of possibility as the general operative nexus, as eidos: a possible pure ego as such standing in relation to an open unending multiplicity of other egos as alien to it, but as standing to it in relationships of empathy and in I-you relationships, in relationships of communicative interaction, reciprocal-ego-determination. Likewise, when I do not carry out the eidetic reduction, I not only pronounce my "I am," but rather, exercising the phenomenological [308] reduction with respect to the factual givenness of alien human-beings, and carrying out phenomenological legitimation by indicating phenomenological *empeiria*, I know myself as pure ego and in addition am empirically certain (in the phenomenological field) of co-being and communicative solidarity with other pure egos. But I also therefore recognize that every truth into which I have insight is intersubjectively valid: It is valid above the empirical, namely, it is not merely dependent upon my empirical ego. It remains if I were to modify myself in thought into a randomly altered ego. But it also holds if I take as a basis any random alien pure ego (that is found in every empirical ego).[113]

---

[112] Editor: Husserl refers here to the lectures, "Grundprobleme der Phänomenologie" [Fundamental Problems of Phenomenology], which are published as Text Nr. 6 in *Husserliana* XIII.

[113] Thus for *a priori* formal truth and for mathematical-natural-scientific truth. No longer completely for empirically intuitive ("perceptual judgments") and for materially *a priori* truths.

INDEX

EDMUND HUSSERL

COLLECTED WORKS

EDITOR: RUDOLF BERNET

---

1. E. Husserl: *Ideas Pertaining to a Pure Phenomenology and to a Phenomenological Philosophy.* Third Book. Phenomenology and the Foundation of the Sciences. Translated by T. Klein and W. Pohl. 1980 ISBN 90-247-2093-1
2. E. Husserl: *Ideas Pertaining to a Pure Phenomenology and to a Phenomenological Philosophy.* First Book. General Introduction to a Pure Phenomenology. Translated by F. Kersten. 1982 ISBN Hb: 90-247-2503-8; Pb: 90-247-2852-5
3. E. Husserl: *Ideas Pertaining to a Pure Phenomenology and to a Phenomenological Philosophy.* Second Book. Studies in the Phenomenology of Constitution. Translated by R. Rojcewicz and A. Schuwer. 1989 ISBN Hb: 0-7923-0011-4; Pb: 0-7923-0713-5
4. E. Husserl: *On the Phenomenology of the Consciousness of Internal Time (1893–1917).* Translated by J.B. Brough. 1991 ISBN Hb: 0-7923-0891-3; Pb: 0-7923-1536-7
5. E. Husserl: *Early Writings in the Philosophy of Logic and Mathematics.* Translated by D. Willard. 1994 ISBN 0-7923-2262-2
6. E. Husserl: *Psychological and Transcendental Phenomenology and the Confrontation with Heidegger (1927–1931).* The *Encyclopaedia Britannica* Article, the Amsterdam Lectures, 'Phenomenology and Anthropology', and Husserl's Marginal Notes in *Being and Time* and *Kant and the Problem of Metaphysics.* Translated by T. Sheehan and R.E. Palmer. 1997 ISBN 0-7923-4481-2
7. E. Husserl: *Thing and Space: Lectures of 1907.* Translated by R. Rojcewicz. 1997 ISBN 0-7923-4749-8
8. E. Husserl: *The Idea of Phenomenology.* Translated by L. Hardy. 1999 ISBN 0-7923-5691-8
9. E. Husserl: *Analyses Concerning Passive and Active Synthesis: Lectures on Transcendental Logic.* Translated by A. J. Steinbock. 2001 ISBN Hb: 0-7923-7065-1; Pb: 0-7923-7066-X

KLUWER ACADEMIC PUBLISHERS – DORDRECHT / BOSTON / LONDON

Printed in the United Kingdom
by Lightning Source UK Ltd.
133381UK00001B/4/A

9 780792 370666